Biomaterials, Bioengineering and Sustainability

Volume 2

There is an urgent need to address the current paradigm shifts in bioengineering for Human Health aiming the creation of breakthrough tissue engineered products, manufacturing technologies and effective regenerative treatments for tackling different diseases/disorders in a personalized manner. Yet, the excessive costs and wastes related to the development and production of biomaterials and advanced therapy medicinal products, and the increasing use of plastics in cell culture methods and animal derived reagents has recognized the importance of decreasing the direct carbon footprint and thus, implement sustainable principles and solutions in the innovation ecosystem. The main goal of the volumes in Biomaterials, Bioengineering and Sustainability series is to catapult and consolidate new concepts and solutions towards the development of the next-generation of sustainable and eco-friendly biomaterials and tissue engineering and regenerative medicine approaches. Each volume will focus on the latest developments dealing with the identification of new sources of sustainable or recycled biomaterials, providing ideas for green technologies and methods that can be applied for biomaterials advanced processing and scaffolding strategies, and applications in biofabrication, tissue engineering, regenerative medicine, and drug delivery systems. It also aims to include the exploitation of renewable and sustainable source of human cells applied for cell therapies or in combination with sustainable biomaterials. The develop complex in vitro 3D/4D models and dynamic cell culture systems will be other subjects to be further explored from a sustainable perspective. This series aims to attract the contributions of leading experts in bioengineering, cell biology, materials engineering, and environmental sciences.

Rishabha Malviya • Sonali Sundram
Editors

Sustainable Nanocomposites with Green Biomaterials

Springer

Editors
Rishabha Malviya
Department of Pharmacy
Galgotias University
Greater Noida, Uttar Pradesh, India

Sonali Sundram
Department of Pharmacy
Galgotias University
Greater Noida, Uttar Pradesh, India

ISSN 2731-751X ISSN 2731-7528 (electronic)
Biomaterials, Bioengineering and Sustainability
ISBN 978-3-031-79109-3 ISBN 978-3-031-79110-9 (eBook)
https://doi.org/10.1007/978-3-031-79110-9

This Springer imprint is published by the registered company Springer Nature Switzerland AG
The registered company address is: Gewerbestrasse 11, 6330 Cham, Switzerland

Foreword

The book "Sustainable Nanocomposites with Green Biomaterials: Applications" is an essential exploration that makes an important contribution to the evolving biomedical research and development using sustainable biomaterials. It demonstrates the development of green biomaterials-based nanocomposites through various techniques that both enhance ecologically friendly objectives and stretch the boundaries of biological breakthroughs. Dr. Rishabha Malviya and the contributors have systematically built a series of chapters that cover significant advances and applications of sustainable nanocomposites.

The exploration begins with the applications of bioinspired green biomaterials in advanced treatments, diagnostics, and the development of biosensors, building the groundwork for the next section of the book. As readers explore deeper, they will come across innovative research and practical applications ranging from tissue-engineered cartilage products to conductive polymers and hydrogels for brain-computer interfaces. The book represents interdisciplinary work by integrating materials science, biology, and engineering, and the chapters provide an in-depth overview of the synthesis and applications of sustainable nanocomposites for personalized biomedical applications.

Underscoring the practical implications of these developments is the inclusion of subjects like biodegradable materials for customized medical devices and antimicrobial biomaterials for dental uses. This book offers concrete instances of how sustainable nanocomposites are currently enhancing lives rather than only focusing on theoretical possibilities. The significance of sustainable innovation is immense, given the global challenges we confront, including but not limited to climate change, resource shortages, and the increased need for healthcare.

Sustainability and environmental impact are the true realm of biomedical engineering, where the combination of cutting-edge nanocomposites and sustainable biomaterials has the potential to transform medical research in novel and environmentally conscious ways.

The book explores the applications of sustainable nanocomposites in biomedical applications, including nano-bioengineered electrochemical biosensors for disease diagnosis and tissue-engineered cartilage products for orthopedic restoration. Innovative drug delivery methods using green biomaterials, selective bio-adsorption of metals like gold and platinum using biomass, and conductive polymers and hydrogels for brain-computer interfaces are among the cutting-edge technologies covered in the book. It also covers the fabrication of customized nanocomposites for injection molding and fused granulate manufacturing, as well as the application of terpenes as natural plasticizers for biomaterial products.

The book also emphasizes the growing importance of biodegradable materials for personalized medicine and as antibacterial material for dental applications. Recent advances in polyhydroxyalkanoates, as well as the controlled release of limonene from alginate-shelled oil in water microemulsion liquid marbles, demonstrate progress toward sustainable solutions.

The book concludes with an exploration of the latest developments in green biomaterials for biomedical implants and tissue engineering, presenting a forward-looking perspective on the future of sustainable biomedicine. Each chapter provides unique insights into the multifaceted applications and benefits of sustainable nanocomposites, making this book an invaluable resource for researchers, clinical practitioners, and academicians.

Happy reading & best wishes.

Professor (HAG Scale) and Former Head
Department of Pharmaceutical
Engineering & Technology
IIT(BHU)
Varanasi, Uttar Pradesh, India

Brahmeshwar Mishra

Preface

In recent years, the search for sustainable and ecologically friendly solutions has significantly grown, especially for biomedical applications. The development of nanocomposites with green biomaterials is at the leading edge of this transformation, offering advances that are both innovative and environmentally friendly. The book, "Sustainable Nanocomposites with Green Biomaterials: Applications," aims to explore the numerous and significant ways in which these materials are revolutionizing the biomedical sciences.

The chapters cover a wide range of applications and breakthroughs, beginning with the fundamental concepts of bioinspired green biomaterials in advanced treatments, diagnostics, and biosensing. The book explores various applications of sustainable biopolymers in tissue-engineered cartilage products for orthopedic restoration, nano-bioengineered platform-based electrochemical biosensors, and plant-derived biomaterials for cardiac tissue regeneration. The use of conductive polymers and hydrogels in brain-computer interfaces demonstrates the innovation in the healthcare sciences. Green biomaterials as carriers for novel drug delivery demonstrate the intersection of sustainability and efficiency in pharmaceutical sciences. The selective biosorption of gold and platinum by *Aspergillus niger* biomass demonstrates the environmental and economic benefits of these sustainable biomaterials.

The book also explores the role of terpenes as natural plasticizers, as well as the fabrication of customized nanocomposites for injection molding and fused granulate manufacturing, demonstrating the versatility and usefulness of the materials for industrial applications. The book also discusses the role of sustainable nanocomposites as platforms for genetic vaccines. The use of biodegradable materials for customized medical devices addresses personalized medicine by providing patient-specific solutions that are both effective and sustainable. The antibacterial characteristics of biomaterials enable it for dental applications.

The book concludes with a forward-looking view of the most recent advances in green biomaterials for biomedical implants and tissue engineering, giving a vision for the future of sustainable biomedicine.

Greater Noida, Uttar Pradesh, India

Rishabha Malviya
Sonali Sundram

Introduction

"Sustainable Nanocomposites with Green Biomaterials: Applications" explores extensively the correlation between sustainability and advanced biomedical technology. To demonstrate the revolutionary potential of nanocomposites associated with eco-friendly biomaterials in the biomedical domain, the book explores their fabrication, characteristics, and diverse applications. The book begins with an overview of bioinspired green biomaterials used in advanced therapeutics, diagnostics, and biosensors. Further, it discusses the fabrication and wide range of uses for sustainable nanocomposites, providing a solid basis for learning more advanced subjects.

The discussion then moves on to tissue-engineered cartilage products for orthopedic regeneration, electrochemical biosensors based on nano-bioengineered platforms for disease recognition, and plant-derived biomaterials for cardiac tissue repair. One of the most innovative uses of these materials in neurotechnology is the combination of hydrogels with conductive polymers for brain-computer interfaces. Modern pharmacology demonstrates the convergence of efficiency and sustainability through the use of green biomaterials as carriers for novel medication delivery methods. The book also explores the environmental and economic advantages of selective bio adsorption of metals like gold and platinum by *Aspergillus niger* biomass.

Subsequent chapters explore the development of customized nanocomposites for injection molding and fused granulate fabrication, the use of terpenes as natural plasticizers for biopolymer enhancement, and the advancement of environmentally friendly biomedicine through sustainable nanocomposites as platforms for genetic vaccines. Biodegradable materials for customized medical devices address personalized medical care, while antimicrobial biomaterials for dental healthcare applications address the practical impact of sustainable biomaterial innovations.

Current developments in polyhydroxyalkanoates show advancement in the commercialization of sustainable solutions by bridging the knowledge gap between laboratory research and real-world applications. Novel studies on the changes in dielectric characteristics of cellulose-based biocomposites underwater aging and the regulated release of limonene by alginate-shelled oil in water microemulsion liquid marbles explain the detailed research leading the breakthroughs.

The book concludes with the most recent advancements in green biomaterials for tissue engineering and biomedical implants, offering an optimistic viewpoint on the future paths of this rapidly evolving topic. This book is an important resource for academics, professionals, and students studying biomedical engineering, materials science, and sustainable technology since each chapter offers a distinct perspective on the many uses and advantages of sustainable nanocomposites.

In an attempt for a more sustainable and appropriate biomaterial advancement, "Sustainable Nanocomposites with Green Biomaterials: Applications" is an extensive resource and an excellent source of direction for continued innovation.

Contents

Bioinspired Green Biomaterials for Advanced Therapeutics, Diagnostics and Biosensors

1

Maurício Dalla Costa Rodrigues da Silva, Daniel Moro Druzian, Tiago Moreno Volkmer, and William Leonardo da Silva

Abstract

Biomaterials are a type of materials that present interaction with biological systems for biomedical applications. The use of biomaterials in devices for tissue engineering, biosensors, drug delivery, and implants has grown significantly over the past decades due to advances in the field. However, the production of biomaterials can be achieved by several methods that raise concerns about environmental burdens. The continuous and increasing awareness related to the environment and sustainability on the planet Earth has boosted the search for innovations to produce biomaterials that align performance and reduction of pollution. In this scope, green methods use less toxic and harmful chemicals in manufacturing, in addition to reducing the waste generation and pollution of water bodies and the atmosphere. One promising strategy is the reuse of agro-industrial wastes to produce green biomaterials. These bioinspired green biomaterials have the potential to be applied in smart and advanced devices for therapeutics, diagnostics, and biosensors. However, besides the benefits for the environment, there are drawbacks related to high costs, control of size and morphology as well as the lack of information about green biomaterials for medical purposes. In this context, this chapter aims to showcase the current status of the use of green biomaterials for advanced biomedical applications, accentuating the innovations in the area.

M. D. C. R. da Silva · D. M. Druzian · W. L. da Silva (✉)
Applied Nanomaterials Research Group (GPNAp), Franciscan University (UFN), Santa Maria, RS, Brazil
e-mail: w.silva@ufn.edu.br

T. M. Volkmer
Technological Development Center, Materials Engineering, Federal University of Pelotas (UFPel), Pelotas, RS, Brazil

R. Malviya, S. Sundram (eds.), *Sustainable Nanocomposites with Green Biomaterials*, Biomaterials, Bioengineering and Sustainability 2,
https://doi.org/10.1007/978-3-031-79110-9_1

Keywords

Biomaterials · Green synthesis · Sustainability · Biosensors · Biomedical applications

Abbreviations

ASSURED	Affordable, Sensitive, Specific, User-friendly, Rapid and robust, Equipment-free, and Deliverable to end-users
AuNCs	Gold nanoclusters
AuNCs/PPI NPs	Gold nanoclusters combined with pea protein isolate
AuNPs	Gold nanoparticles
CdSe QDs	CdSe quantum dots
CMC	Carboxymethylcellulose
ConA	Concanavalin A
CQDs	Carbon quantum dots
DPPH	2,2-Diphenyl-1-picrylhydrazyl
FRET	Fluorescence resonance energy transfer
GCDs	Green carbon dots
GO	Graphene oxide
GOx	Glucose oxidase
KLNP	Kraf liginin nanoparticles
LSPR	Localized surface plasmon resonance
MCDs	Multicolor-emitting carbon dots
MPA	Mercaptopropionic acid
MSCs	Mesenchymal stem cells
MWCNTs	Multiwalled carbon nanotubes
NPs	Nanoparticles
OLNP	Organosolv liginin nanoparticles
PdNPs	Palladium nanoparticles
Pd-rGO	Palladium nanoparticles anchored on reduced graphene oxide nanocomposite
PL	Photoluminescence
POC	Point-of-care
PPI	Pea protein isolate
QDs	Quantum dots
REASSURED	Real-time connectivity, Ease of specimen collection, Affordable, Sensitive, Specific, User-friendly, Rapid and robust, Equipment-free, and Deliverable to end-users
rGO	Reduced graphene oxide
SERS	Surface-enhanced Raman scattering
WHO/TDR	World Health Organization Special Program for Research and Training in Tropical Diseases
ZnS-QDs	ZnS fluorescent quantum dots

1.1 Introduction

Over the last decades, scientific progress in material science has provided discoveries that revolutionized the medical field. Application of new materials for application in treatment, diagnostics, and monitoring flourished as a fresh new field with a wide variety of research opportunities and development of new biomedical devices (Sinha et al. 2022). Thus, due to this significant progress, healthcare was positively impacted and the quality of life of patients affected by different types of illnesses has been improved, in addition to new perspectives for more effective and less aggressive treatments (Himanen et al. 2019).

In this field, biomaterials are defined as materials (synthetic or natural) that interact with biological systems to promote the treatment, replacement, repair, or diagnostics on living bodies (Iqbal et al. 2023). There is a vast variety of naturally available materials that possess these functions and have been used in medical applications. For instance, gypsum (hydrated calcium sulfate) started to be used in orthopedics as bone filling at the end of the nineteenth century (Grado et al. 2018).

With the discoveries in materials science and medicine, the field of biomaterials developed new dispositive for therapeutics, diagnostics, and biosensing based on the different interactions of this class of materials in the body environment. Prominent advances in synthetic biomaterials have been made due to the control of key biocompatibility features, such as morphology, mechanical resistance, chemical composition, purity, and surface functionalization, for instance (Begum et al. 2019). However, the synthesis and manufacture of such biomaterials are often costly, with high risk, and generate environmental impacts due to the high energy consumption and emissions of greenhouse gases (Le Feuvre and Scrutton 2018).

Due to the growing negative effects of environmental pollution on Planet Earth, eco-friendly alternatives for the production of biomaterials for biomedical applications flourish as potential solutions. In this context, the adoption of natural biomaterials has acquired increasing recognition in the medical field, as well as in terms of decreasing the impacts generated in manufacturing (Rajvanshi et al. 2023).

Natural biomaterials are materials synthesized by living organisms (e.g., as fungi, bacteria, plants, and algae) and present significant importance for biomedical applications. Some natural biomaterials include collagen, chitosan, cellulose, and alginate for example (Insuasti-Cruz et al. 2022). These natural biomaterials are produced by renewable resources and are categorized as green biomaterials. Thus, these materials are suitable to manufacture bioinspired structures for biomedical applications, reducing the environmental impacts during the production phases (Biswal et al. 2020).

The rise of green synthesis in the last decades also provides a new perspective on green biomaterials. In green synthesis, bioactive molecules, such as flavonoids, anthocyanins, and vitamins, can be used the replace traditional chemical-reducing agents, which are hazardous and toxic substances (Rahman et al. 2022). These bioactive molecules act as reducers, but also in the nucleation and stabilization of the green particles (Rana et al. 2020).

Another important feature of green biomaterials is the recovery or synthesis of the desired molecules. For instance, agro-industrial residues include a wide variety of plant-derived materials (e.g., peels, pulps, leaves, bagasse, and shells) with low cost and abundant availability (Lopes and Ligabue-Braun 2021). Agro-industrial residues are a promising alternative to recover cellulose fibers, polysaccharides, and hyaluronic acid, as well as substrates for microbial production (Souza et al. 2022; Tsouko et al. 2024). Thus, these agro-industrial residues provided alternative sources of bioactive molecules with valuable use in green synthesis (Osman et al. 2024) and for use in bioinspired materials (Reguengo et al. 2022; Sorrenti et al. 2023).

Therefore, these high-added-value products are recovered for low-cost materials that would cause damages to the environment due to the incorrect disposal or the inefficient transformation of these residues (Freitas et al. 2021). However, despite the mentioned advantages, there are still barriers regarding the usage of green biomaterials for biomedical applications. In this context, this chapter aims to present information related to bioinspired green materials as well as their applications in therapeutics, diagnostics, and biosensors. Moreover, the faced challenges and opportunities for applications of biomaterials are also discussed to promote new outlooks for this topic.

1.2 Bioinspired Green Biomaterials

Green bioinspired materials or biomimetics are some terms used to describe biological structures, functions, and processes to explain and find new solutions to problems related to scientific areas such as biomedicine, and nanobiotechnology (Chen et al. 2019). Additionally, green bioinspired materials vary from the characteristics for imitation presented in Fig. 1.1 such as: (a) particle size, and shape; (b) electronic and molecular interactions; (c) general and specific properties of cells; (d) mechanisms of action and molecular mechanisms; (e) types of organs and biosensors; and (f) functioning of the organism in general (Gou et al. 2014; Li et al. 2019). It is worth mentioning that bioinspiration is a popular trend in many fields and areas such as chemistry, physics, aerodynamics, robotic navigation, materials science, engineering, and medicine (Xiong et al. 2023; Zhang et al. 2023).

In recent years there has been an increase in research based on green bioinspired strategies. However, the projects and reviews presented lacked a holistic interdisciplinary approach mainly lacking an in-depth analysis focused on the performance of results as an example in the areas of energy efficiency of materials and structures, application of the material in civil structures, bioarchitecture, transport systems, and medicinal application involving clinical analysis (Fu et al. 2020; Zhang et al. 2015; Lazarus et al. 2020).

Explaining green bio-inspiration, the nature has existed for centuries, where its wonder has recently been understood in diverse details relating to the area of nanoscience specifically in biological evolution revolutionizing a broad function of materials with new structures providing advantages and solving current problems

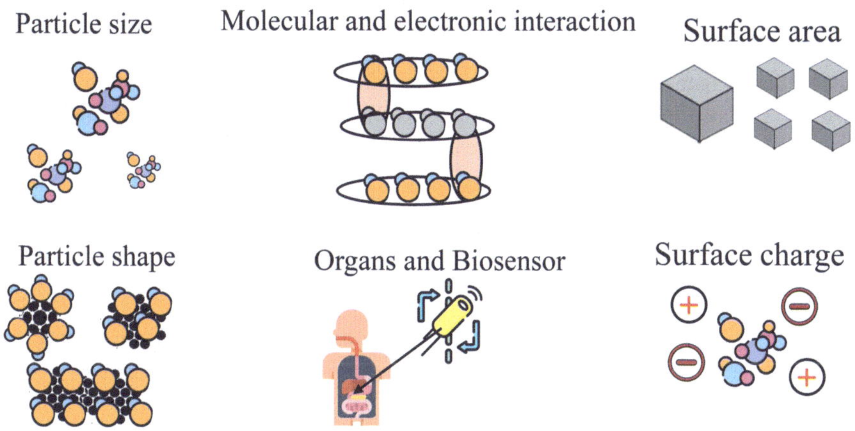

Fig. 1.1 General characteristics of bioinspired materials: shape, particle size, charge, surface area, molecular interactions, and applications in biosensors involving human organs. (Source: Author's construction (2024))

(Zingre et al. 2023). In this sense, innovative properties such as high surface area, structures containing large pores of some materials optimized by nature can be improved using chemical products involving engineering, designing biomimetic devices, and manufacturing biomaterials as singular solutions, and healthcare applications (Al-Obaidi et al. 2017). Emphasizing, nature provides bioinspired structures that are trustworthy, biocompatible, and easily obtained for various extraction methods. Human tissue mimetics, biosensors to simulate the stimulus-responsive activity of natural organisms, and drug delivery devices are some illustrations of bioinspired materials with biomedical applications (Shashwat et al. 2023).

1.3 Applications in Advanced Therapeutics

Bioinspired and biomimetic materials have a high potential for advanced treatments offering an alternative or complement to traditional treatment modalities such as the high capacity to generate oxidative radicals inducing biological effects and inhibit bacterial and tumor cells (Fooladi et al. 2023). It is worth noting that each bioinspired material has distinct properties ranging from particle size, agglomeration, surface charge, porosity, locomotion of particles, and surface area (Lian et al. 2022). Table 1.1 presents the different bioinspired materials, properties, and results for the advanced treatments.

1.3.1 Antibacterial Activity and Surfaces

The biggest challenge currently in pharmaceuticals involving the use of antibiotics is the occurrence of bacterial resistance, where they generally use antibacterial agents with excessive costs and high concentrations of treatments (Johnson et al. 2022). In this way, alternative methods have been researched for the development of antibacterial agents based on nature structures such as certain surfaces of plants and insects (geckos, dragonflies, and cicadas) that can protect against pathogenic bacteria (Cao et al. 2020). Highlighting, the antibacterial activity of agents from nature is mainly due to the superhydrophobic surface of the lamellae structures around 50–250 nm preventing bacterial growth (Tripathy et al. 2017).

Some materials as the diamond cones, titania nanowires, polystyrene, and gold surfaces are based on the characteristics of the Cicada insect wings, where they present antibacterial activity against *Pseudomonas aeruginosa* because of the thickness of the peptidoglycan cell wall (Watson et al. 2015). Furthermore, the plant structures such as Lotus leaves (*Nelumbo nucifera*) are used to produce black silicon based on reactive ionic attack of silicon to inhibit bacterial agents (Ivanova et al. 2013).

One of the most used bioinspired materials for antimicrobial activity are chitosan and chitin peptide, being natural products extracted from the shells of insects and microcrustaceans, presenting biodegradability, absence of toxicity and excellent antibacterial activity on all bacteria, except *Salmonella typhimurium* (Johnson et al. 2022). The main mechanism of action of these structures is the destruction of bacterial cell contents for crossing the membranes and cell lysis due to the reactivity of the ions in these structures (Potbhare et al. 2020).

1.3.2 Anti-inflammatory Activity

The organism's inflammatory response is composed by a series of an intricate biochemical reactions that involves many variables generally cellular, and chemical modulators protecting and responding against different infections (Xu et al. 2024). Additionally, the inflammatory problems involve the rheumatoid arthritis, gastritis, pneumonia, and psoriasis, among others. In this way, bioinspired materials for anti-inflammatory activity mimic the immune cells specifically the cell membrane blocking cellular responses and combating inflammation (Yu et al. 2024).

Some examples such as the surface of a liposome with lymphocyte directional ligands, macrophage adhesion-1 and P-selectin-1 glycoprotein form a proteolipid vesicle targeted and transported to inflammatory sites with lower toxicity for treatment of cancer (Jin et al. 2018). One of the most inspired antioxidants are nanocapsules coated with cytokines containing different macrophage membranes and polysaccharides blocking some specific cytokines and increasing macrophage polarization for tissue regeneration activity (Yin et al. 2020).

1.3.3 Soft and Hard Tissue Engineering

Hard and soft tissues are constituted of clusters of cells, where the tissue production in vivo depends on the adhesion of osteoblasts and fibroblasts to the internal cytoskeleton (Altuntas et al. 2024). In this sense, the introduction of bioinspired structures relating to nanotechnology in the field of tissue regeneration highlighted in heart diseases such as myocardial infarction has achieved remarkable results in repairing injured tissues (Hossain et al. 2024). The use of structures such as graphene oxide, carbon nanotubes, hydrogels, and nanoparticles of silica, iron, and gold mimic the electrophysiological nature and the cardiovascular morphology, allowing successful tissue restoration (Wang et al. 2016).

Examples of bioinspired materials are cardiac tissue adhesives based on the conductive material cryogel and dopamine inspired for mussels (*Mytilus edulis*) inducing myocardial repair (Zhang et al. 2016). Furthermore, silicon tissue with a rough surface is produced by hydrophilic laser in the absence of chemotropic factors for tissue development applications (Stratakis et al. 2011).

1.4 Applications in Diagnostics

Correct diagnosis is important for providing accurate treatment possibilities and prognosis and preventing avoidable patient harm (Centor et al. 2019). Commonly, the expense of early testing is significantly less than that of extensive treatment later (Zhou et al. 2015). Green-inspired biomaterials represent a promising path for creating diagnostic tools that are not only effective but also environmentally considerate. By prioritizing sustainability, they enable the development of diagnostic technologies that reduce environmental impact and enhance healthcare benefits, leading to a new era of responsible innovation in healthcare diagnostics (Hosseini et al. 2021). In recent years, researchers are actively developing various types of prototypes to diagnose multiple diseases. Depending on the application, the decision to use bioimaging or biosensing is often essential for obtaining superior outcomes (Peréz et al. 2019).

Bioimaging provides real-time-non-invasive visualization of biomolecules and biological events, offering significant potential for early diagnostics and therapy (Gu et al. 2019). In optical imaging, targetability is crucial—it enhances signal intensity in specific cells or tissues compared to surrounding areas, providing detailed information about biomarkers and organelles. This is vital for distinguishing between healthy and diseased states. However, poorly targeted optical probes may result in low contrast and accuracy, making small organelles difficult to detect (Zhu et al. 2016). To improve targetability, several design strategies are employed in developing fluorescent probes. These include actively binding to specific organelles such as the cytoplasm (Zheng et al. 2023), mitochondria (Zhu et al. 2020), lysosomes (Ke et al. 2022), and the Golgi apparatus (Zhang et al. 2020).

Among bioinspired biomaterials, the carbon family has been widely used for bioimaging applications. Green carbon dots (GCDs) are particularly notable for their bioimaging applications (Chahal et al. 2021). As zero-dimensional nanomaterials, GCDs display several interesting properties such as biocompatibility, high stability, biocompatibility, and low toxicity. These properties facilitate the creation of multicolor-emitting carbon dots (MCDs) (Sun et al. 2006; Du et al. 2019). Doping carbon dots with transition metal ions enhances their photoluminescence, making them potential fluorescent nanoprobes for bioimaging applications (Pakkath et al. 2018). An eco-friendly method to synthesize nitrogen and sulfur co-doped carbon dots using adenine sulfate as a precursor was reported, where these GCDs were water-soluble and exhibited tunable fluorescence from blue to green and yellow, demonstrating good biocompatibility with HeLa cells (Dang et al. 2023). Another method involves carbonization, where a microwave was used to carbonize chitosan, producing chitosan-derived Carbon Quantum Dots (CQDs) with stable blue fluorescence. By adding ethylenediamine as a dopant, the optical properties of these carbon dots were significantly enhanced, increasing their fluorescence intensity. CQDs penetrated the *Candida albicans* cell membrane, emitting blue, green, and red light according to the confocal laser microscope images (Oliveria et al. 2023). An innovative approach for preparing GCDs using sliced bread from local supermarkets as a precursor, comparing toasting and hydrothermal techniques was also reported (Anpalagan et al. 2023). The samples produced by both methods fluoresced without additional modifications or doping. These GCDs were non-toxic and effectively imaged both human and mouse colon cancer cells, crossing the nuclear membrane and remaining in the cytoplasm.

Other members of the carbon family used for bioimaging include graphene and graphene oxide (GO). Using fruit waste from *Quercus ilex*, potassium-doped graphene oxide (K-doped GO) was synthesized, which featured naturally occurring blue photoluminescence (PL) and a potassium content of 6.8%. The K-doped GO demonstrated low toxicity, intracellular localization, excellent biocompatibility, and robust PL properties (Tewari et al. 2019). Additionally, by combining sulphur/nitrogen-doped graphene with ZnO nanograins through a simple hydrothermal process, a heterostructure with enhanced fluorescence properties was obtained (Sheik et al. 2024). The sulfur/nitrogen-doped graphene, derived from the agricultural waste of cabbage, exhibited red and green fluorescence. In contrast, ZnO nanoparticles displayed green fluorescence.

While carbon-based nanomaterials like graphene and GO have shown promising results in bioimaging, semiconductor quantum dots (QDs) offer distinct advantages due to their unique optical properties. Semiconductor QDs, with their dimensions reduced to the nanoscale, exhibit restricted electron motion, creating an electronic structure similar to atomic with energy levels that depend on the size. This confinement allows the tailoring of nanomaterials that feature broad light absorption that can be tuned along with pure and bright color emission. Moreover, the ability to control electronic transport and customize chemical and physical properties due to their large surface-to-volume ratio enhances their utility in bioimaging applications (Arquer et al. 2021). Tuning the confinement potential in semiconductor quantum dots can effectively manipulate electronic states and optical transitions (Singha et al. 2009), making them very promising for advanced bioimaging techniques. Stabilization of ZnS fluorescent quantum dots (ZnS-QDs) using carboxymethylcellulose (CMC) as a pH-sensitive biopolymer was synthesized and characterized by a simple one-step green aqueous colloidal process (Caires et al. 2020)(. The nanoconjugates (ZnS@CMC) created under varying pH conditions and chemical proportions of precursors exhibited size-dependent photoluminescence properties, showing high potential as active fluorescent nanoprobes for bioimaging malignant glioma cells, suggesting promising applications in cancer nanomedicine. In another study, semiconducting CdSe quantum dots (CdSe QDs) were obtained using a green route. These CdSe QDs presented higher fluorescence in water-soluble systems. After surface modification with streptavidin, biotin was bonded to the CdSe QD aiming for selective cancer cell targeting. Results from cellular imaging and cytotoxicity experiments conducted in HEK-293 and HeLa cells demonstrated that the CdSe QDs were effective for imaging (Vijaya et al. 2020).

As semiconductor quantum dots like ZnS and CdSe revolutionize bioimaging with their size-dependent photoluminescence and tunable optical properties, another category of nanomaterials, metallic nanoparticles—particularly gold nanoparticles (AuNPs)—offers further advancements in this area. Gold nanoparticles are gaining prominence in bioimaging due to their exceptional optical properties. They can be tailored to particular sizes and morphologies, enabling them to demonstrate strong and adjustable phenomena such as localized surface plasmon resonance (LSPR), fluorescence resonance energy transfer (FRET), surface-enhanced Raman scattering (SERS), nonlinear optical properties, and the quantized charging effect (Lee and Jun 2019). Notably, gold nanoparticles are valued for their inherent fluorescence and biocompatible surface features. In a significant development, pea protein isolate (PPI) has been employed as both a reducing and stabilizing agent to produce gold nanoclusters (AuNCs) with intense red fluorescence. When combined with PPI, these AuNCs self-assemble into nanoparticles (AuNCs/PPI NPs) approximately 100 nm in diameter. The protein-based nature of PPI ensures that these nanoparticles exhibit excellent biocompatibility and colloidal stability, making them well-suited for in vivo bioimaging in MCF-7 tumor-bearing nude mice (Li et al. 2018).

1.5 Applications in Biosensors

A biosensor is an instrument that combines a biological component (e.g., nucleic acids, enzymes, antibodies, or aptamers) with an electronic element to produce a quantifiable signal (Naresh and Lee 2021). Due to its ample range of applications, biosensors are widely employed in diagnostics devices, in the last years, for monitoring and treatment of disease progression, drug discovery, and biomedical research (Abolhasan et al. 2019). Typically, a biosensor is an equipment device or probe that consists of 5 parts as presented in Fig. 1.2: (i) an analyte, like an antibody, hormone, or enzyme, (ii) a bioreceptor, a molecule that can bind specifically with the desired analyte, generating a signal that can be recognized by (iii) a transducer, a component that converts that signal into a measurable optical or electrical, with quantity proportional to the amount of signal generated. The transduced signal is processed, converted, and amplified by the (iv) electronics and finally is processed and quantified by the (v) display unit, which generates the output in a way that can be interpreted by the final user (Naresh and Lee 2021).

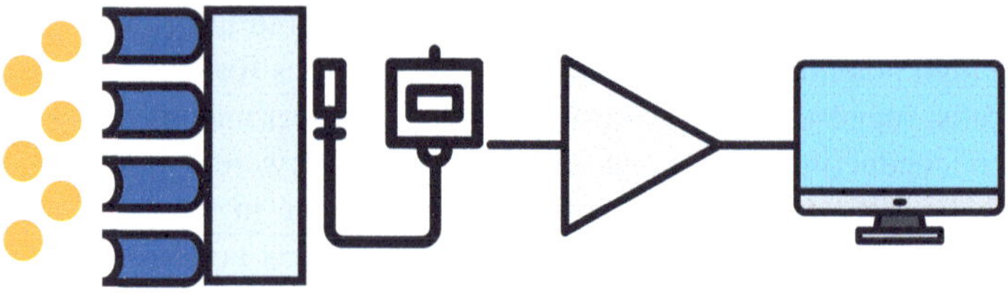

Fig. 1.2 Biosensor components

In many biosensors employed in diverse applications, silicon is the predominant material, along with petroleum-derived polymers, while the electrodes are crafted from various metals. Compared to synthetic alternatives, bio-inspired materials present enhanced potential in terms of efficacy, sensitivity, selectivity, and biocompatibility [60]. The development of eco-friendly biosensors is crucial for ensuring sustainability in environmental monitoring and healthcare. Researchers worldwide are striving to develop devices that offer significant advantages by minimizing the use of toxic materials and reducing waste (Nguyen and Yoon 2016).

The World Health Organization Special Program for Research and Training in Tropical Diseases (WHO/TDR) has endorsed the ASSURED (Affordable, Sensitive, Specific, User-friendly, Rapid and robust, Equipment-free, and Deliverable to end-users) criteria as a pivotal benchmark for assessing the efficacy of diagnostic tests in disease control (Otoo and Schlappi 2022). Meanwhile, with the rapid evolution of digital technology and mobile health (m-health), there is a growing imperative for future diagnostic tools to integrate these principles. This integration has the potential to create REASSURED diagnostic systems (Real-time connectivity, Ease of specimen collection, Affordable, Sensitive, Specific, User-friendly, Rapid and robust, Equipment-free, and Deliverable to end-users). Such systems have the potential to revolutionize disease control strategies in real time, reinforce the efficacy of healthcare systems, and ultimately enhance patient outcomes (Land et al. 2019).

The miniaturization of electronic components has not only spurred the proliferation of 'personal' computing and facilitated the rise of smartphones and other handheld devices, but it has also enabled the development of point-of-care (POC) and wearable biosensors (Brangel et al. 2018; Kim et al. 2019; Bhardan et al. 2023). This chapter will delve into the advancements and applications of point-of-care (POC) highlighting their impact on modern healthcare.

1.5.1 Point-of-Care Biosensors

A point-of-care (POC) biosensor is a type of device designed to be used in medical settings where quick diagnostic results are needed, directly at the site of patient care (Moulahoum et al. 2023). In this context, POC diagnostics serve as decentralized, rapid, highly sensitive, and cost-effective diagnostic tools, enhancing patient care and supporting outbreak prediction capabilities. POC devices facilitate the utilization of more portable and economically viable instrumentation, eliminating the need for sample transportation to centralized clinical laboratories for further analysis, restructuring sample processing procedures, allowing not trained persons to collect samples, and enabling multiplexed detection of diverse analytes (Valera et al. 2021), including viruses (Wang et al. 2022), and antibodies (Brangel et al. 2018). In the scene of the COVID-19 pandemic, the imperative to develop quick, accurate, and portable home-testing point-of-care devices became vital. These innovations were indispensable in promptly identifying infected individuals and curbing the transmission of the virus, emphasizing the critical need for accessible and efficient diagnostic solutions during public health crises (Ertas et al. 2021).

Following the principle of biorecognition, biosensors are classified into two categories: non-catalytic and catalytic biosensors. In catalytic biosensors, the reaction between the analyte and the bioreceptor leads to the formation of a new biochemical product. In this case, enzymes are frequently used as a crucial component. Enzyme-based biosensors provide excellent sensitivity, specificity, portability, cost-effectiveness, and miniaturization for various applications in clinical analysis, food safety, and disease monitoring (Naresh and Lee 2021; Nguyen and Yoon 2016).

For sensitive and selective detection of triglycerides, a green route to synthesize Palladium nanoparticles (PdNPs) anchored on reduced graphene oxide (rGO) nanocomposite (Pd-rGO) was successfully achieved (Singh et al. (2016). Fenugreek seeds were used both as reduction and stabilizing agents for PdNPs and rGO. To create the biosensor, lipase and glycerol dehydrogenase enzymes were co-immobilized on the Pd-rGO platform allowing the enzymes to remain active and stable on the surface of the biosensor. The fabricated Pd-rGO-based biosensor exhibited a wide detection range and excellent sensitivity.

Lignin is a renewable raw material generated as a byproduct of various industries and has also been used for laccase (Capecchi et al. 2019) and tyrosinase (Capecchi et al. 2018). Lignin nanoparticles are used for the immobilization of redox enzymes to ensure the correct distance and orientation between the enzymes for the production of biosensors for glucose detection (Capecchi et al. 2020). Thus, it was tested sulfur-free organosolv liginin nanoparticles (OLNP) and sulfur lignin kraf liginin nanoparticles (KLNP) to modify a gold bare electrode. The modified electrodes were properly assembled with concanavalin A (ConA) and glucose oxidase (GOx), forming a three-layered electrode and applied as biosensing support for glucose (Tortolini et al. 2021). Lignin-derived disposable electrodes for electrochemical lactate biosensors were developed and applied. A green method was demonstrated to convert biomass lignin into graphene electrodes, the conversion involved energy-efficient and localized laser lithography converting lignin into porous laser-induced graphene using water as a green solvent (Meng et al. 2022).

In addition to pursuing the development of bio-inspired receptors, scientists have also been working on developing eco-friendly inks for POC devices. Eco-friendly conductive inks are increasingly important in the development of biosensors due to their reduced environmental impact and potential for innovation in medical diagnostics and environmental monitoring. These inks are formulated from biodegradable and non-toxic materials, which is a shift away from traditional inks that often contain harmful solvents and heavy metals (Camargo et al. 2022).

An innovative approach to fabricate carbon-based conductive inks utilizing carbonized lignin from *Eucalyptus globulus* tree agro-industrial wastes of the paper industry was introduced (Zappi et al. 2021). The new ink formulation involved the use of cyclohexanone/cellulose acetate/carbonized lignin to prepare the electrodes that exhibited an electrochemically active area that outperformed petroleum-based electrodes with similar geometry. The electrodes were tested to quantify uric acid and laccase enzyme, achieving good results for a limit of detection. A water-based conductive ink using chitosan (53.0%) as the polymeric base in the formulation procedure, graphite powder (45.0%), and glycerol (2.0%) was developed (Camargo et al. 2022). The mixture was used in the production of disposable electrodes by screen-printing. Thus, the electrode surface was modified using tyrosinase, dyhexadecyl phosphate, and multiwalled carbon nanotubes (MWCNTs) for the detection of catechol.

1.6 Challenges and Opportunities

Green biomaterials present potential applications for several biomedical applications as stated throughout the prior sections. Their usage for therapeutics, diagnostics, and biosensors has been the focus of diverse studies in the last few years. The advantages of using green biomaterials rely on its availability of sources for production, which could help the development of new healthcare products with less environmental impact. However, this has not been the main trend followed over the years.

Even with the prominent advances in the field of bioinspired green biomaterials, the scale-up from laboratory to industrial manufacturing still faces challenges. These challenges are most associated with the requirement for high production yields and do not count the environmental burdens associated with the manufacturing process. For instance, the production of fibers often uses hazardous organic solvents, such as chloroform. The question asked is simple: Is this procedure environmentally friendly?

In the field of biomaterials, the focus is mainly on the performance of the bioinspired device. In this scope, the carbon emissions, energy consumption, and waste treatment end up not being framed in the overall scope (at least in the priority that should be taken in the 2020s). The preparation of bioinspired biomaterials often generates a large amount of waste with potential for environmental damage. Yet, the main focus is related to the toxicity of the biomaterials (Murphy 2008). Another simple question is raised: Is this the only toxic effect that should be considered?

The bioinspired green biomaterials present inherent advantages to their use in biomedical devices, as presented in the topics of this chapter. Natural polymers, such as chitosan, for instance, present numerous biomedical applications and a natural antibacterial activity (Ul-Islam et al. 2024). Cellulose is another widely available polymer with surprising benefits for biomedical devices (Hasanin 2022). The alternative sources for production of these two polymers, such as agro-industrial residues, could be an eco-friendly approach to add value for the residues and a sustainable method for producing natural polymers in large scale.

The green synthesis of nanoparticles for bioinspired applications is another great example. Substituting traditional reducing agents with naturally available biomolecules present in several agro-industrial residues is a promising alternative for bottom-up methods. Several studies highlighted, not only this aspect but also the new opportunities for the use of these new green biomaterials (Noah and Ndangili 2022).

Another important point is the energy consumption. The main methods for the production of biomaterials often need ultrapure reactants, high temperatures, and ultrahigh vacuum, which requires significant energy demands for these conditions to be reached. Thus, the use or production of toxic or hazardous substances is present (Patwardhan, Manning and Chiacchia2018). Still, these conditions have been embraced for bioinspired biomaterial production despite these drawbacks.

Natural biomaterials present biodegradability, reducing the harmful effects of environmental pollution. Therefore, the efforts in the research for the bio-fabrication of bioinspired structures using green precursors could bring new insights for the scale-up from the laboratory to the industrial fabrication along with the adoption of green manufacturing approaches (Kapoor et al. 2022). However, the principles of green chemistry and engineering have to be encouraged in the initial stages of the research development of bioinspired biomaterials. Thus, the maximization of efficiency and the minimization of negative effects on the environment throughout all the development stages is essential to promote better results. These better results, however, not only focused on the performance of the biomedical devices, but on the longevity and sustainability of our planet.

1.7 Future Remarks

The advances in the biomaterials field promote the development of new alternatives for treatment, diagnostics, and biosensors of a variety of human dysfunctions. The prominent application of such materials enables new paths for healthcare applications. The production of bioinspired biomaterials using naturally available resources promotes significant advances in the biomedical sciences. The use of low-cost agro-industrial residues could bring new horizons in the research field regarding the availability of raw green biomaterials for biomedical applications as well as aligned with the trend for eco-friendly technologies. However, the sustainability and applicability of bioinspired green biomaterials face challenges related to the scale-up from laboratory to industrial level, adoption of green manufacturing approaches, efficiency maximization, and reduction of environmental impacts. The new actions in this field have to focus on the health of the living organisms, but not forgetting that the ecosystem is an important organism that provides resources to promote a better quality of life. Therefore, the future trends in biomaterials for biomedical applications need to tackle their impacts on the environment and stimulate the development of greener alternatives for this field.

Acknowledgements The authors would like to thank the Applied Nanomaterials Research Group (GPNAp) at the Franciscan University (UFN) for the support in carrying out this work. This study was financed by the Conselho Nacional de Desenvolvimento Científico e Tecnológico—Brazil (CNPq/MCTI/FNDCT n° 23/2022)—InovaNióbio (n° 408422/2022-0).

References

Abolhasan R, Mehdizadeh A, Rashidi M et al (2019) Application of hairpin DNA-based biosensors with various signal amplification strategies in clinical diagnosis. Biosens Bioelectron 129:164–174

Al-Obaidi KM, Ismail MA, Hussein H et al (2017) Biomimetic building skins: an adaptive approach. Renew Sust Energ Rev 79:1472–1491

Altuntas U, Coker D, Yavas D (2024) Bioinspired soft-hard interfaces fabricated by multi-material additive manufacturing: a fracture mechanics investigation using essential work of fracture. Mech Mater 195:105035–105035

Anpalagan K, Karakkat JV, Jelinek R et al (2023) A green synthesis route to derive carbon quantum dots for bioimaging cancer cells. Nano 13:2103–2118

Arquer FPG, Talapin DV, Klimov VI et al (2021) Semiconductor quantum dots: technological progress and future challenges. Science 373:eaaz8541

Begum S, Hassan Z, Bräse S et al (2019) Metal–organic framework-templated biomaterials: recent progress in synthesis, functionalization, and applications. Acc Chem Res 52:1598–1610

Bhardan NM, Radisic M, Nurunnabi M (2023) Bioinspired materials for wearable diagnostics and biosensors. ACS Biomater Sci Eng 9:2015–2019

Biswal T, Badjena SK, Pradhan D (2020) Sustainable biomaterials and their applications: a short review. Mater Today Proc 30:274–282

Brangel P, Sobarzo A, Parolo C et al (2018) A serological point-of-care test for the detection of IgG antibodies against Ebola virus in human survivors. ACS Nano 12:63–73

Caires AJ, Mansur AAP, Carvalho IC et al (2020) Green synthesis of ZnS quantum dot/biopolymer photoluminescent nanoprobes for bioimaging brain cancer cells. Mater Chem Phys 244:122716–122730

Camargo JR, Almeida TA, Rivas GA et al (2022) Novel eco-friendly water-based conductive ink for the preparation of disposable screen-printed electrodes for sensing and biosensing applications. Electrochim Acta 409:139968–139976

Cao G, Wang C, Fan Y et al (2020) Biomimetic SIS-based biocomposites with improved biodegradability, antibacterial activity, and angiogenesis for abdominal wall repair. Mater Sci Eng C 109:110538–110552

Capecchi E, Piccinino D, Delfino I et al (2018) Functionalized tyrosinase-lignin nanoparticles as sustainable catalysts for the oxidation of phenols. Nano 6:438–454

Capecchi E, Piccinino D, Bizzarri BM (2019) Enzyme-lignin nanocapsules are sustainable catalysts and vehicles for the preparation of unique polyvalent bioinks. Biomacromolecules 20:1975–1988

Capecchi E, Piccinino D, Tomaino E et al (2020) Lignin nanoparticles are renewable and functional platforms for the concanavalin an oriented immobilization of glucose oxidase-peroxidase in cascade bio-sensing. RSC Adv 10:29031–29042

Centor R, Geha R, Manesh R (2019) The pursuit of diagnostic excellence. JAMA Netw Open 2:e1918040–e1918043

Chahal S, Macairan J, Yousefi N et al (2021) Green synthesis of carbon dots and their applications. RSC Adv 11:25354–25363

Chen S, Matsumoto H, Moro-oka Y et al (2019) Microneedle-array patch fabricated with enzyme-free polymeric components capable of on-demand insulin delivery. Adv Funct Mater 29:1807369–1807377

Dang DK, Nguyen VN, Tahir Z et al (2023) An efficient green approach to constructing adenine sulfate-derived multicolor sulfur- and nitrogen-codoped carbon dots and their bioimaging applications. ACS Appl Mater Interfaces 15:32783–32791
Du J, Xu N, Fan J et al (2019) Carbon dots for in vivo bioimaging and Theranostics. Small 15:1805087–1805102
Ertas YN, Mahmoodi M, Shahabipour F et al (2021) Role of biomaterials in the diagnosis, prevention, treatment, and study of corona virus disease 2019 (COVID-19). Emmergent Mater 4:35–55
Fooladi S, Nematollahi MH, Iravani S (2023) Nanophotocatalysts in biomedicine: cancer therapeutic, tissue engineering, biosensing, and drug delivery applications. Environ Res 231:116287–116300
Freitas LC, Barbosa JR, Costa ALC et al (2021) From waste to sustainable industry: how can agro-industrial wastes help in the development of new products? Resour Conserv Recycl 169:105466–105479
Fu SC, Zhong XL, Zhang Y et al (2020) Bioinspired cooling technologies and the applications in buildings. Energ Buildings 225:110313–110327
Gou M, Qu X, Zhu W et al (2014) Bio-inspired detoxification using 3D-printed hydrogel nanocomposites. Nat Commun 5:3774–3783
Grado GF, Keller L, Idoux-Gillet Y et al (2018) Bone substitutes: a review of their characteristics, clinical use, and perspectives for large bone defects management. J Tissue Eng 9:1–18
Gu K, Zhu W, Peng X (2019) Enhancement strategies of targetability, response and photostability for in vivo bioimaging. Sci China Chem 62:189–198
Hasanin MS (2022) Cellulose-based biomaterials: chemistry and biomedical applications. Starch 74:220060–220069
Himanen L, Geurts A, Foster AS et al (2019) Data-driven materials science: status, challenges, and perspectives. Adv Sci 6:1900808–1900830
Hossain A, Rajput P, Li Z et al (2024) Engineering bioinspired microfluidics; biomimetic self-healing/cleaning coating designs and unique advanced materials. J Chem Eng 489:151336–151367
Hosseini E, Dervin S, Ganguly P et al (2021) Biodegradable materials for sustainable health monitoring devices. ACS Appl Biomater 4:163–194
Insuasti-Cruz E, Suárez-Jaramillo V, Urresta KAM et al (2022) Natural biomaterials from biodiversity for healthcare applications. Adv Healthc Mater 11:2101389–21013411
Iqbal S, Sohali M, Fang S et al (2023) Biomaterials evolution: from inert to instructive. Biomater Sci 11:6109–6115
Ivanova EP, Hasan J, Webb HK et al (2013) Bactericidal activity of black silicon. Nat Commun 41:1–7
Jin K, Luo Z, Zhang B et al (2018) Biomimetic nanoparticles for inflammation targeting. Acta Pharm Sin B 8:23–33
Johnson AP, Sabu C, Nivitha KP et al (2022) Bioinspired and biomimetic micro- and nanostructures in biomedicine. J Control Release 343:724–754
Kapoor RT, Rafatullah M, Qamar M et al (2022) Review on recent developments in bioinspired-materials for sustainable energy and environmental applications. Sustain For 14:16931–16960
Ke Y, Cao J, Gong J et al (2022) A near-infrared naphthalimide fluorescent probe for targeting the lysosomes of liver cancer cells and specifically selecting HAS. Sens Actuators B 352:131015–131023
Kim J, Campbell AS, Ávila BEF et al (2019) Wearable biosensors for healthcare monitoring. Nat Biotechnol 37:389–406
Land KJ, Boeras DI, Chen XS et al (2019) REASSURED diagnostics to inform disease control strategies, strengthen health systems and improve patient outcomes. Nat Microbiol 4:46–54
Lazarus BS, Velasco-Hogan A, Gomez-del-Río T et al (2020) A review of impact resistant biological and bioinspired materials and structures. J Mater Res Technol 9:15705–15738
Le Feuvre RA, Scrutton NS (2018) A living foundry for synthetic biological materials: a synthetic biology roadmap to new advanced materials. Synth Syst Biotechnol 3:105–112

Lee SH, Jun BH (2019) Silver nanoparticles: synthesis and application for nanomedicine. Int J Mol Sci 20:865–888
Li Z, Peng H, Liu J et al (2018) Plant protein-directed synthesis of luminescent gold nanocluster hybrids for tumor imaging. ACS Appl Mater Interfaces 10:83–90
Li Z, Seo Y, Aydin O et al (2019) Biohybrid valveless pump-bot powered by engineered skeletal muscle. Proc Natl Acad Sci 116:1543–1548
Lian M, Shi Y, Chen L et al (2022) Cell membrane and V2C MXene-based electrochemical immunosensor with enhanced antifouling capability for detection of CD44. ACS Sens 7:2701–2709
Lopes FC, Ligabue-Braun R (2021) Agro-industrial residues: eco-friendly and inexpensive substrates for microbial pigments production. Front Sustainable Food Syst 5:589414–589429
Meng L, Chirtes S, Liu X et al (2022) A green route for lignin-derived graphene electrodes: a disposable platform for electrochemical biosensors. Biosens Bioelectron 218:114742–114750
Moulahoum H, Ghorbanizamani F, Beduk T et al (2023) Emerging trends in nanomaterial design for the development of point-of-care platforms and practical applications. J Pharm Biomed Sci 235:115623–115635
Murphy CJ (2008) Sustainability as an emerging design criterion in nanoparticle synthesis and applications. J Mater Chem 18:2173–2176
Naresh V, Lee N (2021) A review on biosensors and recent development of nanostructured materials-enabled biosensors. Sensors 21:1109–1143
Nguyen NS, Yoon HH (2016) Nickel oxide-deposited cellulose/CNT composite electrode for non-enzymatic urea detection. Sens Actuators B 236:304–310
Noah NM, Ndangili PM (2022) Green synthesis of nanomaterials from sustainable materials for biosensors and drug delivery. Sens Int 3:100166–100180
Oliveria BP, Bessa NUC, Nascimento JF et al (2023) Synthesis of luminescent chitosan-based carbon dots for *Candida albicans* bioimaging. Int J Biol Macromol 227:805–814
Osman AI, Zhang Y, Farghali M et al (2024) Synthesis of green nanoparticles for energy, biomedical, environmental, agricultural, and food applications: a review. Environ Chem Lett 22:841–887
Otoo JA, Schlappi TS (2022) REASSURED multiplex diagnostics: a critical review and forecast. Biosensors 12:124–139
Pakkath S, Chetty S, Selvarasu P et al (2018) Transition metal ion (Mn^{2+}, Fe^{2+}, Co^{2+}, and Ni^{2+})-doped carbon dots synthesized via microwave-assisted pyrolysis: a potential nanoprobe for magneto-fluorescent dual-modality bioimaging. ACS Biomater Sci Eng 4:2582–2596
Peréz JAC, Sosa-Hernández JE, Hussain SM et al (2019) Bioinspired biomaterials and enzyme-based biosensors for point-of-care applications with reference to cancer and bio-imaging. Biocatal Agric Biotechnol 17:168–176
Potbhare AK, Umekar MS, Chouke PB et al (2020) Bioinspired graphene-based silver nanoparticles: fabrication, characterization, and antibacterial activity. Mater Today Proc 29:720–725
Rahman A, Chowdhury MA, Hossain N (2022) Green synthesis of hybrid nanoparticles for biomedical applications: a review. Appl Surf Sci Adv 11:100296–100324
Rajvanshi J, Sogani M, Kumar A et al (2023) Biomaterials: a sustainable solution for a circular economy. Eng Proc 59:133–138
Rana A, Yadav K, Jagadevan S (2020) A comprehensive review on green synthesis of nature-inspired metal nanoparticles: mechanism, application and toxicity. J Clean Prod 272:122880–122904
Reguengo LM, Salgaço MK, Sivieri K et al (2022) Agro-industrial by-products: valuable sources of bioactive compounds. Food Res Int 152:110871–110890
Shashwat S, Zingre KT, Thurairajah N et al (2023) A review on bioinspired strategies for an energy-efficient built environment. Energ Buildings 296:113382–113404
Sheik A, Ranjith KS, Ghoreishian SM et al (2024) Green approach for the fabrication of dual-functional S/N doped graphene tagged ZnO nanograins for in vitro bioimaging and water pollutant remediation. Environ Pollut 343:123077–123087
Singh C, Ali M, Sumana G (2016) Green synthesis of graphene based biomaterial using fenugreek seeds for lipid detection. ACS Sustain Chem Eng 4:871–880

Singha A, Pellegrini V, Kalliakos S et al (2009) Optical anisotropy of electronic excitations in elliptical quantum dots. Appl Phys Lett 94:e073114
Sinha A, Simmani FZ, Singh D et al (2022) The translational paradigm of nanobiomaterials: biological chemistry to modern applications. Mater Today Bio 17:100463–100488
Sorrenti V, Burò I, Consoli V et al (2023) Recent advances in health benefits of bioactive compounds from food wastes and by-products: biochemical aspects. Int J Mol Sci 24:2019–2044
Souza MA, Vilas-Boas IT, Leite-da-Silva JM et al (2022) Polysaccharides in agro-industrial biomass residues. Polysaccharides 3:95–120
Stratakis E, Ranella A, Fotakis C (2011) Biomimetic micro/nanostructured functional surfaces for microfluidic and tissue engineering applications. Biomicrofluidics 5:1–31
Sun YP, Zhou B, Lin Y et al (2006) Quantum-sized carbon dots for bright and colorful photoluminescence. J Am Chem Soc 128:7756–7757
Tewari C, Tatrari G, Karakoti M et al (2019) A simple, eco-friendly and green approach to synthesis of blue photoluminescent potassium-doped graphene oxide from agriculture waste for bio-imaging applications. Mater Sci Eng C 104:109970–109976
Tortolini C, Capecchi E, Tasca F et al (2021) Novel nanoarchitectures based on lignin nanoparticles for electrochemical eco-friendly biosensing development. Nano 11:718–733
Tripathy A, Sen P, Su B et al (2017) Natural and bioinspired nanostructured bactericidal surfaces. Adv Colloid Interface Sci 248:85–104
Tsouko E, Pilafidis S, Kourmentza K et al (2024) A sustainable bioprocess to produce bacterial cellulose (BC) using waste streams from wine distilleries and the biodiesel industry: evaluation of BC for adsorption of phenolic compounds, dyes and metals. Biotechnol Biofuels Bioprod 17:40–56
Ul-Islam M, Alabbosh KF, Manan S et al (2024) Chitosan-based nanostructured biomaterials: synthesis, properties, and biomedical applications. Adv Ind Eng Polym Res 7:79–99
Valera E, Jankelow A, Lim J et al (2021) COVID-19 point-of-care diagnostics: present and future. ACS Nano 15:7899–7906
Vijaya BM, Roy N, Moharana P et al (2020) Green synthesis of highly luminescent biotin-conjugated CdSe quantum dots for bioimaging applications. New J Chem 44:16891–16899
Wang L, Jiang J, Hua W et al (2016) Mussel-inspired conductive cryogel as cardiac tissue patch to repair myocardial infarction by migration of conductive nanoparticles. Adv Funct Mater 26:4293–4305
Wang J, Drelich AJ, Hopkins CM et al (2022) Gold nanoparticles in virus detection: recent advances and potential considerations for SARS-CoV-2 testing development. WIREs Nanomed Nanobiotechnol 14:e1754–e1783
Watson GS, Green DW, Schwarzkopf L et al (2015) A gecko skin micro/nano structure—a low adhesion, superhydrophobic, anti-wetting, self-cleaning, biocompatible, antibacterial surface. Acta Biomater 21:109–122
Xiong J, Guo SY, Wu Y et al (2023) Predicting the response of heating and cooling demands of residential buildings with various thermal performances in China to climate change. Energy 269:126789–126805
Xu L, Wei C, Deng L et al (2024) A review of non-biodegradable alloys implantation induced inflammatory and immune cell responses. J Alloys Compd 977:173086–173100
Yin C, Zhao Q, Li W et al (2020) Biomimetic anti-inflammatory nano-capsule serves as a cytokine blocker and M2 polarization inducer for bone tissue repair. Acta Biomater 102:416–426
Yu H, Wu Y, Xu J et al (2024) Neutrophils-mediated bioinspired nanoagents for noninvasive monitoring of inflammatory recruitment dynamics and navigating phototherapy in rheumatoid arthritis. Biomater Adv 158:213764–213773
Zappi D, Varani G, Cozzoni E et al (2021) Innovative eco-friendly conductive ink based on carbonized lignin for the production of flexible and stretchable bio-sensors. Nano 11:3428–3447
Zhang Q, Yang X, Li P et al (2015) Bioinspired engineering of honeycomb structure—using nature to inspire human innovation. Prog Mater Sci 74:332–400
Zhang Y, Ye L, Cui J et al (2016) A biomimetic poly (vinyl alcohol)-carrageenan composite scaffold with oriented microarchitecture. Biomater Sci Eng 2:544–557

Zhang X, Liu C, Cai X et al (2020) A highly specific Golgi-targetable fluorescent probe for tracking cysteine generation during the Golgi stress response. Sens Actuators B 310:127820–127827

Zhang K, Mo CQ, Tang XL et al (2023) Hierarchically porous cellulose-based radiative cooler for zero-energy food preservation. ACS Sustain Chem Eng 11:7745–7754

Zheng BX, Long W, She MT et al (2023) A cytoplasm-specific fluorescent ligand for selective imaging of RNA G-Quadruplexes in live cancer cells. Chem Eur J 29:e202300705

Zhou W, Gao X, Liu DB et al (2015) Gold nanoparticles for in vitro diagnostics. Chem Rev 115:10575–10636

Zhu H, Fan J, Du J et al (2016) Fluorescent probes for sensing and imaging within specific cellular organelles. Accounts Chem Res 49:2115–2126

Zhu L, Fu M, Yin B et al (2020) A red-emitting fluorescent probe for mitochondria-target micro-viscosity in living cells and blood viscosity detection in hyperglycemia mice. Dyes Pigments 172:107859–107866

Zingre KT, Wan MP, Tong S et al (2023) Modeling of cool roof heat transfer in tropical climate. Renew Energy 75:210–223

Maurício Dalla Costa Rodrigues da Silva graduated in Chemical Engineering from Federal University of Santa Maria (UFSM) where he also completed his master's and PhD in Chemical Engineering. He conducted a postdoctoral fellowship in Nanoscience at the Franciscan University (UFN). His expertise lies in Chemical Engineering with focus on process development, machine learning and by-product valorization.

Daniel Moro Druzian graduated in Materials Engineering from the Franciscan University (2013–2018) and a master's degree in Nanosciences from the Franciscan University (2019–2021). He is currently pursuing a doctorate in the Postgraduate Program in Nanosciences at the Franciscan University. His expertise lies in materials and metallurgical engineering, treatment of organic pollutants using nanomaterials, reuse of (agro)industrial waste, material characterization, in-vitro biological tests, nanostructured biomaterials for bone regeneration, and antioxidant activity of different materials.

Tiago Moreno Volkmer graduated in Materials Engineering from the Federal University of Rio Grande do Sul, where he also completed his master's and PhD in Materials Science and Metallurgical Engineering. He conducted part of his PhD research at the Consejo Superior de Investigaciones Científicas (CSIC) in Madrid, Spain, and completed a postdoctoral fellowship at the Georgia Institute of Technology. He was previously a professor at the Franciscan University (UFN) and is now a professor at the Technology Development Center of the Federal University of Pelotas. His expertise lies in Materials and Biomedical Engineering, with a focus on bioceramics and biopolymers.

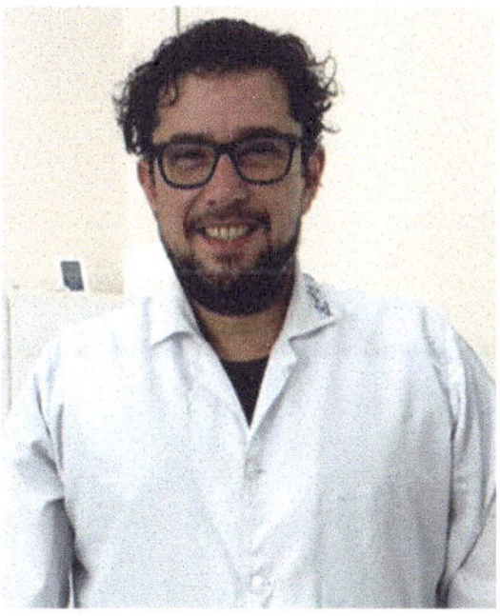

William Leonardo da Silva holds a degree in Chemical Engineering from the Federal University of Santa Maria (UFSM, 2010), a master's degree (2012) and a PhD (2016) in Chemical Engineering from the Federal University of Rio Grande do Sul (UFRGS). He developed research as a sandwich Ph.D. student abroad (PDSE / CAPES-2015) at Ulster University at Nanotechnology and Integrated Bioengineering Centre (NIBEC). Currently, he is an adjunct professor at the Franciscan University (UFN) and a permanent professor of the Graduate Program in Nanosciences, coordinating the Applied Nanomaterials Research Group (GPNAp). He has experience in Green Nanotechnology, with emphasis on green synthesis of metallic nanoparticles, chemical kinetics and catalysis, working mainly on the following topics: removal of organic pollutants in wastewater by Advanced Oxidative Processes (heterogeneous photocatalysis), adsorption process, biomaterials, machine learning and synthesis and characterization of nanocatalysts.

Sustainable Nanocomposites with Green Biomaterials: Synthesis, Properties, and Applications

2

Daniela Marisol Carmona-Govea,
Víctor Adán Cepeda-Tovar, Ana Cecilia Cepeda-Nieto,
Anilú Rubio-Ríos, Sathvik Belagodu Sridhar,
and Jesús Antonio Morlett-Chávez

Abstract

Biomaterials are specialized substances designed to interact with biological systems to diagnose, treat, improve, or replace any part of the human body, from tissue to complete organs or to improve a function. In medicine, nanocomposites are nanoscale materials that unite properties of one or more materials to maximize functionality, performance, and application areas. The synthesis of nanocomposites based on biomaterials is specific and straightforward while maintaining sustainability standards without damaging the compatibility with biological systems. Green biomaterials include bioinspired materials, renewable resources based on ceramics, natural fibers based on composites, biodegradable polymers, and materials obtained from agricultural waste. Their development and synthesis involve techniques such as precipitation, vapor phase deposition, electrochemical synthesis, and emulsions that allow maintaining control over the

D. M. Carmona-Govea · V. A. Cepeda-Tovar · A. Rubio-Ríos
Faculty of Chemical Sciences, Autonomous University of Coahuila,
Saltillo, Coahuila, Mexico

A. C. Cepeda-Nieto
Faculty of Medicine, Autonomous University of Coahuila, Saltillo, Coahuila, Mexico

S. B. Sridhar
RAK College of Pharmacy, RAK Medical & Health Sciences University,
Ras Al Khaimah, United Arab Emirates

J. A. Morlett-Chávez (✉)
Faculty of Chemical Sciences, Autonomous University of Coahuila,
Saltillo, Coahuila, Mexico

Faculty of Medicine, Autonomous University of Coahuila, Saltillo, Coahuila, Mexico
e-mail: antoniomorlett@uadec.edu.mx

R. Malviya, S. Sundram (eds.), *Sustainable Nanocomposites with Green Biomaterials*, Biomaterials, Bioengineering and Sustainability 2,
https://doi.org/10.1007/978-3-031-79110-9_2

nanocomposite's morphology, structure, and composition. Biomaterial-based nanocomposites have properties that can be "tuned" to meet specific requirements for each of the possible medical applications, such as mechanical strength, electrical conductivity, biodegradability, drug loading capacity, and cell binding capacity. This type of nanocomposites is suitable for treatments such as tissue medicine, controlled drug release, medical imaging, biosensors, and bone medicine. Nanocomposite biomaterials are currently used in controlled and targeted drug delivery systems, and they are prevalent in the manufacturing and development diagnostic devices such as biosensors and imaging agents. This chapter will discuss the synthesis and applicability of nanocomposite biomaterials, which represent a promising area of research due to the unique properties of biomaterials and their affinity with the human body.

Keywords

Sustainable · Nanotechnology · Biomaterials · Developments · Polymers

2.1 Introduction

Nanotechnology is a relatively new and developing technology with a wide range of uses. These days, a large range of physico–chemical techniques are employed to synthesize nanoparticles (NPs). In contrast, biogenic reduction of metal precursors to produce corresponding NPs is less costly, environmentally friendly, and free of chemical contaminants for use in biological and medical applications where NP purity is a key consideration. It also explores the role of green chemistry in the contemporary era, offering a comprehensive assessment of the state of affairs and suggesting a variety of ways to employ nanomaterials for a range of purposes, chiefly in sustainable society, with the least amount of hazardous materials.

The fact that metals and metal oxides have a variety of uses in pharmaceutical and biological applications further emphasizes the role that green chemistry plays in the production of these materials. As a key component of upcoming cutting-edge research and development, the chapter integrates green synthesis techniques with nanotechnological characterization techniques. Along with examining green nanomaterials for use in both traditional and cutting-edge biomedicine, it also assesses the potential of the green synthesis route.

2.2 Vision of Nanotechnology and Bionanomaterials Study

Nanotechnology has advanced tremendously in the last few decades. At the nanoscale, materials display novel physicochemical properties. As a result, a wide range of applications, such as tissue engineering, drug delivery, energy storage, electronics, sensors, semiconductors, and imaging of biological agents, are promising for nanomaterials. However, their use and availability are limited due to the need for biological properties (Khandel et al. 2018).

New functional materials are being created in advanced biomedical engineering by simply imitating the chemical, physical, or biological properties of natural materials. This chapter covers the synthesis strategies, acquisition and utilization techniques, and biomedical applications of biomimetic nanocomposites. We give a brief overview of several methods for creating biomimetic nanomaterials, such as electrospinning, gas foaming, and rapid prototyping.

A detailed description was provided of biomimetic drug delivery systems that make use of polymeric nanocomposites, hydrogels, liposomes, micelles, and dendrimers. Tissue engineering can use biomimetic scaffolds based on cell composite nanocomposites due to the right mechanical properties, recapitulation of extracellular matrix, and create an environment that promotes tissue regeneration. Tissue engineering using hydrogels, fibers, biomimetic membranes, and three-dimensional printed nanocomposites will be the main focus of this research. However, more research is required before these materials are used in clinical settings, even with the amazing advancements in biomimetic nanocomposites that have been detailed here (Khandel et al. 2018).

A significant catalyst that nanotechnology has cultivated is the creation of extremely sustainable materials with a myriad of uses, enabling progress in numerous domains. It entails using observation, measurement, manipulation, assembly, control, and manufacturing of matter at the nanoscale to translate the theory of nanoscience into real-world applications. "The understanding and control of matter at dimensions between approximately 1 and 100 nanometers, where unique phenomena enable novel applications across a wide range of fields, is how the National Nanotechnology Initiative (NNI) defines nanotechnology. The two essential requirements for nanotechnology are suggested by this definition. The first is a scale issue: structures are used in nanotechnology by manipulating their size and shape at the nanoscale. The second problem is a novelty: nanotechnology must handle small objects that maximize specific characteristics because of the nanoscale.

A technical-scientific abuse of nanoscience is an innovative, if understudied, idea. Nanoscience, which deals with the manipulation of materials on the atomic and molecular scales, is a synthesis of chemistry, physical science and biology. On the other hand, the ability to observe, measure, control, assemble, control, along with manufacture matter at the nanoscale is known as nanotechnology. Reports detailing the development of nanoscience as well as technology are available. However, there isn't a single report that covers the development of nanoscience and technology from the beginning to the present. To completely comprehend the advancement of nanoscience and technology, it is therefore essential to compile a summary of the key moments in this field (Malik et al. 2023).

2.3 Synthesis and Development of Biomaterials

Nanomaterials encompass materials that are stronger, lighter, more electrically conductive, superparamagnetic, provide tunable optical emission properties, are more porous, have better thermal insulation properties, and are less corrosive than bulk materials. These materials have the potential to address unique biological challenges, which may arise due to the similarity in size domains of the components involved with nanotechnology and those of biological structures. Due to this scale similarity and specific functional properties, nanotechnology is applied in many areas of health-related research, such as synthetic and hybrid nanostructures, diagnostic and therapeutic devices, tissue regeneration matrices, and drug delivery for various biotechnologies. Bionanocomposites that are biologically stable and environmentally degradable can be implemented in multiple medical sciences, such as drug delivery, cancer detection, bone replacement, brain surgery, etc. The literature discusses bionanocomposites, which were created to replace and mend biological tissues. Intelligent and self-assembling nanocomposite biomaterials that react to mechanical, electrical, magnetic, optical, and temperature stimuli are the focus of several novel approaches. The ability of these materials to respond to external stimuli is essential for the development of novel biotechnologies such as biosensors, actuators, in vitro diagnosis, cell culture matrices, contrast substances, and bioassays, as well as intelligent implants and drug delivery systems (Singh et al. 2024).

From the beginning of humanity, plants have been used by humans primarily for the treatment of specific ailments due to the capabilities that plants and herbs can offer on their own. They have been skillfully studied in recent years since they also possess malleable characteristics capable of forming highly biodegradable and biocompatible polymers. This fact marks the difference between materials based on natural extracts, using everything from their biomass to their oils to be processed in different techniques to form polymers capable of forming biofilms, biocapsules, nanoparticles, and fibers, among others. These biomaterials can be used in different phytochemical, pharmacological, and toxicological applications based on their drug delivery, protection, and drug development properties in the medical field (Martínez-Torres et al. 2016).

Green nanotechnology uses naturally occurring bioagents, such as plant microbes, and various biowaste such as agricultural, plant-based wastes, and much more to formulate sustainable nanomaterials (Zhang et al. 2020). It is a low-cost, simple, safe, low-risk, non-toxic, and environmentally friendly approach (Moradi et al. 2021). A crucial element of clean technology for remedication of environment is green nanotechnology, that converts surplus bioactive materials into lucrative and sustainable materials. The increased quantity of biowastes generated by humans is directly correlated with population growth, lead to a significant threat to environment purity.

Some of the central synthesis and development techniques for these materials include specific routes such as (Guerrero et al. 2011):

1. Natural polymers such as starch, cellulose, proteins, and chitin are simply derived from the development of copolymers, which enhance their properties

through biomass extraction or mechanical or chemical processing, modifying their fundamental properties for specific applications.
2. Based on biodegradable polymers such as PHAs (polyhydroxyalkanoates), polyesters, and poly acids, the main synthesis techniques involve monomeric polymerization derived from carbohydrates or fatty acids (Yapias et al. 2022).
3. Composite materials such as nanocrystals, fibers, capsules, and nanoparticles have specific techniques, such as adhesion, combination, reinforcement, and doping. These materials can be entirely natural or have some parts made of synthetic materials (Pontón and Guerrero 2010).
4. Lignocellulosic materials such as wood, lignin, and other woods are based on processing their lignocellulosic biomass to obtain recyclable materials (Tapia et al. 2006).
5. In the case of protein-based materials such as bioplastics, such as wheat gluten, highly reported techniques include extracting and processing film-forming proteins or biodegradable plastics through techniques such as component separation.
6. For materials based on those produced from bacteria and fungi, such as Poly-beta-hydroxybutyrate (PHB), synthesis techniques such as substrate-fed batch culture extract metabolized polymers.

All the above techniques and methods are commonly used to develop and synthesize these materials to maintain sustainability and minimize the impact of contaminants, as is the case with synthetic materials.

Finding a cost-effective and safe key to this issue, that has a significant environmental cost, is therefore imperative (Kalpana and Devi Rajeswari 2018; Khodadadi et al. 2021; Murthy et al. 2020). From this perspective, numerous researchers have recently explored ways to recycle these bioresources into products that are both acceptable and environmentally beneficial (Geetha et al. 2013). Green nanotechnology is one of these techniques, which creates nanomaterials from recyclable bioresources (Iravani 2014; Aravind Kumar et al. 2019; Hassanisaadi et al. 2021). Toxic substances, high pressure, energy, or temperature are not needed. For the production of nanoparticles for use in solar cells, medicines, catalysts, sensors, water desalination, industrial sectors, and air purification, it is an inexpensive and safe process (Bukhari et al. 2021).

Regarded as NextGen biopolymers, polyhydroxyalkanoates (PHAs) are intracellular microbial polymers. They take the place of synthetic plastics because of their important characteristics, which include compatibility, biodegradability, biostability, and excellent physicochemical properties (Kourmentza et al. 2017). In this context, different biocomposites have been presented that have raised expectations in healthcare, exemplifying polyhydroxyalkanoates (PHAs).

Microbial polymers found inside cells are called polyhydroxyalkanoates (PHAs), and they are considered NextGen biopolymers. They take the place of synthetic plastics because of their superior physicochemical qualities, compatibility, biodegradability, and biostability (Tarrahi et al. 2020), due to its biodegradability, as well as thermal, mechanical, and chemical stability, is one of its main characteristics for its immense use in disciplines such as medicine, electronics, agriculture, food, and controlled drug release systems (Gabr 2019). Microorganisms from various domains, including bacteria, archaea, and eubacteria, can be isolated to produce PHA. The biogenesis of PHA is facilitated by a close-knit network of enzymes such as acetyl-CoA acetyltransferases, acetoacetyl-CoA reductases, PHA synthases, and a group of genes that include phaC, phaM, phaZ and many more (Juengert et al. 2018; Pillai et al. 2017; Penkhrue et al. 2020). Understanding the microbial source of production, industrialization, further processing, and application of PHA in PHA and its nanocomposites are critical to its commercialization and success as a sustainable nanocomposite in a variety of medical products and applications (Shen et al. 2019; Gnach et al. 2015).

Nanoparticulate biomaterials are used in many techniques including drug delivery, gene therapy, disease prevention, monitoring, and medical imaging using different modalities, such as positron emission tomography (PET) and nuclear magnetic resonance (NMR) (Tang et al. 2013). Interestingly, these biomolecular interaction methods require chemical alterations on the NP surface before the *in-vitro* and *in-vivo* studies. For instance, in vitro results demonstrated that AgNPs synthesized with fruit extract from Viburnum opulus enhanced the transient secretion of IL-1α in keratinocytes following UVB exposure and reduced the level of IL-6 at both tested intervals, indicating a potential long-term anti-inflammatory effect. In vivo, pre-administration of both AgNPs and fruit extract reduced the amount of cytokines in the soft tissues of the hind foot pad shortly after the induction of inflammation (2 h) and inhibited the inflammatory processes linked to the formation of edema, which in turn decreased the increase in paw volume (Rong et al. 2018; Xie et al. 2021; Farahavar et al. 2019).

Through ligand-receptor interactions, targeting coatings can make it easier for nanoparticles to be incorporated into biological tissues and cells for a variety of therapeutic uses. Additionally, an investigation was conducted into the coating's stability in physiologically simulated conditions. Serum, in fact, contains phosphates that facilitate the biodegradation of metal carboxylate nanoMOFs and is a slightly basic medium (pH ~ 7.4) (Azevedo et al. 2018).

In targeted drug delivery, researchers are involved in finding new biomarkers and ligands that correspond with them. Pathogens, analytes, and biomarkers can all have their signals amplified by NP binding, which makes it easier to diagnose and image them. Functionalization is the process of modifying the framework surface with "bioactive signals" (Gravely et al. 2019). The NP surface has been coupled with a variety of ligands to enable applications such as intracellular administration, disease detection, and the detection of biomolecules and cells.

Because such functionalization is so simple, chemists can easily create functionalities that are appropriate for clinical use (Yu et al. 2021). Since their structural-functional groups make them readily chemically bonded to surfaces, small ligands are frequently used for NP functionalization. The nanogels are among the various surface-modified nanoparticles (NPs) that have been created using ligands/molecules as multifunctional carriers that can combine targeted drug delivery and examination in the arena of pharmaceutical applications (Muthu and Singh 2009). Medications can either be absorbed onto the surface of the NP or be encapsulated within the particle. Additionally, it has been demonstrated that functionalization ceases NPs from aggregation and makes them biocompatible (Gole et al. 2016). Functionalization produces synergistic effects by improving the mechanical, chemical, and physical properties of NPs (Charoensri et al. 2021). Moreover, Hirayama et al. (2021) highlighted the encapsulation and release of quinine, which led to enhanced drug loading capacity due to the combination of silane coupling agents and nanoparticles.

Because they erect a physical barrier around various encapsulates, polymers make good functionalizing agents by preventing the core from direct contact to biological receptors. Although they have a smaller hydrodynamic radius, polymers can act as ligands and produce a barrier. Because polymeric coatings impart characteristics of macromolecular systems, such as biological proteins, to the particle surface, they perform better than small molecule ligands. Coating nanoparticles such as polyethylene glycol (PEG), improves passive targeting in tumor tissue and amplifies the enhanced permeability and retention (EPR) effect. The idea that NPs accumulate significantly more in tumor tissues than in normal tissues is a contentious effect. PEG and other polymeric coatings increase particle uptake in tumor tissues, extend circulation times, and reduce blood serum protein adsorption (Mout et al. 2012).

The properties of biomolecule-coated nanoparticles (NPs), such as their excellent distribution of biomacromolecules with low cytotoxicity, are hard to attain with synthetic substances. One of the earliest applications of bionanotechnology reported by Mirkin and Alivisatos involved the blend of gold-thiol bonds to formulate oligonucleotide-AuNP conjugates (Mout et al. 2012). Kleinfeldt et al. (2019) created a superhydrophobic, biocompatible coating with colloidal stability by stacking layers of anionic polyelectrolyte poly (acrylic acid) and cationic polyallylamine to create gold nanoparticles. Makvandi et al. (2020) examined the functionalization of different nanomaterials and polymers for use in textiles, food packaging, medical applications, and water purification. The study looked into the advantages and disadvantages of functionalizing polymers. Manipulating FNPs opens new pathways for advancing unfamiliar multifunctional tools for biomedical applications (Huang et al. 2020). Nanotechnological uses have sparked great interest since the late 1980s (Li et al. 2016; Nejati et al. 2021; Díez-Pascual 2021).

Choosing no solvents or green solvents (preferably water), alternatives reaction media, one-pot synthesis, multicomponent reactions (MCRs), continuous processing, and processing intensification approaches for atom economy and final waste reduction are just a few ways that the principles of green chemistry (GC) can be fully applied to the green synthesis of pharmaceuticals. By optimizing resource efficiency in the synthesis process, reducing risks and pollution, and designing the active pharmaceutical ingredient (API) life cycle holistically, green synthesis can be executed (Abdul Khalil et al. 2017; Abraham et al. 2017). Other names for green chemistry include benign, clean, and sustainable chemistry. The medical and health care sectors are expanding quickly, and this is being facilitated by the rapid development of other industries, including the pharmaceutical industry. As a result, there are fewer fatalities and cases of suffering (Alexandrescu et al. 2013). The large surface area of nanoparticles confers enhanced properties. The actual process for creating nanoparticles is environmentally dangerous and toxic. The disadvantage of the conventional synthesis method is that byproducts contaminate the colloidal solution. Green nanoparticle synthesis was therefore developed as a solution to this problem (Anirudhan and Rejeena 2014).

These nanoparticles can be used for large-scale production and are both economical and environmentally beneficial. This adheres to some green chemistry principles, including prevention, the synthesis of less hazardous chemicals, the design of safer chemicals, and the real-time prevention of pollution (Barkhordari et al. 2014; Berlioz et al. 2009). Nanotechnology is still in its early stages of development in the pharmacy industry. Six Prior to now, nanoparticles were created using chemical and physical processes. The large-scale production of nanoparticles is a result of growing demand for them. As a result, the commercial process for creating metal nanoparticles was created. However, the need for clean, nontoxic, and environmentally friendly ways to synthesize nanoparticles is becoming more widely recognized due to the use of hazardous solvents or high energy in these processes (Bicu and Mustata 2011).

Ibuprofen's manufacturing process, which produces non-steroidal anti-inflammatory drugs over the counter, is a sophisticated illustration of a high-atom-efficiency process. Researchers set the standard for environmental excellence in chemical processing technology, revolutionizing the production of pharmaceuticals in large quantities. It offers a creative and superb solution to the common issue of the significant waste amounts connected to the conventional stoichiometric use of auxiliary chemicals (Bielska et al. 2013; Bin Choy et al. 2009). With an efficiency of over 99.9%, the anhydrous hydrogen fluoride is recovered and recycled. Because no additional solvent is needed for this process, fugitive emissions are reduced and product recovery is made simpler. This, along with the streamlined process's nearly full atom utilization, makes it a waste-minimizing, environmentally friendly technology that can serve as an example for other pharmaceutical manufacturers (Boyjoo et al. 2014; Brown et al. 2012). Green chemistry and sustainability are thus posing a challenge that is resulting in fundamentally novel advances in organic synthesis, which will eventually benefit society, the environment, and the pharmaceutical sector as well as the chemical and related industries as a whole.

Cellulose, widely abundant in biomass, has garnered increased attention due to energy shortages. Various methods exist for cellulose separation, categorized by solvent types (Brown et al. 2012). Without producing any byproducts, cellulose is directly broken down by ionic liquids and deep eutectic solvents (DES). In biphasic solvent structures and cellulose/ organic solvent-based fractionation, organic solvents are mixed with other phases (Budtova 2019). The authors' group developed a water-based solvent approach called "Cooking with Active Oxygen and Solid Alkali" (CAOSA), which achieves nearly complete cellulose separation with minimal degradation and no strong acids or sulfides. This method presents an environmentally friendly and cost-effective way to utilize biomass.

High temperatures (>200 °C) are needed for traditional methods like steam explosion and organosolv processing, while kraft pulping and sulfite pulping result in acidic and alkaline solutions that are harmful to the environment. Because of its strong inter- and intramolecular hydrogen bonds, cellulose has a limited solubility in water and common organic solvents. This can be addressed by nanomerization. One such nanomaterial is cellulose nanocrystals (CNC), which are divided into three groups according to their width (Cai and Zhang 2005).

Unlike cellulose dissolution, surface modification is more straightforward and widely adopted (Cai and Zhang 2005). Because of its many hydroxyl groups, cellulose is a good material for a variety of processes, such as etherification, esterification, and grafting with functional groups that are active. The most common technique for altering cellulose is esterification, and the degree of activation has a big impact on how much is substituted. By surface esterification with palmitoyl chloride, Berlioz et al. produced a model of cellulose that is easily accessible and has high crystallinity, highlighting the direct relationship between grafting density and the characteristics of the cellulose and the experimental setup. Another common technique for modifying cellulose is etherification. N-oxiranylmethyl-N-methylmorpholinium chloride and 2-oxiranylpyridine were added to cationic cellulose ethers in an alkaline aqueous solution, as shown by Hasani et al. There is a significant relationship between the water content and this cellulose one-pot reaction (Capadona et al. 2009).

Hydrophobic groups and electrostatic forces in aqueous solutions without harsh reaction conditions or solvents facilitate self-assembly of cellulose. This method is essential for capturing hydrophobic particles, like pharmaceuticals and biomolecules. Numerous self-assembly systems have been created based on stimuli-responsive cellulose materials, such as those that react to redox, pH, temperature, light, and multiple stimuli (Capadona et al. 2009; Chen et al. 2010). By changing the molecular structure, these self-assembled cellulose structures have a wide range of uses in functional materials, including drug carriers. This will be elaborated further below (Chen et al. 2010).

Recent advancements in cellulose-based gels have sparked significant interest. Hydrogels, commonly utilized in biomedicine, are typically formed through polymerization reactions involving thoroughly mixed solutions of polymer precursors and cellulose nanocrystals (CNC), nanofibrils (CNF), or microcrystalline cellulose (MCC) or via physical/chemical blending of polymer components. Aerogels, another class of materials with applications in various fields, including biomedicine, are conventionally produced by drying precursing materials or hydrogels, wherein a water phase is replaced by air (Ciobanu et al. 2010).

Materials known as hydrogels have many uses in a variety of industries, including the pharmaceutical and biomedical ones (Ciolacu et al. 2012). Their ability to load various drugs, including hydrophobic ones and biomolecules, and to absorb large volumes of aqueous solutions without losing their mechanical or shape-preserving qualities, among other qualities, demonstrate their versatility and potential market. Similar to this, Kong et al. created multifunctional hydrogels from natural wood, while Qin et al. applied in-situ polymerization. Numerous synthesis routes have been developed due to years of exploration in numerous studies, particularly for chemical/permanent hydrogels. Hydrogels are particularly useful in the biomedical field, but they have many applications in other fields as well (Ciolacu et al. 2012; Dai and Si 2019).

They can be used to create a variety of materials, including contact lenses, blood-contacting hydrogels, scaffolds, wound-healing bioadhesives, artificial kidney membranes, artificial skin, vocal cord replacement, and artificial tendons, due to their adaptable features and distinctive behavior. They are materials of great interest in research because of their relevance and wide range of applications (Dai and Si 2019).

In terms of pharmaceutical applications, the hydrogel structure's pores are highly desirable for drug delivery systems because, among other mechanisms, they can carry active ingredients and release them at a controlled rate through the swelling of the semisolid formulation or even its disintegration. The large variety of polymers that can be obtained is related to the possibility of designing a hydrogel in accordance with the desired application requirements. Therefore, by combining natural and/or synthetic polymers, it is possible to control their rate of biodegradation, mechanical strength, capacity to swell, and response to external stimuli. Hydrogels can also be blended with other materials at the nanoscale to create nanocomposites that have unique qualities like high mechanical strength, antibacterial activity, and magnetic properties.

Free radical polymerization, ionic cross-linking, homogenization, use of natural wood, and freeze-dried processing are examples of common preparation techniques. The sol/gel transition usually happens during thermos- and non-crosslinked gel transitions and is reversible. But if the gel coagulates or becomes anti-solvent overhead the liquid crystal transition temperature (>60 °C), reversibility might be lost (Han et al. 2013).

Synthetic or chemically modified cellulose-based hydrogels frequently have better properties than natural cellulose-based hydrogels. Chemical crosslinking agents create covalent bonds between polymers and cellulose or its derivative chains, which is why they are frequently needed during gelation to guarantee efficient swelling and a stable structure. Cellulose acts as the cross-linker alone in certain situations, but it also interacts with other polymers to form cross-links. For example, a hydrogel based on polyvinyl alcohol (PVA) with good mechanical properties was cross-linked using supramolecular techniques, resulting in a storage modulus increase from 3.8 to 14.3 kPa. Certain cellulose derivatives with -NH2 groups, such as polyamide and glutaraldehyde, are particularly well-suited for certain cross-linkers. After cellulose or its derivatives are shaped into hydrogels, all of these cross-linkers can work efficiently without the need for catalysis, improving the hydrogels' biocompatibility (Dash et al. 2014).

Excellent release characteristics are demonstrated by cellulose-based hydrogels in various environmental circumstances, such as light, temperature, and pH. These hydrogels loaded medications are liberated through modifications to their swelling and shrinking characteristics. Compared to natural hydrogels, incorporating cellulose derivatives alters pore size, enabling the development of "on-off" release devices. For example, grafting carboxymethyl-cellulose sodium onto cellulose creates a thermo-responsive hydrogel network capable of "on" (swelling) or "off" (shrinking) behavior at different temperatures. Composite hydrogels typically exhibit pH-sensitive properties, like those formed with HCl/NaOH aqueous solution, SBF (simulated body fluid), or PBS (phosphate-buffered saline) (Dai and Si 2019).

A unique class of nanostructured materials with highly adjustable physicochemical properties and high porosity is called aerogels. The full potential of aerogels for other technology sectors remains to be evaluated, despite the fact that a few types have already made it to market in the construction materials, textile, and aerospace engineering industries. Based on ongoing efforts to address longevity and quality of life through biotechnological methods and the circular economy approach to the material supply chain, two emerging market opportunities where the use of aerogels needs to be further explored and evaluated in a multidisciplinary manner are environmental and life science applications (Ciolacu et al. 2012).

Plant fiber-derived cellulose nanomaterials have a wide range of possible uses, including biomedical, automotive, and packaging. In the twenty-first century, it is very feasible to use nanocellulose that has been separated from plant fiber—a source that is based on carbohydrates—in biomedicine. The key features of plant fiber-based nanocellulose, such as its mechanical, tensile, and molecular qualities as well as its potential for biodegradation, have been thoroughly investigated for use as functional materials in the creation of aerogel. Aerogels based on plant cellulose nanofiber (CNF) are innovative functional materials that have garnered a lot of attention. The biomedical field has made extensive use of CNF aerogel in recent years because of its biocompatibility, renewability, and biodegradability. Broad applications in biological and medical studies, including biosensing, antimicrobial films for wound healing, scaffold preparation for tissue cultures, sustained antibiotic delivery for wound healing, and drug delivery system development, are influenced by the effective surface area of CNFs (Dash et al. 2014).

Biomaterials based on nanocellulose have low cytotoxicity, high biocompatibility, and biodegradability. Biomaterials that are economically competitive are nanocelluloses, such as cellulose nanocrystals (CNC) and cellulose nanofibrils (CNF), which are typically derived from plants or bacteria. Numerous biomedical uses, including vascular grafts, wound dressings, medication carriers, orthopedic and dental implants, have been proposed for these nanocellulose-based materials.

Thus far, the most promising use of bio-aerogels in pharmaceutical and/or biomedical applications has been proposed as a vehicle for the release of active ingredients, primarily pharmaceuticals. Drugs can be incorporated into bio-aerogels in a variety of ways, such as by mixing with a polysaccharide solution, impregnating a "wet" aerogel precursor, or drying the aerogel. The drug's solubility in the appropriate fluid plays a major role in the route selection. It must be evident that the drug loading must remain high throughout the subsequent processing stages. The majority of the literature describes drug loading in bio-aerogels using adsorption in supercritical carbon dioxide. Aerogels based on alginate, pectin, starch, and chitosan have been reported as drug carriers. The drug's state (crystalline or amorphous) and the matrix's behavior (swelling, contraction, dissolution) during mass transport into and out of the matrix are some of the factors that affect the kinetics of drug release.

It is possible to adjust the release by functionalizing the matrix and creating a composite or hybrid matrix, such as an organic-inorganic or organic-organic matrix (made of various polymers). Bio-aerogels (based on starch, starch/caprolactone, pectin/xanthan, alginate, and alginate/lignin) were also shown to be non-cytotoxic, have good cell adhesion, and some can be used for bone regeneration. Interestingly, despite the fact that many suggest it, very few publications show off the potential use of cellulose II aerogels in biomedical applications. The biocompatibility of cellulose II aerogels with dual porosity—big pores formed by leached porogens and smaller pores resulting from supercritical drying of the coagulated cellulose matrix—was proven. Low-substitution phosphorylated cellulose aerogels demonstrated good hemocompatibility (Dash et al. 2014).

The long-standing use of calcium carbonate ($CaCO_3$) as a delivery system can be attributed to its widespread availability, biocompatibility, and biological degradation. Its low retention capacity, lack of sustained release, and quick dissolution in an acidic stomach environment are common problems, though. Modifying the $CaCO_3$ particles or combining them with other encapsulation systems can solve a number of these problems and create sophisticated $CaCO_3$-derived systems. When the pH is less than 6.5, the tumor microenvironment causes $CaCO_3$ to dissolve quickly and release the drug it contains, even though it can maintain its structural integrity at neutral physiological pHs. Just adjusting the size, shape, composition, and structure of $CaCO_3$ will modulate its delivery performance (Boyjoo et al. 2014).

The porous structure of cellulose/ $CaCO_3$ composites has drawn attention to the possibility of employing mineralized particles in the future. Even though these composites have many benefits, certain difficulties still exist. The difficulty of regulating and controlling the cellulose/$CaCO_3$ particles' size and porosity is one of the main challenges. Ideally, these parameters should be finely tuned to accommodate molecules ranging from tens of nanometers to several micrometers in size. While some progress has been made, such as the development of regular $CaCO_3$ microspheres with diameters ranging from 2 to 3.5 μm based on hollow cellulose nanoparticles by certain research groups, achieving precise control over particle characteristics remains a challenge that requires further exploration and innovation (Boyjoo et al. 2014).

Cellulosic materials, renowned for their excellent physical properties, serve as appealing additives for various polymer nanocomposites. As detailed earlier, they have found utility in biomedicine systems in forms such as hydrogels, aerogels, and biominerals. Furthermore, the use of cellulosic biomimetic materials in the design and assembly of intelligent materials is gaining popularity. Numerous novel delivery systems based on cellulose have been created, such as films, capsules, micro/nanoparticles, and inorganic nanocomposites (Brown et al. 2012). Drawing inspiration from the mechanical reinforcement function of cellulose in plant cell walls, researchers have successfully assembled cellulose nanofibril (CNF)—based capsules at the micrometer scale. Using CNF offers a straightforward but efficient way to create robust and pliable microcapsule shells. These capsules have a lot of potential for encasing poorly soluble medications or protecting molecules that are sensitive to oxygen while being stored. CNF-based capsules provide a novel approach to drug delivery and preservation in biomedical applications by utilizing the inherent qualities of cellulose, such as mechanical strength, sustainability, and biocompatibility.

Additionally, One of the most popular methods for creating nano/microparticles and nano/microfibers is electrospinning, which is achieved by applying a high voltage to the drug-loaded solution. It is possible to fine-tune the functionality-related aspects of the finished product by adjusting the operating parameters, solution properties, or environmental conditions. This produces distinct fiber properties. The latter involves changes in the rate and degree of a drug's solubility, which accounts for the abrupt or gradual start of absorption (Han et al. 2013). Numerous factors, such as material selection, fiber orientation, porosity, surface modification, and tissue application, are frequently taken into account when using electrospun fibers and fiber meshes in tissue engineering applications. Natural and synthetic (recyclable and nondegradable) materials, as well as hybrid combines of the two, are available for use. These blends can offer the best possible balance of mechanical and biomimetic qualities. The fiber orientation (aligned vs. random) and porosity/pore size (cells infiltration) of the electrospun scaffold can be adjusted and optimized for each specific application by changing the previously mentioned processing and solution parameters. Because electrospinning can create aligned scaffolds for anisotropic mechanical and biological properties and can stop smooth muscle cell migration, it has attracted a lot of attention in the field of tissue engineered vascular grafts. Nevertheless, additional studies have looked at how to use the additional properties of electrospun scaffolds to make vascular grafts better than they are now. Using poly(L-lactide-co-e-caprolactone), Inoguchi et al. electrospun a variety of tubular scaffolds for use as tissue-engineered vascular grafts (Han et al. 2013; Dash et al. 2014).

When choosing materials for drug delivery applications, electrospinning offers a great deal of flexibility. Drug release can be regulated by using biodegradable or non-biodegradable materials, and it can happen via diffusion only or diffusion plus scaffold degradation. Furthermore, a variety of medications, such as antibiotics, anticancer medications, proteins, and DNA, can be delivered because of the flexibility in material selection. Additionally, electrospun nanofibres have a high surface area to volume ratio, a multifaceted pore structure, well-regulated surface conformation, and good surface modification. They are also remarkably biocompatible. Since electrospinning is still a simple and straightforward technique for creating intelligent and controlled drug delivery devices, it is impressive that new methods have emerged that take advantage of using greener, biocompatible solvents—like water—instead of more toxic and aggressive ones, like HFIP and chloroform. As an alternative, melt electrospinning, which produces nanofibers without the need for a solvent, might be one of the most promising methods. Nonetheless, it will be essential to shield the medication from the heat and prevent deterioration. Optimized scaffolds have the potential to be a potent tool for patients and clinics in the future, combining precise drug release with non-invasive tissue engineering. Customized nanofibers'distinct qualities and ease of use may be a significant advancement in the field of personalized medicine (Prete et al. 2023).

Fig. 2.1 Distribution of molecular weights of chain polymer

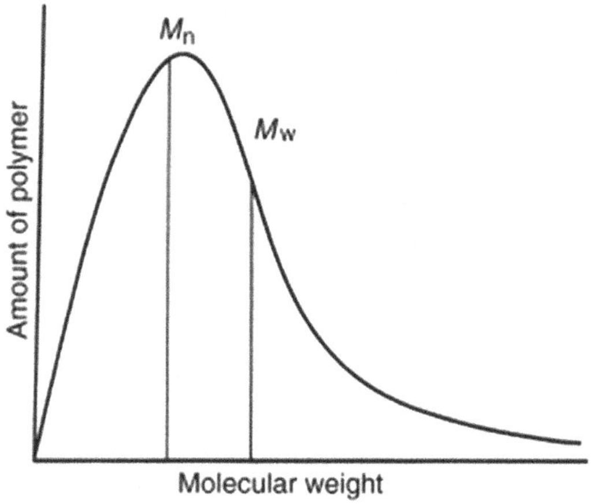

Typical molecular weight distribution of a polymer.

The most remarkable thing about these materials is that some might not change throughout the patient's life, while others might be hydrolytically unstable and break down fast in the body. The distribution of molecular weights used in the synthesis of these polymers typically controls their properties, including crystallinity. The trend for heavier chain sizes of polymers is distributed as shown in the following figure, which is based on the formulas used to determine the weight-average distribution of molecular weight (represented by Mw) and the number-average molecular weight (represented by Mn) (Agrawal et al. 2014).

This graph is generated from the weights of the leading polymer chains used through the formulas:

$Mn = \frac{\sum_i (NiMi)}{\sum_i (Ni)}$, where Mi is the number-average molecular weight of the polymer chains.

$Mw = \frac{\sum_i (NiMi^2)}{Ni\,Mi}$, where we can obtain graphs like the one in Fig. 2.1.

2.4 Subjects Integral to Biomaterials Science

- Biocompatibility of green nanoparticles

A variety of methods are grouped together under the umbrella of "green nanotechnology," which tries to minimize the use of dangerous chemicals, save energy, and support renewable resources. Green nanotechnology in medicine makes it possible to create biocompatible nanoparticles with lower toxicity and higher efficacy for drug delivery, diagnostics, and therapeutics (Kazemi et al. 2023).

Materials created with tissue and cells' biological compatibility in mind are known as biomaterials. They can be used for a range of purposes, including platforms for technological advancement and therapeutic applications, because they are environmentally friendly and biodegradable (Tripathi 2022).

The ability of an implant material to work in vivo without triggering bodily reactions is known as biocompatibility. To ascertain the safety and effectiveness of biomaterials, animal tissues must be used for testing. When an implant is biologically rejected, immune cells trigger an inflammatory response that may require implant removal.

Determining the material's cytotoxicity and biocompatibility requires an understanding of how it interacts with the body. It is crucial to determine the precise procedure and clarify the materials' cytotoxicity and biocompatibility before using them. Green polymeric materials have special qualities and attributes, including being widely available in nature, less harmful, affordable, and having a variety of uses. Bio-based polymers require natural macromolecules and their derivatives, including cellulose, hemicellulose, chitosan, lignin, and starch (Liu et al. 2019). Their distinct structural characteristics include biocompatibility, degradability, environmental friendliness, and richness. Bio-based monomers (rosin, terpene, vegetable oil, furan, and lactic acid) are produced by bio-based polymers, which include ethyl cellulose, starch, and rosin-based polymers. Furthermore, bacterial metabolism may directly synthesize bio-based polymers (Xie et al. 2023).

Biomaterials with varying chemical compositions and physical characteristics have been studied as extracellular environment scaffolds for stem cell differentiation and biological response regulation. For instance, exceptional mechanical properties have been achieved by using genetically engineered natural silk as a biocompatible scaffold material. In vitro, these scaffolds promote cell adhesion, differentiation, and multiplication.

Tissue engineering has made extensive use of nanoparticles and polymer nanocomposite biomaterials. Other nanomaterials have been employed with gelatin (Abar et al. 2024). Choosing an appropriate biomaterial as a scaffold to promote cell adhesion, proliferation, differentiation, and migration with little to no immunological rejection is the primary criterion in bone tissue engineering (Abar et al. 2024).

The most common kind of nanoparticles found in consumer goods are silver ones. But their use has been restricted due to their toxicity (Noga et al. 2023). Conversely, certain in vitro investigations have demonstrated minimal toxicity of silver nanoparticles on diverse cell lines (Brown et al. 2013). One of the more appealing options for drug delivery of macromolecules like proteins, RNA, and DNA is silver nanoparticles. When silver nanoparticles are present, oxidative stress, DNA damage, and apoptosis are the most frequent causes of toxicity. According to an in vivo study, silver nanoparticle exposure causes effects in a number of important organs (Sayed and Soliman 2017). The skin, eyes, kidneys, respiratory, hepatobiliary, and immune systems could all be at risk from it (Noga et al. 2023). In many models of non-mammalian animals, exposure to silver nanoparticles results in morphological deformities, reproductive disorders, and malformations (Du et al. 2018; Noga et al. 2023).

The application of silver nanoparticles has noticeably increased since the COVID-19 pandemic began. Because these methods (laser irradiation, microwave irradiation, ionizing irradiation, and pulse radiolysis) use safe processes to create nanoparticles, they can be used to synthesize silver nanoparticles (Iravani 2014).

Green synthesized silver nanoparticles made from Eucalyptus procera plant extracts have recently been tested as an adjuvant in veterinary rabies vaccines. The European Pharmacopeia 8.0 states that these silver nanoparticles' toxicity was examined in vivo. The findings demonstrated the impact of green silver nanoparticles on boosting the humoral response to rabies vaccination in mice and dogs without causing any in vivo toxicity (Asgary et al. 2016).

In accordance with the size and composition of the nanoparticles, researchers have concentrated on creating materials that reduce the inflammatory hoss effect. Lately, biomimetic nanoparticles—design principles—have been used to synthesize platforms. Biomimetic nanoparticles, which harness the innate tendencies of cell-surface ligands to control cell activation and crosstalk, are the wave of the future for nanomedicine. These nanoparticles imitate the variety of ligands found on the surface of cells because they are based on biological systems and processes. Include membranous proteins or entire cell membranes that have been extracted from tissues, blood, or other biological sources. This term can also be used to describe nanoparticles that have been functionalized with extracted cell membrane proteins or recombinant proteins.

According to Thorp et al. (2020), biomimetic nanoparticles hold promise for next-generation therapies aimed at achieving immune tolerance. Recently, novel biomaterial platforms have been created using alginate-based materials for biomedical applications that aim to lessen the effects of the COVID virus, specially for patients with heart disease, who are significantly more likely to experience complications from the virus (Spoială et al. 2022).

Another cutting-edge therapeutic strategy is the use of gold nanoparticles, which modulate immune responses, control both innate and adaptive immunity, boost antitumor activity during immunotherapy, and reduce excessive inflammation in autoimmune disorders. Green biosynthesis, which produces gold nanoparticles without harming living things, offers a secure way to produce these nanoparticles from plant or microbial extracts (Cai et al. 2022).

Furthermore, zinc oxide nanoparticles have been found to possess antioxidant and antibacterial properties when synthesized using methanolic leaf extracts of the medicinal herb Viscum album, with the ability to combat Escherichia coli, Staphylococcus aureus, and Pseudomonas aeruginosa. These environmentally friendly zinc oxide nanoparticles can be used to treat bacteria that are toxic to reactive oxygen species and resistant to different medications (Mushtaq et al. 2023). Encapsulating or carrier materials are substances that have the ability to encapsulate bioactive agents. These are materials that can be used to form "shell" particles that can trap different compounds, such as cells, liposomes, bioactive agents, drugs, and others, inside them through emulsions, suspensions, and co-precipitations (Miinea et al. 2010).

Various pharmaceutical companies are currently focused on finding new formulations for drug delivery, aiming that new drugs will require fewer doses due to controlled release technology. Being green technology formulations, this also promotes the biodegradation of the capsule within the body. These green materials in the pharmaceutical industry are often called excipients. Manufacturers and distributors of raw materials offer new formulations depending on their use, allowing new

Table 2.1 Green Materials currently used as nanocarriers approved by the FDA

Pharmaceutical or therapeutic agent	Application area
Polyethilene glycol- Adenosin deaminase (Pegademase)	Disease severe immunodeficiency
Doxorubicin liposomes stabilized with polyethylene glycol	Sarcome of Kaposi, ovarian cancer, cervical cancer
Liposomes of Amfotericin B	Treatment of fungical infections
Polyethilene glicol- interferon	Treatment of hepatitis C

dosage forms to be developed. They always seek to improve drug efficiency and reduce the cost of manufacturing biodegradable excipients (Van Arnum 2007). The properties of materials capable of positioning green materials as viable options for replacing synthetic materials include the ability to form individual nano and microparticles, which together initiate the formation of powders that can be administered in various ways within the human body (Hamad et al. 2010).

In the following Table 2.1, the green materials currently used as nanocarriers approved by the FDA are specified (Villafuertes Robles 2011):

Generally, the pharmaceutical structures developed through nanotechnology are known as drug delivery systems. The notion that an excipient must be a completely independent molecule and not covalently bonded to the active ingredient would exclude most products created via nanotechnology. However, there are precedents where components covalently linked to active ingredients have been classified as excipients. Examples include polyethylene glycol in pegylated proteins and albumin in conjugated products like the albumin-doxorubicin complex (Villafuerte Robles 2007).

References

Abar LS, Vandghanooni S, Torab A, Jaymand M, Eskandani M (2024) A comprehensive review on nanocomposite biomaterials based on gelatin for bone tissue engineering. Int J Biol Macromol 254(1):127556. https://doi.org/10.1016/j.ijbiomac.2023.127556

Abdul Khalil HPS, Saurabh CK, Tye YY, Lai TK, Easa AM, Rosamah E et al (2017) Seaweed based sustainable films and composites for food and pharmaceutical applications: a review. Renew Sust Energ Rev 77:353–362

Abraham E, Weber DE, Sharon S, Lapidot S, Shoseyov O (2017) Multifunctional cellulosic scaffolds from modified cellulose nanocrystals. ACS Appl Mater Interfaces 9:2010–2015

Agrawal P, Soni S, Mittal G, Bhatnagar A (2014) Role of polymeric biomaterials as wound healing agents. Int J Low Extrem Wounds 13:180–190

Alexandrescu L, Syverud K, Gatti A, Chinga-Carrasco G (2013) Cytotoxicity tests of cellulose nanofibril-based structures. Cellulose 20:1765–1775

Anirudhan TS, Rejeena SR (2014) Poly(acrylic acid-co-acrylamide-co-2-acrylamido-2-methyl1-propanesulfonic acid)-grafted nanocellulose/poly(vinyl alcohol) composite for the in vitro gastrointestinal release of amoxicillin. J Appl Polym Sci 131: 40699

Aravind Kumar D, Joshua Amarnath S, Jabasingh A et al (2019) One-pot green synthesis of nano magnesium oxide-carbon composite: preparation, characterization and application towards anthracene adsorption. J Clean Prod 237:117691

Asgary V, Shoari A, Baghbani-Arani F, Sadat Shandiz SA, Khosravy MS, Janani A, Bigdeli R, Bashar R, Cohan RA (2016) Green synthesis and evaluation of silver nanoparticles as an adjuvant in rabies veterinary vaccine. Int J Nanomedicine 29(11):3597–3605. https://doi.org/10.2147/IJN.S109098

Azevedo C, Macedo MH, Sarmento B (2018) Strategies for the enhanced intracellular delivery of nanomaterials. Drug Discov Today 23(5):944–959

Barkhordari S, Yadollahi M, Namazi H (2014) pH sensitive nanocomposite hydrogel beads based on carboxymethyl cellulose/layered double hydroxide as drug delivery systems. J Polym Res 21:454

Berlioz S, Molina-Boisseau S, Nishiyama Y, Heux L (2009) Cellulose nanofiber aerogel as a promising biomaterial for customized oral drug delivery. Int J Nanomedicine 12:2020

Bicu I, Mustata F (2011) Cellulose extraction from orange peel using sulfite digestion reagents. Bioresour Technol 102(21):10013–10019

Bielska D, Karewicz A, Kaminski K, Kielkowicz I, Lachowicz T, Szczubialka K, Nowakowska M (2013) Self-organized thermo-responsive hydroxypropyl cellulose nanoparticles for curcumin delivery. Eur Polym J 49:2485–2494

Bin Choy Y, Choi H, Kim K (2009) Uniform ethyl cellulose microspheres of controlled sizes and polymer viscosities and their drug-release profiles. J Appl Polym Sci 112:850–857

Boyjoo Y, Pareek VK, Liu J (2014) Synthesis of micro and nano-sized calcium carbonate particles and their applications. J Mater Chem A 2:14270–14288

Brown EE, Laborie MG, Zhang J (2012) Glutaraldehyde treatment of bacterial cellulose/fibrin composites: impact on morphology, tensile and viscoelastic properties. Cellulose 19:127–137

Brown PK, Qureshi AT, Moll AN, Hayes DJ, Monroe WT (2013) Silver nanoscale antisense drug delivery system for photoactivated gene silencing. ACS Nano 7(4):2948–2959

Budtova T (2019) Cellulose II aerogels: a review. Cellulose 26:81–121

Bukhari SI, Hamed MM, Al-Agamy MH, Gazwi HSS, Radwan HH, Youssif AM (2021) Biosynthesis of copper oxide nanoparticles using Streptomyces MHM38 and its biological applications. J Nanomater 2021:6693302, 16 pages

Cai J, Zhang L (2005) Rapid dissolution of cellulose in lioh/urea and naoh/urea aqueous solutions. Macromol Biosci 5:539–548

Cai F, Li S, Huang H, Iqbal J, Wang C, Jiang X (2022) Green synthesis of gold nanoparticles for immune response regulation: mechanisms, applications, and perspectives. J Biomed Mater Res A 110:424–442

Capadona JR, Shanmuganathan K, Trittschuh S, Seidel S, Rowan SJ, Weder C (2009) Polymer nanocomposites with nanowhiskers isolated from microcrystalline cellulose. Biomacromolecules 10:712–716

Charoensri K, Rodwihok C, Wongratanaphisan D, Ko JA, Chung JS, Park HJ (2021) Investigation of functionalized surface charges of thermoplastic starch/zinc oxide nanocomposite films using polyaniline: the potential of improved antibacterial properties. Polymers 13(3):425

Chen Y, Pang X, Dong C (2010) Dual stimuli-responsive supramolecular polypeptide-based hydrogel and reverse micellar hydrogel mediated by host-guest chemistry. Adv Funct Mater 20:579–586

Ciobanu M, Bobu E, Ciolacu F (2010) In-situ cellulose fibres loading with calcium carbonate precipitated by different methods. Cellul Chem Technol 44:379–387

Ciolacu D, Oprea AM, Anghel N, Cazacu G, Cazacu M (2012) New cellulose-lignin hydrogels and their application in controlled release of polyphenols. Mat Sci Eng C-Mater 32:452–463

Dai L, Si C (2019) Recent advances on cellulose-based nano-drug delivery systems: design of prodrugs and nanoparticles. Curr Med Chem 26:2410–2429

Dash R, Cateto CA, Ragauskas AJ (2014) Synthesis of a co-cross-linked nanocomposite hydrogels from poly(methyl vinyl ether-co-maleic acid)-polyethylene glycol and nanofibrillated cellulose. Cellulose 21:529–534

Díez-Pascual AM (2021) Development of graphene-based polymeric nanocomposites: a brief overview. Polymers 13(17):2978. https://doi.org/10.3389/fimmu.2020.00945

Du J, Tang J, Xu S, Ge J, Dong Y, Li H, Jin M (2018) A review on silver nanoparticles-induced ecotoxicity and the underlying toxicity mechanisms. Regul Toxicol Pharmacol 98:231–239

Farahavar G, Abolmaali SS, Gholijani N, Nejatollahi F (2019) Antibody-guided nanomedicines as novel breakthrough therapeutic, diagnostic and theranostic tools. Biomater Sci 7(10):4000–4016

Gabr GA (2019) Isolation and identification of bacterial strains able to biopolymer polyhydroxybutyrate (Phb) production from soil of Al-Kharj probes, Saudi Arabia. J Pharm Res Int 21:1–11

Geetha AR, George E, Srinivasan A, Shaik J (2013) Optimization of green synthesis of silver nanoparticles from leaf extracts of Pimenta dioica (Allspice). Sci World J 2013:Article ID 362890, 5 pages

Gnach A, Lipinski T, Bednarkiewicz A, Rybka J, Capobianco JA (2015) Upconverting nanoparticles: assessing the toxicity. Chem Soc Rev 44:1561–1584

Gole B, Sanyal U, Banerjee R, Mukherjee PS (2016) High loading of Pd nanoparticles by interior functionalization of MOFs for heterogeneous catalysis. Inorg Chem 55(5):2345–2354

Gravely M, Safaee MM, Roxbury D (2019) Biomolecular functionalization of a nanomaterial to control stability and retention within live cells. Nano Lett 19(9):6203–6212

Guerrero VH, Dávila J, Galeas S, Pontón P, Rosas N, Sotomayor V, Valdivieso C (2011) Nuevos materiales: aplicaciones estructurales e industriales. Ebook: Imprefepp. Quito, Ecuador. ISBN: 978-9978-383-20-9. http://bibdigital.epn.edu.ec/handle/15000/4532

Hamad ML, Bowman K, Smith N, Sheng X, Morris KR (2010) Multi-scale pharmaceutical process understanding: from particle to powder to dosage form. Chem Eng Sci 65(21):5625–5638. https://doi.org/10.1016/j.ces.2010.01.037

Han J, Lei T, Wu Q (2013) Facile preparation of mouldable polyvinyl alcohol-borax hydrogels reinforced by well-dispersed cellulose nanoparticles: physical, viscoelastic and mechanical properties. Cellulose 20:2947–2958

Hassanisaadi M, Bonjar GHS, Rahdar A, Pandey S, Hosseinipour A, Abdolshahi R (2021) Environmentally safe biosynthesis of gold nanoparticles using plant water extracts. Nanomaterials 11(8):2033

Hirayama H, Amolegbe SA, Islam MS, Rahman MA, Goto N, Sekine Y, Hayami S (2021) Encapsulation and controlled release of an antimalarial drug using surface functionalized mesoporous silica nanocarriers. J Mater Chem B 9(25):5043–5046

Huang L, Chen J, Yu Z, Tang D (2020) Self-powered temperature sensor with seebeck effect transduction for photothermal–thermoelectric coupled immunoassay. Anal Chem 92(3):2809–2814

Iravani S (2014) Bacteria in nanoparticle synthesis: current status and future prospects. Int Sch Res Notices 2014:359316, 18 pages

Juengert J, Bresan S, Jendrossek D (2018) Determination of polyhydroxybutyrate (PHB) content in ralstonia eutropha using gas chromatography and Nile red staining. Bio Protoc 8:8

Kalpana V, Devi Rajeswari V (2018) A review on green synthesis, biomedical applications, and toxicity studies of ZnO NPs. Bioinorg Chem Appl 2018:3569758, 12 pages

Kazemi S, Hosseingholian A, Gohari S, Feirahi F, Moammeri F, Mesbahian G, Moghaddam ZS, Ren Q (2023) Recent advances in green synthesized nanoparticles: from production to application. Mater Today Sustain 24:100500. https://doi.org/10.1016/j.mtsust.2023.100500

Khandel P, Yadaw RK, Soni DK, Kanwar L, Shahi SK (2018) Biogenesis of metal nanoparticles and their pharmacological applications: present status and application prospects. J Nanostructure Chem 8:217–254

Khodadadi S, Mahdinezhad N, Fazeli-Nasab B, Heidari MJ, Fakheri B, Miri A (2021) Investigating the possibility of green synthesis of silver nanoparticles using Vaccinium arctostaphlyos extract and evaluating its antibacterial properties. BioMed Res Int 2021:5572252, 13 pages

Kleinfeldt L, Gädke J, Biedendieck R, Krull R, Garnweitner G (2019) Spray-dried hierarchical aggregates of iron oxide nanoparticles and their functionalization for downstream processing in biotechnology. ACS Omega 4(15):16300–16308

Kourmentza C, Plácido J, Venetsaneas N, Burniol-Figols A, Varrone C, Gavala HN, Reis MAM (2017) Recent advances and challenges towards sustainable polyhydroxyalkanoate (PHA) production. Bioengineering 4:55

Li X, Wei J, Aifantis KE, Fan Y, Feng Q, Cui FZ, Watari F (2016) Current investigations into magnetic nanoparticles for biomedical applications. J Biomed Mater Res A 104(5):1285–1296
Liu W, Fang C, Wang S, Huang J, Qiu X (2019) High-performance lignin-containing polyurethane elastomers with dynamic covalent polymer networks. Macromolecules 52(17):6474–6484
Makvandi P, Wang CY, Zare EN, Borzacchiello A, Niu LN, Tay FR (2020) Metal-based nanomaterials in biomedical applications: antimicrobial activity and cytotoxicity aspects. Adv Funct Mater 30(22):1910021
Malik S, Muhammad K, Waheed Y (2023) Nanotechnology: a revolution in modern industry. Molecules 28(2):661
Martínez-Torres AC, Uscanga-Palomeque AC, Rodríguez-Padilla C (2016) Actividad citotóxica in vitro frente a células tumorales. In: RivasMorales C, Oranday-Cardenas MA, Verde-Star MJ (eds) Investigación en plantas de importancia médica. OmniaScience, Barcelona, España, pp 269–312
Miinea L, Farina J, Kallam M, Deorkar N (2010) A new world of excipients for oral solid dosage formulation. Am Pharm Rev White paper, Disponible en: http://www.americanpharmaceuticalreview.com/images/MBI_A_New_World_of_Excipients.pdf. Consultado el 7 de mayo de 2010.
Moradi F, Sedaghat S, Moradi O, Arab Salmanabadi S (2021) Review on green nano-biosynthesis of silver nanoparticles and their biological activities: with an emphasis on medicinal plants. Inorg Nano-Metal Chem 51(1):133–142
Mout R, Moyano DF, Rana S, Rotello VM (2012) Surface functionalization of nanoparticles for nanomedicine. Chem Soc Rev 41(7):2539–2544
Murthy HCA, Desalegn T, Kassa M, Abebe B, Assefa T (2020) Synthesis of green copper nanoparticles using medicinal plant Hagenia abyssinica (Brace) JF. Gmel. Leaf extract: antimicrobial properties. J Nanomater 2020:3924081, 12 pages
Mushtaq W, Ishtiaq M, Maqbool M, Mazhar MW, Casini R, Abd-ElGawad AM, Elansary HO (2023) Green synthesis of zinc oxide nanoparticles using *Viscum album* extracts: unveiling bioactive compounds, antibacterial potential, and antioxidant activities. Plan Theory 12:2130. https://doi.org/10.3390/plants12112130
Muthu MS, Singh S (2009) Targeted nanomedicines: effective treatment modalities for cancer, AIDS, and brain disorders. Nanomedicine 4(1):105–118
Nejati K, Dadashpour M, Gharibi T, Mellatyar H, Akbarzadeh A (2021) Biomedical applications of functionalized gold nanoparticles: a review. J Clust Sci 33:1–16
Noga M, Milan J, Frydrych A, Jurowski K (2023) Toxicological aspects, safety assessment, and green toxicology of silver nanoparticles (AgNPs)-critical review: state of the art. Int J Mol Sci 24(6):5133. https://doi.org/10.3390/ijms24065133
Penkhrue W, Jendrossek D, Khanongnuch C, Pathom-Aree W, Aizawa T, Behrens RL, Lumyong S (2020) Response surface method for polyhydroxybutyrate (PHB) bioplastic accumulation in Bacillus drentensis BP17 using pineapple peel. PLoS One 15:e0230443
Pillai AB, Kumar AJ, Thulasi K, Kumarapillai H (2017) Evaluation of short-chain-length polyhydroxyalkanoate accumulation in Bacillus aryabhattai. Braz J Microbiol 48:451–460
Pontón P and Guerrero VH (2010) Obtención de materiales compuestos de Matriz Poliéster reforzados con Fibra de Abacá mediante Estratificación manual. Revista Tecnológica – ESPOL 23(39):47–54. https://rte.espol.edu.ec/index.php/tecnologica/article/view/62
Prete S, Dattilo M, Patitucci F, Pezzi G, Parisi OI, Puoci F (2023) Natural and synthetic polymeric biomaterials for application in wound management. J Funct Biomater 14(9):455. https://doi.org/10.3390/jfb14090455. PMID: 37754869; PMCID: PMC10531657
Rong L, Qin SY, Zhang C, Cheng YJ, Feng J, Wang SB, Zhang XZ (2018) Biomedical applications of functional peptides in nano-systems. Mater Today Chem 9:91–102
Sayed AE-DH, Soliman HAM (2017) Developmental toxicity and DNA damaging properties of silver nanoparticles in the catfish (Clarias Gariepinus). Mutat Res Genet Toxicol Environ Mutagen 822:34–40
Shen R, Ning Z-Y, Lan Y-X, Chen J-C, Chen G-Q (2019) Manipulation of polyhydroxyalkanoate granular sizes in Halomonas bluephagenesis. Metab Eng 54:117–126

Singh A, Sharma J, Mohanta B, Sood A, Han S, Sharma A (2024) Synthetic and biopolymers-based antimicrobial hybrid hydrogels: a focused review. J Biomater Sci Polym Ed 35:675–716. https://doi.org/10.1080/09205063.2023.2278814
Spoială A, Ilie C-I, Ficai D, Ficai A, Andronescu E (2022) From biomedical applications of alginate towards CVD implications linked to COVID-19. Pharmaceuticals 15(3):318. https://doi.org/10.3390/ph15030318
Tang D, Cui Y, Chen G (2013) Nanoparticle-based immunoassays in the biomedical field. Analyst 138(4):981–990
Tapia C, Paredes C, Simbaña A, Bermúdez J (2006). Aplicación de las Fibras Naturales en el Desarrollo de Materiales Compuestos y como Biomasa, Revista Tecnológica – ESPOL 19(1): Ciencias Naturales y Medio Ambiente
Tarrahi R, Fathi Z, Seydibeyglu MÖ, Doustkhah E, Khataee A (2020) Polyhydroxyalkanoates (PHA): from production to nanoarchitecture. Int J Biol Macromol 146:596–619
Thorp EB, Boada C, Jarbath C, Luo X (2020) Nanoparticle platforms for antigen-specific immune tolerance. Front Immunol 11:945
Tripathi JP (2022) Green polymeric materials for medical implantable and non-implantable devices. In: Baskar C, Ramakrishna S, Daniela La Rosa A (eds) Encyclopedia of green materials. Springer, Singapore. https://doi.org/10.1007/978-981-16-4921-9_98-1
Van Arnum P (2007) Expanding opportunities for specialty excipients. Pharm Tech, abril. Consultado el 3 de septiembre de 2010. Disponible en: http://pharmtech.findpharma.com/pharmtech/Pharma+Ingredients/Expanding-Opportunities-for-Specialty-Excipients/ArticleStandard/Article/detail/415103
Villafuerte Robles L (2007) El diseño de medicamentos al inicio del siglo 21. Los diferentes tipos de medicamentos VI. Nanotecnología Farmacéutica Informacéutico 14(1):15–32
Villafuertes Robles L (2011) The excipients and their functionality in pharmaceutical solid products. Rev Mex Cienc Farm 42(1):18
Xie S, Ai L, Cui C, Fu T, Cheng X, Qu F, Tan W (2021) Functional aptamer-embedded nanomaterials for diagnostics and therapeutics. ACS Appl Mater Interfaces 13(8):9542–9560
Xie Y, Gao S, Zhang D, Wang C, Chu F (2023) Bio-based polymeric materials synthesized from renewable resources: a mini-review. Resour Chem Mater 2(3):223–230. https://doi.org/10.1016/j.recm.2023.05.001
Yapias RM, Astete JA, Uscuchagua YC, Sánchez MCR (2022, 2022) Características fisicoquímicas, composición nutricional y compuestos bioactivos en tres variedades de Mashua (Tropaeolum tuberosum Ruiz y Pavón): Una revisión. Revista Tecnológica – ESPOL 34(2)
Yu Z, Gong H, Li Y, Xu J, Zhang J, Zeng Y et al (2021) Chemiluminescence-derived self-powered photoelectrochemical immunoassay for detecting a low-abundance disease-related protein. Anal Chem 93(39):13389–13397
Zhang D, Ma X-L, Gu Y, Huang H, Zhang G-W (2020) Green synthesis of metallic nanoparticles and their potential applications to treat cancer. Front Chem 8:799

Daniela Marisol Carmona-Govea, M.Sc. Chemical Engineer from the Autonomous University of Coahuila, Mexico, with a master's degree in Materials Science and Technology, specializing in the technology, analysis and synthesis of nanopolymers and nanoceramics from natural sources for medical applications. She has been conducting research and development at the Applied Chemistry Research Center since 2019. Her research focuses on the synthesis of nanometric biomaterials using dispersed media polymerization techniques, as well as the study and kinetic analysis of synthetic and natural polymers within the healthcare and industrial sectors. This work explores alternatives to current drugs and therapies, along with potential applications of these biomaterials. In addition, I have contributed to the development and research of bioactive compounds for use in biopharmaceuticals, particularly in the treatment of chronic and degenerative diseases, through the application of bio-nanoencapsulation delivery techniques.

Víctor Adán Cepeda-Tovar Chemical Engineer from the Autonomous University of Coahuila of Mexico with a Master's in Materials Science and Technology, specializing in technology, analysis, and synthesis of polymers from plant sources. Member of the state research system COECYT since 2022. Carrying out research and development has been at the Center for Research in Applied Chemistry and the Center for Research in Advanced Materials since 2019. The lines of research address the synthesis of biopolymers from plant sources, using polymerization techniques in dispersed media, study and kinetic analysis of synthetic and natural polymers in the industrial sector, extraction processes, and quantification of active biomolecules from agro-industrial waste for the generation of microbial biomasses to obtain high added value products-synthesis and application of biomaterials and accessories in general medicine and biopharmacy for treating chronic or degenerative diseases, using polymerization techniques in dispersed media.

Ana Cecilia Cepeda-Nieto, PhD in Biomedical Sciences. Master's degree in Genetic and Molecular Biology. Research and faculty educator at Medicine School, University Autonomous of Coahuila, Saltillo, Mexico. Goals in Biomedical research and faculty with more than 15 years of experience. Skills in critical thinking and experimental research. Research interests in biomedical sciences oriented to genetic of chronic-degenerative diseases. Member of PRODEP and Mexican Research System CONAHCYT, level 1. ORCID file 0000-0003-2563-6171 E-mail: anacepeda@uadec.edu.mx

Anilu Rubio-Ríos, PhD Anilu Rubio obtained her bachelor's degree in chemical engineering and a master's degree in food science and technology from the Autonomous University of Chihuahua. In 2012, she earned her PhD in Materials Science at the Center for Research in Advanced Materials (CIMAV) in Chihuahua, Mexico.

She worked in the food industry in research and development. In 2015, she joined the Faculty of Chemical Sciences at the Autonomous University of Coahuila. She teaches undergraduate courses in the Chemical Engineering program, as well as in the master's in food science and technology and the Doctorate in Biotechnology programs at the same faculty.

Her research focuses on chemical processes, specifically the production and synthesis of polymeric materials through environmentally friendly emulsion and mini emulsion polymerization techniques, and the use of plant-based monomers for the creation of biopolymeric films.

She is currently an Honorary State Researcher, holds a desirable profile according to PRODEP, and is a Level I member of the National System of Researchers.

Jesús Antonio Morlett-Chávez in Molecular Biology and Genetic Engineering; master's degree in biotechnology. Researcher, Faculty, and Health Research Coordinator at Medicine School, University Autonomous of Coahuila at Saltillo. Senior researcher at the Mexican Social Security Institute. Goals in biomedical research based on proteomics. Faculty with more than 15 years of experience. Membership of the Council National System of Researcher (CONAHCYT), level 1. Program for Faculty Professional Development in Mexico. Skills in molecular biology techniques employed in proteomics and transcriptomics, as well as skills in thinking and experimental research. Research interest in biomedical sciences oriented to proteomics and transcriptomics of chronic-degenerative diseases.

Sathvik Belagodu Sridhar is currently working as an Associate Dean and Professor at RAK College of Pharmacy (RAKCOP), RAK Medical and Health Sciences University located in Ras Al Khaimah, United Arab Emirates. Dr. Sathvik has completed his PhD in Clinical Pharmacy from Rajiv Gandhi University of Health Sciences, Bangalore, India. He has received an FIP fellowship for higher training in renal clinical pharmacy at Royal Adelaide Hospital and Queen Elizabeth Hospital in Adelaide, South Australia. Dr. Sathvik has worked as an Assistant Professor and Clinical Pharmacist at JSS College of Pharmacy and JSS Medical College Hospital, Mysore, India, for eight years and later Joined International Medical University (IMU) Malaysia as a senior lecturer in the Department of Pharmacy Practice from 2008 to 2010. Dr. Sathvik joined as an assistant professor in the Department of Clinical Pharmacy at RAK College of Pharmacy, RAK Medical, and Health Sciences University in 2010 and worked as a professor and chairperson of the Department of Clinical Pharmacy and Pharmacology from 2018 to 2023. Dr. Sathvik has been honored with a prestigious teacher award by the Society of Pharmaceutical Education & Research [SPER]. He has also received training in Pharmacovigilance at the Uppsala Monitoring Centre in Uppsala, Sweden. Dr. Sathvik has 25 years of experience teaching and practicing clinical pharmacy in India, Malaysia, and the UAE. His research interests include Medication Safety, Pharmacovigilance, Pharmacoepidemiology, and Renal and Psychiatry Clinical Pharmacy. Dr. Sathvik has more than 75 research papers published in reputable national and international journals, along with book chapters. He has presented more than 75 research papers at numerous national and international conferences and has been honored with many awards. He also serves as a reviewer for various national and internal pharmacy practice-related journals.

Biomaterials for Tissue Engineered Cartilage Product for Orthopedic (Osteoarthritis) Reconstruction

3

Suraj Kumar, Rishabha Malviya, and Sathvik Belagodu Sridhar

Abstract

The present improved treatment in orthopedics may be attributed mostly to the advancements in biomaterial science, surpassing any other clinical medical specialty. This concise review aims to provide an overview of the current state of biomaterials in the area of orthopedics. It also offers insights into the potential future applications of biomaterials in the treatment of musculoskeletal ailments, with a specific focus on immune regulation. Initially, orthopedic biological materials were developed as passive substitutes for hard tissues like bone and cartilage. However, regenerative medicine is expected to have a more significant impact in both avoiding the loss of tissues due to trauma and addressing the early stages of illnesses. Enhancing our comprehension and control of the immune system's reaction to biomaterials would facilitate the more effective integration of implants into the host tissue or the almost flawless regeneration of the host tissue.

Keywords

Cartilage tissue engineering · Osteoarthritis · Stem cells · Biomaterial · Immune response · Perspective

S. Kumar · R. Malviya (✉)
Department of Pharmacy, School of Medical and Allied Sciences, Galgotias University, Greater Noida, UP, India
e-mail: rishabha.malviya@galgotiasuniversity.edu.in

S. B. Sridhar
RAK College of Pharmacy, RAK Medical & Health Sciences University, Ras Al Khaimah, United Arab Emirates

R. Malviya, S. Sundram (eds.), *Sustainable Nanocomposites with Green Biomaterials*, Biomaterials, Bioengineering and Sustainability 2,
https://doi.org/10.1007/978-3-031-79110-9_3

3.1 Introduction

Articular cartilage is susceptible to a wide range of changes brought about by inflammation, trauma, and the natural aging process. As a result of these challenges, the extracellular matrix, also known as the ECM, of the cartilage undergoes additional deterioration and the cell proliferation of the cartilage continues to deteriorate. Since over a millennium ago, people have been aware of cartilage degradation. "Since Hippocrates and continuing to the present day, it has been widely recognized that the ruptured cartilage is humiliating and that, once destroyed, can not be restored," stated the Scottish physician William Hunter in a dissertation that was published in the Proceedings of the Royal Society in the year 1743. The cartilage of the articular cartilage has a limited capability for regeneration since it cannot vascularize and proliferate chondrocytes. Impairment to cartilage and the breakdown mechanisms that are connected with it are persistent and ultimately result in the occurrence of osteoarthritis (OA) over a prolonged period. Osteoarthritis (OA) is the most prevalent joint ailment that is impacting an increasing proportion of the elderly populace. As people become older, the prevalence of this degenerative condition rises, which is a significant cause for worry in terms of health. It is estimated that osteoarthritis (OA) affects more than 10% of men and 18% of women aged 60 and over worldwide. Osteoarthritis (OA) is characterized by the degradation of the cartilage around the joints, elongation of the subchondral bone, formation of osteophytes, and inflammatory processes of the fluid that surrounds the joint to variable extents. Considering the framework of this discussion, scientific research has been interested in multiple strategies regeneration of cartilaginous tissue that is functional for a considerable amount of time. Nevertheless, the development of transitory fibrocartilaginous tissue is a consequence of procedures involving surgery that have occurred to heal chondro injury. These procedures include abrasive chondroplasty, microfractures, and specialization (Hunziker et al. 2015). Transplantation of material that has chondrogenic qualities, such as transplants of the perichondrium and the perichondrium, as well as osteochondral grafts (mosaicplasty), have been suggested concurrently. The remediation of cartilage injury continues to be challenging even in the modern era, notwithstanding the tremendous progress that has been made in orthopedic surgery. New treatment procedures that are based on cell therapy and tissue engineering have been developed as a result of the various limitations that are connected with the approaches that have been stated above. These drawbacks include graft unpredictability, calcium deposits, and limited applicability to localized lesions. Langer and Vacanti are credited with coining the phrase "tissue remodeling"in 1993. They defined it as "an integrated field that combines engineering and life science approaches to develop naturally existing substitutes that rebuild, continue to or enhance tissues functioning." Tissue engineering is a scientific discipline that originated in the 1990s (Vacanti 2006; Langer and Vacanti 1993).

3.2 Articular Cartilage Tissue Engineering

A relatively new field, tissue engineering integrates engineering and biomedical approaches to create biological replacements that may significantly improve the performance of tissue that has been damaged (Langer and Vacanti 1993). The "tissue engineering tertriad" (Fig. 3.1) refers to the relationship between biomaterials in particular cells as well as either biological or ambient factors. Recent developments in materials science and an increasing understanding of the function underlying substances and stem cells in regenerating tissues have immensely benefited cartilage tissue engineering.

3.2.1 Biomaterials

Cells are often put on scaffolding in the field of tissue engineering. The scaffolding's primary function is to imitate the characteristics of the ECM, or extracellular matrix, of the tissue that is being engineered.. Biomaterials resemble the extracellular matrix (ECM) that provides cells with a supportive framework that promotes growth, survival, and secretive functions. Several ECM-like scaffolds have been developed and evaluated for their potential use in articular tissue engineering applications. The biological materials may be categorized as either synthetic or natural matrices, with a distinction made between those composed of proteins or polysaccharides (Vinatier et al. 2009a) An optimal biological material should possess biocompatibility to mitigate proinflammatory and immune reactions. The environment must be conducive to maintaining the chondrocyte phenotype in three-dimensional and should have sticky properties to allow the cells to bind to the lesion. The material must have permeability to facilitate the flow of chemicals, nutrients, and other

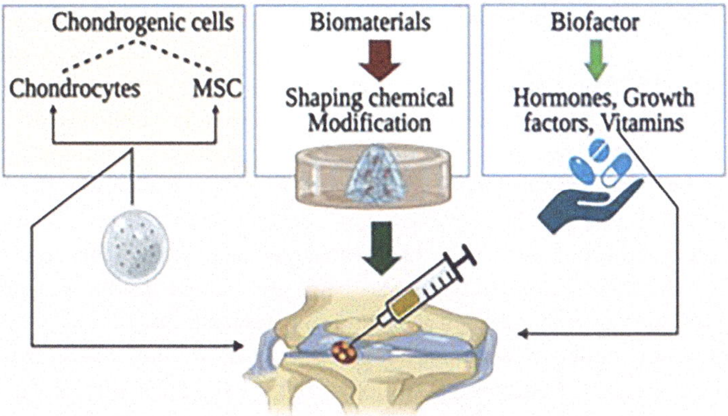

Fig. 3.1 The trio of tissue engineering tools used in the restoration of articular cartilage. To create cartilage tissue-engineering methods, it is essential to combine chondrogenic cells (either enlarged chondrocytes or differentiated MSCs) with biomaterials and biofactors

nurturing factors. Ultimately, the material should possess sufficient biodegradability to be incorporated into the natural tissue remodeling procedures. Ideally, it ought to also be injectable, enabling implantation via fewer invasive surgical procedures. Protein-based matrices, specifically type I and III collagen membranes, are frequently employed in medical applications for autonomous chondrocyte implantation kits including ciated chondrocyte implanting. A specific form of collagen gel called Atelocollagen1, the culture, and implantation of individual allogeneic cartilage from the articulation in three dimensions (Mori et al. 2013) and mesenchymal stem cells derived from bone marrow (MSCs) have been accomplished by applying of this technique (Kuroda et al. 2007) in individual research investigations. Hyalograft1 C is a polysaccharide-based matrix that contains homologous articular cartilage and hyaluronic acid. The acid-containing matrix (Brix et al. 2013) is called HYAFF-111 and is produced by Fidia Advanced Biopolymers in Abano Terme, Italy. This idea has culminated in therapeutic advancements in the treatment of cartilage in articular surfaces in human beings (Nehrer et al. 2006). Due to a negative assessment by the European Medicines Agency, which highlighted a general absence of reliable safety and effectiveness evidence from properly conducted trials, the availability of Hyalograft1 C has been discontinued. BST-CarGel1 is a scaffold composed of chitosan and b-glycerophosphate. It is specifically designed to be embedded in the MACI technique.

The BST-CarGel1 technique had a notable positive impact on therapeutic results, while it had solely a moderate effect on improving the standard of life for patients (Steinwachs et al. 2014). Bio-Seed1-C, developed by BioTissue Innovations in Freiburg, Germany, is a synthetic matrix made of polyglycolic acid (PGA), polylactic acid (PLA), and polydioxanone. This pore biological material serves as a substrate for cartilage cells, which are encased in fibrin glue and deposited onto the matrix (Ossendorf et al. 2007). BioSeed1-C promotes the development of cartilage composed of hyaline and results in enhanced knee functionality in clinical settings. Multifaceted in biological materials have gained popularity. The bilayered scaffolds are composed of two distinct substances: an inner layer with ceramic-like properties that may be inserted to close the gap that exists between the connective tissue below and the bone that is around it, as well as an outermost layer that is adjusted to correspond with the cartilage-like region that is present in osteochondral lesions Clinical experiments have examined different amalgamations of ceramics, polymer synthetics, and organic substances including collagen or hydroxyapatite in bilayered scaffolds (Seo et al. 2014). MaioRegen1, developed by Finceramica in Italy, is a bioceramic scaffolding with many layers arranged in an isolated gradient configuration. Calcium hydroxyapatite which has been supplemented with mg and deantigenated type I equestrian collagen makes up its composition (Filardo et al. 2013). The good being studied is demonstrating encouraging preliminary outcomes that need more research for validation. Although there have been several preliminary and clinical research studies on biological materials for articular tissue manipulation, none of the described matrices are suitable for transdermal injection. Current studies in this field focus on the advancement of self-hardening and injectable materials for the percutaneous transplanting of chondrogenic fibroblasts. Biomaterials

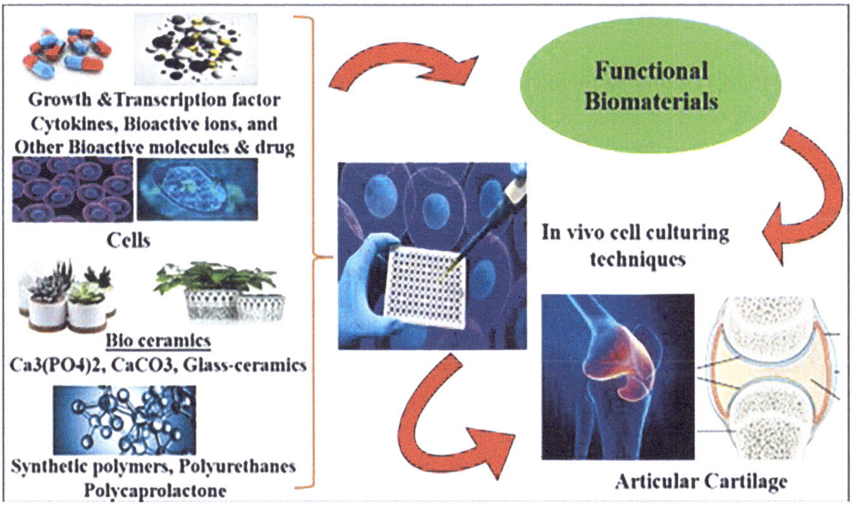

Fig. 3.2 Figure 3.1 illustrates the methodologies used in articular bone tissue engineering

encompass a wide range of substances that vary in arrangement, framework, and physical characteristics. They are characterized by their high compatibility with human beings as well as the capability to integrate with connective tissue or facilitate its regrowth. Examples include hydroxyapatite ceramics, bio-glasses, and improved carbon-based substances (Fig. 3.2).

Hydrogels are substances that possess features and frameworks that lend themselves to them very suitable for implantation by invasive techniques that are as brief as possible (Vinatier et al. 2016). Materials that may develop hydrogels that are include sequencing of macromolecules, which can be either manufactured or naturally occurring and come by physical, sonic, or covalent coupling (Drury and Mooney 2003). They demonstrate a significant degree of hydration that is similar to that of the cartilage in the articular region, enabling them to replicate the 3-D conditions of articular cartilage (Moreira Teixeira et al. 2014). Validating this concept, we have recently utilized a self-setting hydrogel made from cellulose to repair cartilage that has been damaged in the articular area. This is done by employing articular cartilage and mesenchymal stem cells in the rabbit model of osteochondral injuries (Vinatier 2009b; Portron et al. 2015a, b). To summarise, the research we conducted successfully cultured cartilage cells that maintained their physical characteristics and also induced the process of differentiation of mesenchymal stem cells into chondrocytes using various biomaterials. Although there are many different scaffolding available for cartilage engineering, just a few of them have been effectively used in healthcare environments (Smith and Grande 2015).

3.2.2 Biological and Environmental Factors

The synthesis of cartilage in articulations at the early stages of development is accomplished by a process intimately associated with endochondral ossifying. Numerous growth regulators that have a role in the formation of cartilage plates during development have been recognized as variables that have a significant impact Regarding the chondrogenic multiplication and development of mesenchymal stem cells (MSCs). The bulk of the aforementioned factors, namely those belonging to the tumor growth factor b (TGF-b) superfamily, such as transforming growth factor 1 and -b3, are the ones that are used particularly often to produce bone formation in adult mesenchymal stem cells. Another approach that has been extensively studied is the application of bone morphogenetic protein, either in conjunction with or instead of TGF-b (Beane and Darling 2012). To influence the chondrogenic development of mesenchymal stem cells (MSCs), several growth or differentiation agents have been explored, and varied effects have been seen (Vinatier et al. 2009b, c) Through the administration of certain environmental cues, mesenchymal stem cells (MSCs) may be generated or encouraged to differentiate into chondrogenic cells. Among these parameters, the ones that have received the most attention from researchers are undoubtedly three-dimensional culture, inadequate oxygen tension, and mechanical stimulation. As is the case with bone marrow which may be maintained or restored via the use of three-dimensional culture (Vinatier et al. 2005), 3D culture conditions are ideal for promoting mesenchymal stem cell (MSC) chondrogenic development (Merceron et al. 2011). When cells are subjected to high-density agglomeration by centrifugation process they are placed in an in vitro three-dimensional milieu that roughly resembles the cell condensation that often occurs during the early stages of formation of cartilage in vivo (Vinatier et al. 2009c) Interestingly, the biological materials that we propose have been effectively examined for the chondrogenic division of mesenchymal stem cells (MSCs) and 3-D culture (Merceron et al. 2010).

The development of cartilage in the articulation occurs spontaneously in the context of hypoxia that ranges from 2% to 7%. Articular cartilage is considered to be avascular. In addition, hypoxia can increase the manifestations of cartilage extracellular matrix genes via the hypoxia-inducible factors (HIFs), which are responsible for upregulating the chondrocyte phenotypic (Lafont et al. 2008). It is consequently possible to stimulate the bone formation of mesenchymal stem cells (MSCs) by exposing them to hypoxic conditions (Merceron et al. 2010). Furthermore, it has been observed that low oxygen tension in the environment might inhibit the final hypertrophy differentiation of mesenchymal stem cells (Portron et al. 2015b). Therefore, oxygen tension is undeniably an aspect that is both risk-free and readily adjustable, and it is used to influence the chondrogenic development of mesenchymal stem cells (MSCs). In conclusion, the function that mechanical stimulus has been taken into consideration, taking into account the fact that cartilage in articulation is an area of tissue that may be compressed, subjected to deformation forces, or subjected to hydrodynamic tension, among other biomechanical stressors. During the process of articular cartilage formation, physiological loading is a crucial component that helps mesenchymal stem cells (MSCs) differentiate into chondrocytes. Furthermore, evidence has demonstrated that the application of mechanical stimulation to bone marrow cells is completely necessary for the preservation of cartilage integrity. On the other hand, we have studies that are persuasive that mechanical stimulation may be utilized to generate meniscus-engineered structures possessing an extraordinary resemblance to the geometrical arrangement of glycosaminoglycan and collagen observed in genuine cartilage (Huang et al. 2010).

3.3 Possible Future Therapies for Osteoarthritis Based on What We Know About Cartilage Engineering

The potential of mesenchymal stem cells to alleviate osteoarthritis has been the subject of an explosion of study throughout the past decade. The release of immune-modulating, proangiogenic, antiapoptotic, antifibrotic, and antiinflammatory substances is the primary mechanism by which MSCs exhibit their regeneration abilities, as recently shown (English 2013). Multiple randomized controlled studies have shown that MSCs offer therapeutic promise in inflammatory, autoimmune, and auto-immune illnesses (Le Blanc et al. 2008; Connick et al. 2012) preventing T- and B-cell multiplication (English 2013) and shielding against apoptotic (Connick et al. 2012) are two ways in which MSCs might influence adaptive immune cells. In innate immunity, MSCs block the development of dendritic cell populations (Spaggiari et al. 2009) and natural killer (NK) cells (Spaggiari et al. 2006) and change the polarisation of macrophages from pro-inflammatory (M1) to anti-allergic (M2) (Nemeth et al. 2009). Intriguingly, a model of Osteoarthritis was shown to be prevented in animals when MSCs generated from the bone marrow were injected intra-articularly or delivered via a biodegradable scaffold (Murphy et al. 2003). In addition to inhibiting inflammatory processes, adipose-derived MSCs have shown some chondroprotective benefits in Osteoarthritis models (ter Huurne et al. 2012). As a result, there is mounting evidence that MSCs separated from different tissues have a strong protective impact in Osteoarthritis.

Despite these promising findings, there are still significant drawbacks to injecting MSCs into an arthritic joint. These include an elevated incidence of cell death immediately after injection and the possibility of cell leakage outside the joint due to MSC migration (Hunt and Grover 2010). The majority of patients did not show any signs of the infused MSCs 50 days after infusion (von Bahr et al. 2012) Enhancing the clinical efficacy may be achieved by protecting MSCs from immune-mediated detection and prolonging their life in the body, even though MSCs may exert therapeutic benefits via a temporary "hit-and-run" process. The necessity for long-term MSC incorporation and sustained regeneration of tissue in the setting of Osteoarthritis makes this an especially pertinent problem. Cell encapsulating before infusion is being studied as a potential solution to these limitations. The principle of cellular encapsulation is to confine cells inside a biological material sphere or capsule with a semi-permeable membrane. One area where this method has shown promise is in the immuno-isolation of cells that may produce protein therapies for the treatment of illness (Kang et al. 2014). While facilitating the bidirectional passage of oxygen, nutrients, growth hormones, inflammatory chemicals, and waste, this barrier shields imprisoned cells from mechanical stressors. Arthritis, chronic kidney disease, insulin resistance, and cancer are only a few of the numerous long-term illnesses that have prompted discussions for encapsulating cell therapy. The encapsulating cells have a longer half-life after injection, which bodes well for their potential long-term effectiveness. To treat osteoarthritis, encapsulating cells may be the best way to make use of MSCs'secreting capabilities (Fig. 3.3).

In osteoarthritis, as in aging, the breakdown of articular tissue homeostasis leads to an accumulation of catabolism merchandise, dysfunctional organelle-like structures, and molecules corresponding with structural patterns of destruction, which in turn increases inflammation as well as oxidative stress (Haseeb and Haqqi 2013). TM. Autophagic activity is crucial for cartilage cells to remove dysfunctional biomolecules or degenerated organelles, allowing them to keep up homeostatic (Caramés et al. 2010). As we grow older, autophagy slows down in bone marrow cells which means more damaged molecules accumulate. This causes chondrocytes to senescence and more Osteoarthritis to form (Zhou et al. (2014). Autophagic levels of marker proteins (beclin-1, LC3B, and ulk-1) decrease throughout murine Osteoarthritis models, but senescence hallmark concentrations (p16INK4a, P21CIP1, SA-b-Gal) rise, demonstrating a discrepancy among the two processes. Aside from that, rapamycin restored autophagy, and siRNA reduction of p16INK4a decreased osteoarthritic chondrocyte aging and improved their repair activities (Carames et al. 2012) might inhibit the degeneration of articular in both animal models of produced Osteoarthritis and in vitro investigations (Matsuzaki et al.

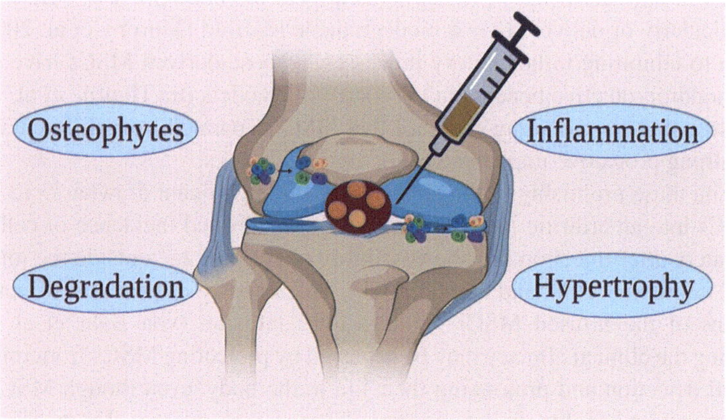

Fig. 3.3 Enclosing mesenchymal stem cells (MSCs) for the intra-articular therapy of osteoarthritis. The administration of mesenchymal stem cells (MSCs) enclosed in biological material capsule-like structures, which enable the controlled release of molecules with anti-inflammatory properties, immune-modulating, and trophic properties, has the potential to prevent the occurrence of inflammatory processes, chondrocyte hypertrophy, cartilage eventually degradation, and osteophyte formation, all of which are characteristic manifestations of osteoarthritis. The shade of blue on the arrow indicates inhibition

2014). Tissue engineering is used in this setting techniques that prevent the breakdown of articular cartilage after surgical induction of osteoarthritis (OA) by regulating the liberation of pro-autophagic medications (rapamycin) from specially engineered biological materials (Chen et al. 2015). This suggests that future tissue-engineering approaches to chondro injury and Osteoarthritis should aim to decrease or prevent senescence while enhancing autophagy of chondrocytes as a potential treatment for OA.

3.4 Regenerative Medicine and Immune Response in Biomaterials

Regenerative therapies will take a lead role in avoiding tissue trauma and early disease phases, while biological material substitutes will keep getting better and used for treating chronic conditions at lower embedded material expenses. These tissue engineering methods will not be possible without stem cells and biological materials. Inserting cells that have been implanted or encased in biological materials has been the primary objective of stem cell-based restorative techniques. Unfortunately, most tissues that have been destroyed due to trauma or degeneration cannot be reliably regenerated with the use of exogenous stem cells. It is anticipated that these foreign cells will be incorporated into the tissue that serves as the host, but unfortunately, the majority of the cells will die off after a brief period. Additionally, the location of implantation is where inflammation occurs. Consequently, immune response is becoming a major concern in the field of tissue engineering (Sridharan et al. 2015). In animals, there is a reverse correlation between regeneration capability and the growth of the immune system as a whole.

The immune system's ability to partially repair tissues that are damaged was lost along with its phylogenetic and growth trajectory (Sridharan et al. 2015; Fukazawa et al. 2009; Mescher et al. 2013). The majority of tissues in mammals are unable to regenerate on their own. Their very well-developed immune system is associated with this (Aurora and Olson 2014). Scarring or the activation of local precursor cells for proliferation occurs in response to tissue injury. Additionally, potentially harmful or immune-stimulating materials are swiftly removed from cellular waste. Immunomodulatory factors are secreted by active phagocytes. Those responsibilities and the main role they serve in immunity generation are carried out by macrophages in mammalian cells. In response to tissue damage, monocytes exhibit interesting polarised and biphasic behavior. Different subsets of macrophages, known as conventionally activated (M1) or activated (M2), with distinct marker/cytokine characteristics, polarise in proinflammatory settings (Gordon and Martinez 2010). As an instance of host defense, M1 cells often release nitric oxide (NO) and inflammatory cytokines, which may harm host tissues. Macrophages with the M2 designation, on the other hand, facilitate resolving inflammation and wound recuperation by secreting immunomodulatory and anti-inflammatory in nature chemicals. Tissue repair depends on the proper recruitment and equilibrium of all macrophage subtypes. Early invasion of M1 macrophages is crucial for clearing necrotic tissue. However, any interruption in the polarisation of macrophages might hinder tissue regeneration.

3.5 Immuno-Modulating Biomaterials

The idea of regulating the immune system to get the best possible therapeutic outcome is going to encounter extensive use in orthopedic biological materials in the years to come. We might generally categorize immuno-modulating biological materials into two types: (1) implanted biological materials that incorporate with the human system and stay there indefinitely resulting in limited inflammation and the development of fibrous tissue; (2) biological materials designed to rejuvenate, which initially stimulate the creation of novel connective tissue and facilitate its development, but whose degradation is regulated and progressive (Brown et al. 2012).

3.5.1 Biomaterials for Replacement

The biological materials used for substitution are usually technologies that may be implanted permanently or that have long-term durability (> 20 years). They exhibit minimal host reaction when implantation and are functionally resilient, being composed of metals, ceramics, or polymers (Gordon and Martinez 2010). Historically, there has been a preference for biologically inactive prostheses that reduce the number of interactions between cells and implantation in the surrounding environment (Brown et al. 2012) As a buffer between the biological materials and the host, the native proteins adsorbed on the surface of these implantations encourage the creation of a transitional matrix. Furthermore, minimal motion between the implantation device and the surrounding tissue is achieved using precision surgical techniques (Sridharan et al. 2015). Alternatively, under certain circumstances, interactions between cells and implants may improve antibody resistance and implantation incorporation into host tissue. When the surface of the titanium bone replacement implant is modified to encourage the migration and attachment of bone marrow cells the rate of osseointegration is increased (Franz et al. 2011; Le Guéhennec et al. 2017) In addition to creating an immunological milieu that is conducive to bone remodeling, these changes may promote a pro-M2 polarisation. To reduce fibrous tissue development and increase the production of regenerative and anti-inflammatory substances, one may change the outermost layer's roughness and chemistry to shift the polarisation towards the M2-type (Rungsiyakull et al. 2010; Anderson et al. 2008; Fink et al. 2008; Engler et al. 2006). The distinction between biological materials that regenerate and those that just serve as a substitute has become more porous as bioengineering has progressed. There are numerous examples of coating technologies that are useful for the field of regenerative medicine and are also employed on replacement implants (Vacanti 2006).

3.5.2 Biomaterials for Regeneration

The goal of regenerative biological materials is to help injured tissues regain their shape and function (Aurora and Olson 2014). Ideally, the aforementioned substances would break down over a few days to a couple of weeks while simultaneously encouraging the host tissue to regrow, which would include the regenerating matrix (Gardner et al. 2013). An essential early step in the recovery of wounds, the first M1 response summons Inflammatory cells to migrate to the site of introduction, which starts the response to the foreign material (Rees 2010; Boehler et al. 2014). On the other hand, cytotoxic reactive oxygen species products are produced when macrophages with the M1 designation are present for an extended period (Spiller et al. 2015; Kearney and Mooney 2013). In addition, biological materials may not be able to encourage tissue growth or disintegrate properly if fibrous capsules form as a result of prolonged inflammation. Therefore to encourage tissue remodeling, a further switch to the M2 type is required (Sridharan et al. 2015). By manipulating the immune system's response with small molecules such as (steroids) or cytokines that are anti-inflammatory (IL-4, IL-10) native signaling may be suppressed and M2 polarisation to macrophage directed (Féréol et al. 2006). Integrating them into automated release systems might accomplish this (McWhorter et al. 2013). It is possible to affect macrophage polarisation by manipulating several biochemical and biophysical features. To generate optimum macrophage polarisation, one may modify the surface topography, which includes the scaffold's pore size (Lotz and Caramés 2011; Caramés et al. 2010; Zhou et al. 2014). Before making any changes to biological material designs to directly influence macrophage behavior, Considering the potential impact on the other kinds of cells involved in tissue regeneration is crucial when assessing these alterations (Sridharan et al. 2015; Carames et al. 2012).

3.6 Orthopedic Biomaterials in the Future

The field of biological materials has been a boon to orthopedics. Beginning with short-lived bone-fixing implants, orthopedic biological materials have progressed to longer-term implantable components and bioabsorbable implants. Both regeneration, as well as substitution of live tissue, will be possible at some point with the help of orthopaedic biomaterials. Biomaterials'intrinsic properties and their function in the regulation of immunity will be better understood as a result of continuing study. This new understanding of the biomaterial-immune responses interaction will pave the way for an entirely novel group of biological materials with the ability to modulate the immune system's response, allowing for enhanced host tissue integration of transplants or nearly flawless host tissue regeneration.

3.7 Conclusion

Over two decades ago, cartilage tissue engineering has seen substantial development, and current clinical investigations indicate that it continues to grow more popular. Joint pain and cartilage integrity are both symptoms of osteoarthritis (OA), which is a complicated disorder caused by several different reasons. Considering tissue-engineering tactics that concentrate on multiple components of the illness is something that should be studied to enhance therapy. Especially in the context of the characteristics of MSCs and the function of autophagy/senescence equilibrium in chondro equilibrium, a tissue-engineering method that combines mesenchymal stem cells (MSCs) with biological materials that eliminate pro-autophagic medicines might be useful. It has to be done to evaluate an enormous amount of different combinations involving biological materials, cells, biofactors, and disease-modifying OA medications while this novel treatment approach can be used regularly. In the future, regenerative therapies will play a more significant role in avoiding the loss of tissues and the initial phases of illness. Greater implant integration or near-perfect regeneration of host tissue may result from an improved comprehension of immune response to biological materials and the ability to modulate that response.

Acknowledgments None.

Availability of Data and Materials Not applicable.

Funding None.

References

Anderson JM, Rodriguez A, Chang DT (2008) Foreign body reaction to biomaterials. Semin Immunol 20:86–100

Aurora AB, Olson EN (2014) Immune modulation of stem cells and regeneration. Cell Stem Cell 15(1):14–25. https://doi.org/10.1016/j.stem.2014.06.009

Beane OS, Darling EM (2012) Isolation, characterization, and differentiation of stem cells for cartilage regeneration. Ann Biomed Eng 40:2079–2097

Boehler R, Kuo R, Shin S, Goodman A, Pilecki M, Leonard J (2014) Lentivirus delivery of IL-10 to promote and sustain macrophage polarization towards an anti-inflammatory phenotype. Biotechnol Bioeng 111:1210–1221

Brix MO, Stelzeneder D, Trattnig S, Windhager R, Domayer SE (2013) Cartilage repair, of the knee with Hyalograft C1: magnetic resonance imaging assessment of the glycosaminoglycan content at midterm. Int Orthop 37:39–43

Brown BN, Ratner BD, Goodman SB, Amar S, Badylak SF (2012) Macrophage polarization: an opportunity for improved outcomes in biomaterials and regenerative medicine. Biomaterials 33:3792–3802

Caramés B, Taniguchi N, Otsuki S, Blanco FJ, Lotz M (2010) Autophagy is a protective mechanism in normal cartilage, and its aging-related loss is linked with cell death and osteoarthritis. Arthritis Rheum 62:791–801

Carames B, Hasegawa A, Taniguchi N, Miyaki S, Blanco FJ, Lotz M (2012) Autophagy activation by rapamycin reduces the severity of experimental osteoarthritis. Ann Rheum Dis 2012(71):575–581

Chen P, Xia C, Mei S (2015) Intra-articular delivery of sinomenium encapsulated by chitosan microspheres and photo-crosslinked GelMA hydrogel ameliorates osteoarthritis by effectively regulating autophagy. Biomaterials 81:1–13

Connick P, Kolappan M, Crawley C, Webber DJ (2012) Autologous mesenchymal stem cells for treating secondary progressive multiple sclerosis: an open-label phase 2a proof-of-concept study. Lancet 11:150–156

Drury JL, Mooney DJ (2003) Hydrogels for tissue engineering: scaffold design variables and applications. Biomaterials 24:4337–4351

Engler AJ, Sen S, Sweeney HL, Discher DE (2006) Matrix elasticity directs stem cell lineage specification. Cell 126:677–689

English K (2013) Mechanisms of mesenchymal stromal cell immunomodulation. Immunol Cell Biol 91:19–26

Féréol S, Fodil R, Labat B, Galiacy S, Laurent VM, Louis B (2006) Sensitivity of alveolar macrophages to substrate mechanical and adhesive properties. Cytoskeleton 2006(63):321–340

Filardo G, Kon E, Perdisa F (2013) Osteochondral scaffold reconstruction for complex knee lesions: a comparative evaluation. Knee 2013(20):570–576

Fink J, Fuhrmann R, Scharnweber T, Franke R (2008) Stimulation of monocytes and macrophages: possible influence of surface roughness. Clin Hemorheol Microcirc 39:205–212

Franz S, Rammelt S, Scharnweber D, Simon JC (2011) Immune responses to implants – a review of the implications for the design of immunomodulatory biomaterials. Biomaterials 32:6692–6709

Fukazawa T, Naora Y, Kunieda T, Kubo T (2009) Suppression of the immune response potentiates tadpole tail regeneration during the refractory period. Development 136:2323–2327

Gardner AB, Lee SK, Woods EC, Acharya AP (2013) Biomaterials-based modulation of the immune system. Biomed Res Int 2013(2013):1–7

Gordon S, Martinez FO (2010) Alternative activation of macrophages: mechanism and functions. Immunity 32:593–604

Haseeb A, Haqqi TM (2013) Immunopathogenesis of osteoarthritis. Clin Immunol 146:185–196

Huang AH, Farrell MJ, Mauck RL (2010) Mechanics and mechanobiology of mesenchymal stem cell-based engineered cartilage. J Biomech 43:128–136

Hunt NC, Grover LM (2010) Cell encapsulation using biopolymer gels for regenerative medicine. Biotechnol Lett 32:733–742

Hunziker EB, Lippuner K, MJB K, Shintani N (2015) An educational review of cartilage repair: precepts and practice—myths & misconceptions—progress & prospects. Osteoarthr Cartil 23:334–350

Kang A, Park J, Ju J, Jeong GS, Lee SH (2014) Cell encapsulation via microtechnologies. Biomaterials 35:2651–2663

Kearney CJ, Mooney DJ (2013) Macroscale delivery systems for molecular and cellular payloads. Nat Mater 12:1004–1017

Kuroda R, Ishida K, Matsumoto T (2007) Treatment of a full-thickness articular cartilage defect in the femoral condyle of an athlete with autologous bone marrow stromal cells. Osteoarthr Cartil 15:226–231

Lafont JE, Talma S, Hopfgarten C, Murphy CL (2008) Hypoxia promotes the differentiated human articular chondrocyte phenotype through SOX9-dependent and -independent pathways. J Biol Chem 283:4778–4786

Langer R, Vacanti JP (1993) Tissue engineering. Science 260:920–926

Le Blanc K, Frassoni F, Ball L, Locatelli F, Roelofs H (2008) Mesenchymal stem cells for the treatment of steroid-resistant, severe, acute graft-versus-host disease: a phase II study. Lancet 372:716

Le Guéhennec L, Soueidan A, Layrolle P, Amouriq Y (2017) Surface treatments of titanium dental implants for rapid osseointegration. Dent Mater 23:844–854

Lotz MK, Caramés B (2011) Autophagy and cartilage homeostasis mechanisms in joint health, aging, and OA. Nat Rev Rheumatol 7:579–587

Matsuzaki T, Matsushita T, Tabata Y (2014) Intra-articular administration of gelatin hydrogels incorporating rapamycin-micelles reduces the development of experimental osteoarthritis in a murine model. Biomaterials 35:9904–9911

McWhorter FY, Wang T, Nguyen P, Chung T, Liu WF (2013) Modulation of macrophage phenotype by cell shape. Proc Natl Acad Sci 110:17253–17258

Merceron C, Vinatier C, Portron S (2010) Differential effects of hypoxia on the osteochondrogenic potential of human adipose-derived stem cells. Am J Physiol Cell Physiol 298:355–364

Merceron C, Portron S, Masson M (2011) The effect of two- and three-dimensional signal cell culture on the chondrogenic potential of human adipose-derived mesenchymal stem cells after subcutaneous transplantation with an injectable hydrogel. Cell Transplant 20:1575–1588

Mescher AL, Neff AW, King MW (2013) Changes in the inflammatory response to injury and its resolution during the loss of regenerative capacity in developing Xenopus limbs. PLoS One 8:80477

Moreira Teixeira LS, Patterson J, Luyten FP (2014) Skeletal tissue regeneration: where can hydrogels play a role? Int Orthop 38:1861–1876

Mori H, Kondo E, Kawaguch Y et al (2013) Development of a salmon-derived crosslinked atelocollagen sponge disc containing osteogenic protein-1 for articular cartilage regeneration: in vivo evaluations with rabbits. BMC Musculoskelet Disord 14:174. https://doi.org/10.1186/1471-2474-14-174

Murphy JM, Fink DJ, Hunziker EB, Barry FP (2003) Stem cell therapy in a caprine model of osteoarthritis. Arthritis Rheum 48:3464–3474

Nehrer S, Domayer S, Dorotka R, Schatz K, Bindreiter U, Kotz R (2006) Three-year clinical outcome after chondrocyte transplantation using a hyaluronan matrix for cartilage repair. Eur J Radiol 57:3–8

Nemeth K, Leelahavanichkul A, PST Y (2009) Bone marrow stromal cells attenuate sepsis via prostaglandin E(2)-dependent reprogramming of host macrophages to increase their interleukin-10 production. Nat Med 15:42–49

Ossendorf C, Kaps C, Kreuz PC, Burmester GR, Sittinger M, Erggelet C (2007) Treatment of posttraumatic and focal osteoarthritic cartilage defects of the knee with autologous polymer-based three-dimensional chondrocyte grafts: 2-year clinical results. Arthritis Res Ther 9:R41

Portron S, Hivernaud V, Merceron C (2015a) Inverse regulation of early and late chondrogenic differentiation by oxygen tension provides cues for stem cell-based cartilage tissue engineering. Cell Physiol Biochem 35:841–857

Portron S, Merceron C, Gauthier O (2015b) Effects of in vitro low oxygen tension preconditioning of adipose stromal cells on their in vivo chondrogenic potential: application in cartilage tissue repair. PLoS One 8:e62368

Rees AJ (2010) Monocyte and macrophage biology: an overview. Semin Nephrol 30:216–233

Rungsiyakull C, Li Q, Sun G, Li W, Swain MV (2010) Surface morphology optimization for osseointegration of coated implants. Biomaterials 31:7196–7204

Seo S-J, Mahapatra C, Singh RK, Knowles JC, Kim H-W (2014) Strategies for osteochondral repair: focus on scaffolds. J Tissue Eng 5:2041731414541850

Smith BD, Grande DA (2015) The current state of scaffolds for musculoskeletal regenerative applications. Nat Rev Rheumatol 2015(11):213–222

Spaggiari GM, Capobianco A, Becchetti S, Mingari MC, Moretta L (2006) Mesenchymal stem cell-natural killer cell interactions: evidence that activated NK cells are capable of killing MSCs, whereas MSCs can inhibit IL-2-induced NK-cell proliferation. Blood 107:1484–1490

Spaggiari GM, Abdelrazik H, Becchetti F, Moretta L (2009) MSCs inhibit monocyte-derived DC maturation and function by selectively interfering with the generation of immature DCs: central role of MSC-derived prostaglandin E2. Blood 113:6576–6583

Spiller KL, Nassiri S, Witherel CE, Anfang RR, Ng J, Nakazawa KR et al (2015) Sequential delivery of immunomodulatory cytokines to facilitate the M1-toM2 transition of macrophages and enhance vascularization of bone scaffolds. Biomaterials 37:194–207

Sridharan R, Cameron AR, Kelly DJ, Kearney CJ, O'Brien FJ (2015) Biomaterial based modulation of macrophage polarization: a review and suggested design principles. Mater Today 18:313–325

Steinwachs MR, Waibl B, Mumme M (2014) Arthroscopic treatment of cartilage lesions with microfracture and BST-CarGel. Arthrosc Tech 3:e399–e402

ter Huurne M, Schelbergen R, Blattes R (2012) Antiinflammatory and chondroprotective effects of intraarticular injection of adipose-derived stem cells in experimental osteoarthritis. Arthritis Rheum 64:3604–3613

Vacanti C (2006) The history of tissue engineering (2006). J Cell Mol Med 1:569–576

Vinatier C, Magne D, Weiss P (2005) A silanized hydroxypropyl methylcellulose hydrogel for the three-dimensional culture of chondrocytes. Biomaterials 26:6643–6651

Vinatier C, Mrugala D, Jorgensen C, Guicheux J, Noël D (2009a) Cartilage engineering: a crucial combination of cells, biomaterials, and biofactors. Trends Biotechnol 27:307–314

Vinatier C, Gauthier O, Fatimi A (2009b) An injectable cellulose-based hydrogel for the transfer of autologous nasal chondrocytes in articular cartilage defects. Biotechnol Bioeng 102:1259–1267

Vinatier C, Bouffi C, Merceron C (2009c) Cartilage tissue engineering: towards a biomaterial-assisted mesenchymal stem cell therapy. Curr Stem Cell Res Ther 4:318–329

Vinatier C, Guicheux J, Daculsi G, Layrolle P, Weiss P (2016) Cartilage and bone tissue engineering using hydrogels. Biomed Mater Eng 16:S107–S113

von Bahr L, Batsis I, Moll G (2012) Analysis of tissues following mesenchymal stromal cell therapy in humans indicates limited long-term engraftment and no ectopic tissue formation. Stem Cells 30:1575–1578

Zhou HW, Lou SQ, Zhang K (2014) Recovery of function in osteoarthritic chondrocytes induced by p16INK4a-specific siRNA in vitro. Rheumatology (Oxford) 43:555–568

Rishabha Malviya completed a B. Pharmacy from Uttar Pradesh Technical University and an M. Pharmacy (Pharmaceutics) from Gautam Buddha Technical University, Lucknow Uttar Pradesh. His PhD (Pharmacy) work was in Novel formulation development techniques. He has 14 years of research experience and has worked as an Associate Professor in the Department of Pharmacy, School of Medical and Allied Sciences, Galgotias University for the past 11 years. His area of interest includes formulation optimization, nanoformulation, targeted drug delivery, localized drug delivery, and characterization of natural polymers as pharmaceutical excipients. He has authored more than 150 research/review papers for national/international journals of repute. He has 51 patents (12 grants, 38 published, 1 filed) and publications in reputed National and International journals with a total of 140 cumulative impact factors. He has also received an Outstanding Reviewer award from Elsevier. He has authored/Edited/edited 91 books (Wily, CRC Press/Taylor and Francis, Springer Nature, Apple Academic Press, IOP Publishing, River Publisher, and OMICS publication) and authored 170 book chapters. His name has been included in Word's top 2% scientist list for the year 2020, 2021, 2022, 2023 and 2024 by Elsevier BV and Stanford University. He is a Reviewer/Editor/Editorial board member of more than 50 national and international journals of repute. He has been invited as an author for "Atlas of Science" and pharma magazine dealing with industry (B2B) "Ingredient South Asia Magazines".

Suraj Kumar completed a B. Pharmacy from SRM University, Chengalpattu, Tamil Nadu, India, and a Pharmacy from Galgotias University, Greater Noida, India. He has 1 year of teaching experience and works as an assistant professor at Pragya College of Pharmaceutical Sciences, Gaya, Bihar. He has attended more than 10 International/National conferences. He has published more than 20 SCI publications with reputed international publishers like Elsevier, Wiley, and Bentham Science with more than 40 cumulative impact factors. He has published 8 chapters, and three books is under consideration for publication.

Sathvik Belagodu Sridhar is currently working as an Associate Dean and Professor at RAK College of Pharmacy (RAKCOP), RAK Medical and Health Sciences University located in Ras Al Khaimah, United Arab Emirates. Dr. Sathvik has completed his PhD in Clinical Pharmacy from Rajiv Gandhi University of Health Sciences, Bangalore, India. He has received an FIP fellowship for higher training in renal clinical pharmacy at Royal Adelaide Hospital and Queen Elizabeth Hospital in Adelaide, South Australia. Dr. Sathvik has worked as an Assistant Professor and Clinical Pharmacist at JSS College of Pharmacy and JSS Medical College Hospital, Mysore, India, for eight years and later Joined International Medical University (IMU) Malaysia as a senior lecturer in the Department of Pharmacy Practice from 2008 to 2010. Dr. Sathvik joined as an assistant professor in the Department of Clinical Pharmacy at RAK College of Pharmacy, RAK Medical, and Health Sciences University in 2010 and worked as a professor and chairperson of the Department of Clinical Pharmacy and Pharmacology from 2018 to 2023. Dr. Sathvik has been honored with a prestigious teacher award by the Society of Pharmaceutical Education & Research [SPER]. He has also received training in Pharmacovigilance at the Uppsala Monitoring Centre in Uppsala, Sweden. Dr. Sathvik has 25 years of experience teaching and practicing clinical pharmacy in India, Malaysia, and the UAE. His research interests include Medication Safety, Pharmacovigilance, Pharmacoepidemiology, and Renal and Psychiatry Clinical Pharmacy. Dr. Sathvik has more than 75 research papers published in reputable national and international journals, along with book chapters. He has presented more than 75 research papers at numerous national and international conferences and has been honored with many awards. He also serves as a reviewer for various national and internal pharmacy practice-related journals.

Bioresorbable Polymeric Scaffold: Advancing Minimally Invasive Surgical Procedure for Cardiovascular and Its Applications.

4

Suraj Kumar, Rishabha Malviya, and Sathvik Belagodu Sridhar

Abstract

Progress in materialistic science and modern medical technology has enabled the creation of minimally invasive interventional techniques for implanting appliances, such as scaffolds used in cardiovascular engineering. Biodegradable materials, such as resorbable polymeric materials, are used in equipment that is only required for a temporary duration. For myocardial stents, the medical equipment is necessary for 6–8 months prior to favorable remodeling occurs. Therefore, the use of biodegradable polymeric stents has been proposed to facilitate this beneficial remodeling and address the problem of persistent confinement of the blood artery. In the context of tissue engineering, the scaffold serves the purpose of facilitating favorable cell-scaffold relations in order to promote the development of functioning tissues. The desired result is for the cells to gradually autonomously generate an individual extracellular framework, ultimately substituting the implantable scaffolding or tissue-engineered construct. Fabricated recyclable materials are the preferred options for scaffolding due to their ability to control breakdown patterns across a wide range of periods and their ease of functionalization. This chapter provides a comprehensive analysis of CAD, highlighting the drawbacks of existing treatments and exploring the prospect of biological materials in overcoming these limits in developing bioresorbable stents,

S. Kumar
Department of Pharmacy, Sarala Birla University Ranchi, Ranchi, Jharkhand, India

R. Malviya (✉)
Department of Pharmacy, School of Medical and Allied Sciences, Galgotias University, Greater Noida, UP, India
e-mail: rishabha.malviya@galgotiasuniversity.edu.in

S. B. Sridhar
RAK College of Pharmacy, RAK Medical & Health Sciences University, Ras Al Khaimah, United Arab Emirates

R. Malviya, S. Sundram (eds.), *Sustainable Nanocomposites with Green Biomaterials*, Biomaterials, Bioengineering and Sustainability 2,
https://doi.org/10.1007/978-3-031-79110-9_4

cardiovascular grafts, and cardiovascular patches. The chapters will cover various material parameters, types of polymeric utilized, present developments, and future difficulties for every application.

Keywords

Bioresorbable scaffolding · Biomaterials · Tissue engineering · Vascular grafts · Patches · Stents

4.1 Introduction

Cardiovascular disease (CVD) is a collection of disorders that affect the internal and external makeup of the circulatory system (Sanchis-Gomar et al. 2016). Among them, coronary heart disease (CHD) is the primary reason for coronary fatalities. Despite significant advancements in treatment as well as oversight of cardiovascular disease (CVD), leading to a 50% decrease in CVD-related deaths in the 1970s (Sanchis-Gomar et al. 2016; Bansilal et al. 2015) cardiovascular disease continues to be a significant cause of fatalities and permanent disabilities worldwide (Bansilal et al. 2015). In 2016, CVD accounted for 31% of global mortality, resulting in approximately 17.9 million fatalities, as reported by the Global Community Health Organisation (WHO 2020). Coronary heart disease (CHD) arises when atherosclerotic plaque expands up on the artery wall, restricting blood circulation to the cardiovascular system and potentially causing myocardial ischemia. For individuals suffering from narrowing ranging from 50% to 70%, stents can be used as a preventative strategy to reclaim blood flow by placing them at the location of the lesion to widen the narrowed area (Karabulut and Cakmak 2015). Nevertheless, the long-term use of a persistent stent can ultimately lead to a delayed catch-up effect, whereas the bioresorbable stent that was beforehand authorized by the Food and Drug Administration encountered problems due to its bulky struts, which in turn caused further troubles. When the SYNTAX score values are high (≥34), stents cannot be used to treat blocked arteries. In these cases, a procedure called coronary artery bypass graft (CABG) will be done (Ramadan et al. 2018). This procedure comprises the utilization of a saphenous vein graft to reroute the bloodstream of blood to the heart muscle. Despite this, the utilization of vein grafts and other artificial channels might not be the best option to be optimal because of a lack of compatibility in terms of dimension and qualities (such as flexibility and rigidity), which can result in the formation of blood clots (Knight et al. 2014).

When individuals have previously had an infarct of the heart (MI), their heart muscles become injured and weakened (Streeter and Davis 2019). Consequently, this will cause the coronary artery to pump inefficiently, potentially resulting in issues with cardiovascular rhythm disorders, for example. Because of advancements in the engineering of materials and modern medical technology, it is now possible to utilize less-invasive surgical procedures for coronary heart disease (CHD), which allows treatment to surpass the limits of conventional methods and alternatives (Williams 2019). In addition, the emergence of tissue engineering and regenerative therapies has stimulated research and development endeavors aimed at creating therapeutic alternatives cardiovascular grafts, bioresorbable stents themselves and myocardial patches are some examples of examples. This review study specifically examines the advancements and present state of Whether they are used in intervention treatments or function as a platform for regeneration procedures, bioresorbable polymeric materials are applied in instances pertaining to heart disease and stroke. Utilization of Synthetic Biomaterials are substances in the Management of Congenital Heart Disease In 1987, the term biological material was initially defined as a non-living substance used in healthcare equipment that interacts with biological mechanisms. Over time, the definition has evolved to include materials that collaborate with biological structures to assess, manage, enhance, either as a replacement for any organ, connective tissue, or functioning within the body. "The" transformation in characterization mirrored the advancement of biological substances, transitioning from a state of biological inactivity to becoming biologically active and compatible with living organisms. Conventionally, materials may be categorized into three main types: metals, ceramics, and elastomeric (Bose and Bandyopadhyay 2013). The three basic categories are distinguished by the nature of the relations that comprise the substance. Metals have metal-based links, ceramics are bound by ionic interactions, and polymeric has covalent connections (Bose and Bandyopadhyay 2013). Biomedical devices typically include three primary categories of biological materials, with metallic and polymeric materials being extensively employed in heart disease rehabilitation (Tibbitt et al. 2015). The use of polymeric biomaterials in cardiovascular applications has resulted in a range of therapeutic choices that promise to enhance the clinical outcome (Song et al. 2018). Polymeric biological materials are utilized in a variety of medical applications, including cardiopulmonary devices, artificial cardiac valves, stents themselves circulatory grafts, patches, and pacemaker devices. Certain criteria, such as non-toxic, must be satisfied by polymeric biomaterials that are used in biological processes, sufficient biomechanical characteristics, appropriate biological compatibility, and blood compatibility. When dealing with decomposed polymer biomaterials, it is important to address factors for example, the correlation between their inflexible nature and the pace of deterioration, and the cytotoxicity and elimination mechanism of by-products that are formed during degrading in the body (Smit et al. 2017; Piskin 1995). To select a particularly appropriate polymeric for a certain application, it is crucial to examine various critical characteristics. This is because making a bad decision in polymer selection might have negative consequences. The next three subsections will cover the context, the factors to consider for selecting a polymer, and a concise assessment of the advances made in the development of bioresorbable cardiovascular stents, cerebrovascular grafts, and cardiac patches.

4.2 Bioresorbable Stents

4.2.1 Development of Coronary Stents

Balloon angioplasty, often known as plain old balloon angioplasty (POBA), revolutionized the management of cardiovascular disease by enabling heart specialists to alter obstructive cardiovascular plaque, therefore reinstating the circulation to the cardiac despite the need for intrusive operations (Tomberli et al. 2018; Cassar et al. 2009). Nevertheless, the effectiveness of POBA was diminished as a result of the emergence of abrupt vascular rebound or vascular segmentation in addition to effects such as late blood vessel reconfiguration resulting from neointimal hypertrophy (Iqbal et al. 2013). The development of coronary stents began in the middle of 1980 to ensure long-term openness of blood vessels following balloon coronary intervention therapy (Iqbal et al. 2013). The Palmaz-Schatz® bare metal stent (BMS), created by Schatz and colleagues, was the inaugural balloon-expandable stent to get FDA approval (Garg and Serruys 2010). Two notable clinical investigations comprised the Belgian Netherland Stent Arterial Revascularization Strategies and the Northeast American stents Restenosis from developing Study. Both of these studies were conducted in the United Kingdom research investigations that played a significant role in determining the adoption of BMS as a typical care practice. These trials clearly showed that BMS is preferable to POBA (Serruys et al. 1994; Fischman et al. 1994). Nevertheless, the extended monitoring of BMS therapy revealed a prevalence of There were as many as 20–30% of cases of in-stent restenosis (ISR), that led to serious consequences or even mortality (Eltchaninoff et al. 1998). Drug-eluting stents, also known as DES later created to address the elevated rates of in-stent restenosis (ISR) observed with bare metal stent (BMS) implants (Chitkara and Gershlick 2010). The initial iteration of DES involved integrating a metal stent platform with a long-lasting polymeric covering and an antiproliferative chemical (Chitkara and Gershlick 2010). These drug-eluting stents (DES) were discovered to elevate the likelihood of subsequent stent thromboses (ST) because of the protracted mending Considering the protective layer of endothelial cells that is present in the bloodstream vessel and the sensitivity to the copolymer coating that is present on the delivery system (Chitkara and Gershlick 2010; Tada et al. 2013). The incorporation of chromium-cobalt alloy in the subsequent versions of DES enabled a reduction in stent component width to 80–90 μm, as opposed to the 130–140 μm thickness of the initial version of DES was constructed out of metallic material 316 L (Chitkara and Gershlick 2010). The use of narrower struts facilitated a faster process of re-endothelialization of the stent, as evidenced by previous research (Sakamoto et al. 2018), hence decreasing the likelihood of stent thrombosis (ST). In the subsequent generations of DES, researchers also investigated the potential of using additional medications, such as everolimus medication and zotarolimus, to lower the occurrence of ST and promote faster re-endothelialization (Chitkara and Gershlick 2010).

The subsequent generations of drug-eluting stent (DES) have achieved notable enhancements in effectiveness and safety for percutaneous coronary intervention

(PCI) therapy as a result of modifications to the drug-polymer framework and the installation of reduced-strut configurations. Although subsequent generations of drug-eluting stents (DES) have achieved significant success in improving treatment outcomes, they nevertheless face some problems, including extremely late stent coagulation, impairment of blood circulation, and delayed vascular repair (Gupta 2016). As a result, biodegradable scaffolding (BRS) was created to avoid these difficulties. Theoretically, if the BRS is fully absorbed, it would allow for recurrent revascularization to the identical spot with reestablished vasomotion (Iqbal et al. 2013), without keeping any residual material behind. This has an opportunity to facilitate good restructuring of the effectively enclosed vessel that circulates blood, the artery has to be addressed and the issue must be resolved (Ormiston and Serruys 2009). To generate a BRS, it is necessary to adhere to certain standards. The optimal duration for BRS to offer assistance in a vessel is estimated to be 6–9 months (Ormiston and Serruys 2009; Berglund et al. 2009). As the BRS degrades, the resulting breakdown of waste products should not elicit any local or systemic adverse reactions (Neamtu et al. 2014). To minimize the formation of blood clots, the substance should possess hem-compatible features. Additionally, it ought to possess favorable mechanical characteristics to limit elastic recoil. The polymeric BRS encountered the difficulty of having weaker mechanical qualities and, therefore required the use of thicker stent struts. This ultimately resulted in unfavorable flow disruptions and other problems during installation (Akinapelli et al. 2017; Soares and Moore 2015).

4.2.2 Polymers in BRS Application

It is widely acknowledged that polymer-based BRS (bioresorbable scaffolds) have inferior mechanical properties in contrast to their metal-based counterparts (Ang et al. 2017). The circumferential stiffness of a stent is closely correlated with its compressive modulus, and bioresorbable scaffolds (BRS) composed of poly-L-lactic acid (PLLA) are approximately 100 times less strong than stents produced from biostable materials such as metallic stainless steel and cobalt, magnesium chromium. In order to address this inherent limitation of the substance, a thickness of these BRS needs to be increased by 240% (Berglund et al. 2009). Polylactic acid (PLA), a kind of polyester-based polymeric material, contains a methyl group that has a chemical association to the primary chain. PDLLA and PLLA have identical chemical-based compositions, but PDLLA lacks a defined crystalline framework, making it ambiguous. In contrast, PLLA has a partially crystalline structure, with crystallinity levels reaching around 70%. The variation in crystal structure is responsible for the disparity in their mechanical qualities. Polyglycolic acid (PGA), in contrast, does not have the extra methyl group that glucose does and is thus more prone to breakdown because of less steric interference. Conversely, the absence of a methyl side group enables Polymethyl methacrylate (PMMA) polymer chains to achieve tighter packing, resulting in increased crystallization, Mechanical attributes, and melting point (Soares and Moore 2015). The enhanced network stiffness of polycaprolactone (PCL) is a consequence of its proportionate thermoplastic material properties and its extended repeating unit. This polymeric material possesses a temperature at which the glass's phase transitions occur (Tg) and the melting temperature (Tm) are both low and demonstrate a significant extension at the point of fracture, along with a comparatively less strong tensile strength in comparison to PLA and PGA. The BRS developed by REVA Medical is made from a polycarbonate, or PC, produced from tyrosine. This PC includes alkyl ester pendant chains of varying lengths that combine to form the carbonate polymerization known as Tyrocore. It consists of counterparts of the amino acid tyrosine. The durability and toughness of Tyrocore are derived from the large pendants group present in the iodinated diphenol, whereas the polymeric material's radiation transparency is due to the chemical bonding of iodide to the tyrosine analog (Leibundgut 2018).

4.3 Vascular Grafts

Vascular grafting, sometimes referred to as CABG, is a surgical procedure that includes using a passageway to bypass a blocked or dysfunctional blood vessel to regain circulation to the area with reduced blood supply (Pashneh-Tala et al. 2016). While PCI has been the preferred method for treating damaged blood arteries, studies have shown that CABG has better over time vascular patency (Antoniou et al. 2013; Mohr et al. 2013; Kindi et al. 2018). The most often used options for vascular grafting are typically obtained from the patient's own body, known as homologous sources. These sources include the interior thoracic outlet artery, and radial artery (Nappi et al. 2021). The saphenous vein is commonly used as a conduit graft since it has a lower risk of problems after elimination and can stay open for at least 10 years (Harskamp et al. 2013). Aside from homologous grafts, synthesized grafts constructed from polytetrafluoroethylene (PTFE) are also a viable substitute for homologous grafts (Ballotta et al. 2003). Synthetic grafts have a high long-time consistency rate of approximately 90% in large-diameter channels (>8 mm) like the femoral artery. They also show favorable integrity levels in medium-sized vessels (6–8 mm). Nevertheless, the utilization of artificial grafts in tiny blood arteries (<6 mm) has led to lower rates of graft patency in comparison to the utilization of grafts derived from the patient's tissues (Brewster et al. 1997). The application of as a consequence of using PTFE grafting in CABG surgeries, the successful healing percentage at one year was around 60% but a saphenous vein graft had a consistency rate of over 95% (Hadinata et al. 2009; Hehrlein et al. 1984). Following a period of 2 years, the patency percentage of PTFE grafts dropped to 32%. Whereas the saphenous venous grafting maintained a patency rate of over 90% (Hadinata et al. 2009; Hehrlein et al. 1984). Additional factors like as patient eligibility, vascular quality, and problems during vascular retrieval continue to hinder the widespread use of CABG procedures for treating CAD. Furthermore, the occurrence of graft failure is not unusual, particularly when artificial grafts are utilized (Greenwald and Berry 2000; Davies and Hagen 1995). Typical causes of unsuccessful grafts include blood clotting, hardening of the arteries, infections, or excessive growth of tissue in

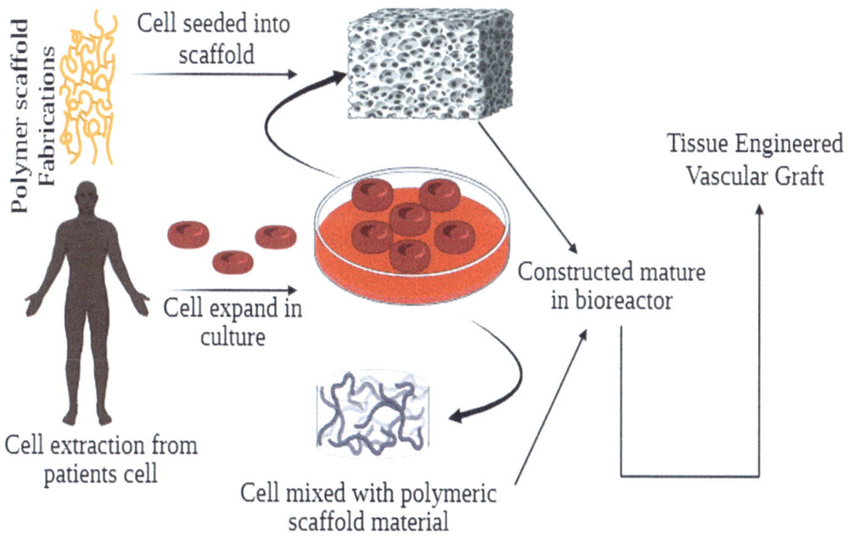

Fig. 4.1 This refers to the procedure of creating a tissue-engineered vascular graft using the scaffold-based technique

the inner lining of the blood vessels (Deutsch et al. 2009). Efforts have been made to enhance the openness of artificial grafts by introducing cells classified as endothelial into the inner surface of the graft (Chlupac et al. 2009). While the enhancements produced were not enough when juxtaposed with natural grafts, this method demonstrated the possibility of an artificially engineered vascular graft for future use in healthcare environments (Stowell and Wang 2018). The idea of creating a tissue-designed vessel that can operate like a natural graft is a very appealing answer to the challenges of obtaining an appropriate graft for individuals who do not have access to their tissue sources. Figure 4.2 several different methods may be utilized to create a cellular tissue-engineered vascular transplant. One of the most common approaches is to use a scaffolding-based technique (Fig. 4.1), in which the scaffolding is constructed employing either artificial, natural, or hybrid resources.

4.3.1 Applications of Polymeric Materials for the Development of Bioresorbable Vasculature Graft Scaffolds

Corresponding to the creation of BRS, scaffolding for cardiovascular grafts can be constructed from somewhat crystalline polymers made from synthetic materials organic materials, or decellularized matrix (Gomes et al. 2013). The production of polyhydroxyalkanoates (PHAs) involves the fermented of food-based carbon resources such as carbohydrates and triglycerides. These PHAs have biomechanical and degrading characteristics that may be tailored to specific needs, based on their molecular masses and the synthesis process (Sudesh et al. 2000; Li et al. 2016). The sections of the various vascular layers include naturally generated polymeric that includes elastin, collagen, and other proteins fibrin, fibronecstresstin, and gelatin (Kakisis et al. 2005; Nicolas and Gagnieu 1997). These polymers contribute to the formation of the vascular layers. To encourage the development of blood vessel cells, these polymers imitate the natural extracellular matrix (ECM) (Nicolas and Gagnieu 1997). One of the most important components of the extracellular matrix (ECM) is type I collagen in the body. The collagen fibers help to restrict huge amounts of stress, so offering the necessary rigidity as well as encouragement for the reliability of the structure of the cardiovascular vessel. During the process of fibrillogenesis, collagen was found to have favorable adherence of cells and low levels of antigenic and inflammatory mediators (Habermehl et al. 2005; Silver et al. 2001). Along the same lines as collagen, elastin is another important component of the extracellular matrix of the circulatory artery, and it is responsible for the flexibility of its framework. During intermittent flow, the presence of conjugated elastin fibers within the vessel's walls works to prevent the vessel from recoiling and undergoing any irreversible deformation. In addition, the generation of smooth muscle cells (SMC) is controlled to inhibit the establishment of neointimal cells (Ibrahim and El-Zairy 2015). Chitosan and silk-derived fibroin are two additional sources of organically produced materials that may be utilized in the production of bioresorbable cardiovascular graft scaffolding. A group of carbohydrates known as chitosan is derived from chitin with crustaceans being the primary source of this monosaccharide. As a result of its biocompatibility and biodegradability, it finds widespread use in medical applications, including reasons related to wound recuperation and antibacterial treatment (Kundu et al. 2013). In terms of its physical qualities, biodegradable qualities, and biological compatibility, silk is a biological material that is extremely adaptable (Zhang et al. 2018). Silk is spontaneously produced by silkworms. Synthetic and natural materials each have their distinct advantages and disadvantages; however, an artificial material that combines the two to maximize the aggregate qualities of the combined material can provide a vascular graft that is acceptable for use.

4.4 Cardiac Patches

The objective of building a cardiovascular patch is to create a proficient protective layer of cardiac muscle to replace cardiac muscle cells that are removed during an ischemia event (Caddeo et al. 2017). The creation of a cardiovascular patch can be achieved by integrating principles from restoration and biological disciplines, giving rise to the traditional tissue engineering model. The desired cells are extracted from the individual's body and placed onto a scaffolding structure that replicates the structure and physical characteristics of the real tissue (Caddeo et al. 2017; Radisic et al. 2008). The implanted scaffolding would thereafter be cultivated in a bioartificial environment that replicates the milieu regarding the knowledge that is contained within the cell, which includes the biochemical procedures as well as the electronic or mechanical motion stimulation (Mei and Cheng 2020). By creating an environment that closely mimics the circumstances of the original tissue, the cells being grown would ideally start to multiply and generate their matrix of extracellular cells. Over time, they would develop into a fully functional tissue, and be capable of carrying out particular duties related to that tissue (Zimmermann et al. 2006). Therefore, numerous attempts have been made to enhance the biological engineering technique for growing cardiac patches that closely mimic the natural myocardial (Eschenhagen et al. 1997). Various factors, including the cell line being studied, culture medium-sized, and scaffolding subject matter, have been found to influence the development and the process of transforming the cells that have been transplanted into heart cells that are functioning (Caddeo et al. 2017; Eschenhagen et al. 1997). A detailed summary of the accomplishments that have been made in the discipline of myocardial patch tissue manipulations is presented in Fig. 4.2. 1997 was the year that Eschenhagen et al. 1997). showed that it was possible to culture juvenile chick cardiomyocytes on a scaffold made from collagen to create a 3D cardiac tissue that exhibited synchronized conjunctions. In the early 2000s, Zimmermann and other researchers demonstrated that incorporating mechanical stresses during the cultivation of tissues and using mixed cultures that included endothelial cells as a component can greatly enhance the structure and function of seeded heart muscle cells (Zimmermann et al. 2002; Gao et al. 2018).

Considering the achievements in cultivating cardiac tissues in a controlled environment, the logical progression would involve utilizing operational cardiac patches to reinstate the heart's functionality after a myocardial infarction. (Gao et al. 2018) successfully enhanced the left ventricular contraction functions in pigs with myocardial infarction by using myocardial patches made from fibrinogen and human pluripotent stem cells that were induced. They achieved proliferation rates of 10.9 ± 1.8% after 4 weeks (Frangogiannis 2017). An essential factor in the

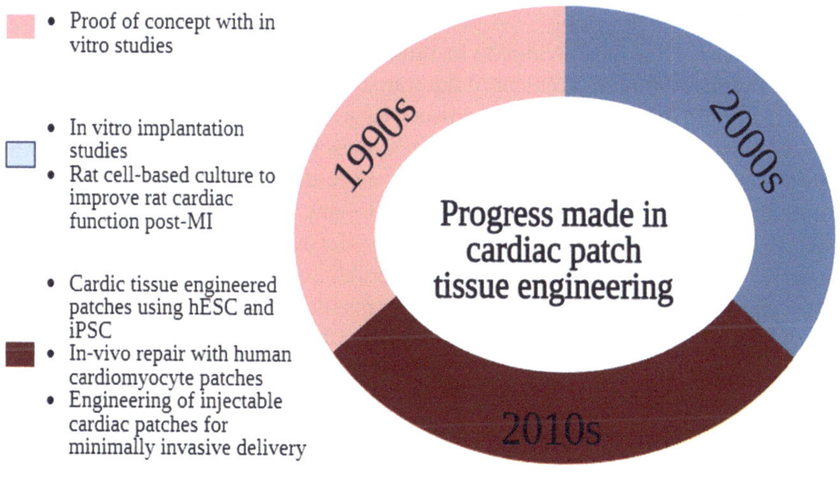

Fig. 4.2 This article offers a detailed account of the developments that have been made in the discipline of myocardial patch tissue engineering

biological regeneration strategy is the selection of scaffolding substance, which directly influences cell-scaffold interaction. The layer of scaffolding is designed to replicate the organic extracellular matrix (ECM) present in the cardiomyopathy. The selection of scaffolding substance will impact the characteristics of the scaffolding. The heart muscle contains primarily fibrillar types I and III collagen, which make up the natural extracellular matrix (ECM). These kinds of collagen are organized in a complex manner to produce the layers of epimysium, the outermost layer of an endomysium. Type I collagen is mostly located in the epimysium and perimysium, where it forms thick rod-like constructions. It serves as a significant part of the framework (Frangogiannis 2019). On the other hand, type III collagen resides primarily when it is found in the endomysium, where it is responsible for the formation of a fragile framework consisting of fibers composed of collagen that wrap surrounding mitochondria in myocytes when stimulated and capillary (LeGrice et al. 2005). The primary function of the collagen network is to offer skeletal reinforcement, maintain ventricle shape, and facilitate the passage of pressure throughout myocardial contraction (Frangogiannis 2017). The choice of material for scaffolding construction is crucial for recreating the framework of the natural extracellular matrix.

However, it is equally necessary to consider the manufacturing process used. By exercising meticulous control over the manufacturing procedure, it is possible to produce anisotropy nanostructures that direct the alignment of the implanted cardiomyocytes. In the following sections, we will go into a more detailed discussion of the appropriateness of numerous biological material options concerning different fabrication procedures and their practicality in creating a heart tissue patch with superior functionality.

4.4.1 Commercially Available Cardiac Patches

Cardiac patches have the potential to be utilized in the restoration of cardiomyopathy by offering both reinforcement of the framework and an optimal setting for the regeneration of native cells within the heart. Regrettably, the fact that mature heart cells are post-mitotic implies that the cardiac muscle has very limited potential to regenerate itself (Cui et al., 2018). Therefore, a healthy heart is incapable of regenerating the billions of destroyed heart cells following a myocardial infarction (MI) event, even if a proper atmosphere is present (McCready et al. 1992). Therefore, existing cardiac patches that are available for purchase only fulfil the main objective of ensuring closure of the pericardium, with a specific emphasis on avoiding hemorrhage along the edge of the suture (Minale et al. 1988). All the myocardial patches listed in Table 4.1 are non-cellular.

This statement reflects the present challenges encountered in the development of a fully functioning cardiac patch. Furthermore, in a subsequent investigation with 110 patients who had implantation of Gore-Tex® surgical membrane, neither cellular ingrowth nor immune-compliant cellular components were detected on the therapeutic membranes (Li et al. 2011). This emphasizes the non-reactive nature of the

Table 4.1 presents a comprehensive inventory of commercially accessible cardiac patches, including details on the brand, material composition, and intended use

Brand	Material	Purpose
GORE-TEX® Cardiovascular Patch	Xenograft; Secretory Submucosa Extracellular Matrix	Preservation and restoration of epicardial tissue
GORE-TEX® Cardiovascular Patch	Extended Polytetrafluoroethylene	To protect and stabilize damaged or deteriorating tissues
Bard Cardiovascular Patch	Extended Polytetrafluoroethylene	This product is recommended for repairing and closing the cardiovascular system.
SteriGraft™—Pericardium	Allograft from the Pericardium	Repair of the dura mater, periodontal reconstruction, and pericardial defects
CardioCel® cardiovascular bio-scaffold	The collagen sheets are derived through bovine cardiac tissue obtained from xenografts and are devoid of cells.	Cardiac problems inside the heart, including septal abnormalities and restorations of the heart's annular structure, are being addressed.
Cryolife: Cardiac Tissue Matrix/ Allograft	Origin of Allograft	Regeneration from birth or as reinforcement material
PB—Bovine Pericardium Patch	Bovine Glutaraldehyde Hemithorax; Allograft	Coronary artery disease repairs and upkeep

selected therapeutic membrane. Xenografts utilizing endogenous cardiovascular extracellular matrix (ECM) may serve as an appealing option due to their superior resemblance to the host's cardiac ECM, facilitating enhanced proliferation and migration of cells (Umashankar et al. 2012). Nevertheless, cardiac extracellular matrix (ECM) produced from natural sources is commonly subjected to a substance called glut treatment, which creates cross-linking in the proteins of both the intracellular and matrix components. This process stabilizes the framework and significantly decreases the immunogenicity of the graft (Ye et al. 2013). Research has shown that the use of a substance called glut therapy leads to cytotoxicity, premature mechanical breakdown, and significant Mineralization in the cardiac tissue of cattle (Benzoni et al. 2016). Therefore, the limitations of the current myocardial patch are far from satisfactory in terms of its capacity to consistently deliver regeneration functionality.

4.4.2 Material Choice for Cardiac Patches

Due to the intricate structure of the cardiac extracellular matrix (ECM), polymeric scaffolds are ideal for creating a highly adjustable milieu that promotes the growth and functionalization of myocardiocytes (Streeter and Davis 2019). There is the possibility of modifying the scaffolding to vary its mechanical durability, deterioration background, and interface chemical characteristics to more exactly resemble the actual characteristics of the original tissue (Streeter and Davis 2019). Hence, the integration of an appropriate biomaterial and cell source is an essential component in the creation of a cardiac patch that can regenerate. The subsequent sections will provide greater details regarding the appropriateness of various biomaterials, as well as the corresponding production methods.

4.4.2.1 Natural Biopolymers

Natural materials refer to compounds that exist within the environment and are derived from biological components. These materials are often well-suited for interaction with cellular and are uncommon to trigger immunological responses or lead to blood clotting (Streeter and Davis 2019). Particularly in the context of a decellularized extracellular matrix (ECM), the internal structure is extensively maintained, greatly Facilitating the proliferation and specialization of stem cells (Streeter and Davis 2019). Myocardial patch construction has utilized many substances that occur naturally such as the protein The substances mentioned are collagen production, a material called fibrous tissue, alginate, additionally Matrigel, another gelatin, consequently, and decellularized ECM (Benzoni et al. 2016). (Benzoni et al. 2016) investigated the use of a chitosan/collagen composite mixture for the creation of a myocardial scaffolding (Benzoni et al. 2016). The chitosan/collagen combination was created by electrophoretic depositing, following a 5:1 ratio. The application process resulted in the formation of cubic and hexagonal rectangular grooves, which intersected to create four and six neighbouring pores (Benzoni et al. 2016). The reason for using chitin in the combination is that glucosamine has a positive charge, which enables it to form relationships with negatively charged glycosaminoglycans (GAG) and proteoglycans (Benzoni et al. 2016). Glycosaminoglycan, whereas an insignificant part of the non-structural ECM (extracellular matrix) is essential in performing a critical function in the myocardial ECM by facilitating the alteration of protein molecules, such as growth hormones. This alteration might result in changes or additions to the functions of these proteins (Denuziere et al. 2000). In addition, the constructed scaffolding was specifically developed with ridges diameters ranging from 30 μm to 80 μm to align the implanted heart muscle cells, hence enhancing their development and functioning (Denuziere et al. 2000). To assess the biological compatibility of the chitosan/collagen scaffolding. Adult heart muscle cells generated from iPSC lines, which are derived from human cells and have the ability to develop into many cell types, were put on the scaffolding and kept in a controlled environment for a period of fourteen days (Umashankar et al. 2012). After 14 days in customs, cardiac muscle cells were shown to be undergoing contractions and producing gene expression related to the heart (Umashankar et al. 2012). In addition, the impact of the artificial channels on the alignment of cardiomyocytes was examined by analyzing the arrangement of Troponin T, a protein found in the cartilage (Umashankar et al. 2012). The study revealed that scaffolding with designed grooves exhibited myocardium growth aligned with the channel orientation, but scaffolding lacking patterns showed random cardiomyocyte proliferation (Umashankar et al. 2012). It is important to mention that the manufacturing of chitosan involves a rigorous treatment of chitin (the raw material) to make it appropriate for biomedical research. Purposes (Rienks et al. 2014). Key steps encompass demineralization, deproteinization, decolorization, and deacetylation (Salick et al. 2014; Sultankulov et al. 2019). Residual proteins, minerals, or color pigments can provoke an immunological reaction when implanted in a living organism, particularly if the chitin used came from crustaceans. Therefore, fungus may serve as an improved substitute source of chitin due to its lower immunogenicity, smaller molecular mass dispersion, and ability to be farmed for a consistent supply (Joseph et al. 2021).

Rashedi et al. researched to examine the impact of a readily accessible type-I bovine connective scaffolding Regarding the ability of the mesenchymal stromal cells to produce myocardial indicators, it was observed that this occurs when MSCs are in combination with rodent cardiomyocytes/fibroblasts (Ghormade et al. 2017). Following 5 days in cultural background, the combination cultivation of MSCs in the extracellular scaffolding initiated the production of cardiac markers, namely MyHC and cTrpT sarcomere proteins (Ghormade et al. 2017). The statement emphasizes the beneficial influence of the milieu of an ECM outside of cells on the maturation and survival of stem cells, as shown by references (Ghormade et al. 2017). Therefore, by carefully considering the processes used to prepare the material, it is feasible to produce a supporting structure that imitates the natural cardiac extracellular matrix. This enables a higher likelihood of achieving a functioning cardiac patch capable of restoring heart functions after a myocardial infarction.

4.4.2.2 Synthetic Materials

Although natural materials are appealing for creating myocardial patches because their chemical constitution is similar to the native extracellular matrix, man-made substances are a feasible option for scaffolding production. Synthetic materials are commonly accessible through repeatable production procedures. Moreover, synthetic materials can provide enhanced flexibility concerning deterioration profiles and stiffness by regulating the ratios of the various components (Rashedi et al. 2017). Cardiovascular tissue technology often utilizes synthetic components including Poly lactic-co-glycolic acid, poly ε-caprolactone, and poly glycerol sebacate are the three components listed. PGS is a recyclable substance that was developed by Robert Langer and flexible polymeric was developed by combining glycerine and sebacic acid by polymer condensation (McMahan et al. 2020). Several research groups have examined the utilization of PGS in the production of myocardial patches. Tallawi et al. utilized a mix of soft-lithographic and electrospinning methods to create the scaffolding con polymer in a weight ratio of 2:1 (McMahan et al. 2020). The electrospun PGS/PCL scaffolds possess porosity, facilitating the circulation of waste and nutrients inside the scaffolding (Wu et al. 2023). The scaffolding elements were constructed using various topographical characteristics, including grooves measuring 10 μm with interstitial miles of 7 μm and 200 μm between them (referred to as "Different-line"), straight grooves with interstitial of Seven and ten micrometers, also known as "Parallel-line" concepts, and round design with a measurement of one hundred micrometers and fifty micrometers between each intermittent square sequence, also known as "Square" patterns (McMahan et al. 2020). An investigation was conducted on cell differentiation and morphological concepts using C2C12 cells, which are a mice myoblast cell line that has been immortalized (McMahan et al. 2020). According to the findings, all the implanted cells showed the expected structure and arrangement of the cellular structure, independent of the topographic aspects of the surrounding environment. Cell realignment was more pronounced in colonies having topographical properties that are aligned to lines, compared to the multi-directional dissemination of individual cells on an unaltered scaffolding composed of randomized fibers. Furthermore, postpartum rat heart muscle cells were utilized to investigate the adherence and aligning of heart muscle cells. The presence of fibronectin-coated scaffolding and glass coverslips resulted in many more cell adhesions in comparison to gelatin-coated glass coverslips. The precise positioning of cells was most pronounced (~60%) in scaffolding that had grooves that were parallel and a consistent spacing between them (McMahan et al. 2020). To summarize, synthetic components are easily accessible and may be used with different manufacturing methods to provide the specific qualities necessary for scaffolds. By precisely adjusting the characteristics of synthetic materials, it is possible to align the breakdown pattern of the scaffolding with the extracellular matrix (ECM) synthesis of the cells that have been introduced. This enables a gradual replacement of the fabricated scaffolding until it is completely replaced as time goes on.

4.5 Future Challenges—The Ideal Cardiac Patch

Several cell therapy research investigations have been conducted, where stem cells were directly injected into the cardiovascular system to replace missing cardiac cells after a myocardial infarction (MI) to safeguard against negative changes in cardiac structure and function (Wang et al. 2002). Although there has been a positive change in medical outcomes, including LVEF-, the proportions of cell integration remain modest. Thus, a potential approach is to generate an extracellular matrix (ECM) that can create an optimal milieu for the growth and specialization of cells, leading to enhanced rates of engraftment (Tallawi et al. 2016). The cardiac patch also offers structural reinforcement to the compromised myocardial wall following a myocardial infarction, enhancing cardiac function and inhibiting the advancement of illness in the infarct zone. Nevertheless, the primary obstacle to address is the avoidance of arrhythmias after the implantation of a functioning myocardial patch onto the heart, which may arise from electrical or mechanical incompatibility (Banerjee et al. 2018). An optimal cardiovascular patch must have three separate Cardiovascular myocytes, congregations of smooth-muscular cells as well as endothelial cells —which work together to give mutually beneficial assistance. In terms of substance and scaffolding design, the scaffolding formulations must be capable of accommodating and promoting the growth of various cell types. When using stem cells, the scaffolding needs the creation of an atmosphere that is quite similar to the natural myocardial. This enables the appropriate differentiation, maturity, and an arrangement of the embryonic stem cells that were seeded. Hence, the scaffolding milieu must include the essential physical and biochemical signals necessary to promote cell adherence and differentiating. Additionally, it should have a sophisticated porosity structure to facilitate the interchange of nourishment and waste, hence ensuring the survival of cells within the scaffolding. Ultimately, after being inserted, the scaffolding must possess the capability to seamlessly merge with the individual's cardiac both in terms of mechanical and electrical integration, to effectively avoid the occurrence of arrhythmia. Once these requirements are fulfilled, the cardiopulmonary patch will become a crucial component for the purpose of cardiac tissue transplantation technology. In 2018, the ESCORT study (NCT03676517) provided evidence that implanting a fibrin patch filled with human embryonic stem cells (ESCs) is acceptable for individuals with significant ischaemic cardiac problems on the left side (Zhang 2015; Wendel et al. 2014; Zheng et al. 2013; Menasché et al. 2018) The ESCORT study was an initial investigation into clinical conditions that administered an intestinal fibrin patch with an average dosage of 8.2 million 10 recipient's epicardium were created using human pluripotent stem cell lines undergoing a CABG surgery (Zhang 2015; Wendel et al. 2014; Zheng et al. 2013; Menasché et al. 2018). All 10 patients did not have any problems, such as irregular heartbeats, and showed enhancements in the Assessment of the proportion of ejection and wall elasticity of the reconciled area in the left ventricle (Zhang 2015; Wendel et al. 2014; Zheng et al. 2013; Menasché et al. 2018). The outcomes of this research investigation emphasized the advancements achieved in the effective implementation of a cardiac patch. Specifically, it has been demonstrated that ventricular fibrillation a significant issue experienced during the deployment of the cardiac patch, has been effectively handled thus far. The findings from the current CARDIOMESH study (NCT03746938, with follow-up expected in October 2020) will Elaborate further to enhance understanding of the prospective use of cardiovascular patches in the treatment of cardiovascular diseases.

4.6 Conclusions

This chapter study has thoroughly examined the principles beneath three primary uses of bioresorbable polymer components in the field of cardiology medicine: bioresorbable scaffolding (BRS), cerebrovascular grafts, and myocardial patches. Biomaterials may be classified Divided into 2 primary classifications: natural and artificial resources. Natural polymeric compounds are often biodegradable and suitable for supporting cells, while artificially produced polymers offer superior mechanical strength. In future decades, biomaterials might potentially combine both of these avenues to create a hybrid substance with tailored qualities that are more suitable for its intended function as implants. The recent progress in fabrication techniques, such as three-dimensional bioprinting, has enabled the printing of intricate composite frameworks. This development holds promise for the effective utilization of tissues-engineered products. Out of the three applications, BRS has achieved the most significant, although temporary, advancement when it comes of healthcare acceptance. The multitude of clinical studies undertaken by multiple BRS firms has enabled healthcare professionals to get comprehensive insights from their clinical backgrounds. The inherent inferior material qualities associated with employing bioresorbable polymeric caused the implants to have wide characteristics, which therefore resulted in challenges in their distribution and disappointing results. Since the discontinuation of the ABSORB BVS in 2017, there have been no recent FDA authorizations for any Bioresorbable Scaffolds (BRS). Future bioresorbable scaffolds (BRS), whether made of polymers or metals, confront the significant challenge of resolving the safety throughout the long run and effectiveness issues that arose throughout the period. To achieve broader clinical acceptance, it is essential to address the problems associated with repeatability, scaling-up, biological compatibility, and preservation for patching of the heart and circulatory grafts/TEVG techniques. The preliminary application of these tissue engineering products in people has resulted in positive results, as outlined in this review. Additional extensive clinical data from meticulously planned controlled experiments will be necessary to authenticate its utilization in therapies for coronary heart disease (CHD). Ultimately, the use of bioresorbable polymeric components in cardiac applications has demonstrated its adaptability and significant benefits, as evidenced through multiple instances of success in clinical studies, such as in technology for creating vascular prostheses products. The growing need for temporary medical equipment and regenerative therapies, fueled by the aging population and increased wealth, will persistently push toward the creation of improved bioresorbable polymers and their more effective utilization. The combination of scientific progress in biological materials engineering and three-dimensional scaffold construction technologies holds great promise for future development and progression in this field.

References

Akinapelli A, Chen JP, Roy K, Donnelly J, Dawkins K, Huibregtse B, Hou D (2017) Current state of bioabsorbable polymer-coated drug-eluting stents. Curr Cardiol Rev 13:139–154. [Google Scholar] [CrossRef] [Green Version]

Ang HY, Huang YY, Lim ST, Wong P, Joner M, Foin N (2017) Mechanical behavior of polymer-based vs. metallic-based bioresorbable stents. J Thorac Dis 9:S923–S934. [Google Scholar] [CrossRef] [PubMed] [Green Version]

Antoniou GA, Chalmers N, Georgiadis GS, Lazarides MK, Antoniou SA, Serracino-Inglott F, Smyth JV, Murray D (2013) A meta-analysis of endovascular versus surgical reconstruction of femoropopliteal arterial disease. J Vasc Surg 57:242–253. [Google Scholar] [CrossRef] [PubMed] [Green Version]

Ballotta E, Renon L, Toffano M, Da Giau G (2003) Prospective randomized study on bilateral above-knee femoropopliteal revascularization: polytetrafluoroethylene graft versus reversed saphenous vein. J Vasc Surg 38:1051–1055. [Google Scholar] [CrossRef] [Green Version]

Banerjee MN, Bolli R, Hare JM (2018) Clinical studies of cell therapy in cardiovascular medicine. Circ Res 123:266–287. [Google Scholar] [CrossRef]

Bansilal S, Castellano JM, Fuster V (2015) Global burden of CVD: focus on secondary prevention of cardiovascular disease. Int J Cardiol 201:S1–S7. [Google Scholar] [CrossRef] [Green Version]

Benzoni P, Ginestra P, Altomare L, Fiorentino A, De Nardo L, Ceretti E, Dell'Era P (2016) Biomanufacturing of a chitosan/collagen scaffold to drive adhesion and alignment of human cardiomyocyte derived from stem cells. Procedia CIRP 49:113–120. [Google Scholar] [CrossRef] [Green Version]

Berglund J, Guo Y, Wilcox JN (2009) Challenges related to development of bioabsorbable vascular stents. EuroIntervention 5(Suppl. F):F72–F79. [Google Scholar] [CrossRef]

Bose S, Bandyopadhyay A (2013) Chapter 1—Introduction to biomaterials. In: Bandyopadhyay A, Bose S (eds) Characterization of biomaterials. Academic, Oxford, pp 1–9. [Google Scholar]

Brewster DC, Hospital FMG, School HM (1997) Current controversies in the management of aortoiliac occlusive disease. J Vasc Surg 25:365–379. [Google Scholar] [CrossRef] [Green Version]

Caddeo S, Boffito M, Sartori S (2017) Tissue engineering approaches in the design of healthy and pathological in vitro tissue models. Front Bioeng Biotechnol:5. [Google Scholar] [CrossRef] [Green Version]

Cassar A, Holmes DR Jr, Rihal CS, Gersh BJ (2009) Chronic coronary artery disease: diagnosis and management. Mayo Clin Proc 84:1130–1146. [Google Scholar] [CrossRef] [PubMed] [Green Version]

Chitkara K, Gershlick A (2010) Second versus first-generation drug-eluting stents. J Interv Cardiol 5:23–26. [Google Scholar] [CrossRef]

Chlupac J, Filova E, Bacakova L (2009) Blood vessel replacement: 50 years of development and tissue engineering paradigms in vascular surgery. Physiol Res 58(Suppl. 2):S119–S139. [Google Scholar] [PubMed]

Cui B, Zheng Y, Sun L, Shi T, Shi Z, Wang L, Huang G, Sun N (2018) Heart regeneration in adult mammals after myocardial damage. Acta Cardiol Sin 34:115–123. [Google Scholar] [CrossRef]

Davies MG, Hagen P-O (1995) Pathophysiology of vein graft failure: a review. Eur J Vasc Endovasc Surg 9:7–18. [Google Scholar] [CrossRef] [Green Version]

Denuziere A, Ferrier D, Domard A (2000) Interactions between chitosan and glycosaminoglycans (chondroitin sulfate and hyaluronic acid): physicochemical and biological studies. Ann Pharm Fr 58:47–53. Google Scholar]

Deutsch M, Meinhart J, Zilla P, Howanietz N, Gorlitzer M, Froeschl A, Stuempflen A, Bezuidenhout D, Grabenwoeger M (2009) Long-term experience in autologous in vitro endothelialization of infrainguinal ePTFE grafts. J Vasc Surg 49:352–362. discussion 362. [Google Scholar] [CrossRef] [PubMed] [Green Version]

Eltchaninoff H, Koning R, Tron MD C, Gupta V, Cribier A (1998) Balloon angioplasty for the treatment of coronary in-stent restenosis: immediate results and 6-month angiographic recurrent restenosis rate. J Am Coll Cardiol 32(4):980–984

Eschenhagen T, Fink C, Remmers U, Scholz H, Wattchow J, Weil J, Zimmermann W, Dohmen HH, Schafer H, Bishopric N et al (1997) Three-dimensional reconstitution of embryonic cardiomyocytes in a collagen matrix: a new heart muscle model system. FASEB J 11:683–694. [Google Scholar] [CrossRef]

Fischman DL, Leon MB, Baim DS, Schatz RA, Savage MP, Penn I, Detre K, Veltri L, Ricci D, Nobuyoshi M (1994) A randomized comparison of coronary-stent placement and balloon angioplasty in the treatment of coronary artery disease. N Engl J Med 331:496–501. [Google Scholar] [CrossRef]

Frangogiannis NG (2017) The extracellular matrix in myocardial injury, repair, and remodeling. J Clin Invest 127:1600–1612. [Google Scholar] [CrossRef] [PubMed] [Green Version]

Frangogiannis NG (2019) The extracellular Matrix in ischemic and nonischemic heart failure. Circ Res 125:117–146. [Google Scholar] [CrossRef]

Gao L, Gregorich ZR, Zhu W, Mattapally S, Oduk Y, Lou X, Kannappan R, Borovjagin AV, Walcott GP, Pollard AE et al (2018) Large cardiac muscle patches engineered from human induced-pluripotent stem cell-derived cardiac cells improve recovery from myocardial infarction in swine. Circulation 137:1712–1730. [Google Scholar] [CrossRef]

Garg S, Serruys PW (2010) Coronary stents: current status. J Am Coll Cardiol 56:S1–S42. [Google Scholar] [CrossRef] [Green Version]

Ghormade V, Pathan EK, Deshpande MV (2017) Can fungi compete with marine sources for chitosan production? Int J Biol Macromol 104:1415–1421. [Google Scholar] [CrossRef] [PubMed]

Gomes M, Azevedo H, Malafaya P, Silva S, Oliveira J, Silva G, João Mano RS, Reis R (2013) 16—Natural Polymers in Tissue Engineering Applications. In: Ebnesajjad S (ed) Handbook of biopolymers and biodegradable plastics. William Andrew Publishing, Boston, pp 385–425. [Google Scholar]

Greenwald S, Berry C (2000) Improving vascular grafts: the importance of mechanical and haemodynamic properties. J Pathol 190:292–299. [Google Scholar] [CrossRef]

Gupta S (2016) Very very late stent thrombosis: 9.5 years after DES implantation. Indian Heart J 68(Suppl. 2):S39–S43. [Google Scholar] [CrossRef] [PubMed] [Green Version]

Habermehl J, Skopinska J, Boccafoschi F, Sionkowska A, Kaczmarek H, Laroche G, Mantovani D (2005) Preparation of ready-to-use, stockable and reconstituted collagen. Macromol Biosci 5:821–828. [Google Scholar] [CrossRef]

Hadinata IE, Hayward PA, Hare DL, Matalanis GS, Seevanayagam S, Rosalion A, Buxton BF (2009) Choice of conduit for the right coronary system: 8-year analysis of radial artery patency and clinical outcomes trial. Ann Thorac Surg 88:1404–1409. [Google Scholar] [CrossRef] [PubMed]

Harskamp RE, Lopes RD, Baisden CE, de Winter RJ, Alexander JH (2013) Saphenous vein graft failure after coronary artery bypass surgery: pathophysiology, management, and future directions. Ann Surg 257:824–833. [Google Scholar] [CrossRef] [PubMed]

Hehrlein FW, Schlepper M, Loskot F, Scheld HH, Walter P, Mulch J (1984) The use of expanded polytetrafluoroethylene (PTFE) grafts for myocardial revascularization. J Cardiovasc Surg 25:549–553. [Google Scholar]

Ibrahim H, El-Zairy E (2015) Chitosan as a biomaterial—structure, properties, and electrospun nanofibers. Concepts Compd Altern Antibact:81–101. [Google Scholar] [CrossRef] [Green Version]

Iqbal J, Gunn J, Serruys PW (2013) Coronary stents: historical development, current status and future directions. Br Med Bull 110:203–221. [Google Scholar] [CrossRef] [PubMed] [Green Version]

Joseph SM, Krishnamoorthy S, Paranthaman R, Moses J, Anandharamakrishnan C (2021) A review on source-specific chemistry, functionality, and applications of chitin and chitosan. Carbohydr Polym Technol Appl 2:100036. https://doi.org/10.1016/j.carpta.2021.100036

Kakisis JD, Liapis CD, Breuer C, Sumpio BE (2005) Artificial blood vessel: the holy grail of peripheral vascular surgery. J Vasc Surg 41:349–354. [Google Scholar] [CrossRef] [Green Version]

Karabulut A, Cakmak M (2015) Treatment strategies in the left main coronary artery disease associated with acute coronary syndromes. J Saudi Heart Assoc 27:272–276. [Google Scholar] [CrossRef] [PubMed] [Green Version]

Kindi HA, Samaan A, Hosny H (2018., [Google Scholar] [CrossRef] [Green Version]) NOBLE and EXCEL: the debate for excellence in dealing with left main stenosis. Glob Cardiol Sci Pract 2018

Knight DK, Gillies ER, Mequanint K (2014) Vascular grafting strategies in coronary intervention. Front Mater 1. [Google Scholar] [CrossRef] [Green Version]

Kundu B, Rajkhowa R, Kundu SC, Wang X (2013) Silk fibroin biomaterials for tissue regenerations. Adv Drug Deliv Rev 65:457–470. [Google Scholar] [CrossRef] [PubMed]

LeGrice I, Pope A, Smaill B (2005) The architecture of the heart: myocyte organization and the cardiac extracellular matrix. In: Villarreal FJ (ed) Interstitial fibrosis in heart failure. Springer, New York, pp 3–21. [Google Scholar]

Leibundgut G (2018) A novel radiopaque, bioresorbable tyrosine-derived polymer for cardiovascular scaffolds. Cardiac Interventions Today, 2018, p. European Featured Technology. Available online: https://citoday.com/articles/2018-july-aug/a-novel-radiopaque-bioresorbable-tyrosine-derived-polymer-for-cardiovascular-scaffolds?c4src=archive:feed. Accessed 29 Feb 2020

Li X, Guo Y, Ziegler KR, Model LS, Eghbalieh SDD, Brenes RA, Kim ST, Shu C, Dardik A (2011) Current usage and future directions for the bovine pericardial patch. Ann Vasc Surg 25:561–568. [Google Scholar] [CrossRef] [Green Version]

Li Z, Yang J, Loh XJ (2016) Polyhydroxyalkanoates: opening doors for a sustainable future. NPG Asia Mater 8:e265. [Google Scholar] [CrossRef]

McCready RA, Siderys H, Pittman JN, Herod GT, Halbrook HG, Fehrenbacher JW, Beckman DJ, Hormuth DA (1992) Delayed postoperative bleeding from polytetrafluoroethylene carotid artery patches. J Vasc Surg 15:661–663. [Google Scholar] [CrossRef] [Green Version]

McMahan S, Taylor A, Copeland KM, Pan Z, Liao J, Hong Y (2020) Current advances in biodegradable synthetic polymer based cardiac patches. J Biomed Mater Res Part A 108:972–983. [Google Scholar] [CrossRef]

Mei X, Cheng K (2020) Recent Development in therapeutic cardiac patches. Front Cardiovasc Med 7:610364. https://doi.org/10.3389/fcvm.2020.610364

Menasché P, Vanneaux V, Hagège A, Bel A, Cholley B, Parouchev A, Cacciapuoti I, Al-Daccak R, Benhamouda N, Blons H et al (2018) Transplantation of Human Embryonic Stem Cell–Derived Cardiovascular Progenitors for Severe Ischemic Left Ventricular Dysfunction. J Am Coll Cardiol 71:429–438. [Google Scholar] [CrossRef] [PubMed

Minale C, Nikol S, Hollweg G, Mittermayer C, Messmer BJ (1988) Clinical experience with expanded polytetrafluoroethylene Gore-Tex surgical membrane for pericardial closure: a study of 110 cases. J Card Surg 3:193–201. [Google Scholar] [CrossRef]

Mohr FW, Morice MC, Kappetein AP, Feldman TE, Stahle E, Colombo A, Mack MJ, Holmes DR Jr, Morel MA, Van Dyck N et al (2013) Coronary artery bypass graft surgery versus percutaneous coronary intervention in patients with three-vessel disease and left main coronary disease: 5-year follow-up of the randomised, clinical SYNTAX trial. Lancet 381:629–638. [Google Scholar] [CrossRef]

Nappi F, Bellomo F, Nappi P, Chello C, Iervolino A, Chello M, Acar C (2021) The use of radial artery for CABG: an update. Biomed Res Int 2021:5528006. https://doi.org/10.1155/2021/5528006

Neamtu I, Chiriac A, Ghilan A, Nita L, Balan V, Nistor M (2014., [Google Scholar] [CrossRef]) Current concepts on cardiovascular stent devices. Mini Rev Med Chem 14:505

Nicolas FL, Gagnieu CH (1997) Denatured thiolated collagen. II. Cross-linking by oxidation. Biomaterials 18:815–821. [Google Scholar] [CrossRef]

Ormiston JA, Serruys PW (2009) Bioabsorbable coronary stents. Circ Cardiovasc Interv 2:255–260. [Google Scholar] [CrossRef] [PubMed] [Green Version]

Pashneh-Tala S, MacNeil S, Claeyssens F (2016) The tissue-engineered vascular graft-past, present, and future. Tissue Eng Part B Rev 22:68–100. [Google Scholar] [CrossRef]

Piskin E (1995) Biodegradable polymers as biomaterials. J Biomater Sci Polym Ed 6:775–795. [Google Scholar] [CrossRef]

Radisic M, Marsano A, Maidhof R, Wang Y, Vunjak-Novakovic G (2008) Cardiac tissue engineering using perfusion bioreactor systems. Nat Protoc 3:719. [Google Scholar] [CrossRef] [Green Version]–738

Ramadan R, Boden WE, Kinlay S (2018) Management of left main coronary artery disease. J Am Heart Assoc 7:e008151. [Google Scholar] [CrossRef]

Rashedi I, Talele N, Wang X-H, Hinz B, Radisic M, Keating A (2017) Collagen scaffold enhances the regenerative properties of mesenchymal stromal cells. PLoS One 12:e0187348. [Google Scholar] [CrossRef] [Green Version]

Rienks M, Papageorgiou A-P, Frangogiannis NG, Heymans SM, Matrix E (2014) Circ Res 114:872–888. [Google Scholar] [CrossRef] [Green Version]

Sakamoto A, Jinnouchi H, Torii S, Virmani R, Finn AV (2018) Understanding the impact of stent and scaffold material and strut design on coronary artery thrombosis from the basic and clinical points of view. Bioengineering 5:71. [Google Scholar] [CrossRef] [Green Version]

Salick MR, Napiwocki BN, Sha J, Knight GT, Chindhy SA, Kamp TJ, Ashton RS, Crone WC (2014) Micropattern width dependent sarcomere development in human ESC-derived cardiomyocytes. Biomaterials 35:4454–4464. [Google Scholar] [CrossRef] [Green Version]

Sanchis-Gomar F, Perez-Quilis C, Leischik R, Lucia A (2016) Epidemiology of coronary heart disease and acute coronary syndrome. Ann Transl Med 4:256. [Google Scholar] [CrossRef] [PubMed] [Green Version]

Serruys PW, De Jaegere P, Kiemeneij F, Macaya C, Rutsch W, Heyndrickx G, Emanuelsson H, Marco J, Legrand V, Materne P (1994) A comparison of balloon-expandable-stent implantation with balloon angioplasty in patients with coronary artery disease. N Engl J Med 331:489–495. [Google Scholar] [CrossRef] [Green Version]

Silver FH, Horvath I, Foran DJ (2001) Viscoelasticity of the vessel wall: the role of collagen and elastic fibers. Crit Rev Biomed Eng 29:279–301. [Google Scholar] [CrossRef]

Smit NW, ten Sande JN, Parvizi M, van Amersfoorth SCM, Plantinga JA, van Spreuwel-Goossens CAFM, van Dongen EMWM, van Dessel PFHM, Kluijtmans SGJM, Meijborg VMF et al (2017) Recombinant human collagen-based microspheres mitigate cardiac conduction slowing induced by adipose tissue-derived stromal cells. PLoS One 12:e0183481. [Google Scholar] [CrossRef] [Green Version]

Soares J, Moore J (2015) Biomechanical challenges to polymeric biodegradable stents. Ann Biomed Eng 44. [Google Scholar] [CrossRef]

Song R, Murphy M, Li C, Ting K, Soo C, Zheng Z (2018) Current development of biodegradable polymeric materials for biomedical applications. Drug Des Dev Ther 12:3117–3145. [Google Scholar] [CrossRef] [Green Version]

Stowell CET, Wang Y (2018) Quickening: translational design of resorbable synthetic vascular grafts. Biomaterials 173:71–86. [Google Scholar] [CrossRef] [PubMed]

Streeter BW, Davis ME (2019) Therapeutic cardiac patches for repairing the myocardium. In: Turksen K (ed) Cell biology and translational medicine, Volume 5: Stem cells: translational science to therapy. Springer, Cham, pp 1–24. Google Scholar

Sudesh K, Abe H, Doi Y (2000) Synthesis, structure and properties of polyhydroxyalkanoates: biological polyesters. Prog Polym Sci 25:1503–1555. [Google Scholar] [CrossRef]

Sultankulov B, Berillo D, Sultankulova K, Tokay T, Saparov A (2019) Progress in the development of chitosan-based biomaterials for tissue engineering and regenerative medicine. Biomol Ther 9:470. [Google Scholar] [CrossRef] [PubMed] [Green Version]

Tada T, Byrne RA, Simunovic I, King LA, Cassese S, Joner M, Fusaro M, Schneider S, Schulz S, Ibrahim T et al (2013) Risk of stent thrombosis among bare-metal stents, first-generation drug-eluting stents, and second-generation drug-eluting stents: results from a registry of 18,334 patients. JACC Cardiovasc Interv 6:1267–1274. [Google Scholar] [CrossRef] [Green Version]

Tallawi M, Dippold D, Rai R, D'Atri D, Roether JA, Schubert DW, Rosellini E, Engel FB, Boccaccini AR (2016) Novel PGS/PCL electrospun fiber mats with patterned topographical features for cardiac patch applications. Mater Sci Eng C 69:569–576. [Google Scholar] [CrossRef] [PubMed]

Tibbitt MW, Rodell CB, Burdick JA, Anseth KS (2015) Progress in material design for biomedical applications. Proc Natl Acad Sci USA 112:14444. [Google Scholar] [CrossRef] [Green Version]

Tomberli B, Mattesini A, Baldereschi GI, Di Mario C (2018) Breve historia de los *stents* coronarios. Rev Española Cardiol 71:312–319. [Google Scholar] [CrossRef]

Umashankar PR, Mohanan PV, Kumari TV (2012) Glutaraldehyde treatment elicits toxic response compared to decellularization in bovine pericardium. Toxicol Int 19:51–58. [Google Scholar] [CrossRef] [Green Version]

Wang Y, Ameer GA, Sheppard BJ, Langer R (2002) A tough biodegradable elastomer. Nat Biotechnol 20:602–606. [Google Scholar] [CrossRef] [PubMed]

Wendel JS, Ye L, Zhang P, Tranquillo RT, Zhang JJ (2014) Functional consequences of a tissue-engineered myocardial patch for cardiac repair in a rat infarct model. Tissue Eng Part A 20:1325–1335. [Google Scholar] [CrossRef] [Green Version]

WHO. Cardiovascular Diseases (CVDs). Available online: https://www.who.int/en/news-room/fact-sheets/detail/cardiovascular-diseases-(cvds). Accessed 9 Mar 2020

Williams DF (2019) Challenges with the development of biomaterials for sustainable tissue engineering. Front Bioeng Biotechnol 7:127. [Google Scholar] [CrossRef] [Green Version]

Wu Z, Li Q, Wang L, Zhang Y, Liu W, Zhao S, Geng X, Fan Y (2023) A novel biomimetic nanofibrous cardiac tissue engineering scaffold with adjustable mechanical and electrical properties based on poly(glycerol sebacate) and polyaniline. Mater Today Bio 23:100798. https://doi.org/10.1016/j.mtbio.2023.100798

Ye L, Zimmermann W-H, Garry DJ, Zhang J (2013) Patching the heart: cardiac repair from within and outside. Circ Res 113:922–932. [Google Scholar] [CrossRef] [PubMed] [Green Version]

Zhang J (2015) Engineered tissue patch for cardiac cell therapy. Curr Treat Options Cardiovasc Med 17:399. [Google Scholar] [CrossRef] [Green Version]

Zhang J, Zhu W, Radisic M, Vunjak-Novakovic G (2018) Can we engineer a human cardiac patch for therapy? Circ Res 123:244–265. [Google Scholar] [CrossRef]

Zheng SX, Weng YL, Zhou CQ, Wen ZZ, Huang H, Wu W, Wang JF, Wang T (2013) Comparison of cardiac stem cells and mesenchymal stem cells transplantation on the cardiac electrophysiology in rats with myocardial infarction. Stem Cell Rev Rep 9:339–349. [Google Scholar] [CrossRef]

Zimmermann W-H, Schneiderbanger K, Schubert P, Didié M, Münzel F, Heubach JF, Kostin S, Neuhuber WL, Eschenhagen T (2002) Tissue engineering of a differentiated cardiac muscle construct. Circ Res 90:223–230. [Google Scholar] [CrossRef] [PubMed] [Green Version]

Zimmermann W-H, Melnychenko I, Wasmeier G, Didié M, Naito H, Nixdorff U, Hess A, Budinsky L, Brune K, Michaelis B (2006) Engineered heart tissue grafts improve systolic and diastolic function in infarcted rat hearts. Nat Med 12:452–458. [Google Scholar] [CrossRef]

Rishabha Malviya completed a B. Pharmacy from Uttar Pradesh Technical University and an M. Pharmacy (Pharmaceutics) from Gautam Buddha Technical University, Lucknow Uttar Pradesh. His PhD (Pharmacy) work was in Novel formulation development techniques. He has 14 years of research experience and has worked as an Associate Professor in the Department of Pharmacy, School of Medical and Allied Sciences, Galgotias University for the past 11 years. His area of interest includes formulation optimization, nanoformulation, targeted drug delivery, localized drug delivery, and characterization of natural polymers as pharmaceutical excipients. He has authored more than 150 research/review papers for national/international journals of repute. He has 51 patents (12 grants, 38 published, 1 filed) and publications in reputed National and International journals with a total of 140 cumulative impact factors. He has also received an Outstanding Reviewer award from Elsevier. He has authored/Edited/edited 91 books (Wily, CRC Press/Taylor and Francis, Springer Nature, Apple Academic Press, IOP Publishing, River Publisher, and OMICS publication) and authored 170 book chapters. His name has been included in Word's top 2% scientist list for the year 2020, 2021, 2022, 2023 and 2024 by Elsevier BV and Stanford University. He is a Reviewer/Editor/Editorial board member of more than 50 national and international journals of repute. He has been invited as an author for "Atlas of Science" and pharma magazine dealing with industry (B2B) "Ingredient South Asia Magazines".

Suraj Kumar completed a B. Pharmacy from SRM University, Chengalpattu, Tamil Nadu, India, and a Pharmacy from Galgotias University, Greater Noida, India. He has 1 year of teaching experience and works as an assistant professor at Pragya College of Pharmaceutical Sciences, Gaya, Bihar. He has attended more than 10 International/National conferences. He has published more than 20 SCI publications with reputed international publishers like Elsevier, Wiley, and Bentham Science with more than 40 cumulative impact factors. He has published 8 chapters, and three books is under consideration for publication.

Sathvik Belagodu Sridhar is currently working as an Associate Dean and Professor at RAK College of Pharmacy (RAKCOP), RAK Medical and Health Sciences University located in Ras Al Khaimah, United Arab Emirates. Dr. Sathvik has completed his PhD in Clinical Pharmacy from Rajiv Gandhi University of Health Sciences, Bangalore, India. He has received an FIP fellowship for higher training in renal clinical pharmacy at Royal Adelaide Hospital and Queen Elizabeth Hospital in Adelaide, South Australia. Dr. Sathvik has worked as an Assistant Professor and Clinical Pharmacist at JSS College of Pharmacy and JSS Medical College Hospital, Mysore, India, for eight years and later Joined International Medical University (IMU) Malaysia as a senior lecturer in the Department of Pharmacy Practice from 2008 to 2010. Dr. Sathvik joined as an assistant professor in the Department of Clinical Pharmacy at RAK College of Pharmacy, RAK Medical, and Health Sciences University in 2010 and worked as a professor and chairperson of the Department of Clinical Pharmacy and Pharmacology from 2018 to 2023. Dr. Sathvik has been honored with a prestigious teacher award by the Society of Pharmaceutical Education & Research [SPER]. He has also received training in Pharmacovigilance at the Uppsala Monitoring Centre in Uppsala, Sweden. Dr. Sathvik has 25 years of experience teaching and practicing clinical pharmacy in India, Malaysia, and the UAE. His research interests include Medication Safety, Pharmacovigilance, Pharmacoepidemiology, and Renal and Psychiatry Clinical Pharmacy. Dr. Sathvik has more than 75 research papers published in reputable national and international journals, along with book chapters. He has presented more than 75 research papers at numerous national and international conferences and has been honored with many awards. He also serves as a reviewer for various national and internal pharmacy practice-related journals.

Plant-Derived Biomaterial and Their Use in Cardiac Tissue Repair

5

Sudipta Choudhury, D. Mutthuraj, M. Maghimaa, Suresh Sagadevan, and Kanthesh M. Basalingappa

Abstract

Globally, cardiovascular disease is the leading cause of death to this date. After an infarction, scar tissue forms because of the incapacity of the heart to heal, which causes the heart malfunction. Therefore, research on heart regeneration is continually in demand. Thanks to the advances in stem cells and biomaterials of the advanced biological tissue design and regeneration medicine, there are promising ways for the engineering of substitute tissues for the cardiovascular system, which may work like the native tissue. Among all types of biomaterials, plant-derived materials have the highest propensity to support cell development because of their mechanical strength, bio composites and inherent biocompatibility. Additionally, compared with common As for the ones derived from plant sources, while collagen and gelatine are perhaps the most commonly derived from animals, they possess substantially fewer immunological properties. Better wettability is another benefit they provide over synthetic fabrics. There is currently a dearth of literature that provides a systematic overview of the development of biomaterials derived from plants for cardiovascular tissue regeneration.

S. Choudhury · D. Mutthuraj · K. M. Basalingappa (✉)
Division of Molecular Biology, School of Life Sciences, JSS Academy of Higher Education & Research, Mysuru, Karnataka, India
e-mail: mutthurajd@jssuni.edu.in; kantheshmb@jssuni.edu.in; https://vidwan.inflibnet.ac.in/profile/84555

M. Maghimaa
Centre for Global Health Research, Saveetha Medical College and Hospital, Saveetha Institute of Medical and Technical Sciences, Chennai, India

S. Sagadevan
Nanotechnology & Catalysis Research Centre, University of Malaya, Kuala Lumpur, Malaysia

R. Malviya, S. Sundram (eds.), *Sustainable Nanocomposites with Green Biomaterials*, Biomaterials, Bioengineering and Sustainability 2,
https://doi.org/10.1007/978-3-031-79110-9_5

The current chapter and analysis highlight the most popular plant-derived biomaterials gotten from both terrestrial and aquatic plants. A fatigue of debate follows on the beneficial properties of these substances in mending damaged tissue. Nonetheless, as summarised using the up-to-date preclinical and clinical evidences, the potential use of plant-derived biomaterials within therapeutic cardiovascular tissue engineering-the tissue engineered scaffolds and the plant derived bio-inks in the course of 3D bio fabrication technique and the bioactive plantation for pharmaceutical utilization-are also briefly incorporated.

Keywords

Cardiovascular disease · Tissue engineering · Regenerative medicines · Biomaterials · Scaffolds · 3D bio fabrication

5.1 Introduction

Cardiovascular diseases are the world's largest cause of death, accounting for 17.9 million deaths yearly (Cattelan et al. 2020). Among the most common cardiovascular disorders are myocardial infarction (MI) and other ischemic heart illnesses. MI is often caused by a momentary or chronic blockage of the cardiac artery, which dramatically reduces the blood supply to the left ventricle (LV). Twenty minutes after the ischemia, the elevated metabolic rate damages the heart irreversibly. The healing process of the heart after MI may be divided into two phases based on the major events that occurred:

1. The inflammatory phase is characterized by intense inflammation brought on by cardiomyocyte necroptosis and the subsequent mobilization of cells from the immune system.
2. The reparative and proliferative phases encompass the inhibition of the inflammatory response, myofibroblast proliferation, formation of scar tissue, and remodelling of the left ventricle function (Dai et al. 2023).

LV dysfunction is caused by structural remodelling of the heart, which includes non-infarcted myocardial enlargement, infarcted wall thinning, and scar development. Congestive heart failure is the end result of a steady decline in heart muscle function. Regretfully, the therapy for ischemic heart disease still presents clinical difficulty, mainly because cardiac tissue and local vasculature have limited ability to regenerate. There is a negligible chance that new cardiomyocytes will naturally arise via stem cell differentiation or cardiomyocyte proliferation, which means that cardiomyocytes will eventually disappear forever (Dai et al. 2023). Relief of symptoms and prevention of the advancement of end-stage cardiovascular illnesses, such as heart attack or stroke, are often the main objectives of current therapeutic therapy. There is currently no alternative proven effective treatment for cardiac arrest aside from a cardiac transplant or left ventricular assist devices (Li et al. 2021). Even in these cases, contraindications, fundamental illnesses, and other concurrent illnesses all contribute to inadequate long-term treatment outcomes (Guglin et al. 2020). A growing percentage of MI patients, a lack of donors (Feingold 2018), and the scarcity of healthcare in remote places make transplantation impractical in the current therapeutic context. Thus, efforts to develop novel therapeutic approaches for ischemic heart disease continue. A crucial component of these approaches is the regeneration or repair of infarcted tissue, intended to maintain or enhance cardiac function and prognostic improvement. Because advanced biomaterials can serve as biomolecules to improve onsite cell activity or transfer cells or bioactive substances, they have drawn interest in the field of cardiac tissue engineering (CTE) research. Currently, various functionalized biomaterials, meaning those produced from plants and animals, have been used in cardiac tissue regeneration. These applications include scaffolds (Tang et al. 2022), delivery vehicles (Ghebouli et al. 2018), bioactive compounds (Qiu et al. 2022), and bio-inks for 3D bio fabrication (Cui et al. 2020). The most researched biomaterials are those obtained from animals, including fibrin, gelatin, laminin, collagen, and extracellular matrix (ECM) formed from animal tissue or decellularized ECM. These biomaterials provide distinct benefits based on their characteristics. However, undesirable traits including xeno-genicity, low scalability, and reproducibility (Pomeroy et al. 2020) lead to erratic therapeutic outcomes.

Ethical issues and the potential existence of endotoxins or virus leftovers further obstruct the therapeutic translation of biomaterials derived from animals in the future (Li et al. 2021). Consequently, there has been a notable surge in the attention of researchers to plant-derived materials, namely plant polysaccharides. Plant-derived biomaterials have several distinct benefits over animal-based biomaterials, including better scalability, lower immunogenicity, and fewer moral conundrums (Pomeroy et al. 2020). Thus, biomaterials generated from plants may hold the key to the future of tissue engineering and regenerative medicine. Several studies have confirmed the application of biomaterials produced from plants in CTE. Depending on where they came from, things made from plants might be classified as either marine or land-based. Nonetheless, a small body of literature offers a thorough and organised analysis of the pertinent topic (Pomeroy et al. 2020).

5.2 Plant-Derived Biomaterials

The utilisation of natural materials for a range of biological purposes has received increasing attention since the development of nanotechnology. The demand for ecologically friendly biomaterials sourced from sources free of ecological load has also been emphasised by this quickly expanding business. Making use of plant-derived biomaterials offers a significant chance to use the biological functions of plants as cell division stimulators and the inherent resilience of plant microarchitecture. Plant-based materials offer improved mechanical stability, bioactivity, biocompatibility, and biodegradability, among other advantages. Because of their facile hydrogel synthesis and chemical manipulation affinity, these materials have also been proposed as potential bio-inks (Wan et al. 2021).

5.3 Marine Plant-Derived Biomaterials

The ocean has a wide variety of marine life and makes up over 71% of the Earth's surface. Alginate, carrageenan, agarose, and fucoidan are just a few of the widely used biomaterials that these aquatic species offer in abundance. Several pharmaceutical, biomedical, cosmetic, and food sectors are drawn to them because of their broad diversity, favourable biocompatibility, renewable nature, low processing costs, and possible bioactivities (Dai et al. 2023). Coralline bone grafts and polysaccharide-based biomaterials are two examples of the many stages of discovery and development that marine biomaterials undergo. The latter are exemplified by collagen obtained from marine sources, chitin and chitosan, and composites made of various marine-derived species (Wan et al. 2021) (Fig. 5.1).

5.3.1 Fucoidan

A polysaccharide derived from sulphated L-fucose is called fucoidan. that is mostly obtained from brown algae. It has been shown to have antibacterial, antioxidant, antiviral, anticoagulant, and anticancer properties (Ghebouli et al. 2018; Oliveira et al. 2019; Niyonizigiye et al. 2019). The fucoidan's molecular weight, sulphate concentration, sulfation structure, and glycosidic branching all have a significant impact on these characteristics. For instance, fucoidan with a low molecular weight frequently has a higher amount of bioactivity (Dai et al. 2023). Endothelial cell adhesion is significantly increased when fucoidan is employed to alter poly (vinyl alcohol) hydrogels, typically not promoting cell adhesion on their own (Torres et al.

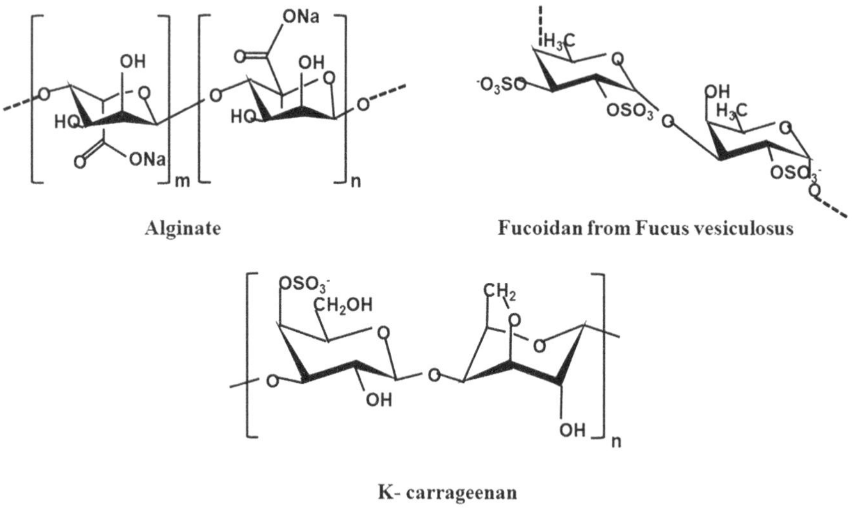

Fig. 5.1 Structures of Marine Biomaterials Derived from Plants—Alginate, Fucoidan, and Carrageenan. (Dai et al. 2023)

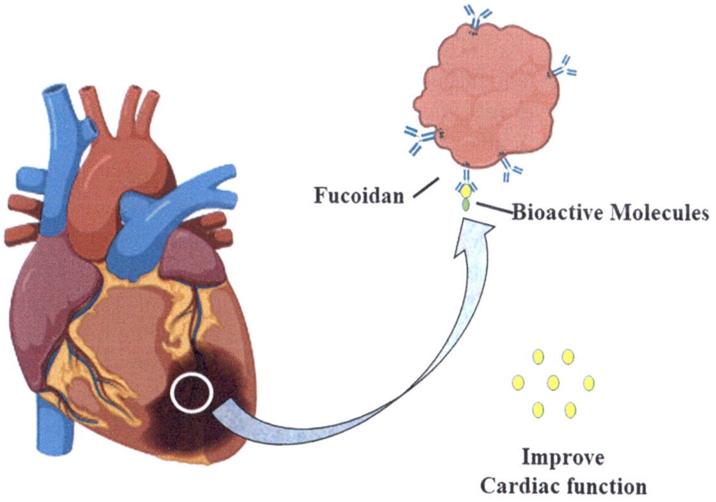

Fig. 5.2 Through its interaction with p-selectin, fucoidan facilitates the targeted delivery of bioactive compounds to thrombus locations. (Dai et al. 2023)

2019). This finding emphasises fucoidan's capacity to promote cell adhesion. Furthermore, fucoidan may bind to P-selectin, an activated platelet and inflammatory endothelium surface glycoprotein that may be used to deliver bioactive compounds precisely (Dai et al. 2023). In addition to its ability to transmit additional bioactive molecules towards the thrombus in a targeted manner, which would alleviate MI, fucoidan itself has positive effects on heart function. To reduce the infarcted area, boost the effectiveness of the antioxidant defence system, and reduce oxidative stress, oral fucoidan was given to rats with isoproterenol-induced MI, according to Thomes et al. (Dai et al. 2023). Furthermore, it has been shown that oral administration of both high-stability fucoxanthin and low-molecular-weight fucoidan improves cardiac function in ageing mice through the GSK3t/CREB and IRS1/GRB2/SOS1 pathways, as well as metabolic pathways (Chang et al. 2019) suggesting a possible molecular mechanism for fucoidan's cardioprotective action (Fig. 5.2).

5.3.2 Alginate

Alginate is a polymer that is produced by brown algae and is also known as alginic acid (Rastogi and Kandasubramanian 2019) It is made up of two primary building blocks: mannuronic acid (M) connected to α-D and guluronic acid (G) linked to α-L. These building blocks often polymerize inside the alginate sequence in an alternating manner. Because of its structural similarity to extracellular matrix (ECM), biocompatibility, and non-thrombogenic qualities, alginate has been used extensively in heart repair. For example, because alginate can form long chains of polysaccharides that help with gelling qualities, it is frequently employed in hydrogel format for medication or cell injection as well as to give structural aid for the MI area in CTE (Fig. 5.3) (Dai et al. 2023). Additionally, by adjusting the ratio of guluronic acid (G) to mannuronic acid (M) with alginate, the elasticity modulus for the alginate-based hydrogel may be adjusted (from 5 to 55 kPa). Higher G alginate hydrogels are more pliable than higher M hydrogels. Alginate in CTE is made possible by its customisable elasticity. Interestingly, alginate binds to divalent ions involving Ca2+, Sr2+, Mg2+, and Ba2+ to create a gel. By unleashing these divalent ions, alginate hydrogel stability may be jeopardised. Thus, covalently cross-linking the alginate hydrogels using additional particles, like poly-(ethylene glycol) diamines, would be one method of stabilising them. Alginate's hydrophilicity, which also restricts its capacity to absorb proteins, contributes to its non-cell-adhesive properties (Dai et al. 2023). Alginate has therefore been molecularly altered to improve cell attachment by covalent bonding with certain cell adhesion ligands, including the arginine-glycine-aspartate (RGD) ligand. Through their interaction with cellular integrin, RGD peptides can increase a cell's inclination towards extracellular matrix (ECM). This can trigger cell signalling that prevents apoptosis and promotes proliferation. It was also shown in vivo studies using MI rats that neovascularization may be enhanced by in situ administrations of RGD-coupling alginate into the infarcted region. Thus, one potential approach for using alginate in CTE is RGD alteration It has been demonstrated that alginate hydrogel enhances stem cell retention and delivery, while fuller-enol nanoparticles provide antioxidant

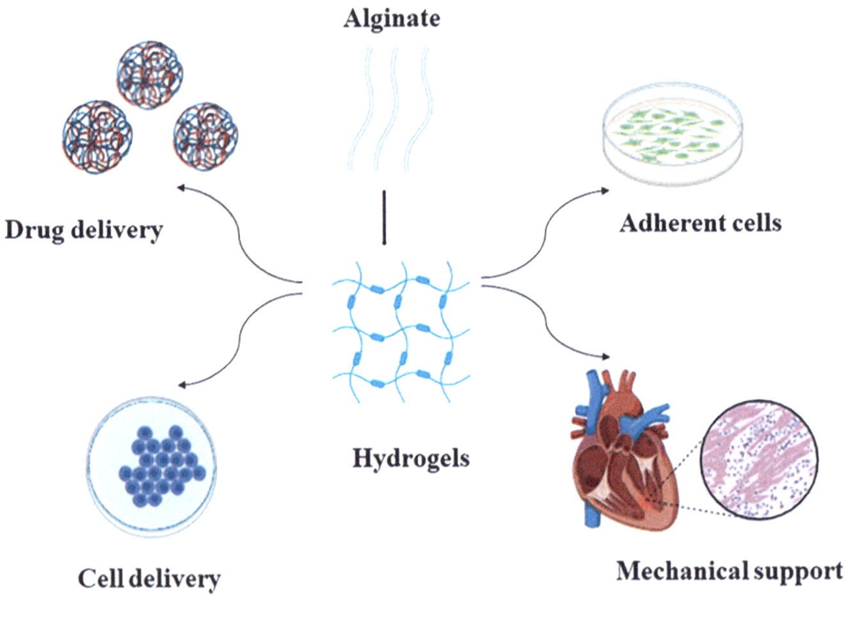

Fig. 5.3 The potential of biomaterials generated from plants for heart regeneration and repair. Alginate can give the affected area mechanical support and is readily gel-able. It is frequently used as a hydrogel for medication or cell administration. (Dai et al. 2023)

qualities that lessen ROS damage to cells. By penetrating cells, these nanoparticles have the ability to further control the JNK, p38, and ERK MAPK signalling pathways. These pathways are crucial in order to survive proliferation, death, and cardiomyogenic stem cell differentiation(Dai et al. 2023).

5.3.3 Carrageenan

Seaweeds produce carrageenan, a class of sulfated polysaccharides. It consists of lengthy straight chains with ester sulphates in between that alternate between D-galactose (α-1, 3) and 3-anhydro-galactose (β-1, 4). Along with being biodegradable, biocompatible, and nontoxic, Carrageenan possesses beneficial bioactivities, including anti-inflammatory, antibacterial, antiviral, antihyperlipidemic, anticoagulant, and immunomodulatory qualities (Pacheco-Quito et al. 2020). It can also encourage cell proliferation and adhesion (Ocampo et al. 2019). When Carrageenan with a negative charge is combined with cationic chitosan, an insoluble mixture can be created without the need for harmful cross-linkers. This method has been employed for several bio-applications, such as medication delivery and tissue engineering scaffolds (Yu et al. 2018). κ-carrageenan, when paired with chitosan, can provide synthesized hydrogels withs more physical strength than other polymers like collagen, gelatin, n and polymethyl methacrylate (Pourjavadi et al. 2019).

5.3.4 Agarose

The saccharide polymer agarose originates from marine algae and is composed of repeating units of D 'Galactose and 3,6-anhydro-L-galactose. It can be broken down enzymatically by α-galactosidase or phagocytized by macrophages. Thermo-reversible gelation behaviour, porosity, plasticity, temperature sensitivity, and biocompatibility are characteristics of agarose (Zou et al. 2021). In order to accommodate the various CTE requirements, the agarose gel's pore size may also be adjusted by varying the agarose concentration. Furthermore, agarose is frequently mixed with other hydrogels to increase temperature responsiveness and biomechanical strength (Dai et al. 2023). Agarose is often used in cell encapsulation and distribution because its high water absorption property typically promotes cell proliferation, differentiation, and proliferation (Zarrintaj et al. 2018). Nevertheless, agarose lacks natural ligands that facilitate its association with anchorage-dependent mammalian cells, similar to alginate. Therefore, surface modification Covalent interaction with cell adhesion peptides, the production of nano/microstructures, stiffness modulation, and combining with other polymers is required to promote cell adhesion across agarose substrates. However, agarose's non-cell-adhesive characteristic could also be useful in some circumstances. Agarose microwell plates, for example, might be applied to pluripotent stem cells to produce aggregation and embryoid entities that may eventually grow into the heart lineage (Dai et al. 2023).

Agarose

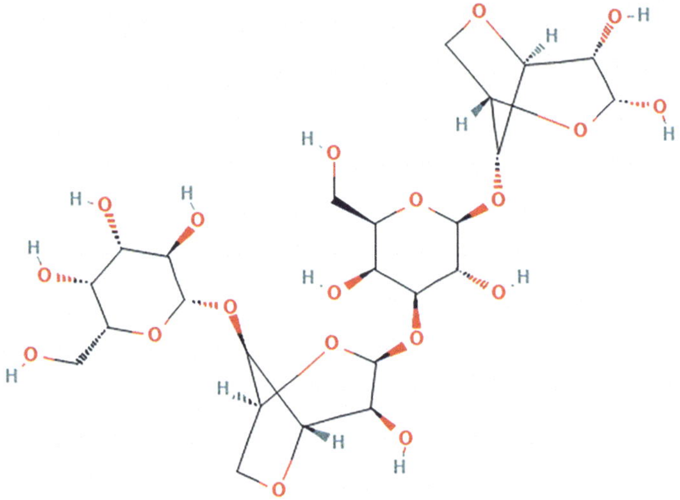

Fig. 5.4 Structure of Marine Plant-Derived Biomaterials- Agarose. (Dai et al. 2023)

Marine Biomaterials produced from plants have demonstrated be promising for heart regeneration and repair. They also often have a variety of advantageous physicochemical characteristics and bioactivities. In clinical practice, their cost-effectiveness and capacity for commercial manufacturing are preferred. However, a barrier to their therapeutic translation is a lack of thorough knowledge of their behaviour in vivo, including the underlying processes. Future clinical practice will be facilitated by the growing amount of research that has been conducted in this area (Fig. 5.4).

5.3.5 Land Plant-Derived Biomaterials

There have been few investigations on the application of these biomaterials in order to restore heart tissue, and only a handful of them have been successfully created for biomedical uses. The terrestrial plant-derived biomaterials that have been documented for various uses are mentioned below, along with a summary of their special benefits and difficulties (Fig. 5.5).

5.3.5.1 Starch

Starch, the second most common natural biomaterial, is present in many plant sections, including the roots, stalks, and seeds of crops (Dai et al. 2023). Its two main structural forms are amylose, which is a linear polymer of glucose-D-glucose units linked by glucose-1,4 glycosidic connections, and amylopectin, which is a branched polymerization of glucose-D-glucose units linked by glucose-1,4 and glucose-1,6 glycosidic linkages. Amylose produces the amorphous portion of starch, whereas

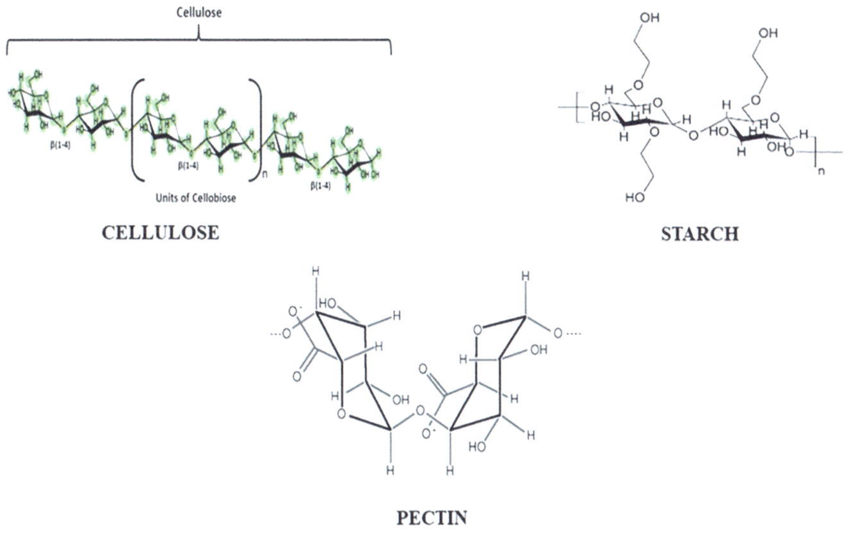

Fig. 5.5 Structures of Land Plant-Derived Biomaterials—Cellulose, Starch & Pectin. (Dai et al. 2023)

amylopectin contributes to its crystalline property. It is noteworthy that starch exhibits inadequate cell attachment when left unmodified (Zhuang et al. 2021). Starch can function as a vital energy source by breaking down into glucose without causing an inflammatory reaction, making it a potentially affordable replacement for in vivo polysaccharides (Wu et al. 2017). Starch has several active hydroxyl groups that simplify modification. Thus, starch has great promise in the biomedical domain. It has been extensively studied in the fields of tissue engineering and drug delivery, and it has demonstrated great promise for the treatment of heart problems (Fig. 5.6). When using starch for medication delivery, it is usually combined with another polymer or used on its own. Several investigations have demonstrated that the presence of starch in alginate hydrogels can modify their structural properties (such as the size and porosity of the generated network) and regulate the drug delivery profile (Mohammadi et al. 2018; Okunlola and Adewusi 2019). Hydrogels composed of synthetic polyelectrolytes may operate better when starch is crosslinked with extra biopolymers. Alginate and starch esterification together may facilitate site-specific, regulated acidic drug release.

Making smart solid dosages may be accomplished by esterifying and cross-linking starch nanoparticles to improve their surface. If one wants to modify the beneficial qualities of starch on its own, it has to be chemically changed. The modification processes of Commonly utilized processes include oxidation, etherification, esterification, acid hydrolysis, and cross-linking (Lemos et al. 2021). Starch serves as a carrier for delivery and has also been utilized in biological engineering. The electrospinning technology has been used in several studies to build scaffolds for skin and bone tissue engineering. Furthermore, starch enhances hydrophilicity and mediates in vitro breakdown.

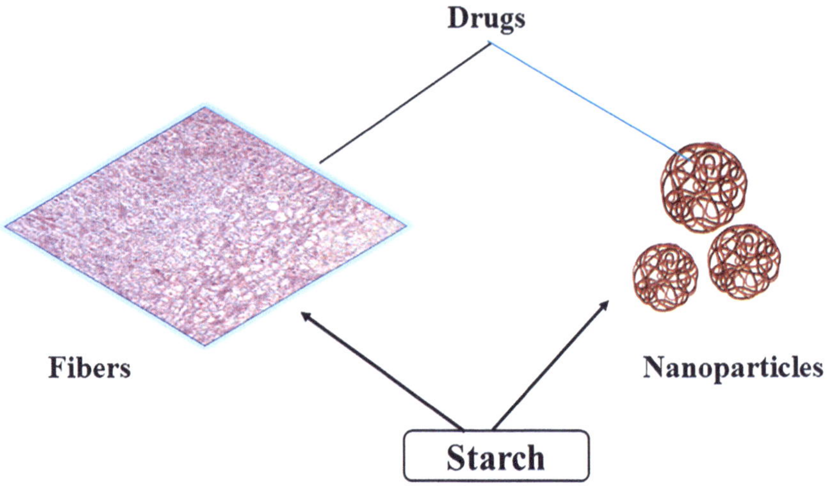

Fig. 5.6 The many active hydroxyl groups found in starch make it simple to alter potential use in medicine delivery and tissue engineering. (Dai et al. 2023)

5.3.5.2 Cellulose

Lycosidic (1–4) connections bind glucose (d-glucopyranose) to form cellulose, a linear homopolysaccharide (Simeoni et al. 2021). It may be obtained from several bacteria or plants, such as Sarcina ventriculi, Pseudomonas, Acetobacter, and Azotobacter (Torres et al. 2019). One example of a plant is pineapple. Cell adhesion and proliferation may be aided by cellulose (Kiiskinen et al. 2019). It's interesting to note that cellulose obtained from various plants can vary greatly in its degree of polymerization (Sultan and Mathew 2019). Regarding cellulose's possible use in CTE, Entcheva et al. effectively cultivated fibroblasts from cardiac tissue and cardiomyocytes in vitro on scaffolds made of cellulose acetate. Cardiomyocyte network development with improved cell packing, connection, and electromechanical activity was encouraged by these scaffolds, which also helped cardiac fibroblasts proliferate (Souza et al. 2018). We may find more instances of the fibre in CTE in the section on property-tailored 3D porous scaffolds. Owing to its advantageous properties, nanocellulose, or nanocellulose, has been receiving a lot of attention lately. The three further different forms of nanocellulose are nanocrystalline cellulose, nano fibrillated cellulose, and bacterial cellulose (Li et al. 2019). The latter of which is mostly made by breaking down plant fibres (Chen et al. 2018; Liu et al. 2021; Ling et al. 2019). Because the β-1, 4-glucose in its molecular chain contains three active hydroxyl groups, nanocellulose, like nanoscale cellulose (Luo et al. 2019), may create a hydrogen bond network. Its surface area and aspect ratio are also high (Curvello et al. 2019). Because of its superior mechanical strength, greater hydrophilicity, tunable surface chemistry, and increased compatibility, nanocellulose is a substance that might be used to biological purposes (Luo et al. 2019). Take into consideration the various potential uses of nanocellulose-based scaffolds in tissue engineering—vascular, for example hepatic, ocular, and cutaneous tissue

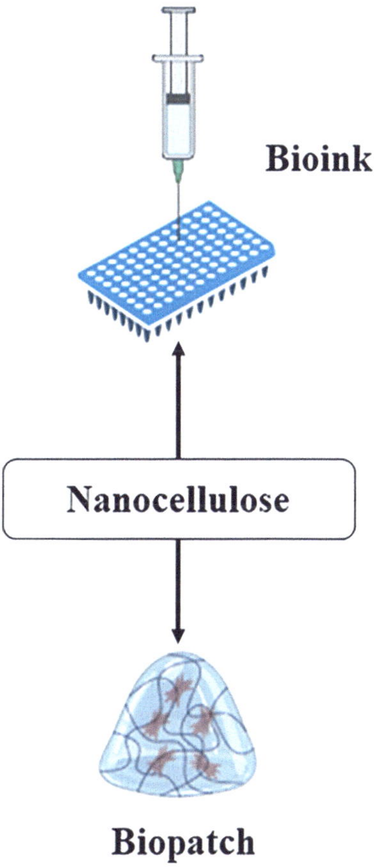

Fig. 5.7 Diagrammatic representations of a cardiac biopatch made of nanocellulose printed using 3D printing. (Dai et al. 2023)

engineering (Luo et al. 2019). Furthermore, it offers lots of potential for application as a bio-ink for extrusion-based three-dimensional bioprinting (Fig. 5.7) (Sultan and Mathew 2019; Markstedt et al. 2017; Leppiniemi et al. 2017).

5.3.5.3 Pectin

In industrial processes, pectin is extracted using citrus peel and apple pomace. It is a structural polysaccharide that primarily takes place in the fundamental plant cell wall. Pectin is a complex substance made up of several neutral sugars, including rhamnose, galactose, and arabinose, as well as (1,4)-linked α-d-galacturonic acid residues (Noreen et al. 2017). It can be readily functionalized by chemical modification because of several polysaccharide domains inside its branched molecule structure (Pereira et al. 2018). A number of other beneficial characteristics of pectin include its anionic nature, hydrophilicity, biological compatibility, and biological degradation (Noreen et al. 2017). When in contact with divalent cations, such as Ca2+, pectin can create a hydrogel structure. Pectin has a lot of potential uses in the biomedical sciences because of all these characteristics, including tissue engineering and controlled drug release. Involves the use of tissue engineering and regulated medication release (Kumar et al. 2013). Pectin's poor cell-adhesion feature is partly due to its inability to adsorb proteins, just as alginate does (Xu et al. 2020) Therefore, To alter or enhance the functionality of pectin-based biomaterials, peptides or polymers are frequently added (Noreen et al. 2017; Pereira et al. 2018; Lapomarda et al. 2019; Gawkowska et al. 2018; Fares et al. 2018).

In conclusion, cellulose, starch, and pectin are terrestrial plant-derived biomaterials with significant promise for tissue engineering. Pectin and starch can be used as medication delivery platforms. In addition, modified citrus pectin may be used as a bioactive medication to treat heart problems (Table 5.1).

5.4 Using Biomaterials Derived from Plants to Mend Cardiac Tissue

Ischemic cardiac damage can result from several cardiac diseases, such as MI and valvular heart disease (McKirnan et al. 2019). Because of restricted blood flow, the infarcted tissue receives less oxygen, which leads to myocardial necrosis, apoptosis, and inflammation (Dai et al. 2023). Either way, reperfusion or not, it results in a significant production of ROS, TNF-α, and damage-associated molecular patterns. In addition, the action of fibroblast-secreted matrix metalloproteinases promotes the breakdown of collagen fibres, which thin and dilate the ventricular wall. Consequently, aberrant stresses are produced by the myocardium to counteract the reduced cardiac activity, leading to structural changes in the heart (Andreadou et al. 2019) The restricted ability of heart tissue to regenerate itself presents the biggest treatment difficulty. Tissue engineering and regenerative medicine approaches that utilize biomaterials have emerged as viable treatment options for myocardial infarction (MI) in the last few years. The ways in which these techniques include increased onsite cell migration and proliferation (Smagul et al. 2020). decreased cellular mortality and inflammatory processes, and improved angiogenesis and electrical mobility (Fig. 5.8). The biological material can also provide skeletal support by expanding the ventricular wall and reducing wall stress, all while preserving heart function. Because of their better integration with the host tissue, they can also increase the rates of cell survival and retention (Wang et al. 2018). However, because the myocardium is contractile, there are very few alternatives for scaffold materials for myocardial regeneration. In response to these limitations, plant-derived biomaterials with biomechanical and elastic characteristics have become well-liked options for imitating the naturally occurring extracellular matrix (ECM) and promoting the cardiogenic differentiation and electromechanical integration of transplanted mesenchymal stem cells (MSCs). Biomaterials generated from plants have also been employed as delivery systems for medications, growth hormones, and short RNAs intended for heart repair (Dai et al. 2023; Choe et al. 2019a).

In terms of bioactivity, cardiac scaffolds should be able to decrease myocardial fibrosis and the inflammatory response, enhance angiogenesis, and improve cell migration and proliferation (e.g., by blocking the TLR4/MyD88/NF-κB signalling

Table 5.1 An overview of biomaterials generated from plants, especially their general uses in cardiac tissue engineering (nanocrystalline cellulose, nanofibrillated cellulose, extracellular matrix, embryoid bodies, and IPN, interpenetrating polymer network)

Plant-based biomaterials	Source	Attributes	Drawbacks	Utilisation and associated forms	References
Fucoidan	Brown algae	Antioxidant, antiviral properties, anticoagulant, antimicrobial, and antitumor biological activities; specific interaction with P-selectin.	Batch-to-batch variability in molecular weight, sulphate concentration, sulfation pattern, and glycosidic branch-dependent bioactivities	Bioactive molecules: Targeted delivery of medications and additional chemicals	Ghebouli et al. (2018), Oliveira et al. (2019) and Niyonizigiye et al. (2019)
Alginate	Brown algae	Simple gelation in the presence of divalent ions, structural similarity to the extracellular matrix, biocompatibility, and non-thrombogenic quality.	Lack of cell adhesion, instability, and batch-to-batch variations.	Hydrogel: Drug and cell delivery	Rastogi and Kandasubramanian (2019)
Carrageenan	Seaweeds	Antiviral, antibacterial, antihyperlipidemic, anticoagulant, antioxidant, anticancer, and immunomodulating bioactivities; biodegradability, biocompatibility, and non-toxicity.	Fluctuation from batch to batch and the paucity of research on carrageenan's use in CTE.	Nanosphere: Intelligent medication administration; scaffold: Tissue engineering	David et al. (2018), Pacheco-Quito et al. (2020), Yu et al. (2018), Pourjavadi et al. (2019) and Geyik and Işıklan (2021)

(continued)

Table 5.1 (continued)

Plant-based biomaterials	Source	Attributes	Drawbacks	Utilisation and associated forms	References
Agarose	Sea algae	Thermo-reversible gelation behaviour, reliability, high water absorbance, porosity, plasticity, biocompatibility, and transparency	Batch-to-batch variability and non-cell adhesive nature.	Hydrogel: Encapsulation and transport of cells; agarose microwell plates: Production of EBs and aggregation by the use of pluripotent stem cells and subsequent differentiation	Zou et al. (2021) and Zarrintaj et al. (2018)
Starch	Various plant elements, such as roots, stalks, and crop seeds	Easy to modify, inexpensive alternative to in vivo polysaccharides, and a vital source of energy as it breaks down into glucose without causing inflammation.	Batch-to-batch variability.	Functional additives or nanoparticles can change the structure's characteristics or allow for regulated medication release. Scaffolds: The fabrication of tissues	Wu et al. (2017), Mohammadi et al. (2018), Okunlola and Adewusi (2019), Lemos et al. (2021) and Simeoni et al. (2021)
Cellulose	Different plants, like pineapples	Better hydrophilicity, improved compatibility, remarkable mechanical strength, and customizable surface chemistry.	There are variations in cellulose polymerization levels across batches that are produced from various plants.	Bioink: Extrusion-based bioprinting; scaffolds: Tissue engineering	Chen et al. (2018), Liu et al. (2021), Ling et al. (2019), Luo et al. (2019), Sultan and Mathew (2019), Markstedt et al. (2017) and Leppiniemi et al. (2017)
Pectin	Primary plant cell wall	Among its easy-to-modify chemical properties are hydrophilicity, biocompatibility, biodegradability, nontoxicity, anionic nature, and gelling behavior in the presence of divalent cations.	Low cell adhesion and batch-to-batch variation.	Bioink: IPN and semi-IPN hydrogels: Tissue engineering; scaffold or excipient: Drug delivery; bioactive molecule: Cardioprotective property; 3D printing of complex, changeable porous scaffolds	Noreen et al. (2017), Pereira et al. (2018), Lapomarda et al. (2019), Gawkowska et al. (2018), Fares et al. (2018), Kumar et al. (2013) and Xu et al. (2020)

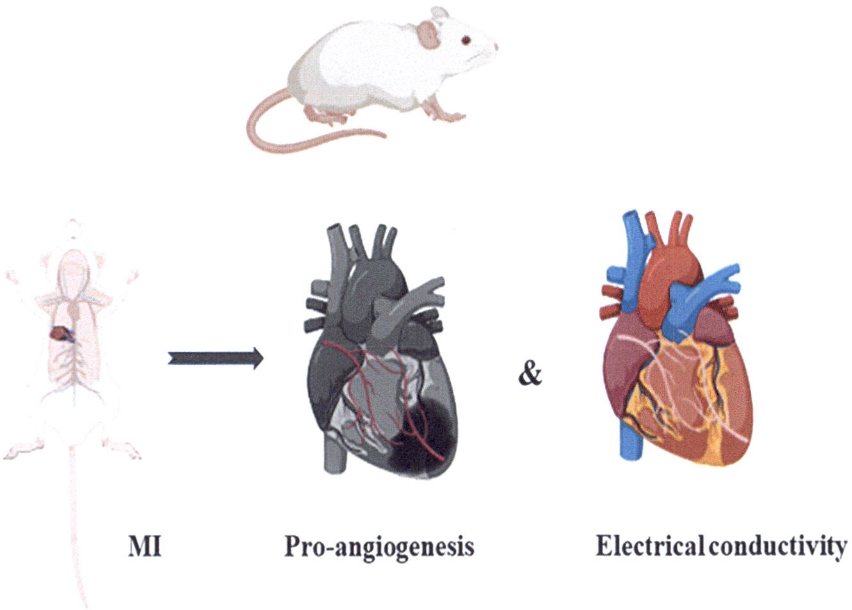

Fig. 5.8 Biomaterials generated from plants have intrinsic properties that offer particular advantages in CTE. Both electrical conductivity and angiogenesis can be induced by them. These are the methods by which biomaterials generated from plants help with heart repair. (Dai et al. 2023)

cascade). When creating a coronary scaffold that allows the heart to contract freely while supporting the cyclic loads caused by ventricular expansion and ejection, elasticity must be carefully taken into account.. To provide structural support without impeding contraction, the scaffold's flexibility must be matched to the properties of the myocardium. It is still early to determine the best elasticity properties to fit this. Furthermore, as the biofunction of encapsulated cells and the printing resolution would be affected, bio-printability and tailorable viscosity must also be taken into consideration when using plant-derived biomaterials as bio-ink (Dai et al. 2023) (Table 5.2).

5.4.1 Using Scaffolds to Transport Cells and Bioactive Components

CTE aims to regenerate and build engineered cardiac tissue suits to replace or repair infarcted myocardium. A combination of framework and cells is frequently employed to accomplish this purpose, and the choice of scaffolding components is crucial. The right material should act as a carrier for proteins, drugs, or other chemicals that support cell growth, in addition to offering a matrix to improve cell preservation and aid in cell growth in order to support and aid in the growth and operation of transplanted cells. Furthermore, given the ischemia sensibility and elasticity of

Table 5.2 A summary of current advancements in the application of biomaterials produced from plants in CTE

Material	Properties	Setting	Significance	References
Alginate	Hydrogel	Anterior MI in rat	Provide mechanical assistance to damaged heart tissue.	Dai et al. (2023).
	Hydrogel (RGD-modified)	hMSCs CMs	Encourage the release of development and the survival of stem cells factors linked to heart regeneration in the presence of H2O2-induced oxidative stress.	Bushkalova et al. (2019)
	Composite hydrogel	Acute MI in rat	Alginate-chitosan composite has the ability to stimulate endogenous cardiomyocyte proliferation and attract cardiac stem cells, as well as inhibit LV remodelling following MI.	Dai et al. (2023).
	Hydrogel	MI in rat	After MI, local sequential administration of PDGF-BB and VEGF-A using alginate hydrogels causes an increase in function and an angiogenic impact that is larger than that of single factors.	Dai et al. (2023).
	Scaffold	In vitro: MSCs In vivo: Skeletal muscle implantation model of rat	The PEC scaffold, consisting of alginate and chitosan, enhanced MSC retention, long-term survival, and FGF-2 secretion.	Nemati et al. (2017)
	Patch	MI in rat	The host myocardium exhibited both structural and electrical integration of the vascularized cardiac patch.	Dai et al. (2023)
	Bioink	Primary feline adult cardiomyocytes (HL1) cardiac muscle cell line	A functioning heart pseudo tissue with a hierarchical architecture that balances porosity for mass transfer and structural support might be created using the ink jet bio-prototyping technique.	Dai et al. (2023)

(continued)

Table 5.2 (continued)

Material	Properties	Setting	Significance	References
	Microcapsule	In vitro: hUVECs In vivo: Immunocompromised mouse implant model	Both in vivo and in vitro settings allow the alginate-gelatin encapsulation to trigger the angiogenic response.	Frost et al. (2019)
Fucoidan	Composite hydrogel	In vitro: hMVECs In vivo: Mice	The interaction between FGF-2 and the chitosan/fucoidan complex-hydrogel significantly extended FGF-2's biological half-life and increased its activity.	Dai et al. (2023)
	Scaffold	hESCs	In order to regulate cardiac differentiation from hESCs, fucoidan scaffolds seem to be a highly promising biomaterial. They might also be paired with mechanical stress to encourage the creation of sarcomeres during the terminal phases of differentiation.	Dai et al. (2023)
Cellulose	Scaffold	H9C2 rat cardiac myoblasts cells	Ten weight percent of naturally organised cellulose was added to PU by electrospinning it into a three-dimensional scaffold. This scaffold has high mechanical properties and is biocompatible, allowing it to facilitate and sustain the ongoing contraction and expansion cycles of the heart's tissue.	Dai et al. (2023).
	Bioink		Show that adjusting the CNC content of bioinks may effectively adjust their mechanical characteristics.	Lin et al. (2019)
Starch	Patch	MI in rats	The gel-point adhesive patch reversed the left ventricular remodelling following a MI and allowed for the myocardium's cyclic deformation.	Gershlak et al. (2017)

(continued)

Table 5.2 (continued)

Material	Properties	Setting	Significance	References
Agarose	Microcapsule	MSCs	Overcome the present restrictions of inadequate cell retention in cardiac cell treatment.	Dai et al. (2023)
	Microcapsule	CSCs	In order to promote the healing of cardiac injuries and avoid detachment-induced cell death, single-cell encapsulation promotes the mechanical retention of CSCs.	Dai et al. (2023).
	Scaffold	hMSCs hPSCs	Demonstrating that it is feasible to create CTE scaffolding sustainably by using decellularized plants.	Cernencu et al. (2019)

cardiomyocytes, materials with proper mechanical stability, versatility, and elasticity, need to be developed (Dobos et al. 2018). These conditions can be met by plant-derived biomaterials, which have various uses in CTE, such as cardiac patches, 3D porous scaffolding, and injectable hydrogels (Dai et al. 2023).

5.4.1.1 3D Porous Scaffolds with Property-Tailoring

Porous scaffolds are implantable scaffolds that are sufficiently flexible, mechanically strong, and biocompatible. Unlike hydrogels, porous scaffolds have interconnecting pores that allow seed cells to move, mature, and communicate with one another. This process forms the extracellular matrix (ECM) across the scaffold. Additionally, this enables the efficient supply of minerals, vitamins, and metabolic waste (Zhao et al. 2018). Materials and preparatory methods such as combination moulding, 3D printing, and electrospinning, can control the porosity and measurement of scaffolds in accordance with the specific requirements of various tissues or organs regarding the mechanical properties, porosity, and rate of degradation of scaffolds (Iravani and Varma 2019). The existing limitations of animal-derived and synthetic biomaterials as scaffolds for tissue engineering are their scarcity, poor biocompatibility, and high cost. On the other hand, plant-derived biomaterials are comparatively cheap and have favourable durability and biocompatibility to promote cell development, in addition to their inherent biodegradability and medicinal qualities. These qualities make them ideal for use as scaffold materials (Choe et al. 2019a, b).

Apart from the traditional scaffold generated from plants, Ger Shlak et al. used decellularized tissues from plants as perfusable pre-vascularized scaffolds by utilizing the comparable differences in vascular structure between plant and animal tissues. The scaffold maintained a working vascular network that was capable of transferring microparticles even after it was decellularized. It's noteworthy to highlight that after 21 days, the scaffold showed constrictile activity as human MSC heart muscle cells and endothelial cells colonized the deepest layer of plant capillaries, and linked to its exterior were human pluripotent stem cells (Cernencu et al. 2019).

By combining chitosan and alginate to create a polyelectrolyte complex, Bushkalova et al. created a macroporous framework that was then freeze-dried and cross-linked. The advantage of this strategy was that it allowed for the improvement of the polymer's rheological characteristics while preserving its original biocompatibility to accommodate soft tissues' mechanical resistance. It is noteworthy that This work's primary objective was not to differentiate MSCs into specific types of cells within the scaffolds, but rather to ensure cell viability and paracrine action to promote vascularization. Consequently, in vitro tests showed their ability to secrete FGF-2, retain MSCs (above 90%), and maintain viability over the long term. The scaffold design of alginate–chitosan combined with its angiogenic capacity, made it an intriguing option for improving the results of MSC therapy in ischemic heart conditions (Nemati et al. 2017).

5.4.1.2 Hydrogels with Cell Instructive Injectables

Hydrogels, as opposed to other synthetic biomaterials, are hydrophilic polymer networks in three dimensions that are capable of absorbing, swell, and release biofluids. These biofluids are more similar to natural extracellular matrix biology and live tissues (Dai et al. 2023). In particular, injectable hydrogels may be administered as a solution and then quickly and precisely gel at infarcted areas. The hydrogel that forms in situ can provide structural support while also increasing cell proliferation, adhesion, and operation and inhibiting dilatation of the heart. Additionally, it can serve as a scaffold to increase cell retention rates or give the wounded area's cardiac cells biological cues, or cell instructive signals (Dai et al. 2023).

Alginate is among the most widely utilized injectable hydrogels.. Alginate is frequently treated with adhesion peptides to enhance cell binding because it lacks cell adhesion. For instance, Yu et al. discovered When compared to the untreated group using non-modified alginate, alginate modification with RGD peptides enhanced the adhesion and proliferation of human umbilical vein endothelial cells (hUVECs).. To promote in situ hydrogel production, calcium chloride and alginate were administered separately and concurrently to rats (no cells were present). After 5 weeks, both groups displayed more arterioles throughout the infarcted area, with the RGD-modified group exhibiting a greater density. This suggests that adhesion peptide alteration can enhance angiogenesis following myocardial infarction (Dai et al. 2023). Furthermore, A significant amount of study has been done on injectable hydrogels for the administration of extracellular vesicles, proteins, and cells. For example, Choe and colleagues synthesised graphene oxide/alginate microgels by the electrospray process and employed the microgels to co-culture MSCs. Using in situ injection, the composite was introduced into the centre of the infarcted region. Both in vitro and in vivo studies demonstrated a high cytoprotective capacity accompanied by higher MSC survival along with survivability in MI-related hostile microenvironments (such as oxidative stress), decreased infarction area, and improved heart function (Liang et al. 2019). Ventricular failure is worsened by the nonconductivity of damaged tissue, which develops along the border of ischemic tissue, is a serious problem that frequently arises after MI. Liang et al. addressed this problem by grafting 2-aminopyridine-5-thiocarboxamide and tetraaniline onto partially oxidized algae, which simultaneously made the system conductive and released H2S. The suspension that resulted from combining the solution with adipose-derived stem cells was applied to three different areas surrounding the affected region. The electrically conductive properties of hydrogel allowed electrical pulses to be transmitted between cardiomyocytes, and the proangiogenic and anti-inflammatory characteristics of H2S improved the function of the area of infarction in the rats (Mazzola and Di Pasquale 2020). The insufficiency of injected hydrogels in providing appropriate physical support for wounded hearts can be attributed to their mechanical characteristics not matching those of the myocardium. This presents another issue. Research on enhancing the mechanical characteristics of injectable hydrogels has been spurred by this issue.

5.4.1.3 Cardiac Tissue-Aiding Patches

In addition to giving the heart mechanical support, epicardial patches, known as cardiac patches, can be utilized to distribute tissues or other restorative materials in situ to provide therapeutic advantages. The cells that reside in the myocardial patch can create an internal lamellar morphology by placing it in a specific spatial order and allow for precision targeting of the lesion site. Additionally, this configuration increases cell survival and persistence (Mir and Nakamura 2017). Proper adhesion, favourable biocompatibility, and an extracellular matrix environment that encourages cell proliferation and differentiation, or the release of paracrine factors, are necessary for the perfect cardiac patch (Dai et al. 2023). To produce cardiac cells that are both functional and well-designed in order to repair or restore damaged myocardium, Dvir et al. implanted cardiac flecks composed of tissue from the heart along with alginate within the infarcted area of the rat myocardium for 7 days following infarction. Their pre-vascularized cardiac patches were shown to fuse with the host myocardium 4 weeks following implantation. Its main objective was to prevent ventricular dilatation and dysfunction while fortifying the myocardium's structure (Dai et al. 2023).

5.4.2 Utilising Bio-inks for 3D Bio Fabrication

3D bioprinting is a layer-upon-layer additive manufacturing process that helps recreate biological heterogeneity and complex hierarchy or denization in three dimensions by accurately depositing cells that are alive and substances in accordance with specified spatial patterns (Birla and Williams 2020). Even with recent advancements in the three-dimensional printing of functional cardiac tissue, it remains challenging to create the entire heart in the field of cardiac tissue engineering while maintaining the integrity of its structure, variability in cells, and cardiac rhythm (Indurkar et al. 2021). One of the most important things is choosing the right bio-inks, since a bio functional 3D manufacturing process needs one with the right rheological, physic, mechanical, and bio functional characteristics (Cui et al. 2020). With the benefits of gelation capacity, rheological characteristics, and biocompatibility, plant-derived biomaterials are a popular choice for bio inks. They show significant promise for using the biological role of plant microstructures as growth promoters and their inherent toughness. Furthermore, there is a lot of interest in 3D bioprinting research since plant-derived bio-inks are readily available, affordable, and renewable (Gu et al. 2020; Indurkar et al. 2021).

One of the main obstacles in CTE is the development of vascularized tissues. The survival of new tissues in vivo depends on a sufficient blood supply, and the therapeutic use of artificial tissue constructions is restricted if this problem is not resolved. It is now feasible to bio-print perfusable vascular networks with regulated dimensions because of developments in 3D bio fabrication (Nishimura et al. 2018). For instance, human lumbar vein smooth muscle cells were encapsulated in calcium cross-linked alginate and printed as vascular conduits using a coaxial deposition method. The histological findings demonstrated that the conduit-encapsulated cells were proliferating well, and acceptable extracellular matrix deposition was observed on both the luminal and peripheral surfaces (Dai et al. 2023). Furthermore, to promote Nishimura et al. used a microfluidic approach in conjunction with the biological production of alginate microfibers connected to theta-glass capillaries to achieve vascularized tissue engineering. Vascular linked branch/chain structures were formed as a result (Thomas et al. 2021; Gu et al. 2020).

5.4.3 Delivery Vehicles

The improvement of patient outcomes for various medical illnesses depends on the creation of efficient drug delivery methods. Nonetheless, the creation and use of novel materials will greatly influence the development of medicine in the future. This is a further development that affects diagnostic tools, implantable devices, regenerative medicine, and medication delivery systems. Novel materials are driving a revolution in medical technology characterised by a multidisciplinary approach that combines nanotechnology, biotechnology, and materials science. In order to create biocompatible and bioresorbable scaffolds for tissue engineering, drug delivery systems, and medical devices, polymers have emerged as key actors. Biobased and biodegradable polymers provide a wealth of engineering opportunities, including regulated rates of breakdown, customised mechanical attributes, and surface qualities that complement human physiology (Pires et al. 2023). The evolution of medication delivery systems is one of the most important uses of biomaterials (Jovic et al. 2019). Pharmaceutical substances can be contained and discharged with the help of these materials in a targeted and regulated way. This strategy has several benefits, including increased medication efficacy, fewer adverse effects, and extended-release for long-term illnesses. To ensure a prolonged release of medication at the site of action, biodegradable polymer nanoparticles are capable of being loaded with medication and injected into the body. Polymers, lipids, and other substances commonly used to encapsulate and regulate the release of medications can be used as biomaterials for drug-delivery systems. With their amazing capabilities at the nanoscale, nanomaterials are essential for the advancement of medical technology. Their uses are numerous and include the development of novel diagnostic

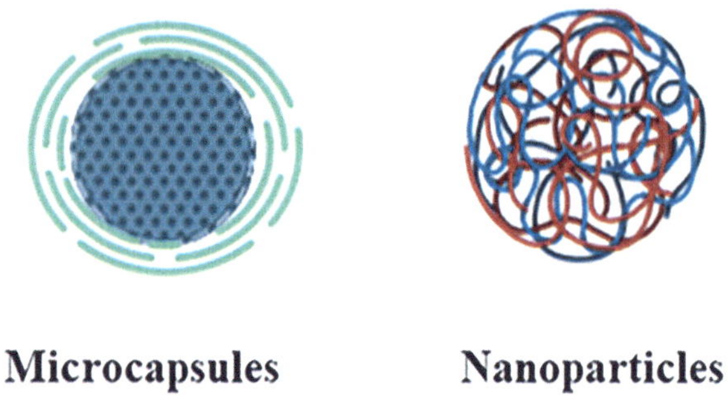

Fig. 5.9 Structure of Delivery Vehicles. (Dai et al. 2023)

instruments, imaging agents, and targeted medication administration (Pires et al. 2023) (Fig. 5.9).

5.4.3.1 Microcapsules

The variable mechanical characteristics, biodegradability, and biocompatibility of microcapsules provide further benefits. Hydrogel-based microcapsules' high hydrophilicity and porous shape, which support cell growth, enable them to effectively transport nutrients, oxygen, and metabolic waste products. Additionally, the surface of the microcapsules has been altered to improve their immune-modulating capacity, retention capacity, and targeting ability (Dai et al. 2023). Blocki et al. packaged MSCs with microcapsules for later intramyocardial transplantation in rats in order to overcome the shortcomings of current cardiac cell-based therapies, such as poor cell retention. Dextran sulphate, a glycosaminoglycan mimic, was added to the microcapsules along with small amounts of collagen, agarose, and fibrin in order to preserve cell viability. MRI showed that the infarcted area retained cell-loaded microcapsules for a notably longer period of time than The group suspended in cells. Likewise, Mayfield et al. proved that single-cell differentiation promotional materials (agarose hydrogel microcapsules nourished along with fibronectin and fibrinogen) preserved Cardiovascular Stem Cell Viability according to hypoxic stress conditions in contrast to suspended cardiac stem cells and prevented detachment-induced cell death to aid in the restoration of a damaged heart valve (myocardium) (Dai et al. 2023).

5.4.3.2 Nanoparticles

Biopolymers generated from plants, such starch and carrageenan, have been thoroughly studied as nanoparticles for applications involving controlled medication delivery because of their improved biological compatibility, biodegradability, and low immunogenicity. Thomas and colleagues, for instance, developed starch-modified alginate nanoparticles for use in drug delivery applications. Theophylline and bovine serum albumin, the model pharmaceuticals, were completely dispersed in simulated intestinal fluid (pH 7.4) and well shielded from gastric fluid (pH 1.2). In vitro tests further confirmed that L929 fibroblast cells absorbed the nanoparticles (Benyerbah et al. 2019). In a similar vein, Benyerbah et al. investigated the viability of using polyelectrolytic and ampholytic starch to provide delayed medication release. It was shown. That ampholytic trimethylamine-carboxymethyl starch along with co-processed trimethylamine-carboxymethyl starch (formed by spray drying) starch polymeric excipients produced even with extremely high drug loading, the optimal drug release characteristics. This may be explained by the presence of amine and carboxylic groups at various locations throughout the chains, which attract opposing charges to improve polymer stability and limit tablet hydration. In addition, ionic interactions support structural stability (Scarpella 2017; Thomas et al. 2021).

5.4.4 Bioactive Molecules

Owing to their distinct biological properties, a number of biomaterials generated from plants can be utilized as bioactive supplements or as drugs to treat cardiovascular diseases. enhance the therapeutic advantages, as well as acting as delivery systems, scaffolds, and bio-inks. For example, it has been shown that fucoidan possesses cardioprotective qualities. In a rat model of MI caused by isoproterenol, it was discovered to decrease the infarcted area and cellular oxidative stress and improve the antioxidant defence system (Qiu et al. 2022).. Likewise, fucoidan has been shown by Krishna Murthy et al. to have cardioprotective benefits through the preservation of minerals and membrane-bound ATPases, which in turn lessened the size of the infarct. Modified citrus pectin is a tiny bioactive molecule that is produced by specifically modifying citrus pectin with a low molecular weight of fucoidan. This is an additional biomaterial derived from plants. Thus, further research is needed before this may be fully implemented in clinical settings (Dai et al. 2023). A distinct biomaterial generated from plants, modified citrus pectin has a molecular weight of less than 15 kDa. It is a very small bioactive compound. Because of its antifibrotic, anti-inflammatory, and anticancer bioactivities, it has drawn increasing amounts of interest. For example, Xu et al. demonstrated that by blocking the synthesis of galectin-3 and the TLR4/MyD88/NF-κB signalling pathway, modified citrus pectin decreased cardiac inflammation and corrected isoproterenol-induced heart failure in a rat model (Xu et al. 2020).

Danshen, or Radix salvia miltiorrhiza, is a popular herb in China that has been demonstrated to have medicinal benefits for coronary heart disease, namely for MI, in long-term clinical applications. In order to optimize its mobility, chemical stability, and accessibility and hence maximize its efficiency, danshen is frequently produced as nanoparticles. However, there are two problems with storing herbal nanoparticles: the first is that they easily aggregate because of the substantial amount of surface energy and unsaturated bonding within the cellular structure; the second is that the active ingredients quickly oxidize when exposed to heat and light. Su et al. created Danshen nanoparticles using gelatin and carboxymethyl cellulose as roofing materials, devised a spray-drying preparation method for Danshen microcapsules, and detailed the physicochemical properties of Danshen microcapsules to identify the optimal manufacturing parameters in order to address these problems (Dai et al. 2023).

5.5 Future Perspectives and Challenges

As an electromechanical organ, the heart is contractile. The creation of a strong, conductive biomaterial is essential to the supply of a stable transient framework, or scaffold, for the treatment of MSC transplantation, to withstand the hostile local microenvironment that follows MI. CTE scaffolds are often produced using synthetic or animal-derived biomaterials, which have a high cost of manufacture and a detrimental effect on the environment (Beckwith et al. 2021). Collagen and chitosan, two biomaterials originating from animals, are linked to the extensive use of fodder for feeding and the management of the ensuing enormous waste. Furthermore, there are several ethical restrictions on the collection of animal organs or tissue for commercial manufacturing. Conversely, when non-renewable resources are chemically processed to create synthetically produced biomaterials, multiple dangerous by-products are typically created (Dai et al. 2023).

To minimise the abuse of animals and chemical reagents, scientists have thus hastened the development of environmentally friendly biomaterials with the goal of substituting conventional biomaterials with renewable resources. The good news is that plant-derived biomaterials usually possess the required tissue engineering qualities, are sufficiently flexible to be fabricated into various forms and dimensions, are sustainable and renewable for large-scale manufacturing, and are significantly reduced in terms of host immune response risk. Comparatively speaking, synthetic hydrogels also considerably lower the production of hazardous by-products during the manufacturing phase. Biomaterials originating from plants are therefore perfect substitutes for synthetic and animal-based materials due to these qualities.

More significantly, combining tissue from plants with the newest green technology can help overcome some of the major challenges in the scaffold production process. For instance, one of the most important fields of study within the discipline of CTE is advancements in vascularization techniques. With plants' vascular structures resembling human tissues, perfusion-based decellularization methods provide a possible substitute for pre-vascularized scaffolds in order to prevent cancer that might arise from the stimulation of angiogenesis (Cernencu et al. 2019).

Another work by Norotte et al. aimed to recreate perfusable small-diameter vessels using agarose rod-like structures precisely as the lumen template. This study developed a scaffold-free tissue production approach. Although conventional agriculture is significantly less harmful to the environment than animal husbandry, there is still a need to increase the output of biomaterials generated from plants. It is encouraging to know that plant materials with adjustable mechanical qualities may be generated in the lab, according to scientific research (Dai et al. 2023). In one experiment, MIT researchers gave regular plant cells characteristics similar to those of stem cells. Zinnia elegans is a blooming plant whose leaves were used to harvest cells, which were then preserved in liquid media. Following treatment with a gel-based medium that was high in nutrients along with hormones, the plant cells multiplied, as evidenced by the observations made. The outcomes demonstrated that the quantity of hormones in the form of gel medium affected the characteristics of the proliferating cells (e.g., greater amounts of hormones created stiffer plant material).

Significantly, 3D printing techniques were employed to create personalized formations from the gel-cultivated cells (Fig. 5.10a) (Mohammadinejad et al. 2019). Although the field is still in its infancy, lab-grown plant material indicates that innovations at the intersection of science and technology may be able to address environmental problems by lowering deforestation and offering an energy-saving, low-waste substitute. Moreover, improvements The poor generation of physiologically active compounds in various plant suspension cell cultures has been addressed in the field of gene transfer. For example, foreign genes can be incorporated, synthesized, and passed on into plant tissues using recombinant DNA technology, resulting in the creation of unique plants (genetically modified versions of plants). Plant genetic engineering techniques are employed to modify critical genes implicated in their molecular biosynthesis.

This is done to change the plant biomaterials' component ratio or to enhance the biological function and secretion of active chemicals, which eventually boosts the effectiveness of the compound intended as a secondary metabolic by-product (Fig. 5.10b). Furthermore, novel studies are being conducted that employ plant scaffolds for therapeutic gene delivery (Choe et al. 2019a, b).

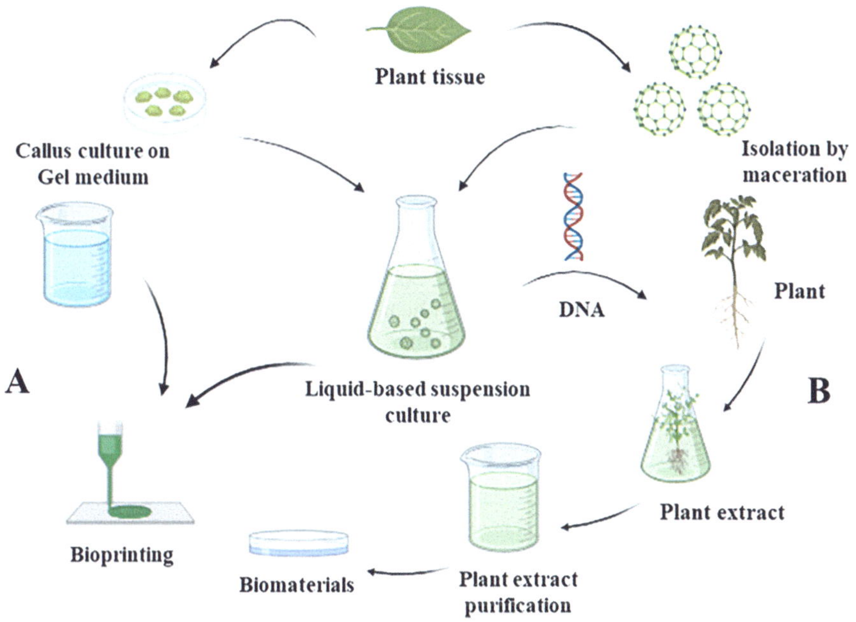

Fig. 5.10 The advancement of non-destructive plant samples through (**a**) biological printing and (**b**) large-scale genome editing-based biomaterials manufacture in a selective tissue-like manner. A small amount of plant material can be used in the laboratory to create ex vivo tissue-like structures. The method used to separate cells from plant samples differs depending on the kind of plant. Certain species allow for the direct extraction of cells by the process of maceration of leaves, but others necessitate the callus culture phase, as seen above. (Dai et al. 2023)

To impede the post-transcriptional expression of important genes, for instance, small interfering RNAs (siRNAs) can specifically knockdown complementary messenger RNA (mRNA) (gene silence). However, RNases can degrade siRNAs in vivo. Krebs et al. responded by encasing therapeutic siRNAs in plant-derived hydrogels and using the hydrogels' varying release rates to achieve locally controllable gene silencing. To be more precise, siRNA was continually released from collagen for approximately 2 weeks and from alginate for approximately 1 week (Trucillo 2024). The overall quantity of siRNA released by the various alginate cross-linking processes varied, with photo-crosslinking and calcium crosslinking releasing the most. This siRNA delivery technology produced from plants has the potential to be highly helpful in treating CTE since it can decrease the production of particular genes at particular bodily locations. Furthermore, as the majority of plant-derived biomaterials lack some crucial bio-properties (such as poor cell attachment, unwanted mechanical behaviour, low stability, and so on), the biofunctionalization of these materials is still an area that needs to be explored. Techniques including blending multiple biomaterials, optimizing the cross-linking process, and modifying the topography and/or components are frequently used. In addition, the transfer of stem cells is a common use for plant-derived biomaterials (Oduk et al. 2018), which may result in contamination by organisms as well as host proinflammatory and immunological responses. Acellular approaches (such as replacing stem cells with extracellular vesicles or using recombinant proteins (Cong et al. 2022, Jovic et al. 2019), may offer a way to do so while removing any danger.

For example, because they having the ability to be applied in the delivery of drug vesicles and in biomedicine, exosome-like nanoparticles (ELNs) extracted from plants represent an intriguing alternative for tissue reconstruction. Compared to extracellular vesicles, ELNs are more widely available and need a simpler separation procedure. Their pharmacological effects are typically comparable in comparison to those of their origin, and they can alternatively be conceptualised as lipid-like drug vesicles (Trucillo 2024; Pires et al. 2023). To sum up, several plant-derived biomaterials have been researched for application in healing damaged myocardium through several applications because of their inherent benefits. Although certain challenges persist, innovative approaches have been devised to address possible problems. Plant-derived biomaterials will become more significant in the field of CTE as related research improves.

References

Andreadou I, Cabrera-Fuentes HA, Devaux Y, Frangogiannis NG, Frantz S, Guzik T, Liehn EA, Gomes CP, Schulz R, Hausenloy DJ (2019) Immune cells as targets for cardioprotection: new players and novel therapeutic opportunities. Cardiovasc Res 115(7):1117–1130

Beckwith AL, Borenstein JT, Velásquez-García LF (2021) Tunable plant-based materials via in vitro cell culture using a Zinnia elegans model. J Clean Prod 15:288

Benyerbah N, Ispas-Szabo P, Sakeer K, Chapdelaine D, Mateescu MA (2019) Ampholytic and polyelectrolytic starch as matrices for controlled drug delivery. Pharmaceutics 11(6):253

Birla RK, Williams SK (2020) 3D bioprinting and its potential impact on cardiac failure treatment: an industry perspective. APL Bioeng 4(1):010903. https://doi.org/10.1063/1.5128371. PMID: 32095736; PMCID: PMC7028435

Bushkalova R, Farno M, Tenailleau C, Duployer B, Cussac D, Parini A, Sallerin B, Fullana SG (2019) Alginate-chitosan PEC scaffolds: a useful tool for soft tissues cell therapy. Int J Pharm 571:118692

Cattelan G, Guerrero Gerbolés A, Foresti R, Pramstaller PP, Rossini A, Miragoli M, Caffarra MC (2020) Alginate formulations: current developments in the race for hydrogel-based cardiac regeneration. Front Bioeng Biotechnol 8:414

Cernencu AI, Lungu A, Stancu IC, Serafim A, Heggset E, Syverud K, Iovu H (2019) Bioinspired 3D printable pectin-nanocellulose ink formulations. Carbohydr Polym 220:12–21

Chang PM, Li KL, Lin YC (2019) Fucoidan–fucoxanthin ameliorated cardiac function via IrS1/Grb2/SOS1, GSK3β/CREB pathways and metabolic pathways in senescent mice. Mar Drugs 17:69

Chen S, Song Y, Xu F (2018) Highly transparent and hazy cellulose nanopaper simultaneously with a self-cleaning superhydrophobic surface. ACS Sustain Chem Eng 6(4):5173–5181

Choe G, Kim SW, Park J, Park J, Kim S, Kim YS, Ahn Y, Jung DW, Williams DR, Lee JY (2019a) Antioxidant activity reinforced reduced graphene oxide/alginate microgels: mesenchymal stem cell encapsulation and regeneration of infarcted hearts. Biomaterials 225:119513

Choe G, Park J, Jo H, Kim YS, Ahn Y, Lee JY (2019b) Studies on the effects of microencapsulated human mesenchymal stem cells in RGD-modified alginate on cardiomyocytes under oxidative stress conditions using in vitro biomimetic co-culture system. Int J Biol Macromol 123:512–520

Cong M, Tan S, Li S, Gao L, Huang L, Zhang HG, Qiao H (2022) Technology insight: plant-derived vesicles—how far from the clinical biotherapeutics and therapeutic drug carriers? Adv Drug Deliv Rev 182:114108

Cui X, Li J, Hartanto Y, Durham M, Tang J, Zhang H, Hooper G, Lim K, Woodfield T (2020) Advances in extrusion 3D bioprinting: a focus on multicomponent hydrogel-based bioinks. Adv Healthc Mater 9(15):e1901648

Curvello R, Raghuwanshi VS, Garnier G (2019) Engineering nanocellulose hydrogels for biomedical applications. Adv Colloid Interf Sci 267:47–61

Dai Y, Qiao K, Li D, Isingizwe P, Liu H, Liu Y, Lim K, Woodfield T, Liu G, Hu J, Yuan J et al (2023) Plant-derived biomaterials and their potential in cardiac tissue repair. Adv Healthc Mater 12:20

David S, Levi CS, Fahoum L, Ungar Y, Meyron-Holtz EG, Shpigelman A, Lesmes U (2018) Revisiting the carrageenan controversy: do we really understand the digestive fate and safety of carrageenan in our foods? Food Funct 9(3):1344–1352

Dobos A, Grandhi TS, Godeshala S, Meldrum DR, Rege K (2018) Parallel fabrication of macroporous scaffolds. Biotechnol Bioeng 115(7):1729–1742

Fares MM, Sani ES, Lara RP, Oliveira RB, Khademhosseini A, Annabi N (2018) Interpenetrating network gelatin methacryloyl (GelMA) and pectin-g-PCL hydrogels with tunable properties for tissue engineering. Biomater Sci 6(11):2938–2950

Feingold B (2018) The challenges of donor-derived risk, donor shortage and waitlist mortality in children: time for a new measuring stick? J Heart Lung Transplant 37(3):317–318

Frost BA, Sutliff BP, Thayer P, Bortner MJ, Foster EJ (2019) Gradient poly (ethylene glycol) diacrylate and cellulose nanocrystals tissue engineering composite scaffolds via extrusion bioprinting. Front Bioeng Biotechnol 7:280

Gawkowska D, Cybulska J, Zdunek A (2018) Structure-related gelling of pectins and linking with other natural compounds: a review. Polymers 10(7):762

Gershlak JR, Hernandez S, Fontana G, Perreault LR, Hansen KJ, Larson SA, Binder BY, Dolivo DM, Yang T, Dominko T, Rolle MW (2017) Crossing kingdoms: using decellularized plants as perfusable tissue engineering scaffolds. Biomaterials 125:13–22

Geyik G, Işıklan N (2021) Design and fabrication of hybrid triple-responsive κ-carrageenan-based nanospheres for controlled drug delivery. Int J Biol Macromol 192:701–715

Ghebouli R, Loyau S, Maire M, Saboural P, Collet JP, Jandrot-Perrus M, Letourneur D, Chaubet F, Michel JB (2018) Amino-fucoidan as a vector for rtPA-induced fibrinolysis in experimental thrombotic events. Thromb Haemost 118(01):42–53

Gu Z, Fu J, Lin H, He Y (2020) Development of 3D bioprinting: from printing methods to biomedical applications. Asian J Pharm Sci 15(5):529–557

Guglin M, Zucker MJ, Borlaug BA, Breen E, Cleveland J, Johnson MR, Panjrath GS, Patel JK, Starling RC, Bozkurt B (2020) Evaluation for heart transplantation and LVAD implantation: JACC council perspectives. J Am Coll Cardiol 75(12):1471–1487

Indurkar A, Pandit A, Jain R, Dandekar P (2021) Plant-based biomaterials in tissue engineering. Bioprinting 21:e00127

Iravani S, Varma RS (2019) Plants and plant-based polymers as scaffolds for tissue engineering. Green Chem 21(18):4839–4867

Jovic TH, Kungwengwe G, Mills AC, Whitaker IS (2019) Plant-derived biomaterials: a review of 3D bioprinting and biomedical applications. Front Mech Eng 5:19

Kiiskinen J, Merivaara A, Hakkarainen T, Kääriäinen M, Miettinen S, Yliperttula M, Koivuniemi R (2019) Nanofibrillar cellulose wound dressing supports the growth and characteristics of human mesenchymal stem/stromal cells without cell adhesion coatings. Stem Cell Res Ther 10:1–3

Kumar PS, Ramya C, Jayakumar R, Lakshmanan VK (2013) Drug delivery and tissue engineering applications of biocompatible pectin–chitin/nano CaCO3 composite scaffolds. Colloids Surf B Biointerfaces 106:109

Lapomarda A, De Acutis A, Chiesa I, Fortunato GM, Montemurro F, De Maria C, Mattioli Belmonte M, Gottardi R, Vozzi G (2019) Pectin-GPTMS-based biomaterial: toward a sustainable bioprinting of 3D scaffolds for tissue engineering application. Biomacromolecules 21(2):319–327

Lemos PV, Marcelino HR, Cardoso LG, de Souza CO, Druzian JI (2021) Starch chemical modifications applied to drug delivery systems: from fundamentals to FDA-approved raw materials. Int J Biol Macromol 184:218–234

Leppiniemi J, Lahtinen P, Paajanen A, Mahlberg R, Metsa-Kortelainen S, Pinomaa T, Pajari H, Vikholm-Lundin I, Pursula P, Hytonen VP. (2017) 3D-printable bioactivated nanocellulose–alginate hydrogels. ACS applied materials & interfaces

Li J, Cha R, Mou K, Zhao X, Long K, Luo H, Zhou F, Jiang X (2019) Nanocellulose-based antibacterial materials. Adv Healthc Mater 7(20):e1800334

Li H, Bao M, Nie Y (2021) Extracellular matrix–based biomaterials for cardiac regeneration and repair. Heart Fail Rev 26:1231–1248

Liang W, Chen J, Li L, Li M, Wei X, Tan B, Shang Y, Fan G, Wang W, Liu W (2019) Conductive hydrogen sulfide-releasing hydrogel encapsulating ADSCs for myocardial infarction treatment. ACS Appl Mater Interfaces 11(16):14619–14629

Lin X, Liu Y, Bai A, Cai H, Bai Y, Jiang W, Yang H, Wang X, Yang L, Sun N, Gao H (2019) A viscoelastic adhesive epicardial patch for treating myocardial infarction. Nat Biomed Eng 3(8):632–643

Ling Z, Edwards JV, Guo Z, Prevost NT, Nam S, Wu Q, French AD, Xu F (2019) Structural variations of cotton cellulose nanocrystals from deep eutectic solvent treatment: micro and nano scale. Cellulose 26:861–876

Liu L, Chen HG, Li Y, Li H, Li J, Wang Y, Yao S, Qin C, Tong S, Yuan X, Luo X (2021) Temporal profiles of antibody responses, cytokines, and survival of COVID-19 patients: a retrospective Cohort in Wuhan, China. Engineering (Beijing) 7(7):958–965

Luo H, Cha R, Li J, Hao W, Zhang Y, Zhou F (2019) Advances in tissue engineering of nanocellulose-based scaffolds: a review. Carbohydr Polym, vol 224, p 115144

Markstedt K, Escalante A, Toriz G, Gatenholm P (2017) Biomimetic inks based on cellulose nanofibrils and cross-linkable xylans for 3D printing. ACS Appl Mater Interfaces 9(46):86

Mazzola M, Di Pasquale E (2020) Toward cardiac regeneration: combination of pluripotent stem cell-based therapies and bioengineering strategies. Front Bioeng Biotechnol 8:455

McKirnan MD, Ichikawa Y, Zhang Z, Zemljic-Harpf AE, Fan S, Barupal DK, Patel HH, Hammond HK, Roth DM (2019) Metabolomic analysis of serum and myocardium in compensated heart failure after myocardial infarction. Life Sci 221:212–223

Mir TA, Nakamura M (2017) Three-dimensional bioprinting: toward the era of manufacturing human organs as spare parts for healthcare and medicine. Tissue Eng Part B Rev 23(3):245–256

Mohammadi N, Ehsani MR, Bakhoda H (2018) Development of caffeine-encapsulated alginate-based matrix combined with different natural biopolymers, and evaluation of release in simulated mouth conditions. Flavour Fragr J 33(5):357–366

Mohammadinejad R, Shavandi A, Raie DS, Sangeetha J, Soleimani M, Hajibehzad SS, Thangadurai D, Hospet R, Popoola JO, Arzani A, Gómez-Lim MA (2019) Plant molecular farming: production of metallic nanoparticles and therapeutic proteins using green factories. Green Chem 21(8):1845–1865

Nemati S, Rezabakhsh A, Khoshfetrat AB, Nourazarian A, Biray Avci Ç, Goker Bagca B, Alizadeh Sardroud H, Khaksar M, Ahmadi M, Delkhosh A, Sokullu E (2017) Alginate-gelatin encapsulation of human endothelial cells promoted angiogenesis in in vivo and in vitro milieu. Biotechnol Bioeng 114(12):2920–2930

Nishimura K, Morimoto Y, Mori N, Takeuchi S (2018) Formation of branched and chained alginate microfibers using theta-glass capillaries. Micromachines 9(6):303

Niyonizigiye I, Ngabire D, Patil MP, Singh AA, Kim GD (2019) In vitro induction of endoplasmic reticulum stress in human cervical adenocarcinoma HeLa cells by fucoidan. Int J Biol Macromol 137:844–852

Noreen A, Akram J, Rasul I, Mansha A, Yaqoob N, Iqbal R, Tabasum S, Zuber M, Zia KM (2017) Pectins functionalized biomaterials; a new viable approach for biomedical applications: a review. Int J Biol Macromol 101:254–272

Ocampo JI, de Paula MM, Bassous NJ, Lobo AO, Orozco CP, Webster TJ (2019) Osteoblast responses to injectable bone substitutes of kappa-carrageenan and nano hydroxyapatite. Acta Biomater 83:425–434

Oduk Y, Zhu W, Kannappan R, Zhao M, Borovjagin AV, Oparil S, Zhang J (2018) VEGF nanoparticles repair the heart after myocardial infarction. Am J Phys Heart Circ Phys 314(2):278–284

Okunlola A, Adewusi SA (2019) Development of theophylline microbeads using pregelatinized breadfruit starch (artocarpus altilis) as a novel co-polymer for controlled release. Adv Pharm Bull 9(1):93

Oliveira C, Granja S, Neves NM, Reis RL, Baltazar F, Silva TH, Martins A (2019) Fucoidan from Fucus vesiculosus inhibits new blood vessel formation and breast tumor growth in vivo. Carbohydr Polym 223:1–10

Pacheco-Quito EM, Ruiz-Caro R, Veiga MD (2020) Carrageenan: drug delivery systems and other biomedical applications. Mar Drugs 18(11):583

Pereira RF, Barrias CC, Bártolo PJ, Granja PL (2018) Cell-instructive pectin hydrogels crosslinked via thiol-norbornene photo-click chemistry for skin tissue engineering. Acta Biomater 66:282–293

Pires PC, Mascarenhas-Melo F, Pedrosa K, Lopes D, Lopes J, Macário-Soares A, Peixoto D, Giram PS, Veiga F, Paiva-Santos AC (2023) Polymer-based biomaterials for pharmaceutical and biomedical applications: a focus on topical drug administration. Eur Polym J 187:111868

Pomeroy JE, Helfer A, Bursac N (2020) Biomaterializing the promise of cardiac tissue engineering. Biotechnol Adv 42:107353

Pourjavadi A, Doroudian M, Ahadpour A, Azari S (2019) Injectable chitosan/κ-carrageenan hydrogel designed with au nanoparticles: a conductive scaffold for tissue engineering demands. Int J Biol Macromol 126:310–317

Qiu F, Yuan Y, Luo W, Gong YS, Zhang ZM, Liu ZM, Gao L (2022) Asiatic acid alleviates ischemic myocardial injury in mice by modulating mitophagy-and glycophagy-based energy metabolism. Acta Pharmacol Sin 43(6):1395–1407

Rastogi P, Kandasubramanian B (2019) Review of alginate-based hydrogel bioprinting for application in tissue engineering. Biofabrication 11(4):042001

Scarpella E (2017) The logic of plant vascular patterning. Polarity, continuity and plasticity in the formation of the veins and of their networks. Curr Opin Genet Dev 45:34–43

Simeoni RB, Mogharbel BF, Francisco JC, Miyague NI, Irioda AC, Souza CM, Souza D, Stricker PE, da Rosa NN, Souza CF, Franco CR (2021) Beneficial roles of cellulose patch-mediated cell therapy in myocardial infarction: a preclinical study. Cells 10(2):424

Smagul S, Kim Y, Smagulova A, Raziyeva K, Nurkesh A, Saparov A (2020) Biomaterials loaded with growth factors/cytokines and stem cells for cardiac tissue regeneration. Int J Mol Sci 21(17):5952

Souza SF, Mariano M, Reis D, Lombello CB, Ferreira M, Sain M (2018) Cell interactions and cytotoxic studies of cellulose nanofibers from Curauá natural fibers. Carbohydr Polym 201:87–95

Sultan S, Mathew AP (2019) 3D printed porous cellulose nanocomposite hydrogel scaffolds. J Vis Exp 24:(146). https://doi.org/10.3791/59401. PMID: 31081812

Tang J, Cui X, Zhang Z, Xu Y, Guo J, Soliman BG, Lu Y, Qin Z, Wang Q, Zhang H, Lim KS (2022) Injection-free delivery of MSC-derived extracellular vesicles for myocardial infarction therapeutics. Adv Healthc Mater 11(5):e2100312

Thomas D, Mathew N, Nath MS (2021) Starch modified alginate nanoparticles for drug delivery application. Int J Biol Macromol 173:277–284

Torres FG, Arroyo JJ, Troncoso OP (2019) Bacterial cellulose nanocomposites: an all-nano type of material. Mater Sci Eng 98:1277–1293

Trucillo P (2024) Biomaterials for drug delivery and human applications. Materials 17(2):456

Wan MC, Qin W, Lei C, Li QH, Meng M, Fang M, Song W, Chen JH, Tay F, Niu LN (2021) Biomaterials from the sea: future building blocks for biomedical applications. Biomed Mater 6(12):4255–4285

Wang Z, Lee SJ, Cheng HJ, Yoo JJ, Atala A (2018) 3D bioprinted functional and contractile cardiac tissue constructs. Acta Biomater 70:48–56

Wu D, Samanta A, Srivastava RK, Hakkarainen M (2017) Starch-derived nanographene oxide paves the way for electrospinnable and bioactive starch scaffolds for bone tissue engineering. Biomacromolecules 18(5):1582–1591

Xu GR, Zhang C, Yang HX, Sun JH, Zhang Y, Yao TT, Li Y, Ruan L, An R, Li AY (2020) Modified citrus pectin ameliorates myocardial fibrosis and inflammation via suppressing galectin-3 and TLR4/MyD88/NF-κB signaling pathway. Biomed Pharmacother 126:110071

Yu HC, Zhang H, Ren K, Ying Z, Zhu F, Qian J, Ji J, Wu ZL, Zheng Q (2018) Ultrathin κ-carrageenan/chitosan hydrogel films with high toughness and antiadhesion property. ACS Appl Mater Interfaces 10(10):9002–9009

Zarrintaj P, Manouchehri S, Ahmadi Z, Saeb MR, Urbanska AM, Kaplan DL, Mozafari M (2018) Agarose-based biomaterials for tissue engineering. Carbohydr Polym 187:66–84

Zhao P, Gu H, Mi H, Rao C, Fu J, Turng LS (2018) Fabrication of scaffolds in tissue engineering: a review. Front Mech Eng 13:107–119

Zhuang P, Greenberg Z, He M (2021) Biologically enhanced starch bio-ink for promoting 3D cell growth. Adv Mater Technol 6(12):2100551

Zou Q, Tian X, Luo S, Yuan D, Xu S, Yang L, Ma M, Ye C (2021) Agarose composite hydrogel and PVA sacrificial materials for bioprinting large-scale, personalized face-like with nutrient networks. Carbohydr Polym 269:118222

Ms. Sudipta Choudhury is an M.Sc. student at the School of Life Sciences, JSS Academy of Higher Education and Research, Mysuru, where she is pursuing her postgraduate degree in Molecular Biology. In 2022, she received her UG degree from Jyoti Nivas College Autonomous in Bangalore with a triple major in Biotechnology, Biochemistry, and Genetics.

The author was able to complete one project during her undergraduate studies that dealt with the pathogenesis of obesity research by inducing obesity in Danio rerio with the administration of a high protein diet and were able to give an oral presentation at the 18th International Conference organised by the Department of Animal Science & Indian Association of Applied Microbiologists (IAAM) on "Converging Microbiological Innovation for Application in Animal, Plant, Environment, and Health Care (ICCMIA – 2023)" on 27th and 28th January, 2023.

During her undergraduate graduation, the author did an internship at EDUFABRICA in association with PULSE, AIIMS DELHI on "Food Microbiology" in 2021. She had taken some other short-term courses and been certified in Medical Laboratory Technology, Clinical Research, Nutrition and Dietetics. As part of her M.Sc. Internship, she was trained briefly at Mediomix Diagnosis and Bioresearch Pvt. Ltd. and gained experience in the transcriptomic approach using various bioinformatics tools.

Mr. D. Mutthuraj currently a PhD research scholar under the guidance of Dr. Kanthesh BM, Associate Professor, Division of Molecular Biology, JSS Academy of Higher Education & Research, Mysuru, India. He has been actively involved in research since post graduation with specialization in Biochemistry & Molecular Biology. Identifying various biomarkers are studied involved in autoimmune disorders and their pathways.

He received his undergraduate degree from the prestigious University of Mysore with First class with distinction, later he joined JSS AHER where he worked on auto-immune disorders like Rheumatoid arthritis titled "inhibition of pro-inflammatory molecules by ginger (Zingiber officinale Roscoe) and its anti-inflammatory effects on arthritis patients" – where he checked for the inhibitory molecules and their effects on arthritis patients and also published 5 review paper and one original paper in peer-reviewed journals and also worked as guest lecturer in various colleges in Mandya before joining as PhD Research Scholar at JSS AHER. The author graduated from the JSS AHER (Deemed to be university) with first class with distinction "A" Grade. Currently focused on ovarian cancer and its chemopreventive studies.

Dr. M. Maghimaa, is working as a Professor at the Centre for Global Health Research, Saveetha Medical College and Hospital, Saveetha Institute of Medical and Technical Sciences, Chennai, India. With two decades of teaching and research experience, her primary focus lies within the realm of Microbiology, particularly delving into cancer biology, nanomaterials, drug discovery, nanomedicine, drug resistance, and phytomedicine. Her contributions to the field have garnered significant recognition, including winning the prestigious National level first prize in National Science Teachers Congress (NSTC) at the 6th India International Science Festival (IISF-2020), organized by the Ministry of Science & Technology, Government of India in December 2020. She was honored with the Best Researcher award from Periyar University among the 116 affiliated colleges in September 2022. Dr. Maghimaa's research endeavors have been supported by various funded projects, notably from Department of Biotechnology (DBT) (New Delhi) major research project grants (8 Lakhs), five TNSCST (Chennai, India) SPS projects (0.4 Lakhs), Science & Technology Project, TNSCST-DST Grant, (4.2 Lakhs) played a seat anchor role in DBT star college scheme (NewDelhi) (104 Lakhs), DST–FIST (NewDelhi) (42 Lakhs) and CRS Project, UGC-DAE-CSR, Indore, Madhya Pradesh (4.54 Lakhs). Her research output includes over 70 papers published in international peer-reviewed, Science Indexed (SCI) journals, along with contributions to book chapters and two books. She has also obtained patents from Australia, Germany, and India, reflecting the innovative nature of her work. In addition to her research and academic achievements, Dr. Maghimaa has demonstrated a commitment to nurturing future talent in the field. She has supervised and mentored 2 M.Phil and 5 Ph.D. scholars, contributing to their academic and professional development. Recognized for her excellence, Dr. Maghimaa has received ten accolades at both national and international levels, including the Young Scientist Award, Best Young Researcher award, and awards for Best Oral Presentation. Her dedication to scientific outreach is evident through her organization of 34 DBT sponsored workshops & twinning programmes & a national level seminar. She has participated in the brainstorming session on DBT–state partnership programme on Skill Vigyan in life sciences and biotechnology in July 2018 at TNSCST, Chennai.

Dr. Suresh Sagadevan, CChem, FRSC is working as an Associate Professor at the Nanotechnology & Catalysis Research Centre, University of Malaya. He has published numerous papers in internationally reputed journals. He is a Guest Editor and Editorial board member of many reputed ISI journals. He is a member of many professional bodies at the national/international level. He has been a recognized reviewer for many reputed journals. Indeed, he is working in various fields such as Nanofabrication, functional Materials, graphene, polymeric nanocomposite, glass materials, thin films, bio-inspired materials, drug delivery, tissue engineering, cell culture, supercapacitor, optoelectronics, photocatalytic, green chemistry, and biosensor applications.

Dr. Kanthesh M. Basalingappa is an Associate Professor of Molecular Biology at the JSS AHER, Mysuru, India. His goal of research is to determine the role of RNA Binding proteins in tumor progression and metastasis. Post-transcriptional regulation of gene expression by RNA binding protein is a crucial mechanism in regulating the timing and the amount of expression of genes. Growing evidence indicates that the alteration of the expression and function of RNA binding proteins could potentially play a role in inflammation and cancer.

Dr.Kanthesh BM did his PhD from University of Madras (2005). He also did postdoctoral research at the University Malaya, Kuala Lumpur, Malaysia (2007–2009); West Virginia University, Morgantown, USA (2090–2011) and University of Oklahoma Health Sciences, Oklahoma, USA(2011–2014).

He received Malaysia Prestigious Bio-Malaysia Gold medal Award (2008). His research area is Arbovirus infections, in that they done patented work on "Early detection of BK virus using molecular methods". He also Received Dr. Wilson Aruni's "Best Research Mentor and Teacher Gold medal Award" from the Indian Association of Applied Microbiology (IAAM) (2018).

He has been engaged in teaching and research in Microbiology & Molecular Biology for the past 21 years. He has published over 90 original research papers, 20 book chapters, and 20 review articles. He is also Professional and Scientific Memberships in, American Association for Cancer Research (AACR), Life Member of Indian Association of Applied Microbiology (IAAM), Life Member of Indian Association of Biomedical Scientists (IABMS), Indian Association of Medical Microbiologist (IAMM). He Received Fellowship Award from Indian Association of Applied Microbiology (FIAAM). At present he is collaborating with Royal Research Foundation, a research institute in India.

Conductive Polymers and Hydrogels for Brain-Computer Interfaces

6

Shahab Ahmadi Seyedkhani, Azam Irajizad, and Raheleh Mohammadpour

Abstract

Brain-Computer Interfaces (BCIs) have emerged as revolutionary technologies in neuroscience, providing bidirectional communication between the brain and external devices. However, the development of BCIs faces significant challenges, particularly in creating sensitive and biocompatible interfaces. Recently, conductive polymers (CPs) and hydrogels have attracted many attentions due to their unique properties in addressing these challenges. In this chapter, we provide an overview on CPs and hydrogels, discussing their different synthesis methods and properties required for neural interfacing applications. Also, we highlight the synergy of CPs and hydrogels in composite and hybrid systems for improving the performance of neural interfaces. Afterwards, we delve into their various applications, including signal recording and stimulation devices, tissue engineering and neural regeneration, smart drug delivery, artificial synapses and muscles, molecular bioelectronics, and biodegradable implants. Finally, we mention the main challenges for fabrication of high-performance BCIs based on CPs and hydrogels, and conclude with insights in future directions, emphasizing the essential strategies to overcome the current limitations and unlock the full potentials of these materials in advancing the BCIs.

Keywords

Conductive Polymer · Hydrogel · Biomedical Applications · Neural Interface · Brain-Computer Interfaces

S. Ahmadi Seyedkhani (✉) · A. Irajizad · R. Mohammadpour
Center for Nanoscience and Nanotechnology, Institute for Convergence Science & Technology, Sharif University of Technology, Tehran, Iran
e-mail: ahmadiseyedkhani@gmail.com; sh.ahmadi@sharif.edu

R. Malviya, S. Sundram (eds.), *Sustainable Nanocomposites with Green Biomaterials*, Biomaterials, Bioengineering and Sustainability 2,
https://doi.org/10.1007/978-3-031-79110-9_6

Abbreviations

BCI	Brain-Computer Interface
CP	Conductive Polymer
WHO	World Health Organization
Pt	Platinum
Au	Gold
FBR	Foreign Body Response
CIC	Charge Injection Capacity
CVD	Chemical Vapor Deposition
MEAs	Microelectrode Arrays
μ-ECoG	Micro-Electrocorticography
3D	Three-dimensional
ECM	Extracellular Matrix
Redox	Reduction-Oxidation
PANi	Polyaniline
APS	Ammonium Persulfate
EP	Electrochemical Polymerization
CV	Cyclic Voltammetry
PEDOT	poly(3,4-ethylenedioxythiophene)
PPy	Polypyrrole
BMIMPF6	1-butyl-3-methylimidazolium hexafluorophosphate
LiTFSI	lithium bis(trifluoromethosulfonyl)amide
VPP	Vapor Phase Polymerization
oCVD	Oxidative Chemical Vapor Deposition
PECVD	Plasma-Enhanced Chemical Vapor Deposition
FESEM	Field-Emission Scanning Electron Microscopy
SAED	Selected Area Electron Diffraction
TEM	Transmission Electron Microscopy
PEG	Polyethylene Glycol
HRP	Horseradish Peroxidase
DNA	Deoxyribonucleic Acid
EDOT	3,4-Ethylenedioxythiophene
Py	Pyrrole
CTP	Catalyst-Transfer Polycondensation
Ni	Nickel
Pd	Palladium
KCTP	Kumada Catalyst Transfer Polymerization
SCTP	Suzuki Catalyst Transfer Polymerization
StCTP	Stille Catalyst Transfer Polymerization
NPs	Nanoparticles
Ru	Ruthenium
DArP	Direct Arylation Polymerization
OLEDs	Organic Light-Emitting Diodes
OFETs	Organic Field-Effect Transistors

EGOFETs	Electrolyte-Gated Organic Field-Effect Transistors
ADMET	Acyclic Diene Metathesis
S	Sulfur
PEC	Polyelectrolyte Complex
Ca	Calcium
PCPP	Poly-di(carboxylatophenoxy) Phosphazene
CS	Chitosan
K	Potassium
PVA	Polyvinyl Alcohol
UV	Ultraviolet
UV-Vis	Ultraviolet-visible
CPH	Conductive Polymer-based Hydrogel
PAA	Polyacrylamide
LED	Light-Emitting Diode
AA	Acrylic Acid
ICOHs	Ion Conducting Organohydrogels
ICHs	Ion-Conducting Hydrogels
DES	Dee Eutectic Solvent
IPN	Interpenetrating Polymer Network
PI	Polyimide
PU	Polyurethane
CSC	Charge Storage Capacity
DNH	Dual Network Hydrogel
LBL	Layer-by-Layer
CBD	Chemical Bath Deposition
PVD	Physical Vapor Deposition
SILAR	Successive Ionic Layer Adsorption and Reaction
CNT	Carbon Nanotube
QD	Quantum Dot
AgQDs	Silver Quantum Dots
SNR	Signal-to-Noise Ratio
SEM	Scanning Electron Microscopy
LFP	Local Field Potential
OECT	Organic Electrochemical Transistor
R-COOH	Carboxylic Acid
PGLu	Polyglutamic Acid
NGF	Nerve Growth Factor
MOF	Metal-Organic Framework
ECRAM	Electrochemical Random-Access Memory
AC	Alternating Current
BQ	Benzoquinone
HQ	Hydroquinone
PLCL	Poly(L-lactide-co-ε-caprolactone)

6.1 Introduction

According to reports from the World Health Organization (WHO), brain stroke is the 2nd leading cause of death, responsible for approximately 11% of global deaths (more than 6,500,000 deaths in 2023). In addition, Alzheimer's disease and other forms of dementia rank as the 7th leading reason of human death. Numerous other diseases, such as epilepsy, ataxia, brain tumors, hearing and visual impairments, and chronic headaches, also originate from neurological disorders. Fortunately, many of these diseases can be controlled or treated, especially in their early stages. Diagnosing these diseases and finding a correct understanding of the mechanisms behind their emergence and development require accurate and rapid detection of disease signs by establishing an efficient connection with the brain. Therefore, making a direct communication between the brain and external diagnostic devices for real-time and precise detection of neuronal activities could be the primary motivation for developing the brain-computer interfaces (BCIs). In fact, BCIs are interfaced with the brain to extract precise medical data and to interfere with its function by providing external stimuli. Moreover, bypassing traditional neuromuscular challenges, BCIs allow us to control devices, interact with surrounding environments, and communicate just through brain activity. Development of BCIs holds insightful implications in various fields, including healthcare, assistive technology, neuroscience research, and human-machine communications (Ahmadi Seyedkhani et al. 2024). BCIs offer new therapeutic approaches for individuals having neurological disorders or movement impairments (Karikari and Koshechkin 2023; Chai et al. 2024; Holt et al. 2024). For instance, BCIs can facilitate motor recovery following stroke or spinal cord injury through providing real-time feedback and promoting neural plasticity (Levett et al. 2023). They provide researchers unparalleled tools to understanding the brain dynamics and functional connectivity mapping. These opportunities made by BCIs can help study the neural correlates of perception, cognition, and behavior, leading to our better understanding of learning, memory, and consciousness.

Despite noteworthy progresses, the development of BCIs faces several challenges. One of the main challenges is achieving reliable and high-resolution neural recording and stimulation while maintaining biocompatibility and long-term stability (Yadav et al. 2020; Saha et al. 2021). Traditional electrode materials, such as platinum (Pt) and gold (Au), have limitations in flexibility, tissue integration, and chronic responses, as well as improper electrochemical properties (Ahmadi Seyedkhani and Mohammadpour 2022; Tang et al. 2023). The foreign body responses (FBRs) caused by implanted rigid electrodes lead to tissue damages and signal declining over time (Campbell and Wu 2018; Carnicer-Lombarte et al. 2021; Wu et al. 2021; Tang et al. 2023). In addition, due to their high electrochemical impedance and relatively low charge injection capacity (CIC), the metals are usually used where large electrodes are applicable. Hence, many efforts have been focused on developing new materials for fabrication of high-efficiency BCIs.

Recently, conductive polymers (CPs) have attracted many attentions for neural interfacing due to combination of synergistic electron/ion conductivity, flexibility, and biocompatibility (Nezakati et al. 2018; Uz and Mallapragada 2019; Krauhausen et al. 2023; Ziai et al. 2023). These organic materials can conform to the complex morphology of neural tissues, reducing FBR and improving signal transferring processes (Ohayon and Inal 2020; Ahmadi Seyedkhani and Mohammadpour 2022). The CPs can be synthesized *via* various methods, including chemical polymerization, electrochemical deposition, chemical vapor deposition (CVD), biocatalytic production and other approaches (Poddar et al. 2021; Wang and Feng 2022a). By controlling the synthesis conditions and additives, researchers can modify the properties of CPs to meet the specific application requirements. On the other hand, hydrogels are three-dimensional (3D) networks of hydrophilic polymers that possess biocompatibility, high water storage capacity, ion permeability and flexibility. The hydrogels can mimic the extracellular matrix (ECM), encouraging cell adhesion, migration, growth and proliferation (Ware et al. 2013; Sheng et al. 2019; Fu et al. 2020; Correa et al. 2021). Furthermore, the hydrogel's mechanical properties can be tuned to adapt with the stiffness of brain tissue, minimizing mechanical mismatch-induced FBRs (Sheng et al. 2019; Ahmadi Seyedkhani and Mohammadpour 2022; Sagdic et al. 2023).

The integration of CPs and hydrogels can also be promising approach for fabrication of electrode materials with novel characteristics and performances for neural interfacing. The hybrid CPs-hydrogel compositions can facilitate synergistic ion-electron transmitting at tissue-electronic interface, promoting high quality and stable neural signals and reducing inflammatory responses (Ahmadi Seyedkhani and Mohammadpour 2022). Hence, the future investigations should be focused on production of new CPs/hydrogel composites, as well as designing the novel neural interfaces based on these materials. In this regard, the convergence of nanotechnology, materials science, biotechnology, and neuroscience can drive the progression of BCIs towards less invasive, more reliable, and more accessible devices. In this chapter, we will delve into the recent advancements, challenges, and opportunities of CPs and hydrogels for BCIs, providing insights in design principles, fabrication techniques, and potential applications of these materials.

6.2 Conductive Polymers (CPs)

Organic electronics in neural interfaces are recognized for enhancing compatibility with biological tissues (Berggren and Richter-Dahlfors 2007; Rivnay et al. 2014). This is mainly due to the flexible interfaces developed based on organic materials, reducing FBRs and improving communication with biological tissues. CPs are characterized by a repeating sequence of single and double bonds known as π bonds. Once appropriately doped, these polymers establish a continuous pathway for charge propagation along the their chain (Bredas and Street 1985; Shirakawa 2001). This configuration enables electron delocalization and the presence of dynamic charge carriers, permitting electrical conductivity within CPs through traversing the π-bonded carbon atoms (Roncali 1992). It is noteworthy that the doping process can be reversible, while the dopant concentration and location dictate the CPs' conductivity, spanning from insulating to metallic properties. In addition, CPs exhibit appropriate capabilities to encapsulate drugs and biomolecules within their structure, allowing for controlled release during reduction-oxidation (redox) reactions. This ability arises from the inherent electroactivity of CPs, enabling them to undergo reversible redox reactions. In a redox process, the polymer can alternately accept and donate electrons, triggering changes in its oxidation state. This redox activity serves as a mechanism to precisely release the entrapped drugs or biomolecules in response to specific stimuli, such as changes in the surrounding environment. This controlled release mechanism enhances drug delivery systems, where the CPs can serve as smart carriers, which respond to physiological conditions or external stimuli to release the therapeutic agents, minimizing side effects and optimizing treatment outcomes (Abidian et al. 2006; Thompson et al. 2006, 2010; Isaksson et al. 2007; Richardson et al. 2009; Simon et al. 2009; Green et al. 2010). Moreover, the CPs can be functionalized by biological species like ECM proteins and growth factors, which enhances their biocompatibility and induces specific cellular responses. This high functionalization capability of CPs expands their applications for biosensing, tissue engineering, and neural interfacing, where the modified CPs promote cell adhesion, proliferation, and differentiation (De Giglio et al. 2000; George et al. 2006). Moreover, CPs exhibit a dual charge conducting mechanism, possessing both ionic and electronic conductivities (Ahmadi Seyedkhani and Mohammadpour 2022). In fact, CPs such as PEDOT directly convert between electronic and ionic charges *via* electrochemical doping and dedoping processes. Figure 6.1 illustrates a schematic representation of mixed ion-electron conductivity mechanism in PEDOT. It has been demonstrated that the holes' chemical potential modulated through changes in doping level, can be a significant driving force for synergic charge transfer (Keene et al. 2024). Within such systems, despite the great charge densities, a competition between diffusion and hole drift can result to diffuse space charge regions. This phenomenon is unique to synergic conductive systems, in which the mobile ionic charges can compensate the depletion or accumulation of electronic charges, thereby screening the electrostatic driving forces (Keene et al. 2024). As we know, the biological systems transfer the electrophysiological signals through ion movements, while the electronic devices work by electron conduction. Therefore, the capability of CPs for mixed ion-electron charge conversion enables high-efficient signal transduction at tissue-electronic interfaces, making a wide range of applications in neuromorphic computing, bioelectronics, BCIs (Deslouis et al. 1996; Heeger 2001; MacDiarmid 2001).

6.2.1 Fabrication of CPs

The discovery of CPs in 1970s made a different conception in polymer science, providing the basis for the emergence of organic electronics (Shirakawa et al. 1977). Significant progress in production of CPs has been made, accompanying with improvements in their charge carrier mobility. However, there's still a need for enhanced performance in their stability, mechanical behavior, and biocompatibility

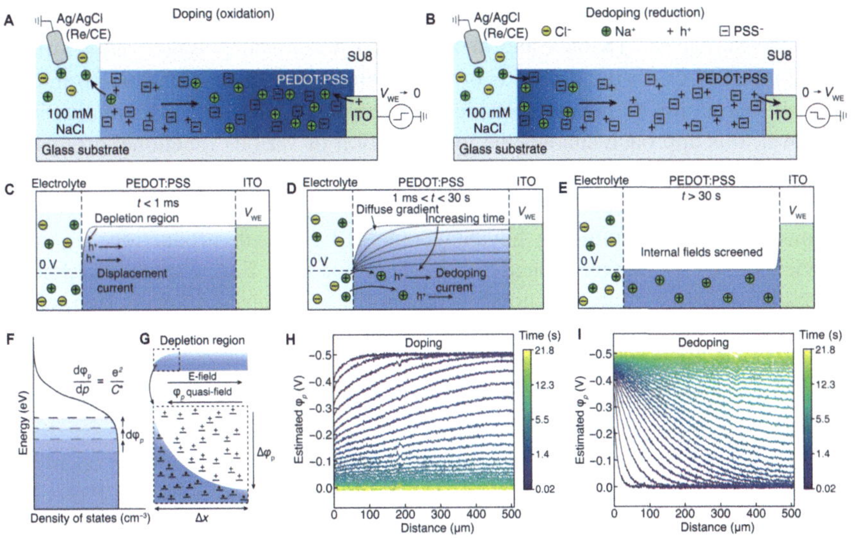

Fig. 6.1 Schematic illustration for propagation of electrochromic transition through electrochemical (**a**) doping and (**b**) dedoping of PEDOT:PSS. (C-I) Driving forces for electrochemical (de) doping of PEDOT:PSS. (**c**) Once the potential is applied to the ITO, a hole depletion region is created at the electrolyte interface. (**d**) The electric field leads to drifting the cations towards the PEDOT:PSS, dedoping PEDOT. (**e**) When the internal potential reaches the electrolyte's potential, the dedoping finishes. (**f**) The hole chemical potential (ϕ_p) changes during dedoping affect the ultimate hole concentration. (**g**) A hole quasi-electric field, which drives diffusion in an inverse direction of the drift, made by the concentration-dependent hole chemical potential. (**h**) Doping and (**i**) dedoping processes (V_{WE} from 0 to −0.5 V vs Ag/AgCl). (Keene et al. 2024)

to meet wide-ranging application demands. Here, we will discuss some main fabrication processes of CPs in more details.

6.2.1.1 Chemical Polymerization

Chemical polymerization of a conductive polymer conducts by the formation of covalent bonds between its monomer units, leading to construction of the polymer chains. This process typically requires chemical initiators or catalysts to facilitate the polymerization reactions. For CPs, the monomers possess π-electron-rich structures that undergo oxidative or reductive polymerization to form extended conjugated polymer chains. Oxidative polymerization is a generally employed method for the synthesis of CPs, in which the polymer monomers are oxidized to produce radical species that subsequently initiate the polymerization. This process typically utilizes oxidizing agents or dopants to facilitate the formation of the chains. For example, during polymerization of polyaniline (PANi), the aniline monomers can be oxidized by the ammonium persulfate (APS, $[NH_4]_2S_2O_8$) or ferric chloride ($FeCl_3$) to form the polymer backbone. Otherwise, reductive polymerization can also be utilized to synthesize CPs *via* reducing monomers to radical species, forming the polymer chains through coupling reactions. This method is particularly appropriate to monomers having electron-deficient structures, such as pyrrole or thiophene derivatives (Kitada and Ozaki 1995). Chemical polymerization offers several advantages for the synthesis of CPs, including precise control on produced polymer's structure, highly pure products, and capability for large-scale production. Moreover, this approach enables the incorporation of various functional groups or dopants into the polymer backbone, allowing for tailored properties and applications (Poddar et al. 2021). Nevertheless, it should be noted that although chemical polymerization is economically advantageous for large-scale production, it generally yields to CPs with constrained electrical conductivity. Furthermore, the chemical oxidation often relies on non-biocompatible oxidants, imposing significant limitations on this method for biomedical applications.

6.2.1.2 Electrochemical Polymerization

The electrochemical polymerization (EP) of CPs follows the chemical oxidation mechanism, except that it utilizes the electrical potential instead of the chemical oxidants to polymerize the monomers. Actually, the synthesis of CPs *via* EP process involves a simultaneous precipitation and oxidation of polymer monomers on an electrode' surface. Hence, this process generally requires a three-electrode system, consisting of the working electrode, counter electrode, and a reference electrode, alongside an electrolyte solution. By tuning the process's parameters, different copolymerization can be also carried out through EP synthesis methods. In this regard, the monomers' oxidation and reduction potentials, which can be characterized by electrochemical analysis such as cyclic voltammetry (CV), determine the favored EP techniques. Besides primary precursors such as polymer monomers, the electrolyte solution generally includes supplementary substances essential for preparing the electrolyte and conducting oxidation process. In recent years, great attention has been focused towards electrochemical deposition of poly(3,4-ethylenedioxythiophene) (PEDOT), with its inception dating back to 1988 (Nie et al. 2021). During electrochemical deposition of PEDOT, the electrolyte anions enter the PEDOT structure as counterions, making charge stabilization along the polymer chains. Moreover, these anions affect significantly the morphology, stability, photoelectric properties, and mechanical characteristics of the PEDOT films (Zotti et al. 2003; Poverenov et al. 2010; Culebras et al. 2014). Therefore, optimizing the PEDOT's electrochemical properties is feasible by employing diverse counterions and altering the electrolyte composition. Some commonly electrolytes for PEDOT deposition include lithium perchlorate ($LiCLO_4$), 1-butyl-3-methylimidazolium hexafluorophosphate (BMIMPF6), and lithium bis(trifluoromethosulfonyl)amide (LiTFSI) (Zotti et al. 2003; Culebras et al. 2014). Recent studies have demonstrated a notable enhancement in electrical conductivity of the PEDOT layers (reaching up to 2000 $S.cm^{-1}$) just by optimization electrolyte composition (Culebras et al. 2014). Finally, although the EP methods are mostly limited to synthesis of CPs on the conductive substrates, they have great potentials for producing a variety of CPs and their composites, as well as the high capability for integration with microscale manufacturing technologies of BCIs.

6.2.1.3 Vapor Phase Synthesis

Vapor phase polymerization (VPP) techniques, such as oxidative chemical vapor deposition (oCVD), plasma-enhanced chemical vapor deposition (PECVD), and pulsed laser deposition, are distinguished due to their ability to generate high quality thin films of both soluble and insoluble CPs. In 1986, researchers deposited conductive polypyrrole (PPy) through VPP (Mohammadi et al. 1986), initiating a method that continues to progress for developing various CPs. Initially, the VPP employed a sequential deposition process, where a substrate already containing an oxidizing agent was exposed to vapor of the monomers, providing a desired place for polymerization reactions. Over the years, VPP processes have been developed *via* advances in oxidant agents, vacuum technology, and additives. However, upscaling VPP remains challenging because of the need for intricate equipment, and high solvent-substrate compatibility. During VPP process, oxidative polymerization occurs at the liquid-vapor interface, affected by various parameters like monomer and oxidant concentrations, temperature, and pressure. Actually, these parameters determine the properties of resulting CP layer. Hence, by controlling the VPP inputs and conditions, the final properties of CPs, such as crystallinity, electrical conductivity and optical characteristics, can be engineered for specific applications (Bhattacharyya et al. 2012; Lawal and Wallace 2014; Brooke et al. 2017). Recently, the single-crystal PEDOT nanowires with ultrahigh conductivity (up to ~8800 S/cm) was developed using liquid-bridge-mediated nanotransfer printing with VPP method (Cho et al. 2014). Figure 6.2 illustrates fabrication process and characteristics of the single-crystal PEDOT nanowire films.

Moreover, there are various oxidants with different oxidation strengths that impact polymerization rates, resulting in different morphologies and functionalities (Bhattacharyya et al. 2012). On the other hand, additives like sulfonate-based anions, pyridine and polyethylene glycol (PEG) derivatives can influence doping levels, polymerization kinetics, and subsequent electrical conductivity of CPs (Wang and Feng 2022a). Many studies have revealed that the presence of anion post-polymerization can confer higher conductivity to CPs based on ion exchange mechanisms (Brooke et al. 2017).

Furthermore, the VPP processes can surpass the traditional solution-based coating methods, where the substrates show poor surface wettability. For example, CP-based sensors on fabrics have great potential for developing versatile applications, particularly where the fabric or ready-made wearable can be directly coated. To this end, the solution-based coating approaches indicate limitations in preparing adhesive and uniform thin films due to the fabrics' surface hydrophobicity. While, the VPP methods have demonstrated appropriate capability for fabrication of patterned polymeric thin films with controllable thicknesses, while maintaining the breathability and mechanical stability of the fabrics. Recently, utilizing the oCVD technique, researcher developed binder-free printed PEDOT wearable sensors on commercially available disposable fabric gloves and masks (Clevenger et al. 2024). Figure 6.3 shows fabrication setup and characterization of the prepared PEDOT coatings on fabric substrates.

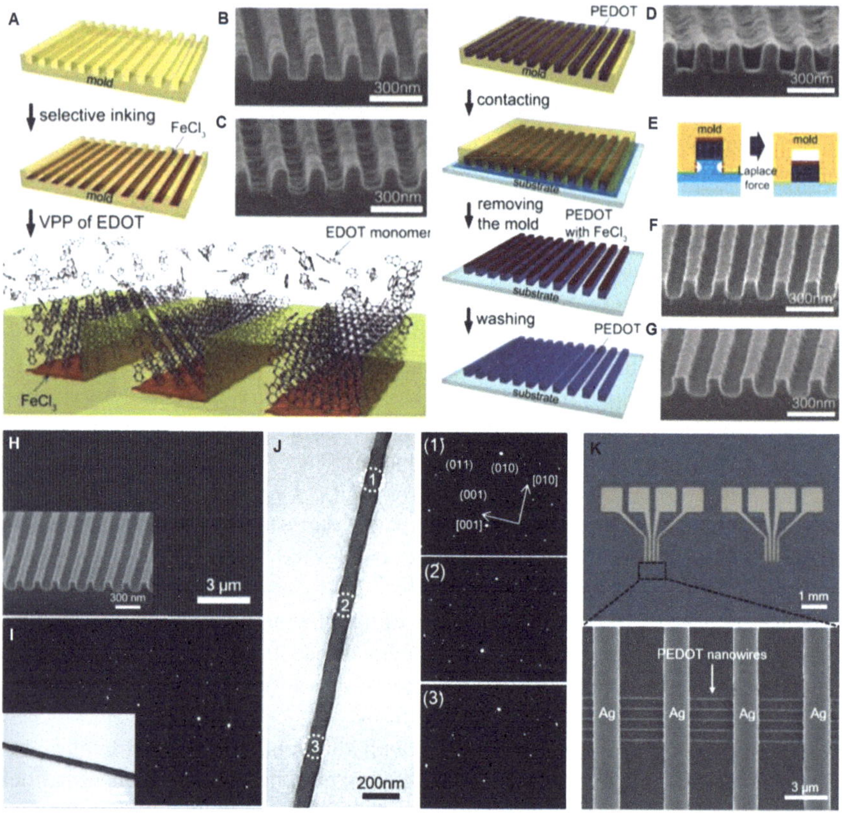

Fig. 6.2 (**a**-**g**) Schematic illustration for VPP fabrication of single-crystal PEDOT nanowires and field-emission scanning electron microscopy (FESEM) images of the developing nanowires related to each step. (**h**) FESEM and (**i**) selected area electron diffraction (SAED) pattern of the prepared PEDOT nanowire film. (**j**) Transmission electron microscopy (TEM) image of a PEDOT nanowire and corresponding SAED patterns from three different areas of the PEDOT nanowire. (**k**) Top: optical microscopy and bottom: electron microscopy images of a device made by single-crystal PEDOT nanowires for four points resistivity measurements. (Cho et al. 2014)

6.2.1.4 Biocatalytic Synthesis

Biocatalytic synthesis of CPs utilizes biological organisms such as enzymes or cells to catalyze polymerization reactions, offering several advantages over traditional chemical methods. Firstly, the biocatalytic synthesis allows for controlled kinetics, enabling precise regulation of polymerization rates and molecular weight distributions. Enzymes can serve as catalysts in these reactions, facilitating the formation of polymer chains in a controlled manner. This control is critical for producing polymers with desired properties, such as electrical conductivity, which is essential for applications in neuroelectronics and BCIs. Moreover, the conventional chemical polymerization processes typically require highly acidic conditions and toxic oxidants. These conditions not only pose environmental concerns but also generate

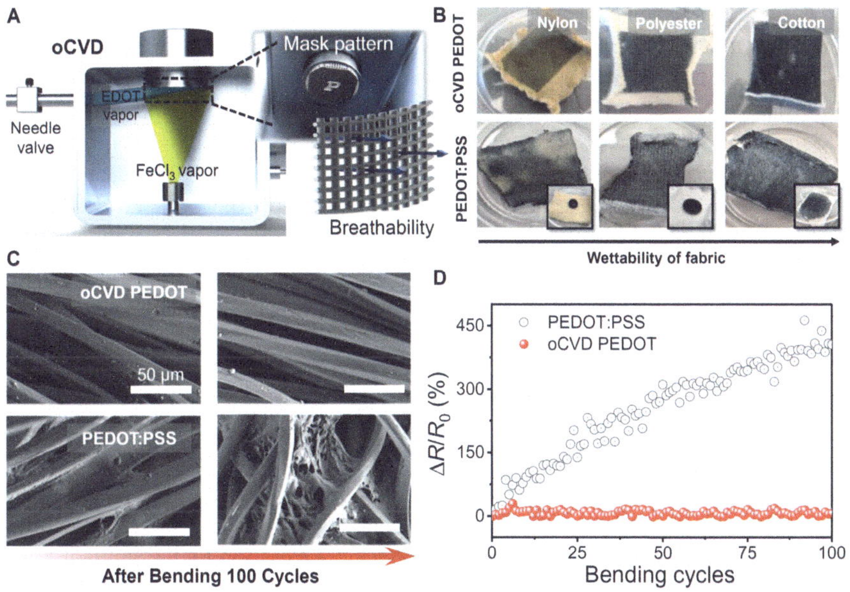

Fig. 6.3 (**a**) Schematic illustration for fabrication of PEDOT thin films on fabric *via* oCVD process. (**b**) Optical images of fabrics (cotton, polyester, and nylon) coated with solution-processed PEDOT:PSS and oCVD PEDOT, demonstrating surface hydrophobicity of the fabrics (insets shows PEDOT solution droplets on fabrics). (**c**) Electron microscopy images of the solution-processed PEDOT:PSS and oCVD PEDOT on polyester before and after bending 100 cycles. (**d**) Resistance changes for solution-processed and oCVD PEDOT coatings during the bending test, indicating highly stable electrical performance of the oCVD PEDOT. (Clevenger et al. 2024)

hazardous by-products, complicating multistep purification and disposal processes. In contrast, biocatalytic synthesis performs under slighter conditions, typically at moderate pH and biological temperature. On the other hand, unlike EP and VPP processes that require energy-intensive conditions, biocatalytic synthesis can be carried out by living organisms, reducing energy consumption.

Enzymatic catalysis is environmentally friendly, energy efficient, non-toxic, readily scalable, and produces high yields of products (Bouldin et al. 2010). Among the enzymes utilized to synthesis CPs, horseradish peroxidase (HRP) stands out as a commercially available enzyme that has found widespread applications. The HRP has been applied in various strategies for production of chiral PANi (Bruno et al. 2001; Fernandes et al. 2003). One approach involves utilization of the HRP in combination with different templates, which guide the polymerization process and affecting the final properties of the polymer. For instance, small deoxyribonucleic acids (DNA) can be employed as templates in conjunction with HRP to direct the formation of PANi nanowires (Patrick Nickels et al. 2004). This template-assisted synthesis not only facilitates the control of polymer morphology but also enables the incorporation of chiral elements, resulting in polymers with specific optical and electronic properties. Furthermore, pH-stable peroxidases obtained from soybeans and palm leaves have been studied as alternative catalysts in the synthesis of PANi. These enzymes exhibit advantages such as robust activity under a wide range of pH conditions, making them compatible for synthesis of CPs in different environmental conditions. By harnessing the catalytic capabilities of pH-stable peroxidases alongside their appropriate templates, researchers have produced chiral PANi with tailored properties and enhanced stability (Otrokhov et al. 2013; Wang and Feng 2022a). Chiral CPs demonstrate unique optical and electronic properties arising from their helical or spiral structures, making them appropriate candidates for advanced sensors and optoelectronics. Moreover, the polymer's chirality opens up opportunities for enantioselective sensing and separation, where the ability to distinguish between enantiomers is critical in various analytical and pharmaceutical processes.

Fungal laccases, which are known by their high redox potential, are phenol oxidases belonging to the group of multi-copper enzymes that catalyze the oxidation of a great variety of CPs (Walde et al. 2019). Unlike peroxidase-catalyzed polymerization, which generally necessitates external oxidants, laccase-catalyzed polymerization uses ambient oxygen as the primary oxidant, streamlining the synthesis process and reducing the reliance on additional chemicals. Operating under ambient temperature and pressure, is one the prominent advantages of employing fungal laccases, particularly in the synthesis of PANi. Furthermore, laccases produced by basidial fungi exhibit strong stability and activity at acidic pH values. This pH compatibility can ensure ideal enzyme performance throughout the polymerization process, leading to efficient polymerization and high-quality PANi formation. Many works have been done to study the mechanisms underlying both chemical and enzymatic synthesis of PANi to distinguish their differences. Chemical polymerization typically involves an induction period, while enzymatic synthesis exhibits controlled kinetics with no induction period. Actually, the enzymatic polymerization results in the growth of PANi chains through the production of halfoxidized forms of the aniline oligomers. However, for a long time, it was widely believed that the enzymes alone could not oxidize monomers like 3,4-Ethylenedioxythiophene (EDOT) and pyrrole (Py), due to their high oxidation potentials. Hence, the redox mediators were applied for enzymatic oxidation and subsequent polymer chain growth. For example, CPs like PPy were synthesized using HRP and laccases based on enzyme–mediator systems (Bhoyar et al. 2024). But, it is worth noting that the first enzymatic synthesis of PEDOT using HRP alone was reported in 2007 (Rumbau et al. 2007), demonstrating the great potential of biocatalytic synthesis in producing different CPs (Wang and Feng 2022a).

6.2.1.5 Other Synthesis Methods

Catalyst-Transfer Polycondensation (CTP)

Catalyst-Transfer Polycondensation (CTP) is utilized for the controlled synthesis of CPs. It involves transition metal-catalyzed cross-coupling polycondensation reactions, which can proceed through a controlled chain-growth mechanism (Yokozawa and Yokoyama 2009; Kiriy et al. 2011; Nezakati et al. 2018). Unlike conventional polycondensation, where the catalyst detaches after each step, in CTP, the catalyst transfers from monomer to monomer, facilitating controlled chain growth (Aplan and Gomez 2017). The choice of functional groups and cross-coupling agents is crucial in CTP. Transition metal catalysts such as nickel (Ni) and palladium (Pd) are utilized to boost cross-coupling reactions between bifunctional monomers containing reactive groups that complement each other (Kiriy et al. 2011; Wang and Feng 2022a). Different types of monomers, such as difunctional and electron-rich or electron-deficient monomers, can undergo CTP *via* chain-growth or step-growth mechanisms, respectively. The selection of appropriate transmetalation reagents influences the polymerization kinetics and functional group tolerance. The Ni catalysts, such as $Ni(d_{ppp})Cl_2$, which are often used with bidentate phosphine ligands, show effective performance in Kumada Catalyst Transfer Polymerization (KCTP). They facilitate chain growth by promoting oxidative addition and transmetalation steps. On the other hand, the Pd catalysts, including Suzuki and Stille catalysts, are employed in Suzuki Catalyst Transfer Polymerization (SCTP) and Stille Catalyst Transfer Polymerization (StCTP), respectively. These reactions also proceed through controlled chain-growth mechanisms, although they may be less controlled than KCTP. The phosphine ligands play a key role in CTP by regulating the electronic and steric environment around the metal center. Monodentate and bidentate phosphines offer different advantages of reactivity and stability, depending on the polymerization requirements (Kiriy et al. 2011; Wang and Feng 2022a). However, despite the great achievements in CTP, further advancements require new monomers and catalysts to expand the range of conjugated building blocks and improve polymerization efficiency. Overall, the CTP represents a competent approach for the controlled synthesis of CPs, offering opportunities for tailoring their molecular weight, dispersity, and chain structure to achieve desired physical and electronic properties.

Photopolymerization

Photopolymerization offers advantages over chemical and electrochemical methods for synthesizing CPs, such as applying coating on insulating substrates, and avoiding unstable organic compounds. The CPs produced *via* photopolymerization exhibit enhanced thermal stability and structural properties compared to those from chemical polymerization (Woods et al. 2021; Wang and Feng 2022a). Photopolymerization can be conducted through two main approaches, including (i) direct photoexcitation and (ii) photopolymerization *via* photocatalytic systems. Through direct photoexcitation, a self-synthesized photopolymerization of CPs is carried out *via* the photoexcitation of the monomers without photosensitizers (Heydarnezhad et al. 2018). Actually, using photo-irradiation polymerization with no additional toxic chemical compounds, the direct photopolymerization can be classified in the green and environmentally friendly methods. However, the CPs produced by this approach usually exhibit weak electrical conductivity, which is attributed to the small dopant molecules, low concentrations, and large distances between polymer chains (Wang and Feng 2022a). On the hand, during photopolymerization *via* photocatalytic systems, the photosensitizers such as ruthenium (Ru) complexes or metallic nanoparticles (NPs) catalyze CP production. For instance, the Ru complexes, like tris (bipyridine) ruthenium (II), facilitate photopolymerization by inducing photo-reduction or photo-oxidation processes (Wang and Feng 2022a). Transition metal salts or metallic NPs, such as Cu and Au NPs, can also catalyze photopolymerization processes (Heydarnezhad et al. 2018). Overall, photopolymerization methods can be utilized for preparing a wide range of CP nanostructures and hybrid systems, which are discussed further in future sections.

Direct Arylation

Direct arylation polymerization (DArP) progresses *via* cross-coupling polymerization between dihaloarene monomers and non-substituted arenes. In this method, contrary to conventional cross-coupling polymerizations like Suzuki and Stille reactions, no prior preparation of arene monomers with organometallic functionalities is required (Nakabayashi 2018). Initial works for development of DArP were focused on three main purposes, including (i) enhancing site selectivity, (ii) eliminating structural faults, and (iii) transitioning from simple homopolymers to more complex donor–acceptor copolymers. The CPs synthesized *via* DArP find applications in organic electronics such as organic light-emitting diodes (OLEDs), organic photovoltaics, and organic field-effect transistors (OFETs). Traditional coupling systems for (hetero)aryl-(hetero)aryl couplings produce toxic by-products and require multistep purification of organometallic monomers. While, DArP utilizes low catalyst loadings, produces high molecular weight CPs, yielding high production efficiencies. Actually, the DArP stands out due to its efficiency in producing desired aromatic derivatives while minimizing the need for expensive or environmentally hazardous reagents. Moreover, the absence of organometallic species in DArP simplifies the subsequent purification processes, and reduces potential of the side reactions, enhancing the overall efficiency of this practical synthesis method (Nakabayashi 2018; Wang and Feng 2022a).

Acyclic Diene Metathesis (ADMET) Polymerization

Acyclic Diene Metathesis (ADMET) polymerization is a technique for creating CPs with precise structures and diverse properties. The ADMET resembles the olefin metathesis and traditional polycondensation reactions, initiating with the metathesis of terminal olefins to form a metal alkylidene and ethylene. When the ADMET process is done under vacuum conditions, high molecular weight CPs can be attained *via* high conversion of monomers, which is attributed to the removal of ethylene. However, undesired reactions like depolymerization and cyclization can occur, affecting the final CPs microstructures. The conjugated ADMET polymers synthesized using chromophore-bearing monomers offer unique structures with a high capability for optimization. In ADMET, appropriate solubility in high-boiling solvents and catalyst compatibility are crucial for producing moderate to high molecular weight CPs. Considering important factors like the negative neighboring effect can guide the selection of ideal monomers to optimize the synthesis process of CPs (da Silva et al. 2017). Typically, ADMET polymerization presents an efficient approach for producing π-conjugated polymers with several promising advantages, including:

(a) *Defect-Free Polymers*: ADMET polymers possess inherent defect-free characteristics, which is attributed to the absence of conjugated unit terminations or the inclusion of detrimental impurities like sulfur (S) or halogens. Consequently, these conditions can ensure the purity and integrity of the resultant CPs.
(b) *Highly Trans Olefinic Double Bonds*: The ADMET polymerization proceeds *via* a metallacycle intermediate, resulting in polymers with highly trans olefinic double bonds. This trans configuration enhances the conjugation along the polymer backbone, leading to improved electronic properties and structural stability.
(c) *Well-Defined Polymer Chain Ends*: Polymers synthesized using Ru catalysts in ADMET reactions typically exhibit well-defined chain ends, often terminating with vinyl groups. These terminal chains make opportunities for further modification and functionalization of the conjugated materials, enabling tailored properties and applications.

Table 6.1 summarizes advantages and challenges of different synthesis methods for producing CPs.

Table 6.1 Advantages and challenges of different synthesis processes of CPs

Method	Advantages	Challenges
Chemical Polymerization	Works by a wide variety of monomers for producing diverse CPs. Tuning molecular weight and doping level *via* reaction parameters. Well-established and scalable. Requires simple, inexpensive equipment.	Producing hazardous waste from organic solvents and reagents. Works by severe pH conditions (usually acidic). Needs multistep complementary purification processes. Uses toxic and flammable chemicals. Exhibits weak reproducibility across synthesis batches (Batch-to-Batch variability). Synthesis of adhesive CP coatings is challenging. Requires efficient recycling systems for solvents. Long reaction times in catalyst-free processes. Difficult removal of residual catalysts.

(continued)

Table 6.1 (continued)

Method	Advantages	Challenges
EP	Direct film formation on electrodes. Control the CP properties *via* tuning electrical parameters. Eliminating chemical initiators and catalysts, reducing contamination. Fabrication of various CPs from a wide range of monomers. Producing few hazardous by-products, reducing environmental impact. Fabricating highly pure polymers *via* one-step technique, eliminating multiple purification steps. *in situ* doping and modification for simultaneous polymerization and properties tuning. Suitable for scalable production of thin films, ideal for microelectronics and sensors.	Limited to conductive substrates. Requires highly optimized electrolytes and solvents for efficient polymerization. Making inconsistent film properties on non-uniform substrates. Possibility of side reactions from improper EP techniques and parameters. Challenging to precisely control doping levels High initial costs for electrochemical setup.
VPP	Producing highly pure and single crystal CPs. Uniform CP coatings on complex surfaces. Efficient for large-scale coating processes. Producing high molecular weight CPs with enhanced electrical and mechanical performances. Appropriate substrate-coating adhesion for durability. Capability for *in situ* doping and functionalization.	High costs, expensive equipment, and complex processes. Requires monomers with adequate volatility and thermal stability. Highly sensitive process, small deviations can affect polymer quality. Challenging reaction kinetics control to avoid defects. Limited monomers suitable for vaporization. Uniform coating on porous substrates is challenging. Requires substrates with high thermal stability to avoid degradation. Handling volatile and potentially toxic monomers necessitates stringent safety protocols.

(continued)

Table 6.1 (continued)

Method	Advantages	Challenges
Biocatalytic Synthesis	Mild reaction conditions (ambient temperature, pH) avoid harsh chemicals and reduce energy consumption. Generating few toxic by-products and waste. Enzymes offer high selectivity and specificity, resulting in fewer side reactions and by-products. *In situ* functionalization during polymerization. Using biodegradable catalysts (Enzymes), reducing environmental impacts and side effects. Compatible with aqueous systems, eliminating organic solvents. Sustainable manufacturing processes by renewable biological sources. Low energy consumption and spontaneous polymerization and functionalization by living species.	Stability and activity of enzymes can be affected by temperature, pH, and inhibitors. High costs for large-scale production and purification of enzymes. Limited number of enzymes available for CP synthesis. Scaling up while ensuring similar and consistent enzyme performance at large scale can be challenging. Low reaction rates. Enzyme inactivation due to by-products. Some biocatalytic mechanisms are still not fully understood.
CTP	Controlled chain-growth mechanisms. Working by different monomers. Rapid polymer chain growth by using catalysts. Precise control over molecular weight, dispersion, and chain structure of CPs to achieve desired physical and electronic properties.	Removing catalyst residues is challenging. Poor process repeatability. Providing proper functional groups and cross-coupling agents is difficult. Transmetalation reagents can influence polymerization kinetics and functional group tolerance. Producing by-products, which can reduce yield and purity. CTP requires new monomers and catalysts to expand the range of conjugated building blocks.

(continued)

Table 6.1 (continued)

Method	Advantages	Challenges
Photopolymerization	Capability for CP coating on insulating substrates. Rapid polymerization reactions with high spatial and temporal control. Using few chemical compounds, leading to high purity and low contamination. Avoiding unstable organic compounds. Photopolymerized CPs exhibit thermal stability. Self-synthesis of CPs without photosensitizers is possible. Using no additional toxic chemicals, it is a green and environmentally-friendly method. Capability for using catalysts, facilitating photopolymerization *via* inducing photo-reduction or photo-oxidation processes. *In situ* production of CP nanostructures and hybrid systems like interpenetrating polymer networks.	Large distance between polymer chains leads to CPs with poor electrical conductivity. Low penetration of light into thick or opaque substrates leads to limited depth of cure and reduces polymerization efficiency. Significant polymer shrinkage and stress during curing. Low compatibility of photoinitiators with specific monomers and substrates. Expensive equipment and light sources (such as UV lamps and lasers). Not all monomers are suitable for photopolymerization. Surface sensitivity to oxygen inhibition or impurities can result in incomplete polymerization. Precise control over crosslinking density and network structure is difficult.
DArP	Eliminating the need for pre-functionalized monomers. Utilizing low catalyst loadings, producing high molecular weight CPs. Simple purification processes due to absence of organometallic species. Synthesis of CPs suitable for organic electronics like OLEDs and OFETs. Creating desired aromatic derivatives. Minimizing the need for expensive or environmentally hazardous reagents. High production efficiencies.	Requiring highly active and selective catalysts to avoid side reactions (limited available effective catalysts for DArP). Not all monomers are suitable for DArP, limiting the range of CPs that can be synthesized. The need for high purity of starting materials to prevent catalyst poisoning and ensure high reaction efficiency.

(continued)

Table 6.1 (continued)

Method	Advantages	Challenges
ADMET	Precise control on polymer structure, leading to uniform polymer properties. Working by a wide range of functional groups, which enables the synthesis of different structures for CPs. Performing without solvents, reducing environmental impact and solvent wastes. Synthesis of high molecular weight CPs with proper mechanical properties. Producing few toxic by-products (eco-friendly polymerization method)	Highly pure monomers are required to avoid defects. Strict reaction conditions (specific temperature and pressure settings). ADMET catalysts are highly sensitive to impurities, leading to simple catalyst deactivation. Risk of oligomer formation, resulting in lower molecular weight and poor electrical conductivity. Slow process with extended reaction times, increasing energy consumption and costs. Requiring efficient removal systems to continuously remove by-products and drive desired reactions. Limited monomers desired for ADMET. Challenging scaling up

6.3 Hydrogels

Hydrogels are 3D cross-linked networks composed of hydrophilic polymers, which can absorb and retain large amounts of water or biological fluids. Hence, these gel-like materials can exhibit significant flexibility and ion transferring capability, making them appropriate for various biomedical applications (Yang and Suo 2018; Sheng et al. 2019; Yuk et al. 2019; Dreiss 2020; Liu et al. 2020). Hydrogels are produced by natural or synthetic polymers through different crosslinking processes. Through evolution of polymerization and crosslinking processes, the polymers' chains can be covalently bonded to create a 3D network configuration, which ensures appropriate hydrogel's mechanical behavior and prevents its rapid dissolution within aqueous environments (Lake and Thomas 1967). We have summarized various strategies for synthesis of the hydrogels below.

6.3.1 Physical Crosslinking

The physical crosslinking can be attributed to non-covalent intermolecular interactions, including hydrogen bonds (Zhang et al. 2017), ionic bonds (Yuan et al. 2018), electrostatic forces (Papagiannopoulos et al. 2023), polymer entanglement (Alves et al. 2011), crystallite formation, and van der Waals forces (Rebers et al. 2021).

Hence, physical crosslinked hydrogels (or physical hydrogels) show reversible viscosity and elasticity. The physical crosslinking is simple and conducted at mild conditions, resulting in low toxicity and high biocompatibility compared to chemical crosslinking processes, since they need no external chemical crosslinking agents. But, it should be noted that the physical hydrogels are unstable, which may lead them to be degraded at high temperatures, where the chemical crosslinked hydrogels might be stable (Zhao et al. 2023). However, this instability and consequent degradation can make the physical hydrogels appropriate for short-term applications such as temporary scaffolds and bioresorbable neural interfaces.

6.3.1.1 Crosslinking Based on Hydrogen Bonding

Hydrogen bonds and their combination with other interactions like electrostatic forces, can result in self-assembly cross-linked hydrogels containing non-covalent bonds. These hydrogels exhibit shear-thinning properties and superior elasticity, making them particularly appropriate for 3D bioprinting (Appel et al. 2015; Ding et al. 2016). Having low shear viscosity, stress-relaxation behavior, and rheological properties, these hydrogels can also be directly injected by a very fine needle or catheter, leading to protect the cells from severe mechanical strains (Umashankar et al. 2012; Peak et al. 2018). Consequently, they are extensively applied for scaffolds, cell delivery, and embolic biomaterials (Avery et al. 2016; Liu et al. 2018). Furthermore, hydrogen bonding or electrostatic interactions can lead to the formation of polyelectrolyte complex (PEC) hydrogels. The PECs contain repeating units that possess electrolyte groups of polyanions and polycations. The connection between these poly-ions causes the formation of interpolymer complexes that are directed by the hydrogen bonding between different polymer chains (Tsuchida and Abe 2005). Nonetheless, due to the low crosslinking density and the fragile nature of hydrogen bonds that break easily, the mechanical strength of the hydrogels crosslinked *via* hydrogen bonding is unsatisfactory. Moreover, because of high gel concentration, injecting of these hydrogels are still challenging (Guvendiren et al. 2012; Ye et al. 2017).

6.3.1.2 Ionic Crosslinking

Ionic crosslinking is a kind of physical crosslinking that relies on ionic interactions between oppositely charged groups on polymer chains. In fact, the crosslinking process is performed based on electrostatic attractions, and does not involve the covalent bonds. The mechanism of ionic crosslinking process typically involves (i) preparation of polyelectrolyte solution, (ii) introduction of crosslinking agents, and (iii) formation of ionic bonds. Through the ionic crosslinking, polycations can naturally form crosslinked networks with anions. Alginate is a well-known example of polymers that undergoes ionic crosslinking based on calcium (Ca^{2+}) ions at room temperature and physiological pH (Gacesa 1988; Gacesa and Russell 1990). However, the resulting network can be degraded *via* the releasing of Ca^{2+} ions by chelation. Poly-di(carboxylatophenoxy) phosphazene (PCPP) is also crosslinked with Ca^{2+} ions. The crosslinking process is conducted through spraying an aqueous

PCPP solution into a calcium chloride solution, creating gel microbeads that dissolve under physiological conditions (Andrianov et al. 1994). The chitosan (CS)-based hydrogels, which are usually composed of CS and glycerol-phosphate disodium salt, crosslinked *via* ionic interactions, where the chitosan and salt solution remains liquid at room temperature and forms gels by heat treatment (Chenite et al. 2000). It is worth noting that in some cases, the ions can encourage hydrogel formation with no need for ionic groups in the polymer. For example, the dextran that lacks ionic binding sites, can form a hydrogel in the presence of potassium (K^+) ions. It has been demonstrated that the K^+ ion can fit perfectly into the cage formed by glucose's oxygen atoms in the polymer chains, creating a hydrogel microstructure (Watanabe et al. 1996). It should be noted that the metal ions can also enhance hydrogel strength (Zhang et al. 1999), so that the hydrogels formed based on metal ions exhibit more proper mechanical behavior compared to those prepared in salt-free conditions (Hossain et al. 2001). Overall, the ionic crosslinking is reversible, and often occurs under mild conditions. Hence, the ionic crosslinked hydrogels can be used in various biomedical applications, including drug delivery systems, wound dressing, and tissue engineering.

6.3.1.3 Crystallite Formation

Polymer crystallization progresses *via* alignment of the polymer molecular chains (Carraher Jr 2003). The crystallite formation, which might be also recognized as the freeze-thaw technique, conducted by freezing a polymer solution, and then rapidly melting (or drying) it at room temperature. The mechanism of crystallite formation includes (i) *ice crystal formation*, in which the water molecules in hydrogel form ice crystals during the freezing step. The created ice crystals exclude the polymer chains, concentrating them within an unfrozen region. The second step is done *via* (ii) *polymer chain alignment*, where the polymer chains align themselves in an arranged manner around the ice crystals. The last step involves (iii) *melting and crystallite growth*, when the crystals melt upon thawing, making small voids. In this stage, the aligned polymer chains form physical crosslinks (crystallites). This cycle is frequently repeated until a 3D gel-like network is formed (Wahab et al. 2019). For instance, the polyvinyl alcohol (PVA), which is a water soluble polymer with a high hydrolysis degree, can gradually form a gel with low mechanical strength within aqueous solutions. However, one the aqueous solutions of the PVA undergo a freeze-thawing process, a highly elastic a strong gel can be created (Yokoyama et al. 1986). Generally, the crystallite formation avoids the incorporation of toxic chemicals. Also, the hydrogel's characteristics can be optimized *via* tuning the primary polymer solution concentration and the freeze-thaw cycles (Wahab et al. 2019). For instance, multistep cycles lead to a denser gel network, thereby improving mechanical strength (Alves et al. 2011). Simultaneously, a higher initial polymer solution concentration can enhance the final mechanical properties of the hydrogel (Chee et al. 2018), as the polymer's molecular chains interact more in the frozen state. The freeze-thaw method usually improves the mechanical properties, thermal stability, and water storage capability of the hydrogels.

6.3.2 Chemical Crosslinking

Chemical crosslinking of hydrogels entails the formation of covalent bonds between polymer chains. The hydrogels synthesized based on chemical crosslinking processes demonstrate high chemical and thermal stability, as well as proper mechanical behaviors due to their covalent bonding (Mushtaq et al. 2022). The chemical crosslinking includes various strategies, such as crystallite formation (Suo et al. 2018), radiation crosslinking (Huang et al. 2013), thermal crosslinking, and enzymatic crosslinking (Badali et al. 2021). In the following, we will describe some of the most widely used methods of chemical crosslinking.

6.3.2.1 Thermal Crosslinking

Thermal crosslinking involves heat treatment of polymer solution to facilitate the formation of covalent bonds between polymer chains. This process can be performed to polymer solutions that can sustain temperature changes during the synthesis process. Different mechanisms of thermal crosslinking include (i) *polymerization and crosslinking* that involve free radical polymerization. In the mechanism, thermal energy induces the generation of free radicals that initiate the polymerization of monomers and subsequently form crosslinks between polymer chains. The second mechanism entails (ii) *condensation reactions*, in which the thermal energy promotes the elimination of small molecules (such as water and alcohol) as by-products. The next mechanism uses (iii) *Diels-Alder reactions*, which is a thermally reversible cycloaddition between a diene and a dienophile. This mechanism can lead to the formation and breakage of crosslinks with temperature changes, providing self-healing properties. Overall, the thermal crosslinking is simple, however, its gelation times can be considerably longer than the other methods, and a well-controlled over degree of crosslinking by this method is challenging. Moreover, the high temperatures in heating steps required for thermal crosslinking of polymer chains can adversely affect biological species, limiting its applications for bioinks and living electrodes (GhavamiNejad et al. 2020).

6.3.2.2 Photo-Crosslinking

Photo-crosslinked hydrogels are synthesized through irradiating the polymer aqueous solution using ultraviolet-visible (UV-Vis) wavelengths, leading to curing the gel process. This approach conducts *in situ* formation of hydrogel with no additional hazardous chemicals. Also, the photo-crosslinking represents a cost-effective method as it can be carried out at room temperature, consuming less energy compared to other approaches. The photo-crosslinking has special importance in 3D printing applications, due to its facile operation, spatiotemporal, and remote control. The principle of 3D printing through photo-crosslinking is based on photocurable polymer bioinks in the presence of photoinitiators that can be crosslinked *via* different mechanisms such as step-growth mechanism, chain-growth process, and redox-based reactions (GhavamiNejad et al. 2020). Moreover, this method provide rapid gelation process, followed by producing transparent hydrogels appropriate for optical microscopy of biological species (Carvalho and Mansur 2017). Furthermore, the photo-crosslinking can be an ideal process due to its fast gelation times (several minutes) with no by-product output (Gajewski et al. 2021). By using the UV-Vis irradiation and low energy consumption for polymerization at low temperatures, the photo-crosslinking is an environmentally friendly process (Langevin et al. 2019).

6.3.2.3 High-Energy Radiation Crosslinking

Radiation crosslinking process deals with the high energy sources, including the microwaves, gamma, and UV wavelengths, which can provide free radicals necessary for the hydrogel formation (Zhao et al. 2023). Consequently, the radiation crosslinking can be classified as one of the radical polymerization methods. The wide radiation area as well as high penetration length of the high-energy electron beam, which can facilitate breaking down of the water molecules and crosslinking of the polymers simultaneously, leading to intermolecular radical recombination. This results to a rapid polymer-to-hydrogel conversion, accompanying with a more uniform gel microstructure (Huang et al. 2013). It is worth noting that the radiation crosslinking processes are relatively green methods that can be utilized for synthesis the hydrogels with no chemical initiators and crosslinkers. However, working with high-energy radiation sources can be challenging and requires assured safety protocols. In general, the hydrogels produced *via* the radiation crosslinking are pure and nontoxic, making them significantly appropriate for biomedical applications. However, at the present state, the high costs of radiation crosslinking processes limit it to small-scale laboratory production.

6.3.2.4 Enzymatic Crosslinking

Enzymatic hydrogels are covalently bonded through reactions that are catalyzed by enzymes. In comparison to the other crosslinking methods, the enzymatic crosslinking offers a highly controlled process benefits from mild synthesis conditions and rapid gelation stage (Luo and Lai 2019). Principally, due to the utilization of enzymes instead of toxic chemical compounds during the synthesis process, the enzymatic crosslinked hydrogels demonstrate proper biocompatibility (Ryu et al. 2022). Also, the injectable (or spray) hydrogels produced through *in situ* enzymatic crosslinking reactions can be utilized in various applications such as nerve regeneration and neural interfaces (Raia et al. 2017; Xiang et al. 2020).

6.3.3 Microfluidic Techniques and 3D Printing

Microfluidic-assisted gelation is a cutting-edge technique for precisely controlling the formation of hydrogels by using microscale fluid dynamics. In fact, by mixing and delivery of polymer precursors and crosslinking agents at the micro- and nanoscales, the microfluidic-based synthesis methods allow for a precise control over gelation kinetics, morphology, and spatial distribution of the crosslinks. Moreover, optimizing the process parameters such as flow rates, channel geometries, and chemical compositions, can improve the properties of hydrogels with high reproducibility. Microfluidic-assisted gelation presents several advantages, including (i) rapid and homogeneous gel formation, (ii) the ability to create complex and gradient structures, and (iii) compatibility with high-throughput screening for hydrogel formulation optimization. According to the nature of the polymers' backbone and their functional groups, hydrogel bioinks can be crosslinked *via* different methods, including physical, chemical, or a combination of them. The chemical crosslinking of hydrogel bioinks occurs via non-reversible covalent bonding between polymer chains followed by network formation, usually through addition of chemical crosslinkers such as thrombin (Pimentel et al. 2018) and sodium bicarbonate (Ng et al. 2018), or *via* chemical reactions, including enzymatic crosslinking (Carvalho et al. 2018), thiolene coupling (Bertlein et al. 2017), hydrazide-aldehyde coupling (Kageyama et al. 2016), azide-alkyne cycloaddition (Madl et al. 2016), and Schiff base chemistry (Du et al. 2017). Additionally, the chemical crosslinking of hydrogel bioinks can be facilitated using visible (Xu et al. 2018), ultraviolet (UV) (Hu et al. 2018), and near-infrared (Dinh et al. 2017) radiations. As we discussed, these chemically crosslinked hydrogels exhibit adequate mechanical strength to provide appropriate shape stability, however, the crosslinking kinetics should be controlled to prevent the 3D printer nozzle blockage. On the other hand, the physical hydrogels although represent poor mechanical properties, they can supply appropriate cell-friendly environments compared to the chemically crosslinked hydrogels. Nonetheless, to address the mechanical shortcomings of physical hydrogels, chemical functionalities or nanoscale reinforcements can be introduced to hydrogel structures to improve its stability required for 3D bioprinted constructs (Hribar et al. 2014). However, it is worth noting that not only the method of crosslinking is important for 3D bioprinting of hydrogels, but also the formed crosslinks concentration should be considered as a key parameter. In fact, achieving a desired balance between the crosslinking degree and printability is essential. Reducing the crosslinking degree might lead to rapid flow of the bioink, while increasing it may result to higher stiffness that hamper printability. Overall, the microfluidic-assisted techniques for fabrication of hydrogels holds great potential in various biological applications such as tissue engineering, drug delivery, biosensors, and neural interfaces, where precise control over hydrogel's properties is critical for achieving desired functionalities and performance.

6.4 CPs and Hydrogels Composites and Hybrid Systems

The primary motivation for creating composites and hybrid systems of CPs and/or hydrogels for neural interfaces is improving their electrical conductivity, mechanical behavior, and biological responses. These strategies have particularly attracted significant attentions for developing neural interfaces that mimic neural tissues' properties. In fact, CPs are ideal materials for integrating with hydrogels to create composites due to several reasons. Firstly, some similar soft and highly flexible mechanical behaviors of CPs and hydrogels prevent dynamic mismatches, reducing additional stresses on the neural implant. Moreover, their synthesis methods are generally carried out *via* similar conditions and parameters, such as process temperature, formation media, and the same fabrication equipment. Next, we will introduce state-of-the-art CPs/Hydrogel composites and hybrid systems developed for neural interfaces.

6.4.1 Conductive Polymer-Based Hydrogels (CPHs)

Among various hydrogels, the CPHs are the most promising candidates for neural interfaces and BCIs. The CPHs can be fabricated *via* different synthesis methods to obtain desired characteristics, ranging from molecular structures to microscale configuration. One of the main strategies for producing CPHs is copolymerization of the CP's monomers with non-conductive polymers. However, it should be noted that the most of CPHs synthesized by this route show non-elastic behavior and suffer from inadequate stimuli-responsiveness because of the constrained mobility of their chains, which is originated from the robust mechanical interlocking or chemical interactions between the polymer matrix and fillers (Haraguchi and Li 2005). In fact, although many kind of hydrogel materials have shown promising properties for bioelectronics because of their unique advantages like excellent biocompatibility, water storage capability, and tissue-like softness, there are still significant challenges in their homogeneity, electrochemical performance, and seamless integration with flexible devices. Accordingly, many efforts have been made to improve the electrical properties and mechanical behavior of hydrogels simultaneously. In this regard, researcher developed homogeneously CPH interfaces comprised of conductive polyacrylamide (PAA) hydrogel incorporated by PANi-decorated PEDOT:PSS networks (Shin et al. 2024). The prepared CPHs demonstrated low impedance and high electrical conductivity. Moreover, the interfaces exhibited highly robust adhesion and reliable integration with flexible devices *via* hydrogen and covalent bonding as well as on-device polymerization. Figure 6.4 shows detailed configuration, properties and performance of these CPH interfaces.

In CPHs, a hydrophilic non-conductive polymer matrix can serve as a fragment for the final copolymer's chains or the crosslinking unit to link the CPs. Hence, for integrating the CP's chains into the non-conductive polymer matrix and obtaining a through conductive network within CPHs, the copolymerization process can be carried out either in a two-step fabrication method (Fig. 6.5a, b), or *via* a one-step self-assembly approach (Fig. 6.5c).

6.4.1.1 Two-Step Synthesis Processes

The two-step synthesis method includes two main routes for preparing the CPHs networks:

1. *Creating CP networks within a hydrogel matrix*. In this method (Fig. 6.5a), a hydrophilic non-conductive hydrogel is initially formed through UV radiation-based curing or free radical polymerization (Chen et al. 2019b; Wei et al. 2021; Zhang et al. 2022). Afterwards, by adding the CP's precursor into the hydrogel matrix (or soaking the hydrogel in a solution containing the CP's monomer), the

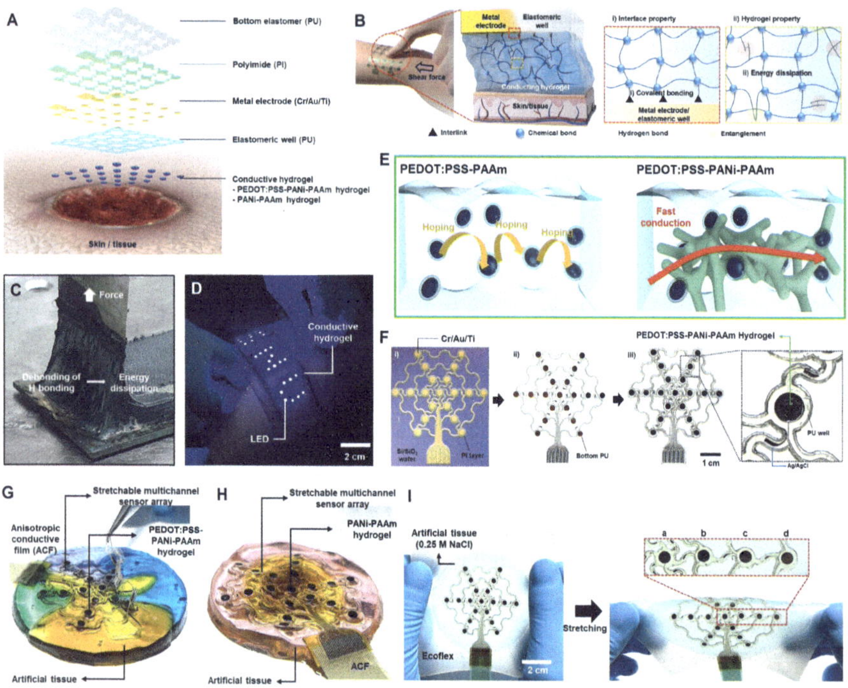

Fig. 6.4 (**a**) Expanded configuration of the CPH interface based on PEDOT:PSS-PANi-PAAm hydrogel. (**b**) Schematic illustration for showing the mechanisms of (left) seamless and reliable adhesion between CPH interfaces and flexible device, and (right) robust bonding *via* surface covalent bonding and bulk hydrogen bonding. (**c**) Optical image of the prepared CPH interface undergoing a peeling test. (**d**) An image depicting appropriate electrical conductivity of CPH interfaces to turn on a wearable light-emitting diode (LED) array. (**e**) Conduction mechanisms before and after treatment of CPHs with PANi. (**f**) Fabrication steps for a flexible multichannel sensor array using PEDOT:PSS-PANi-PAAm hydrogels. (**g, h**) Stretchable multi-channel sensors made by the prepared CPH interface on the artificial tissue having different ion concentrations and pH values, respectively. (**i**) Optical images of the stretchable multi-channel sensors on the artificial tissue (left) before, and (right) after stretching, demonstrating excellent mechanical performance of the fabricated CPH interfaces (Shin et al. 2024)

electron transferring pathways are created *via* absorbing or diffusing the CP's monomers into the porous hydrogel matrix. This step follows by *in situ* redox polymerization to form a through conductive CPH network (Cong et al. 2021; Wang et al. 2021b; Sun et al. 2022). Figure 6.5a shows the fabrication process of CPHs *via* creating CP networks in a hydrogel matrix. Using this method, researchers fabricated a conductive PPy-PAM/CS hydrogel by using a molecular chitosan template for controlling the *in situ* polymerization of PPy nanorods in the hydrogel network (Gan et al. 2018). They demonstrated that the hydrophobic PPy nanorods were uniformly distributed within the hydrogel matrix and strongly bound to the hydrophilic polymer phase within the PPy-PAM/CS,

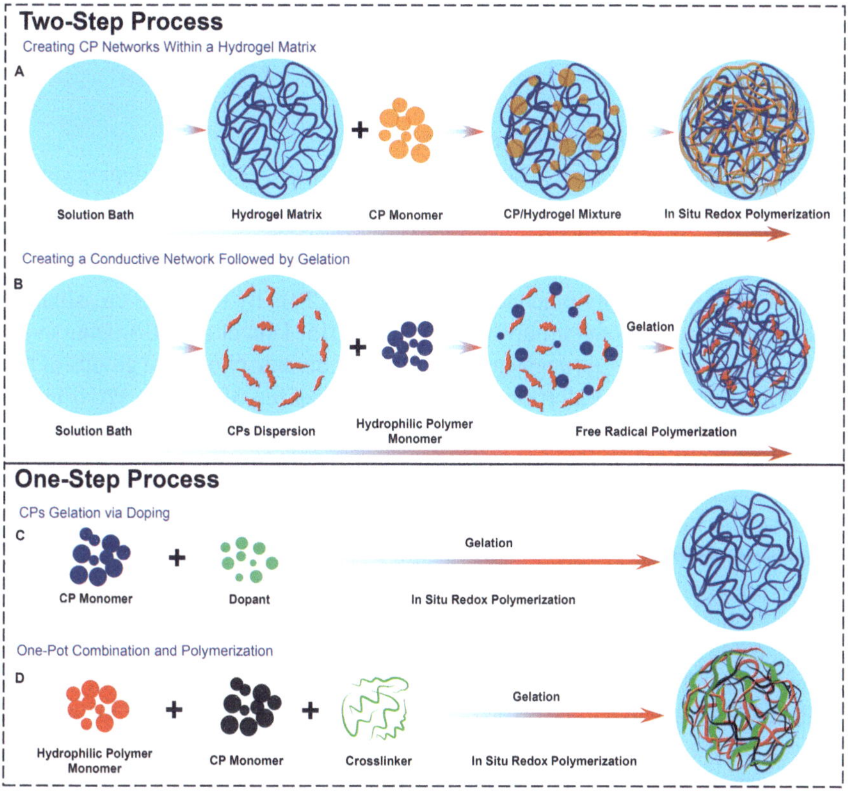

Fig. 6.5 Schematic diagram of the main synthesis methods of CPHs. (**a**) Creating CP networks within a hydrogel matrix, (**b**) creating a conductive network followed by gelation, (**c**) CPs gelation *via* doping, and (**d**) one-pot combination and polymerization. (Liu et al. 2023)

providing excellent mechanical strength accompanying with appropriate electrical conductivity.

2. *Creating a conductive network followed by gelation.* In this method, for enhancing the water solubility and increasing the capacity of CPs to make uniform mixtures with other monomers, the CPs are modified with surfactants or amphiphilic molecules. Subsequently, the free radical gelation process is conducted to form the CPHs. Figure 6.5b illustrates the fabrication of CPHs *via* this approach. Recently, some research groups fabricated a CPH made by strand-entangled supramolecular PANi/PAA hydrogels through the "doping then gelling" strategy (Liu et al. 2022). For this, they doped the PANi nanorods with an acrylic acid (AA) monomer. Their results indicated that the AA-doped PANi nanorods not only possessed auspicious dispersion capability, but also enhanced electrical conductivity. In 2020, researchers developed different CPHs *via* doping the various CPs (PEDOT, PPy, and PANi) by catechol-quinone containing of sulfonated lignin to improve water solubility of the CPs, facilitating the CPHs formation

(Gan et al. 2020). Through well-constructed electrical pathways, the synthesized CPHs exhibited excellent electrical conductivity and long-term adhesion properties.

6.4.1.2 One-Step Synthesis Processes

(a) *CPs Gelation via Doping*. The One-step gelation of CPs through doping can be carried out *via* directly crosslinking of the CP's chains with no insulating (non-conductive) polymer matrix. Figure 6.5c illustrates a schematic representation for the one-step gelation of CPs based on doping for preparing of the CPHs. This method is highly practical for producing CPHs by using the crosslinkers having multifunctional groups (Wang et al. 2015; Dou et al. 2016; Ding et al. 2018). For example, through employing the six anionic phosphate groups of the phytic acid as gelling dopant, researchers synthesized 3D porous PANi-based CPHs (Ding et al. 2018). Also, the plant-derived polyphenol tannic acid, which contains multi-phenolic hydroxyl groups, can be used as dopant and crosslinker for fabricating the PPy-based conductive CPHs through making plenty of inter-molecular interactions (Zhou et al. 2018a).

(b) *One-pot combination and polymerization*. The CPHs can also be produced *via* one-pot combination of the CPs' monomers and hydrophilic polymer precursors, following by gelation through *in situ* polymerization, as is illustrated in Fig. 6.5d (Lee et al. 2015; Wu et al. 2017; Li 2021; Rani et al. 2022). Different approaches have been developed for one-pot synthesis method of CPHs using different CPs, including PEDOT, PANi, and PPy (Moussa et al. 2019; Zhao et al. 2020). For instance, through a one-pot low temperature polymerization process, researchers fabricated stretchable PANi/PVA CPHs for high-performance wearable strain sensors (Zhao et al. 2020). The prepared CPH sensors showed improved electrical conductivity, proper mechanical behavior, and excellent electrochemical properties, as well as ability for monitoring the human movements in real time. Furthermore, one of the main advantages of one-pot synthesis of CPHs is facilitating the self-assembly process of CPHs using solution-processing CPs, such as PEDOT:PSS (Kayser and Lipomi 2019; Ren et al. 2021). It should be noted that the blending of PEDOT with other non-conductive polymers usually lakes suitable electrical and/or mechanical properties. To address these issues, researchers fabricated a pure PEDOT with interconnecting nets of the PEDOT nanofibrils (Lu et al. 2019). Contrary to conventional methods, they applied a simple technique based on mixing aqueous PEDOT solutions with volatile organic solvents, which was followed by optimized dry-annealing and rehydration steps. This method resulted in producing the CPHs having excellent mechanical behavior, superior electrical conductivity, and tunable isotropic/anisotropic swelling within physiological environments.

6.4.2 Ion Conducting Organohydrogels (ICOHs)

Conductive hydrogels find applications in various fields, including artificial muscles, ionic skin, artificial axons, and central nervous system connections. For neural interfaces, the frequently employed hydrogels include ion-conducting hydrogels (ICHs), ion-conducting organohydrogels (ICOHs), and hydrogel composites (Ahmadi Seyedkhani and Mohammadpour 2022). The ICHs are characterized by their prominent hydrophilicity and possess a 3D hydrated network structure, leading to great ionic conductivity and mimic the ionic transferring mechanism of biological tissues. The ionic conduction in ICHs primarily entails a non-faradaic (capacitive) charge transfer mechanism, ideal for transmission high-frequency electrical signals over long distances. However, ICHs containing salts may face limitations in biocompatibility and stability due to the excessive ion releasing. Despite their potential for neural interfaces, ICHs suffer from losing the electrical conductivity, time-dependent reduction of flexibility, and morphological changes due to their dehydration. To address this issue, different strategies, including (i) adding the dehydration reagents (Zhang et al. 2020b; Chen et al. 2021), (ii) binding with sealing compounds (Zhang et al. 2020a), (iii) mixing by polymeric liquid gels (Sun et al. 2014), (iv) using deep eutectic solvents (DESs), and (v) adding organic solvents into hydrogels to make ICOHs have been suggested. By constraining dehydration, the ICOHs can maintain the hydrogels' benefits, but face challenges originated from high electrochemical impedance and inadequate long-term adhesion to biological tissue, limiting their neuroelectronics applications. Many works have been conducted to improve the ICOHs' performance *via* different methods, such as solvent replacement and gelation in binary systems (Ahmadi Seyedkhani and Mohammadpour 2022).

6.4.3 Interpenetrating Polymer Networks (IPNs) and Semi-IPNs

Interpenetrating polymer networks (IPNs) are blends formed *via* entanglement of at least two polymers within a network. Unlike conventional polymeric blends, the IPNs lack covalent bonds between their components, resulting in that they remain intertwined until their chemical bonds are broken (Banerjee et al. 2010). These polymer networks are typically classified into two types, including (i) full IPN that comprise two crosslinked polymer networks, and (ii) semi-IPN that consist of only one crosslinked polymer network. The IPNs are typically prepared through initially formation of one network followed by formation of the second one through external stimuli such as light, heat, or chemicals. It is worth noting that the formation steps occur sequentially rather than simultaneously. In fact, the formation of first rigid crosslinked network, which possesses basis mechanical properties, is followed by the formation of a network with a lower crosslinking degree to fill gaps and disperse external stress. Therefore, the IPNs can be applied to enhance the mechanical properties of hydrogels, which leads to development of their biomedical applications such as mimicking ECMs and cell culture scaffolds (Jenkins et al. 1996). Moreover, the IPN films exhibit appropriate adhesion to substrates compared to the conventional polymer networks. In fact, despite the great capabilities of CPs and hydrogels for different biomedical applications, the weak and unstable adhesion to substrates within the wet physiological environment has significantly limited their performances. To address this issue, introducing the hydrophilic polymers into the CPs to form IPN intermediate layers between the CPs and underlying substrates has been proposed (Inoue et al. 2024). Figure 6.6 shows a schematic illustration for making IPN intermediate layer between CPs and different substrates and related mechanisms for improving its mechanical adhesion.

Using the proposed method, as illustrated in Fig. 6.6a, the adhesive layer can be applied to various substrates *via* different coating methods, including dip coating, spin coating, and spray. Then, different CPs can be fabricated on the modified substrate to form IPNs through different techniques such as electropolymerization of monomers (e.g., EDOT) and solvent casting of CP solutions (e.g., PEDOT, PANi, and PPy). It is worth noting that although the proposed method can significantly improve the mechanical behaviors of CP coatings, it does not compromise their electrical properties, making it promising approach for bioelectronics such as neural interfacing and BCIs (Inoue et al. 2024). Figure 6.7 illustrates the strong adhesion of CPs on different bioelectronic devices through creating IPN intermediate layers.

Generally, the IPNs can be produced *via* three primary approaches, including:

(a) **Direct Mixing:** This approach involves a straightforward blending of CPs with hydrogel precursors. By this route, the CPs are evenly dispersed within the hydrogel matrix, forming a uniform interconnected network. This approach is relatively simple and enables the incorporation of CPs into the hydrogel structure without the need for additional processing steps.

(b) ***In situ* Polymerization:** In this method, polymerization of the CPs occurs within the hydrogel matrix. This process contains introducing of the polymer monomers or semi-polymerized CPs into the hydrogel precursor solution, followed by initiation of polymerization within the hydrogel network. After the polymerization process, the CPs integrate into the hydrogel network, resulting in an IPN hydrogel with improved mechanical and electrical properties.

(c) **In-growth Polymerization:** In this method, CPs are polymerized directly within the pre-formed hydrogel matrix. Initially, monomers or pre-polymerized CPs are introduced into the hydrogel, where they diffuse and polymerize to form a conductive network throughout the hydrogel matrix. This approach allows for precise control over the distribution and density of CPs within the hydrogel, leading to enhanced electrical conductivity and mechanical strength.

Figure 6.8 illustrates schematic representations of the IPNs production processes.

However, it should be noted that the fabrication of IPN electrodes faces substantial challenges, primarily focused on integrating CPs and hydrogel into a consistent configuration. Actually, not every amalgamation of CPs and hydrogels necessarily ends up producing IPNs, which exhibit both soft and electroactive characteristics

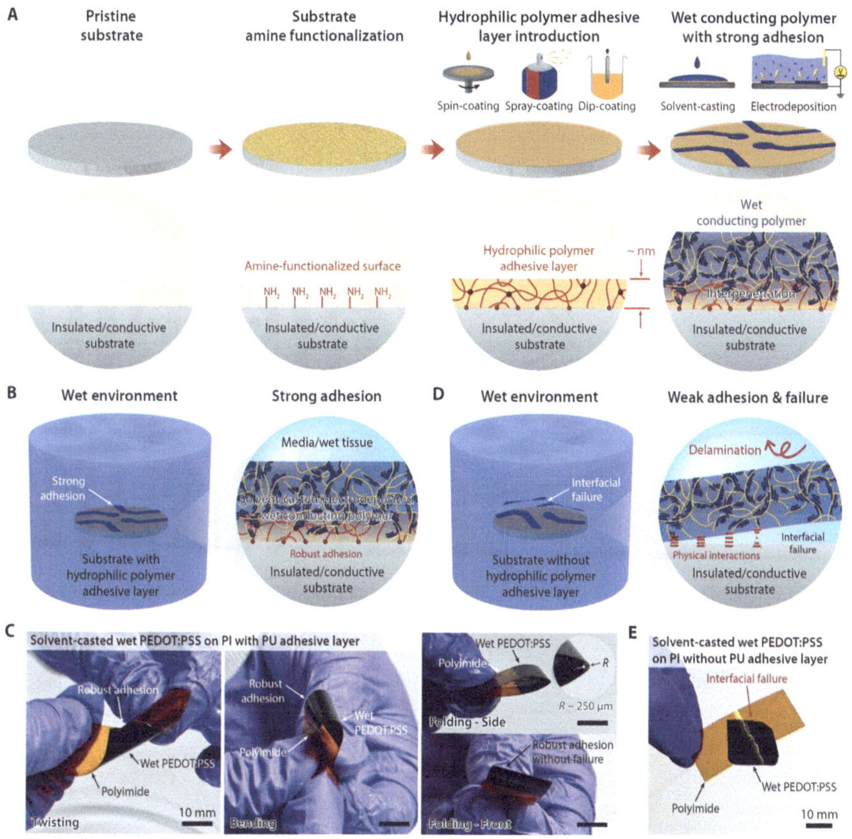

Fig. 6.6 (**a**) Schematic illustration for preparing a wet PEDOT:PSS coating with high adhesion strength on an amine-functionalized substrate due to formation of an IPN intermediate layer based on the PEDOT:PSS and a hydrophilic polymer. (**b**) Robust adhesion of the PEDOT:PSS layer to the functionalized substrate in wet environment. (**c**) The wet PEDOT:PSS coating on a polyimide (PI) substrate with polyurethane (PU) adhesive layer fabricated *via* solution casting technique. (**d**) Unstable coating with poor adhesion strength on a substrate without the adhesive layer. (**e**) The PEDOT:PSS layer on PI substrate without the adhesive layer (no IPN intermediate layer) exhibits interfacial separation upon substrate's mechanical deformation. (Inoue et al. 2024)

simultaneously. To address this issue, researchers developed a novel hydrogel system designed, by elucidating the role of immobilized dopants in the formation of conductive hydrogels, to facilitate the precise placement of covalently immobilized dopant groups within the hydrogel matrix (Goding et al. 2017). A key finding of the study was about the significance of interdopant spacing in the formation of IPNs. They found that the smaller interdopant spacing, as a determinant factor, can leads to the formation of IPNs with superior charge storage capacity (CSC) and lower electrochemical impedance (Goding et al. 2017).

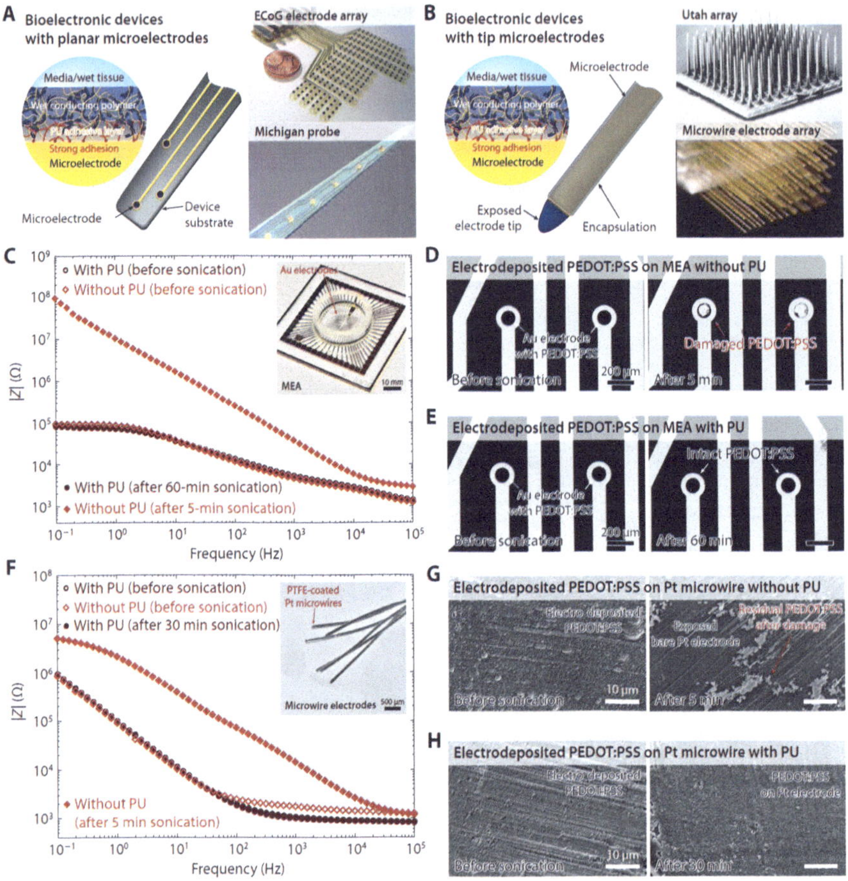

Fig. 6.7 Robust adhesion of CPs on different bioelectronic devices through IPN intermediate layer: (**a**) Wet CPs on (**a**) planar and (**b**) tip microelectrode arrays (MEAs). (**c**) Impedance changes and (**d**, **e**) optical images of PEDOT:PSS coating layers on Au MEAs with and without the adhesive layer, demonstrating better electrochemical properties and improved mechanical behavior of the CP layers *via* IPN intermediate layers. (**f**) Impedance changes and (**g**, **h**) SEM images of wet PEDOT:PSS layer PTFE-coated Pt microwires with and without the adhesive layer. (Inoue et al. 2024)

6.4.4 Dual Network Hydrogels (DNHs)

Dual network hydrogels (DNHs) are a class of interpenetrating configurations composed of two distinct polymer networks within a single hydrogel matrix, which can be considered as a special kind of IPNs. However, the DNHs consist of two independent crosslinked networks: (1) a rigid and brittle network with a highly crosslinked structure that provides initial strength and structural integrity, while the

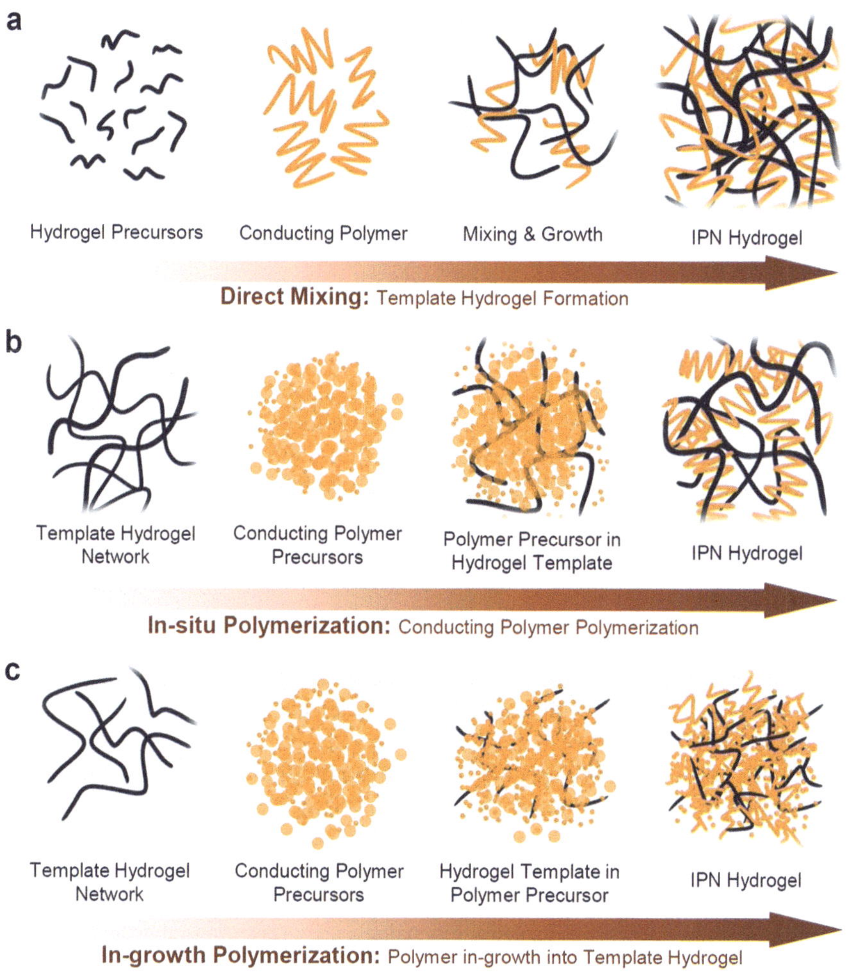

Fig. 6.8 The primary approaches for creating conductive IPN hydrogels include: (**a**) direct mixing that contains blending of CPs with the hydrogel's precursors, (**b**) *in situ* polymerization, which involves polymerizing CPs within the hydrogel matrix, and (**c**) in-growth polymerization through integrating CPs *via* polymerization processes into the hydrogel matrix. (Ahmadi Seyedkhani and Mohammadpour 2022)

another is (2) a more flexible and loosely crosslinked network that can interpenetrate the primary network and deform significantly without breaking. Having this two-phase interpenetrating structure, the DNHs exhibit typically superior mechanical properties and unique functionalities that surpass those of single-network hydrogels. The DNHs can be generally categorized into three classes, including (i) the hybrid hydrogels comprising of (first) a physical crosslinked polymer network and (second) a covalent bond polymer network, (ii) the physical hydrogels containing dual crosslinked polymer networks, and (iii) the covalent hydrogels with double covalent bond polymer networks. Hence, through combining two distinct polymer networks to obtain appropriate mechanical behavior, including high Young's modulus, and proper compressive and tensile properties, the DNHs have found promising capabilities for dynamic biomedical applications such as soft robotics, contact lenses, wound dressing, and artificial muscles (Chen et al. 2019a).

6.4.5 Layer-by-Layer (LBL) Assembly and Thin Films

Layer-by-Layer (LBL) assembly is an approach for synthesis multi-layered thin films through highly precise control over their properties at molecular level. The LBL assembly methods are particularly beneficial for combining CPs and hydrogels, which can be engineered to have appropriate unique electrical and mechanical characteristics for numerous applications, including biosensors and neural interfaces.

The conventional LBL assembly processes involve the sequential adsorption of alternating layers of oppositely charged materials onto underlying substrate. The process can be summarized as following steps:

(a) *Surface Preparation*. To ensure adequate interactions between the substrate's surface and materials of the first layer, surface cleaning and modification procedures should be carried out. This can also involve surface functionalization approaches to introduce specific chemical groups onto each component.
(b) *Adsorption of the First Layer*. There are different methods for precipitation of the first layer, including deep coating, spin coating, chemical bath deposition (CBD), CVD, physical vapor deposition (PVD), successive ionic layer adsorption and reaction (SILAR), and electrochemical methods (Richardson et al. 2016). In this regard, the substrate is usually exposed to an environment containing the material of the first layer (e.g., a positively charged CP monomers present within solution). Therefore, the desired material adsorbs onto the substrate's surface because of hydrogen bonding, electrostatic interactions, or van der Waals forces.
(c) *Rinsing*. After the first layer precipitation, the substrate should be rinsed to remove any unbound material and undesired impurities, ensuring a pure uniform layer.
(d) *Adsorption of the Second Layer*. This process can be carried out by the techniques offered in step b, which of course are selected according to material of the second layer (e.g., a negatively charged hydrogel). This material forms a layer over the first due to their complementary interactions.

(e) *Rinsing and Repetition*. The substrate, which now has a double layer coating, is rinsed again to remove undesired materials. The steps b to d are repeated to create multiple layers, achieving a desired thickness and composition.

As we found out, the LBL assembly processes allow for precise control over the material, thickness and composition of each layer, enabling the building up of complex multilayer structures. Moreover, a wide range of materials, including CPs, hydrogels, proteins, and nanoparticles, can be used to form each layer. Hence, different functionalities, such as electrical conductivity, mechanical strength, and biological responses, can be integrated into each layer. Nevertheless, ensuring the long-term stability of LBL films, especially in biological environments, is a significant challenge. Also, developing efficient processes for the large-scale production of LBL films can facilitate their widespread applications (Richardson et al. 2016; Liu et al. 2023).

6.4.6 Hybrid Nanocomposites

As we discussed, the charge transfer mechanisms within hydrogels and CPs are mainly based on the movement of ions. Therefore, integration of electron-conducting nanostructures, including metal nanoparticles, nanowires and nanomeshes, carbon nanotubes (CNTs), graphene, and quantum dots (QDs) into the CPs and hydrogels can enhance their synergistic electron-ion conductivity (Shahab Ahmadi Seyedkhani 2024b; Ahmadi Seyedkhani and Mohammadpour 2022). Moreover, fabricating hybrid nanocomposites using these materials can improve mechanical behaviors, electrochemical properties, and biological responses of neural electrodes. For instance, carbon-based materials are especially advantageous for blending with hydrogels and CPs due to their outstanding conductivity, large specific surface area, proper chemical stability in biological environments, and the ability to form covalent bonds with different polymer groups. Consequently, the CP- and hydrogel-based nanocomposites reinforced with carbon-based materials can exhibit improved neural cell-electrode interactions, thereby facilitating more efficient signal transmission between biological tissues and external electronics. To creating nanocomposites, different fabrication methods can be employed, including (i) direct formation of hydrogels and CPs within nanoparticle suspensions, (ii) physical embedding of nanoparticles into the structure of CPs and hydrogel after gelation, (iii) synthesis of nanoparticles within the CPs and/or hydrogel precursors, (iv) utilization of nanoparticles for crosslinking and polymerization (Thoniyot et al. 2015). Figure 6.9 shows some hybrid nanocomposites fabricated by different additives to CPs and hydrogels.

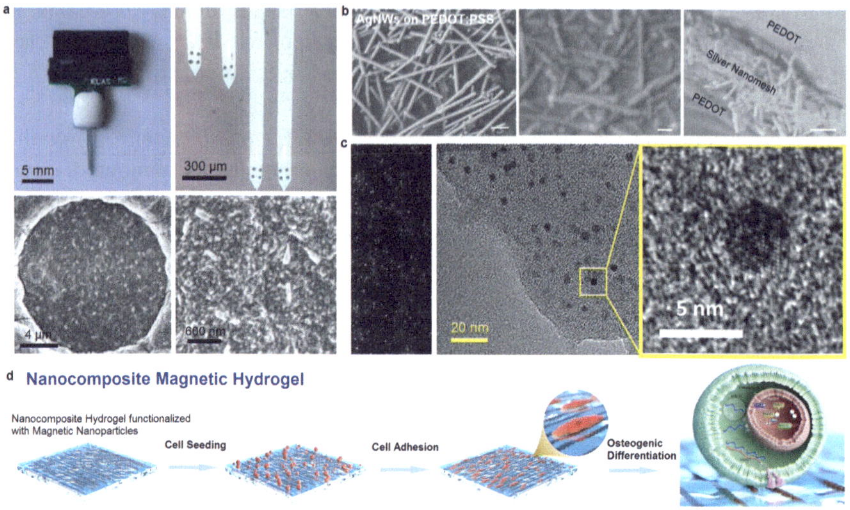

Fig. 6.9 Different Composites and hybrid systems based on CPs and hydrogels. (**a**) Design and fabrication of MEAs comprising of PtNP/PEDOT:PSS nanocomposite sensing pads (Lu et al. 2023), (**b**) surface and cross-sectional FESEM micrographs of PEDOT:PSS-AgNWs composites with interpenetrating networks (Namkoong et al. 2022), (**c**) FESEM and TEM images of a nanocomposite prepared using silver quantum dots (AgQDs)-embedded PEDOT through VPP (Shi et al. 2022), and (**d**) nanocomposite magnetic hydrogels with dual anisotropic properties for inducing osteogenesis. (Tang et al. 2024)

6.5 Applications

CPs and hydrogels have attracted significant attentions in biomedical researches, particularly in development of neural interfaces. In this section, we will delve into utilization of these materials in various neural applications, including recording and stimulating devices, neural regeneration, drug delivery, tissue engineering, artificial synapses, molecular bioelectronics, artificial muscles, and biodegradable electronics.

6.5.1 Recording and Stimulation Devices Based on CPs and Hydrogels

- *Recording Devices*

The primary goals behind using CPs and hydrogels in neural recording electrodes are improving bioactivity, electrochemical impedance, and more sensitivity, which allow reducing the electrode size followed by the detection of weak electrophysiological signals and individual cell's activities (Green and Abidian 2015). Furthermore, the unique molecular structure of CPs activates ion-electron synergistic charge transfer mechanisms, enhancing signal-to-noise ratio (SNR) in recording processes (Ahmadi Seyedkhani and Mohammadpour 2022).

Recently, CPs have attracted many attentions to fabricate transparent and flexible MEAs, particularly for monitoring brain electrical signals and optical imaging of neuronal networks simultaneously. Among CPs, the PEDOT stands out due to its appropriate transparency, making it excellent for applications requiring both optical access and electrical functionality. However, it should be noted that the fabrication of PEDOT interfaces requires precise procedures to prevent organic material degradation, which could compromise their performance and lifespan. Recent advancements have introduced a technique for incorporating transparent PEDOT:PSS electrodes for neural recording and study the intricate brain's neural circuitry (Dijk et al. 2022). Figure 6.10 shows the fabricated transparent and flexible PEDOT:PSS MEAs. Using these MEAs, researchers carried out advanced imaging based on

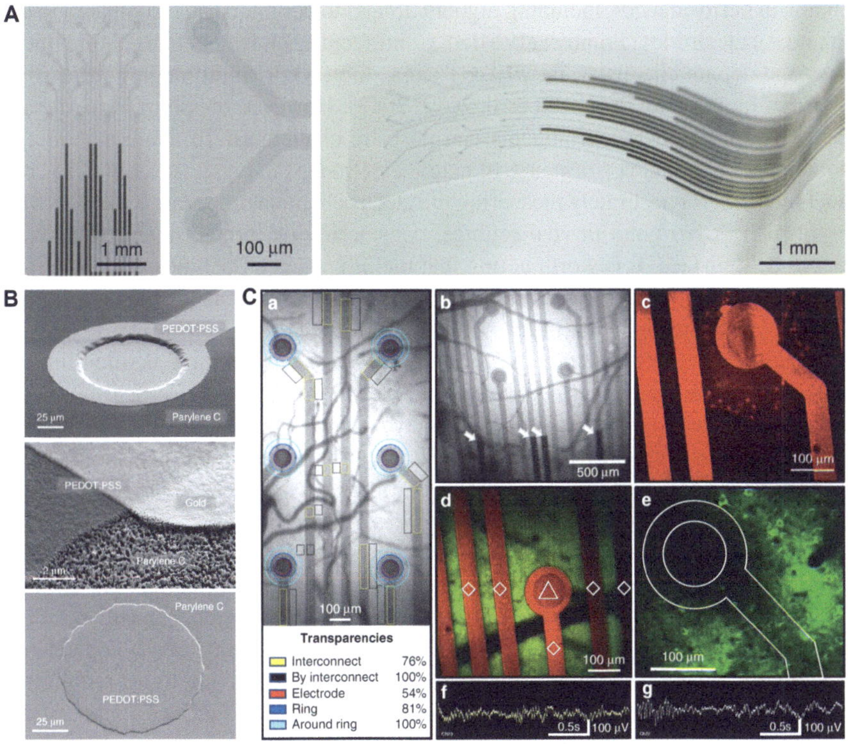

Fig. 6.10 Transparent and flexible PEDOT:PSS MEA: (**A**) Optical image of (left) the MEA featuring 16 PEDOT:PSS sensing electrodes, (middle) enlarged view of two sensing electrodes, (right) the MEA positioned on a curved agarose brain model. (**B**) Scanning electron microscope (SEM) images of the PEDOT:PSS sensing pad. (**C**) Imaging and recording of local field potential (LFP) of the PEDOT:PSS neural interfaces on the mouse' cortex: (a, b) infrared images, (c-e) 2-photon images. (f, g) LFP recording from the neural interfaces in (d) and (e), respectively. (Dijk et al. 2022)

optical techniques such as 2-photon microscopy, and recorded the local field potentials from the mice's cortex. Also, the electrodes were able to maintain adequate transparency to visualize neurons expressing GCaMP6f (a genetically encoded calcium indicator) (Dijk et al. 2022). These achievements underscore the capability of CPs for developing transparent MEAs to facilitate *in vivo* studies of neuronal activities with exceptional spatial and temporal resolution.

- *Stimulating electrodes*

Hydrogels and CPs also hold potential in development of stimulating devices, such as neuroprostheses, by addressing the need for smaller electrodes while ensuring safety and efficacy. Traditionally, Pt and Au electrodes were used in such devices, however, their CIS pose limitations, resulting in adverse chemical reactions at the tissue-electrode interface, followed by tissue damage and electrode destruction (Green et al. 2014). While, the CPs and hydrogels offer appropriate electrochemical performances, including higher CIS and lower impedance compared to the Pt and Au electrodes (Pranti et al. 2018). Contingent upon factors like coating thickness and dopant chemistry, the PEDOT can exhibit a CIC limit ranging from 10 to 30 times higher than Pt (Green et al. 2012, 2013). However, obtaining these advantages within *in vivo* environments can pose by challenges. To attain a thorough understanding about performance of neural electrodes *in vivo*, it is essential to conduct methods for accurately measuring the electrochemical impedance and CIS limits in both *in vitro* and *in vivo* settings, considering the protein interactions with electrode's surface. It is worth noting that the *in vivo* electrical measurements face challenges due to spatial constraints, the complex biological milieu, and tissue-implant mechanical disparities. Actually, a reliable three-electrode system is required for accurate electrical performance assessment, but there is often limited space for a reference electrode *in vivo*. Furthermore, *in vitro* electrical measurements conducted in physiological saline may not fully represent the protein interactions that occur with electrodes *in vivo*. These protein interactions can significantly impact electrode's performance, creating differences between *in vitro* and *in vivo* data. Studies have shown that CP coatings can be mechanically compromised during electrical stimulation, highlighting the need for robust CP coatings that can withstand such conditions. For instance, making mechanical interlocking through surface roughening of substrate electrode prior to CP deposition, can enhanced the CP coating durability. Furthermore, the detection of weak electrophysiological signals and/or the application of accurate currents to targeted living tissue require the advanced hydrogel and CP electrodes with appropriate design and composition. Recently, organic electrochemical transistors (OECTs) based on CPs have been developed to overcome this issues (Kim et al. 2018). Figure 6.11 illustrates a high performance electrochemical transistor based on crystalized PEDOT:PSS with long-term stability. Due to the synergistic electron-hole and ion transfer mechanisms within aqueous environment, the OECTs show great potentials for nanobioelectronics and BCIs.

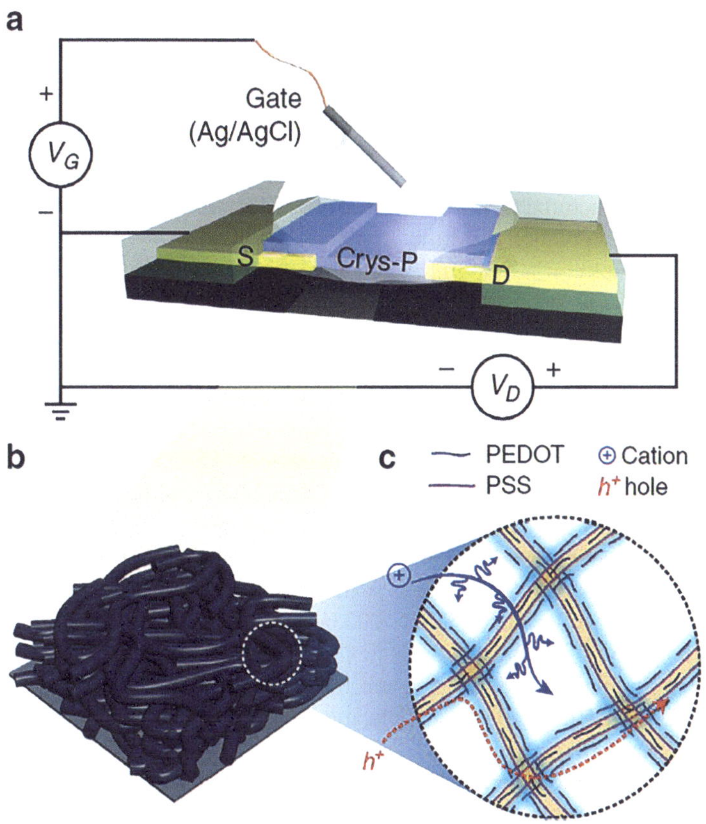

Fig. 6.11 (**a**) An illustration of the device configuration of Crystalized PEDOT:PSS organic electrochemical transistor. (**b**) A schematic diagram of the crystalized PEDOT:PSS film microstructure and (**c**) the hole and ion transport in that. (Kim et al. 2018)

6.5.2 Tissue Engineering and Neural Regeneration

In nerve tissue engineering, fabricating biomaterials based on CPs and hydrogels presents a diverse array of opportunities for creating advanced scaffolds that support cell adhesion and proliferation, followed by tissue regeneration (Petty et al. 2020; Rocha et al. 2021). Hydrogels and CPs can be used as substrate to guide and facilitate axonal outgrowth, which is critical for functional recovery of the nerve system damages. However, a successful tissue regeneration is often confronted with significant challenges arising from the post-injury intricate biological milieu. Nonetheless, there are some approaches to promote axonal outgrowth, including electrical stimulation (Schmidt et al. 1997; Kotwal and Schmidt 2001; Gomez et al. 2007; Park et al. 2007), topographic contact guidance (Lee et al. 2009a; Abidian et al. 2010; Ho et al. 2015), growth factor delivery (Thompson et al. 2006, 2010), and biomolecule doping (Gomez and Schmidt 2007; Green et al. 2010), which can appropriately conducted using CPs and hydrogels. For example, the PSS-doped PPy electrodes have shown capability for stimulating the neurite outgrowth (Schmidt et al. 1997). The results revealed that the electrical stimulation by PPy films leads to a significant increase in neurite length, demonstrating the great potential of PPy and other CPs for electrical guidance of the axons. Also, topographic contact guidance through micro and nanostructured CPs has also been studied by linear PEDOT:PSS nanoparticles arrays, which guided dendritic growth and enhanced cell area, approving the effect of topography in promoting neurite outgrowth (Gomez et al. 2007; Abidian et al. 2010; Ho et al. 2015). Moreover, the CPs doped with biomolecules, such as laminin peptides and growth factors, can enhance neurite outgrowth and promote axonal regeneration, although they may exhibit reduced electrical performance and mechanical stability (Green et al. 2009, 2010). Recent studies have indicated *in vivo* axonal regeneration using electrochemically polymerized PEDOT embedded within a tubular agarose gels for an animal nerve-gap model (Abidian et al. 2012).

Moreover, due to the inherent properties of polymers and their simple processing, fabricating CPs and hydrogels in the form of nanofibers is comparatively easier than producing nanofibers from metals and semiconductors. By mimicking the ECM nanoscale structure, nanofibrous scaffolds can promote cell proliferation and differentiation (Barjasteh et al. 2022, 2023a, b, 2024; Ahmadi Seyedkhani et al. 2023; Dehnavi et al. 2023). Hence, the flexible conductive nanofibers can provide a 3D microenvironment to cell growth and function, making them promising candidates for neural tissue engineering applications (Bertuoli et al. 2019). For instance, the cardiac tissue engineering utilizes CP containing nanofibers to create anisotropic scaffolds that promote cardiomyocyte alignment and synchronous beating through electrical stimulation. Accordingly, the CPs nanofibers designed for cardiac tissue engineering, can further modulate the heartbeat frequency, enhancing tissue development. Moreover, covalently CP-biomolecules attachment, containing reactive groups such as $-NH_2$ or carboxylic acid (R–COOH) reduces biomolecule denaturation and leaching (Berezhetska et al. 2015). For example, doping PPy with polyglutamic acid (PGlu) provides R–COOH groups for supplementary functionalization, enabling cell adhesion and facilitating the incorporation of various biomolecules, leading to complex 3D substrates for tissue engineering and controlled protein releasing (Jin et al. 2016; Qin et al. 2016; Zhou et al. 2018b). Similarly, nerve growth factor (NGF) can be immobilized on CPs films using intermediate photocrosslinkers, demonstrating the potential of CPs that can provide covalent bonds for tissue engineering scaffolds, which offers both electrical and biological stimulation simultaneously (Kim et al. 2007; Lee et al. 2009b; Wang and Feng 2022b). Furthermore, studies have exhibited controlled proliferation and differentiation of stem cells cultured on the conductive substrates (Wang and Feng 2022b), demonstrating high capability of CPs nanofibrous to boost tissue engineering.

6.5.3 Drug Delivery

Smart drug delivery methods aim to boost treatment efficacy and reduce side effects. These goals can be achieved by suitable therapeutic agent carriers, which require appropriate material design and composition. In this regard, the CPs enable "dry" drug delivery, localized implantable devices, controlled electrical releasing, and personalized treatment based on closed-loop systems (Raj et al. 2023). The CPs and hydrogels coatings undergo volume changes through redox reactions, facilitating drug incorporation and release processes. Various therapeutic agents, including cationic and anionic drugs, neutral ions, anti-oxidants, and antimicrobial materials can easily be integrated into the polymer and/or hydrogel structures *via* chemical or electrochemical polymerization processes. In addition, the creating composites of CPs and hydrogels using nanomaterials such as graphene, CNT, and metal-organic frameworks (MOFs) increases the specific surface area for drug storage, greater ion exchange and faster redox cycling reactions, and capability to controlled drug releasing *via* external stimuli, allowing for more efficient treatment (Bansal et al. 2020; Raj et al. 2023). Many CPs undergo reversible redox processes that help the ions mobility for releasing drugs. Drug releasing mechanisms by CPs and hydrogels can be categorized in different types, including chemical or electrical approaches. The chemical release can be conducted by redox reagents or pH changes, while the electrical release utilizes external potentials or currents to oxidize or reduce the coating carriers. However, it should be noted that the requirement for external source stimuli might limit *in vivo* applications.

The multilayer CP systems inhibit spontaneous release until they are fully degraded, providing controlled drug releasing process (Wang and Feng 2022b). As mentioned, the CPs can offer precise control over drug release through reversible redox reactions. The anionic drugs are released during reduction, while the cationic drugs are released during oxidation (Alshammary et al. 2016; Puiggalí-Jou et al. 2019). Hydrogels and CPs can also trap drugs or utilize ion-electron synergistic conducting mechanisms for ion delivery using electrophoretic ion pumps. These pumps offer spatial and temporal control over drug distribution without damaging tissues, which provide a promising method for highly precise drug delivery in biological applications.

For neural interfaces, there is a growing interest in utilizing CPs and hydrogels for therapeutic drug delivery (Pernaut and Reynolds 2000; Abidian et al. 2006; Thompson et al. 2006, 2010; Simon et al. 2009; Svirskis et al. 2010; Xiao et al. 2012; Su et al. 2014). The potential of these materials lies in their capacity to support neural cells regeneration, promote integration with neurons, and modulate tissue-electronic interactions through anti-inflammatory and antibiotic agents. The CPs and hydrogels are also highly appropriate materials to integrate with biological organisms to make "living electrodes" (Serruya et al. 2018; Adewole et al. 2024). However, creating bioactive electrodes using CPs and hydrogels faces important challenges due to their limited space available for incorporating large biomolecules and additives. To overcome this challenge, several approaches, including (i) immobilization within the polymer matrix (Green et al. 2009, 2010), (ii) electrochemical delivery (Massoumi and Entezami 2002; Wadhwa et al. 2006), and (iii) entrapment within the polymer matrix (Thompson et al. 2006, 2010) have been suggested. The selection of each strategy relies on specific biofunctional requisites of the drug. The components that do not necessitate cellular internalization, can be integrated in a manner that renders them accessible but immobilized on the polymer surface. For example, the cell adhesion/promoting agents such as laminin peptides (Green et al. 2009, 2010) and RGD sequences (Cui et al. 2001) have been utilized for this purpose, facilitating cell integration with the polymer. However, *in vivo* studies have shown that these factors may not be adequate to maintain the cell attachment due to inflammatory reactions, which lead to glial cell dominance and neuronal death near the implant interface (Polikov et al. 2005). Instead, for drugs that require uptake by cells, the incorporation within the polymer matrix is necessary. Some therapeutic agents like anti-inflammatory molecules, can be doped into the CPs and electrochemically delivered (Stevenson et al. 2010). However, maintaining the same releasing levels over extended periods remains a challenge. For electrochemically drug delivery, the presence of anions in the electrolyte solution can also affect drug doping *via* replacing the released drug (Green and Abidian 2015). On the other hand, the big drugs or those with negligible electrical charges, such as growth factors, need to be encapsulated within the CP or hydrogel matrix (Green et al. 2008). Nevertheless, drug delivery by CP or hydrogel carriers in the neural interface has generally two primary challenges, (i) achieving sustained therapeutic levels and (ii) overcoming the scar tissue formation reactions *in vivo*. The current approaches involve combining CPs and hydrogels with complementary systems to improve the drug storage capacity and controlled release processes.

6.5.4 Artificial Synapses: Computer-Neuron Dialog

Artificial synapses are a cutting-edge approach to mimic the complicated functions of natural neural networks, aiming to replicate the dynamic interactions of electrical impulses at the synaptic space. CPs are the key materials to develop the artificial synapses, through mimicking the synaptic transmission processes that are detected in biological systems. In fact, the CPs exhibit incomparable properties, such as mixed ion-electron charge conduction, as well as storage and release of ions that is critical for neuronal pulse transmission, making them appropriate for artificial synapse applications. This CPs' capability allows them to mimic the releasing processes of neurotransmitters, in which the ions are released into the synaptic cleft for propagating signals between neurons. When exposed to electrical stimulation, the CPs undergo ion exchange processes, which lead to releasing ions following by neuronal firing, therefore, enabling the synaptic activity replication. Figure 6.12 shows the concepts of artificial synapses and suggested materials and systems for creating artificial synapses based on CPs and hydrogels.

Neural networks, which are celebrated for their capability in problem-solving, suffer from the challenge of determining the optimal network structures. In biology, the natural phenomena such as neurogenesis and structural plasticity can provide

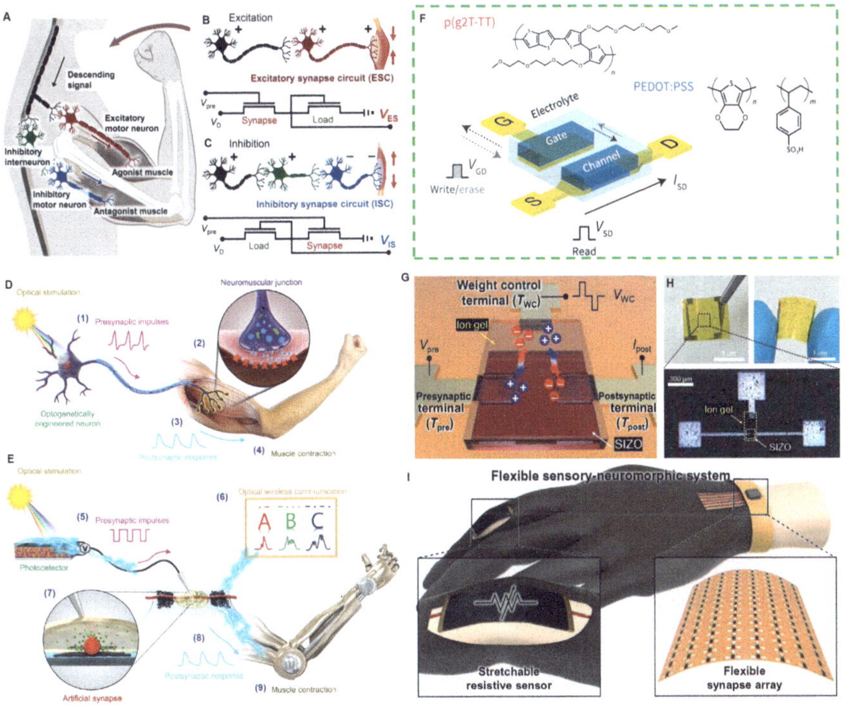

Fig. 6.12 Schematic representation of (**a**) biological reciprocal inhibition and (**b**, **c**) biological excitatory and inhibitory synapse pathways with their corresponding circuits, as the principle concepts for development of the artificial synapses (Kim et al. 2024). (**d**, **e**) Biological and organic optoelectronic artificial synapses and neuromuscular electronic system: (1–4) within biological systems, light stimulates motor neuron, which has photosensitive protein expression, thus action potentials are generated. (5–9) Within organic artificial synapses, light triggers a photodetector to generate output voltage spikes (Lee et al. 2024). (**f**) Schematic illustration for an electrochemical random-access memories (ECRAMs) device composed of PEDOT:PSS and p(g2T-TT) channel/gate developed for artificial synapses (Melianas et al. 2024). (**g**) Schematic illustration for a flexible synaptic device comprised of a SIZO/ion gel hybrid structure, and (**h**) optical microscopy images of the prepared synaptic device array. (**i**) Design of a sensory-neuromorphic system for translation the sign language, comprising of flexible artificial synapses and stretchable resistive sensors. (Oh et al. 2024)

valuable insights into addressing this challenge. Although the advanced neural network algorithms mainly rely on mechanisms such as synaptic plasticity and learning, their integration with living biological systems faces the absence of appropriate neural interfaces capable to replicate the complex biological neural networks. To address this issue, researchers have demonstrated that the dendritic electropolymerization of PEDOT:PSS nanofibers through alternating the electrical current densities can replicate the structural plasticity required for natural formation of the neural networks (Janzakova et al. 2023). This approach that is in alignment with Hebbian principles—a foundational concept in neurobiology demonstrating that neurons that fire together wire together—facilitates the creation of network topologies that

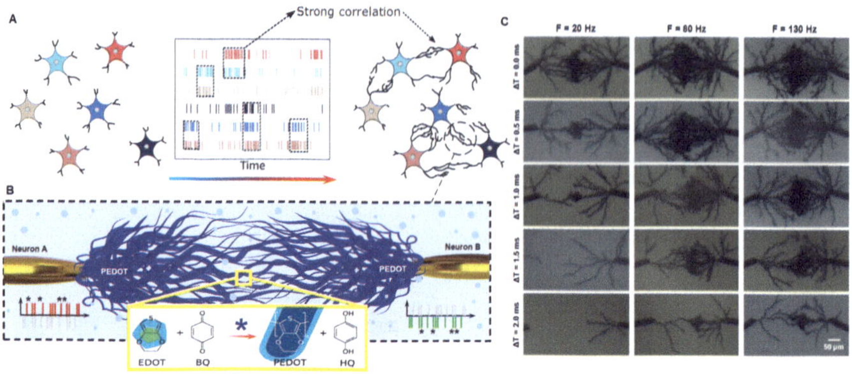

Fig. 6.13 Emulating structural plasticity using PEDOT:PSS dendritic fibers. (**a**) The growth of dendrites between neurons is driven by activity correlation, resulting in the formation of physical topologies resembling artificial neural networks. (**b**) Dendritic electropolymerization occurs through the application of alternating current (AC) electrical signals between two conductive Au wires (illustrated by neurons A and B). The applied electrical potential at each node facilitates the oxidation of EDOT monomers into PEDOT as well as reduction of benzoquinone (BQ) into hydroquinone (HQ). Node activity is represented by voltage pulses at each terminal, which can induce overpotential (depicted by a blue star), thereby triggering electropolymerization. (**c**) Dendritic PEDOT electropolymerization induced by spike timing modulation at different time-frame and frequencies. (Janzakova et al. 2023)

enhance computational performance by promotion sparse synaptic connectivity, particularly beneficial for addressing complex tasks. By this approach, researchers have indicated a significant improvement in network sparsity of up to 61% for classification tasks, revealing a substantial reduction in network connections. Furthermore, in signal reconstruction tasks, the prepared PEDOT:PSS interfaces made a notable 50% enhancement, demonstrating their capability for improving the network's capacity to accurately reconstruct signals (Janzakova et al. 2023). Figure 6.13 shows the prepared structural plasticity emulation based on PEDOT:PSS dendritic fibers.

6.5.5 Artificial Muscles

When exposed to electrical fields, the CPs such as PPy and PANi show changes in volume and shape, allowing them to generate mechanical strains. These polymers also possess electrical conductivity while also being flexible and lightweight. Moreover, although they face limitations in efficiency compared to alternative materials like CNTs, the CP actuators offer advantages like low operating voltage (Wang and Feng 2022b). These characteristics allow the CPs to deform in response to electrical stimuli like the biological muscles, making them the eminent materials for development of artificial muscles. Hydrogels, on the other hand, are water-swollen 3D polymer networks that can absorb and retain large amounts of water. Hence, they show significant changes, especially in their volume, when exposed to external stimuli such as pH, temperature, or electrical potentials. These volume changes result in mechanical deformation, which is essential for development of artificial muscles. Therefore, it can be concluded that the combination of CPs with hydrogels can make several compositional synergies that significantly enhance the performance of artificial muscles. Firstly, the CPs' electrical conductivity lets accurate control on the stimulation of the muscle, enabling rapid and efficient actuation. Also, the flexibility and biocompatibility of hydrogels make them appropriate for dynamic biological applications, such as prosthetics and soft robotics. However, there are still challenges that need to be addressed in utilization of hydrogels for artificial muscles. One such challenge can be reaching a balance between mechanical strength and flexibility, as well as optimizing the reactions of materials to electrical stimuli. Furthermore, long-term stability and durability of the artificial muscles remains an important concern, particularly where the reliability is paramount. Overall, the integration of CPs and hydrogels can be a promising approach for development of advanced artificial muscles. Through replicating the complex behavior of natural muscles, these materials have great potential to advance fields such as robotics, cyborg technology, and artificial muscles.

6.5.6 Molecular Bioelectronics

Molecular bioelectronics denotes an interdisciplinary field of chemistry, biology, and electronics, in which the CPs and hydrogels can interact with cell membranes, affecting the electron and ion charge transfer processes within living systems. This interaction can lead to enhancement of cellular activities or making new functionalities, which can open up new strategies for development of bioelectronic applications. One of the advantages of CPs for molecular bioelectronics is development of neural electrodes that can serve as non-specific bioelectronic sensors, which are capable to record the local potential changes of biological systems. Through direct connection to cellular membranes, the CP-coated electrodes can facilitate high-quality electrophysiological signal recordings, resulting in valued insights in cells' dynamics and communication. In some cases, for obtaining high sensitivity and long-term reliability within the complex biological environments, the CP electrodes are able to incorporate with antifouling agents. There are three main antifouling polymers, including PEG/OEG-based, zwitterionic, and others like glycans or hydrophilic peptides (Saxena et al. 2024). These polymers can be incorporated into the CPs *via* direct synthesis or surface modification techniques to create antifouling CPs. For instance, the electrodes comprised of antifouling CPs, like PEGylated PANi nanofibers and zwitterionic PSBEDOT, exhibit high sensitivity for detecting biomarkers such as BRCA1 and glucose, with improved stability *in vivo*. Furthermore, the controlled cell capture and release by antifouling CPs can be achieved through techniques like redox-responsive. For example, the hydroquinone-functionalized EDOT enable regulated cell adhesion and detachment, essential for cancer cell separation (Wang and Feng 2022b; Saxena et al. 2024). In recent years, CP-based transistors have been developed that show advanced functionalities for amplification and transduction of the biological signals. Organic transistors, such as OFETs and electrolyte-gated organic field-effect transistors (EGOFETs) provide highly appropriate platforms for interfacing with biological systems at molecular levels. These transistors can be functionalized *via* multiple methods to upgrade their capability for specific detection with high sensitivity, which enables precise recording and modulation of the biological signals. Furthermore, the CP-based transistors exhibit significant compatibility with biological environments, allowing for seamless integrations with different living microorganisms. This biocompatibility accompanying with the tunable properties of the CPs, makes them excellent candidates for tissue-electronics interfaces applications, biosensors, and BCIs (Kergoat et al. 2012; Wang et al. 2016, 2021a; Wang and Feng 2022b).

6.5.7 Biodegradable Implants

Bioresorbable electronics necessitate the development of devices that are designed to be gradually absorbed within the body, bypassing the surgery required to remove the implant post-treatment. These devices are composed of biodegradable materials, providing temporary medical monitoring and therapeutic interventions. The controlled degradation of the materials ensures gradually absorption, which reduces inflammatory responses and adverse reactions. Among biodegradable materials, CPs and hydrogels have attained a prominent position for fabricating temporary neural electrodes. For example, a biodegradable electrode containing of PPy nanoparticles can promote cell growth *via* electrical stimulation (Shi et al. 2004). Biodegradable conductive membranes composed of PPy nanoparticles and chitosan can also facilitate Schwann cell growth and enhanced neuron regeneration *via* electrical stimulation (Huang et al. 2010). On the other hand, the conductive biodegradable composites made by PANi combined with different materials such as gelatin and poly(L-lactide-co-ε-caprolactone) (PLCL) demonstrated enhanced cell adhesion and proliferation under electrical stimulation (Wang and Feng 2022b). Moreover, controlled biodegradation is particularly beneficial for drug carrier electrodes, where the CPs and hydrogels degrade and release the encapsulated drug through a controlled rate (Shahab Ahmadi Seyedkhani 2024a; Kwon and Furgeson 2007; Hardy et al. 2014). Recently, bioresorbable OECTs comprised of biodegradable PEDOT:PSS conductive channels, as illustrated in Fig. 6.14, have been developed for temporary BCIs (Wu et al. 2023).

6.6 Challenges and Future Directions

Although CPs and hydrogels can be extensively used to develop neural interfaces and especially BCIs, benefiting from their maximum potential faces significant challenges. Producing high quality CPs that can exhibit appropriate electrical conductivity is still challenging. This problem is typically attributed to low crystallinity and improper chains configuration of the fabricated CPs. Using different chemical

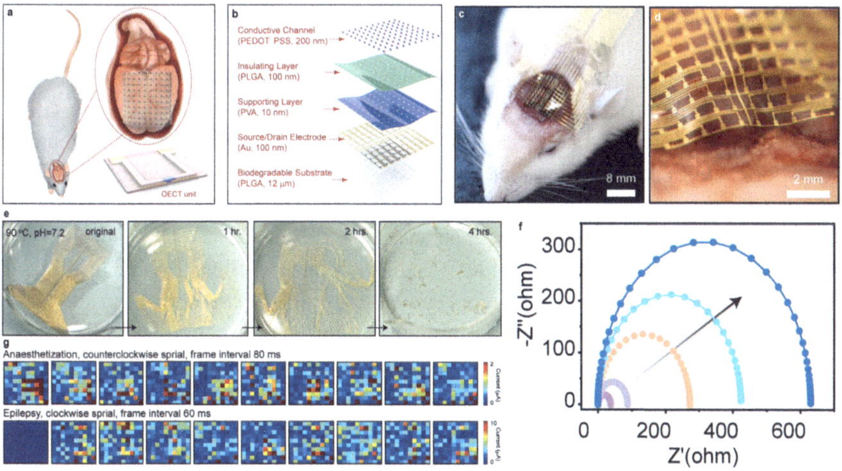

Fig. 6.14 Bioresorbable ultrathin and flexible transient OECT array as a core platform for a high-fidelity brain-machine interface: (**a**) Schematic illustrations depicting the setup of the transient OECT array positioned on the cerebral cortex of an animal model. (**b**) A detailed schematic representation shows the transient OECT with a fully biodegradable framework. (**c**, **d**) Optical images of the fabricated flexible OECT on the brain of a rat. (**e**) Degradation of the OECT in PBS, and (**f**) electrochemical impedance changes of the PEDOT:PSS versus degradation time. (**g**) micro-Electrocorticography (μ-ECoG) signals recorded under (top) anesthesia and (bottom) during seizures using OECT arrays. (Wu et al. 2023)

compounds and additives, liquid-based synthesis methods of CPs generally show impurities and toxic byproducts, requiring multiple modification processes. Moreover, low crystallinity and undesired polymer chain configuration for CPs can result in low conductivity and high electrochemical impedance. On the other hand, the vapor phase synthesis techniques such as VPP usually conducted by complex equipment. Additionally, these methods are not suitable for applying CPs coatings to complex or porous substrates. The biocatalytic synthesis of CPs, such as enzymatic polymerization, should also be optimized and developed for the mass production. While, the hydrogels' applications can be limited due to their high degradation rates and highly soft gel-like mechanical properties. Also, hydrogels gradually lose their structural water over time, leading to change their electrical conductivity. Consequently, the conductivity changes may compromise the reliability of the recorded data and limit the long-term applications of hydrogels. To address these problems, different strategies, including creating hybrid systems such as CPHs, IPNs, ICOHs, DNHs, and nanocomposites have been suggested. Improving the selectivity and specific detection capability of neural interfaces made by CPs and hydrogels need to be considered, demanding advancements in material compositions, electrode design, and interface configurations.

Nonetheless, the future directions in CPs and hydrogels applications for neural interfaces and BCIs can involve highly promising multifaceted approaches. Design and fabrication of biomimetic electrodes using CPs and hydrogels that closely replicate the neural tissue's properties can facilitate seamless integration and reduce adverse reactions at tissue-electrode interfaces. In this regard, novel fabrication methods such as 3D printing and microfluidics offer great potentials for creating complex electrode structures with tailored functionalities. Furthermore, integrating advanced sensing and actuation capabilities within CPs and hydrogels can enable closed-loop neural modulation and wireless signal transmitting, which facilitate real-time feedback and adaptive responses in BCIs. Also, having appropriate capability for combination with biological species, the CPs and hydrogels can facilitate the development of "living electrodes". Hence, interdisciplinary collaboration of biology, materials science, neuroscience, and electronics can drive desired advancements in BCIs and translate them into practical solutions for neural applications.

6.7 Conclusions

In this chapter, we investigated fabrication methods and properties of CPs and hydrogels, focusing on their applications in neural interfaces and especially BCIs. We found that although the electrical conductivity of CPs and hydrogels cannot compete with metals, their lower electrochemical impedance and ion-electron synergistic charge transfer processes, which can mimic biological signal transmitting, make them highly appropriate materials for creating neural electrodes. Moreover, through lower mechanical mismatch with biological tissues, appropriate flexibility, and excellent biocompatibility, which lead to reducing FBRs and subsequent inflammations, the CPs and hydrogels can overtake other materials in neural interfacing. Through controlled synthesis conditions, modification processes, or creating hybrid systems, the CPs and hydrogels find different degradation rates within biological environments, which make them widely appropriate for various temporary implants and bioresorbable electronics. We also indicated that the controlled proliferation and differentiation of the stem cells can be conducted on the electroactive substrates, demonstrating high capability of conductive polymer and hydrogel templates to boost tissue engineering and nerve regeneration. Some CPs like PEDOT show high transparency characteristics, making them appropriate materials for creating transparent and flexible MEAs, particularly for monitoring brain electrical signals and optical imaging of neuronal networks simultaneously. The capability for integrating with wireless tissue-electronics interfaces, serving as highly compatible materials for building the living electrodes, and creating the proper carriers for smart drug delivery systems are additional advantages of CPs and hydrogels, making them appropriate candidates for advanced multifunctional neural interfacing applications. Future investigations should be focused on optimizing the fabrication processes, developing new composites and hybrid systems to obtain high chemical stability, improved mechanical behavior, and adequate electrical conductivity for CPs and hydrogels, which provide a more reliable long-term two-way communication between biological systems and external electronic devices. Finally, it can be concluded that the advances in CPs and hydrogels can be great promising for developing nanobioelectronics and especially BCIs, through offering enhanced biocompatibility, improved signal transfer processes, and the great potentials for seamless integration with neural tissues.

Conflict of Interest The authors declare no conflict of interest.

References

Abidian MR, Kim D, Martin DC (2006) Conducting-polymer nanotubes for controlled drug release. Adv Mater 18:405–409

Abidian MR, Corey JM, Kipke DR, Martin DC (2010) Conducting-polymer nanotubes improve electrical properties, mechanical adhesion, neural attachment, and neurite outgrowth of neural electrodes. Small 6:421–429

Abidian MR, Daneshvar ED, Egeland BM et al (2012) Neural tissue engineering: hybrid conducting polymer–hydrogel conduits for axonal growth and neural tissue engineering (Adv. Healthcare Mater. 6/2012). Adv Healthc Mater 1:681

Adewole DO, Struzyna LA, Burrell JC et al (2024) Development of optically controlled "living electrodes" with long-projecting axon tracts for a synaptic brain-machine interface. Sci Adv 7:eaay5347. https://doi.org/10.1126/sciadv.aay5347

Ahmadi Seyedkhani Sh, Mohammadpour R (2022) Tissue-electronics interfaces. In: Elnashar PM (ed). IntechOpen, Rijeka., p Ch. 1

Ahmadi Seyedkhani Sh, Dehnavi SM, Barjasteh M (2023) A comprehensive study on a novel chitosan/Ag-MOFs nanocomposite coatings for bone implants: physico-chemical, biological and electrochemical properties. Mater Chem Phys 308:128268. https://doi.org/10.1016/j.matchemphys.2023.128268

Ahmadi Seyedkhani Sh, Mohammadpour R, Irajizad A (2024) Principles and advancements of microelectrode arrays in brain-machine interfaces. In: Villarreal-Gómez DLJ (ed). IntechOpen, Rijeka., p Ch. 6

Alshammary B, Walsh FC, Herrasti P, Ponce de Leon C (2016) Electrodeposited conductive polymers for controlled drug release: polypyrrole. J Solid State Electrochem 20:839–859

Alves M, Jensen BEB, Smith AAA, Zelikin AN (2011) Poly (vinyl alcohol) physical hydrogels: new vista on a long serving biomaterial. Macromol Biosci 11:1293–1313

Andrianov AK, Payne LG, Visscher KB et al (1994) Hydrolytic degradation of ionically cross-linked polyphosphazene microspheres. J Appl Polym Sci 53:1573–1578

Aplan MP, Gomez ED (2017) Recent developments in chain-growth polymerizations of conjugated polymers. Ind Eng Chem Res 56:7888–7901

Appel EA, Tibbitt MW, Webber MJ et al (2015) Self-assembled hydrogels utilizing polymer–nanoparticle interactions. Nat Commun 6:6295

Avery RK, Albadawi H, Akbari M et al (2016) An injectable shear-thinning biomaterial for endovascular embolization. Sci Transl Med 8:365ra156-365ra156

Badali E, Hosseini M, Mohajer M et al (2021) Enzymatic crosslinked hydrogels for biomedical application. Polym Sci Ser A 63:S1–S22

Banerjee S, Ray S, Maiti S et al (2010) Interpenetrating polymer network (IPN): a novel biomaterial. Int J Appl Pharm 2:28–34

Bansal M, Dravid A, Aqrawe Z et al (2020) Conducting polymer hydrogels for electrically responsive drug delivery. J Control Release 328:192–209

Barjasteh M, Mohsen Dehnavi S, Ahmadi Seyedkhani S et al (2022) Synergistic Wound Healing by Novel Ag@ZIF-8 Nanostructures. Int J Pharm 629:122339. https://doi.org/10.1016/j.ijpharm.2022.122339

Barjasteh M, Dehnavi SM, Ahmadi Seyedkhani S et al (2023a) Improved biological activities of dual nanofibrous chitosan/bacterial cellulose wound dressing by a novel silver-based metal-organic framework. Surf Interfaces 36:102631. https://doi.org/10.1016/j.surfin.2023.102631

Barjasteh M, Dehnavi SM, Ahmadi Seyedkhani S, Akrami M (2023b) Cu-vitamin B3 donut-like MOFs incorporated into TEMPO-oxidized bacterial cellulose nanofibers for wound healing. Int J Pharm 646:123484. https://doi.org/10.1016/j.ijpharm.2023.123484

Barjasteh M, Dehnavi SM, Ahmadi Seyedkhani S et al (2024) Hybrid nanocomposite wound dressings by a novel nanorod vitamin-B3-Ag metal-organic framework and bacterial cellulose nanofibers. J Drug Deliv Sci Technol 95:105532. https://doi.org/10.1016/j.jddst.2024.105532

Berezhetska O, Liberelle B, De Crescenzo G, Cicoira F (2015) A simple approach for protein covalent grafting on conducting polymer films. J Mater Chem B 3:5087–5094. https://doi.org/10.1039/C5TB00373C

Berggren M, Richter-Dahlfors A (2007) Organic bioelectronics. Adv Mater 19:3201–3213

Bertlein S, Brown G, Lim KS et al (2017) Thiol-Ene Clickable Gelatin: A Platform Bioink for Multiple 3D Biofabrication Technologies. Adv Mater 29. https://doi.org/10.1002/adma.201703404

Bertuoli PT, Ordono J, Armelin E et al (2019) Electrospun conducting and biocompatible uniaxial and Core–Shell fibers having poly (lactic acid), poly (ethylene glycol), and polyaniline for cardiac tissue engineering. ACS Omega 4:3660–3672

Bhattacharyya D, Howden RM, Borrelli DC, Gleason KK (2012) Vapor phase oxidative synthesis of conjugated polymers and applications. J Polym Sci Part B Polym Phys 50:1329–1351

Bhoyar SS, Sangale VB, Desai MA et al (2024) Enzymatic synthesis of biocompatible polypyrrole with enhanced electrochemical properties for supercapacitors. J Power Sources 607:234599. https://doi.org/10.1016/j.jpowsour.2024.234599

Bouldin R, Kokil A, Ravichandran S et al (2010) Enzymatic synthesis of electrically conducting polymers. In: Green polymer chemistry: biocatalysis and biomaterials. American Chemical Society, pp 23–315

Bredas JL, Street GB (1985) Polarons, bipolarons, and solitons in conducting polymers. Acc Chem Res 18:309–315

Brooke R, Cottis P, Talemi P et al (2017) Recent advances in the synthesis of conducting polymers from the vapour phase. Prog Mater Sci 86:127–146

Bruno FF, Nagarajan R, Stenhouse P et al (2001) Polymerization of water-soluble conductive polyphenol using horseradish peroxidase. J Macromol Sci Part A 38:1417–1426

Campbell A, Wu C (2018) Chronically implanted intracranial electrodes: tissue reaction and electrical changes. Micromachines 9:430

Carnicer-Lombarte A, Chen S-T, Malliaras GG, Barone DG (2021) Foreign body reaction to implanted biomaterials and its impact in Nerve Neuroprosthetics. Front Bioeng Biotechnol 9:622524

Carraher CE Jr (2003) Seymour/Carraher's polymer chemistry. CRC Press

Carvalho IC, Mansur HS (2017) Engineered 3D-scaffolds of photocrosslinked chitosan-gelatin hydrogel hybrids for chronic wound dressings and regeneration. Mater Sci Eng C 78:690–705

Carvalho CR, Costa JB, da Silva MA et al (2018) Tunable enzymatically cross-linked silk fibroin tubular conduits for guided tissue regeneration. Adv Healthc Mater 7:1800186. https://doi.org/10.1002/adhm.201800186

Chai X, Cao T, He Q et al (2024) Brain–computer interface digital prescription for neurological disorders. CNS Neurosci Ther 30:e14615

Chee BS, de Lima GG, Devine DM, Nugent MJD (2018) Investigation of the effects of orientation on freeze/thawed Polyvinyl alcohol hydrogel properties. Mater Today Commun 17:82–93

Chen F, Yang K, Zhao D, Yang H (2019a) Thermal-and salt-activated shape memory hydrogels based on a gelatin/polyacrylamide double network. RSC Adv 9:18619–18626

Chen J, Peng Q, Thundat T, Zeng H (2019b) Stretchable, injectable, and self-healing conductive hydrogel enabled by multiple hydrogen bonding toward wearable electronics. Chem Mater 31:4553–4563

Chen L, Wang Z, Zhan Z et al (2021) 3D printed super-anti-freezing self-adhesive human-machine interface. Mater Today Phys 19:100404

Chenite A, Chaput C, Wang D et al (2000) Novel injectable neutral solutions of chitosan form biodegradable gels in situ. Biomaterials 21:2155–2161

Cho B, Park KS, Baek J et al (2014) Single-crystal poly (3, 4-ethylenedioxythiophene) nanowires with ultrahigh conductivity. Nano Lett 14:3321–3327
Clevenger M, Kim H, Song HW et al (2024) Binder-free printed PEDOT wearable sensors on everyday fabrics using oxidative chemical vapor deposition. Sci Adv 7:eabj8958. https://doi.org/10.1126/sciadv.abj8958
Cong J, Fan Z, Pan S et al (2021) Polyacrylamide/chitosan-based conductive double network hydrogels with outstanding electrical and mechanical performance at low temperatures. ACS Appl Mater Interfaces 13:34942–34953
Correa S, Grosskopf AK, Lopez Hernandez H et al (2021) Translational Applications of Hydrogels. Chem Rev 121:11385–11457. https://doi.org/10.1021/acs.chemrev.0c01177
Cui X, Lee VA, Raphael Y et al (2001) Surface modification of neural recording electrodes with conducting polymer/biomolecule blends. J Biomed Mater Res An Off J Soc Biomater Japanese Soc Biomater Aust Soc Biomater Korean Soc Biomater 56:261–272
Culebras M, Gómez CM, Cantarero A (2014) Enhanced thermoelectric performance of PEDOT with different counter-ions optimized by chemical reduction. J Mater Chem A 2:10109–10115
da Silva LC, Rojas G, Schulz MD, Wagener KB (2017) Acyclic diene metathesis polymerization: history, methods and applications. Prog Polym Sci 69:79–107
De Giglio E, Sabbatini L, Colucci S, Zambonin G (2000) Synthesis, analytical characterization, and osteoblast adhesion properties on RGD-grafted polypyrrole coatings on titanium substrates. J Biomater Sci Polym Ed 11:1073–1083
Dehnavi SM, Barjasteh M, Ahmadi Seyedkhani S et al (2023) A novel silver-based metal-organic framework incorporated into nanofibrous chitosan coatings for bone tissue implants. Int J Pharm 640:123047. https://doi.org/10.1016/j.ijpharm.2023.123047
Deslouis C, El Moustafid T, Musiani MM, Tribollet B (1996) Mixed ionic-electronic conduction of a conducting polymer film. Ac impedance study of polypyrrole. Electrochim Acta 41:1343–1349
Dijk G, Kaszas A, Pas J, O'Connor RP (2022) Fabrication and in vivo 2-photon microscopy validation of transparent PEDOT:PSS microelectrode arrays. Microsyst Nanoeng 8:90. https://doi.org/10.1038/s41378-022-00434-7
Ding X, Gao J, Awada H, Wang Y (2016) Dual physical dynamic bond-based injectable and biodegradable hydrogel for tissue regeneration. J Mater Chem B 4:1175–1185
Ding Q, Xu X, Yue Y et al (2018) Nanocellulose-mediated electroconductive self-healing hydrogels with high strength, plasticity, viscoelasticity, stretchability, and biocompatibility toward multifunctional applications. ACS Appl Mater Interfaces 10:27987–28002
Dinh N, Luo R, Christine MTA et al (2017) Effective light directed assembly of building blocks with microscale control. Small 13:1700684
Dou P, Liu Z, Cao Z et al (2016) Rapid synthesis of hierarchical nanostructured Polyaniline hydrogel for high power density energy storage application and three-dimensional multilayers printing. J Mater Sci 51:4274–4282
Dreiss CA (2020) Hydrogel design strategies for drug delivery. Curr Opin Colloid Interface Sci 48:1–17
Du Z, Li N, Hua Y et al (2017) Physiological pH-dependent gelation for 3D printing based on the phase separation of gelatin and oxidized dextran. Chem Commun 53:13023–13026
Fernandes KF, Lima CS, Pinho H, Collins CH (2003) Immobilization of horseradish peroxidase onto polyaniline polymers. Process Biochem 38:1379–1384
Fu F, Wang J, Zeng H, Yu J (2020) Functional conductive hydrogels for bioelectronics. ACS Mater Lett 2:1287–1301
Gacesa P (1988) Alginates. Carbohydr Polym 8:161–182
Gacesa P, Russell NJ (1990) The structure and properties of alginate. In: Pseudomonas infection and alginates: biochemistry, genetics and pathology. Springer, pp 29–49
Gajewski P, Lewandowska A, Szcześniak K et al (2021) Optimization of the properties of photocured hydrogels for use in electrochemical capacitors. Polymers (Basel) 13:3495

Gan D, Han L, Wang M et al (2018) Conductive and tough hydrogels based on biopolymer molecular templates for controlling in situ formation of polypyrrole nanorods. ACS Appl Mater Interfaces 10:36218–36228

Gan D, Shuai T, Wang X et al (2020) Mussel-inspired redox-active and hydrophilic conductive polymer nanoparticles for adhesive hydrogel bioelectronics. Nano-Micro Lett 12:1–16

George PM, LaVan DA, Burdick JA et al (2006) Electrically controlled drug delivery from biotin-doped conductive polypyrrole. Adv Mater 18:577–581

GhavamiNejad A, Ashammakhi N, Wu XY, Khademhosseini A (2020) Crosslinking strategies for 3D bioprinting of polymeric hydrogels. Small 16:2002931

Goding J, Gilmour A, Martens P et al (2017) Interpenetrating conducting hydrogel materials for neural interfacing electrodes. Adv Healthc Mater 6:1601177

Gomez N, Schmidt CE (2007) Nerve growth factor-immobilized polypyrrole: Bioactive electrically conducting polymer for enhanced neurite extension. J Biomed Mater Res Part A 81:135–149

Gomez N, Lee JY, Nickels JD, Schmidt CE (2007) Micropatterned polypyrrole: a combination of electrical and topographical characteristics for the stimulation of cells. Adv Funct Mater 17:1645–1653

Green R, Abidian MR (2015) Conducting polymers for neural prosthetic and neural interface applications. Adv Mater 27:7620–7637

Green RA, Lovell NH, Wallace GG, Poole-Warren LA (2008) Conducting polymers for neural interfaces: challenges in developing an effective long-term implant. Biomaterials 29:3393–3399

Green RA, Lovell NH, Poole-Warren LA (2009) Cell attachment functionality of bioactive conducting polymers for neural interfaces. Biomaterials 30:3637–3644

Green RA, Lovell NH, Poole-Warren LA (2010) Impact of co-incorporating laminin peptide dopants and neurotrophic growth factors on conducting polymer properties. Acta Biomater 6:63–71

Green RA, Hassarati RT, Bouchinet L et al (2012) Substrate dependent stability of conducting polymer coatings on medical electrodes. Biomaterials 33:5875–5886

Green RA, Matteucci PB, Hassarati RT et al (2013) Performance of conducting polymer electrodes for stimulating neuroprosthetics. J Neural Eng 10:16009

Green RA, Matteucci PB, Dodds CWD et al (2014) Laser patterning of platinum electrodes for safe neurostimulation. J Neural Eng 11:56017

Guvendiren M, Lu HD, Burdick JA (2012) Shear-thinning hydrogels for biomedical applications. Soft Matter 8:260–272

Haraguchi K, Li H-J (2005) Control of the coil-to-globule transition and ultrahigh mechanical properties of PNIPA in nanocomposite hydrogels. Angew Chem Int Ed 44:6500–6504. https://doi.org/10.1002/anie.200502004

Hardy JG, Mouser DJ, Arroyo-Currás N et al (2014) Biodegradable electroactive polymers for electrochemically-triggered drug delivery. J Mater Chem B 2:6809–6822

Heeger AJ (2001) Nobel Lecture: semiconducting and metallic polymers: the fourth generation of polymeric materials. Rev Mod Phys 73:681–700

Heydarnezhad HR, Pourabbas B, Tayefi M (2018) Conducting electroactive polymers via photopolymerization: a review on synthesis and applications. Polym Plast Technol Eng 57:1093–1109

Ho D, Zou J, Chen X et al (2015) Hierarchical patterning of multifunctional conducting polymer nanoparticles as a bionic platform for topographic contact guidance. ACS Nano 9:1767–1774

Holt MW, Robinson EC, Shlobin NA et al (2024) Intracortical brain-computer interfaces for improved motor function: a systematic review. Rev Neurosci 35:213–223

Hossain KS, Miyanaga K, Maeda H, Nemoto N (2001) Sol– gel transition behavior of pure ι-carrageenan in both salt-free and added salt states. Biomacromolecules 2:442–449

Hribar KC, Choi YS, Ondeck M et al (2014) Digital plasmonic patterning for localized tuning of hydrogel stiffness. Adv Funct Mater 24:4922–4926

Hu D, Wu D, Huang L et al (2018) 3D bioprinting of cell-laden scaffolds for intervertebral disc regeneration. Mater Lett 223:219–222

Huang J, Hu X, Lu L et al (2010) Electrical regulation of Schwann cells using conductive polypyrrole/chitosan polymers. J Biomed Mater Res Part A An Off J Soc Biomater Japanese Soc Biomater Aust Soc Biomater Korean Soc Biomater 93:164–174

Huang X, Zhang Y, Zhang X et al (2013) Influence of radiation crosslinked carboxymethyl-chitosan/gelatin hydrogel on cutaneous wound healing. Mater Sci Eng C 33:4816–4824

Inoue A, Yuk H, Lu B, Zhao X (2024) Strong adhesion of wet conducting polymers on diverse substrates. Sci Adv 6:eaay5394. https://doi.org/10.1126/sciadv.aay5394

Isaksson J, Kjäll P, Nilsson D et al (2007) Electronic control of Ca2+ signalling in neuronal cells using an organic electronic ion pump. Nat Mater 6:673–679

Janzakova K, Balafrej I, Kumar A et al (2023) Structural plasticity for neuromorphic networks with electropolymerized dendritic PEDOT connections. Nat Commun 14:8143. https://doi.org/10.1038/s41467-023-43887-8

Jenkins AD, Kratochvíl P, Stepto RFT, Suter UW (1996) Glossary of basic terms in polymer science (IUPAC recommendations 1996). Pure Appl Chem 68:2287–2311

Jin J, Huang Z, Yin G et al (2016) Preparation of polypyrrole-protein composite films and the electrochemically controlled release of proteins. J Nanosci Nanotechnol 16:2283–2290

Kageyama T, Osaki T, Enomoto J et al (2016) In situ cross-linkable gelatin-CMC hydrogels designed for rapid engineering of perfusable vasculatures. ACS Biomater Sci Eng 2:1059–1066

Karikari E, Koshechkin KA (2023) Review on brain-computer interface technologies in healthcare. Biophys Rev 15:1351–1358

Kayser LV, Lipomi DJ (2019) Stretchable conductive polymers and composites based on PEDOT and PEDOT: PSS. Adv Mater 31:1806133

Keene ST, Rao A, Malliaras GG (2024) The relationship between ionic-electronic coupling and transport in organic mixed conductors. Sci Adv 9:eadi3536. https://doi.org/10.1126/sciadv.adi3536

Kergoat L, Piro B, Berggren M et al (2012) Advances in organic transistor-based biosensors: from organic electrochemical transistors to electrolyte-gated organic field-effect transistors. Anal Bioanal Chem 402:1813–1826

Kim D, Richardson-Burns SM, Hendricks JL et al (2007) Effect of immobilized nerve growth factor on conductive polymers: electrical properties and cellular response. Adv Funct Mater 17:79–86

Kim S-M, Kim C-H, Kim Y et al (2018) Influence of PEDOT:PSS crystallinity and composition on electrochemical transistor performance and long-term stability. Nat Commun 9:3858. https://doi.org/10.1038/s41467-018-06084-6

Kim S, Kim S, Ho DH et al (2024) Neurorobotic approaches to emulate human motor control with the integration of artificial synapse. Sci Adv 8:eabo3326. https://doi.org/10.1126/sciadv.abo3326

Kiriy A, Senkovskyy V, Sommer M (2011) Kumada catalyst-transfer polycondensation: mechanism, opportunities, and challenges. Macromol Rapid Commun 32:1503–1517

Kitada K, Ozaki S (1995) Reductive polymerization of halothiophene. Polym J 27:1161–1166

Kotwal A, Schmidt CE (2001) Electrical stimulation alters protein adsorption and nerve cell interactions with electrically conducting biomaterials. Biomaterials 22:1055–1064

Krauhausen I, Coen C, Spolaor S et al (2023) Brain-inspired organic electronics: merging neuromorphic computing and bioelectronics using conductive polymers. Adv Funct Mater 2307729

Kwon GS, Furgeson DY (2007) Biodegradable polymers for drug delivery systems. In: Biomedical polymers. Elsevier, pp 83–110

Lake GJ, Thomas AG (1967) The strength of highly elastic materials. Proc R Soc London Ser A Math Phys Sci 300:108–119

Langevin SA, Tan B, Freeman AW et al (2019) UV-cured gel polymer electrolytes with improved stability for advanced aqueous Li-ion batteries. Chem Commun 55:13085–13088

Lawal AT, Wallace GG (2014) Vapour phase polymerisation of conducting and non-conducting polymers: a review. Talanta 119:133–143

Lee JY, Bashur CA, Goldstein AS, Schmidt CE (2009a) Polypyrrole-coated electrospun PLGA nanofibers for neural tissue applications. Biomaterials 30:4325–4335

Lee JY, Lee J-W, Schmidt CE (2009b) Neuroactive conducting scaffolds: nerve growth factor conjugation on active ester-functionalized polypyrrole. J R Soc Interface 6:801–810

Lee Y-Y, Kang H-Y, Gwon SH et al (2015) A strain-insensitive stretchable electronic conductor: PEDOT: PSS/acrylamide organogels. Adv Mater 28:1636–1643

Lee Y, Oh JY, Xu W et al (2024) Stretchable organic optoelectronic sensorimotor synapse. Sci Adv 4:eaat7387. https://doi.org/10.1126/sciadv.aat7387

Levett JJ, Elkaim LM, Niazi F et al (2023) Invasive brain computer interface for motor restoration in spinal cord injury: a systematic review, vol 27. Neuromodulation Technol Neural Interface, pp 597–603

Li C (2021) Towards conductive hydrogels in e-skins: a review on rational design and recent developments. RSC Adv 11:33835–33848. https://doi.org/10.1039/D1RA04573C

Liu K, Zang S, Xue R et al (2018) Coordination-triggered hierarchical folate/zinc supramolecular hydrogels leading to printable biomaterials. ACS Appl Mater Interfaces 10:4530–4539

Liu X, Liu J, Lin S, Zhao X (2020) Hydrogel machines. Mater Today 36:102–124

Liu D, Zhou H, Zhao Y et al (2022) A strand entangled supramolecular PANI/PAA hydrogel enabled ultra-stretchable strain sensor. Small 18:2203258. https://doi.org/10.1002/smll.202203258

Liu D, Huyan C, Wang Z et al (2023) Conductive polymer based hydrogels and their application in wearable sensors: a review. Mater Horiz 10:2800–2823. https://doi.org/10.1039/D3MH00056G

Lu B, Yuk H, Lin S et al (2019) Pure PEDOT:PSS hydrogels. Nat Commun 10:1043. https://doi.org/10.1038/s41467-019-09003-5

Lu B, Fan P, Li M et al (2023) Detection of neuronal defensive discharge information transmission and characteristics in periaqueductal gray double-subregions using PtNP/PEDOT:PSS modified microelectrode arrays. Microsyst Nanoeng 9:70. https://doi.org/10.1038/s41378-023-00546-8

Luo L-J, Lai J-Y (2019) Amination degree of gelatin is critical for establishing structure-property-function relationships of biodegradable thermogels as intracameral drug delivery systems. Mater Sci Eng C 98:897–909

MacDiarmid AG (2001) "Synthetic metals": a novel role for organic polymers (Nobel lecture). Angew Chem Int Ed 40:2581–2590

Madl CM, Katz LM, Heilshorn SC (2016) Bio-orthogonally crosslinked, engineered protein hydrogels with tunable mechanics and biochemistry for cell encapsulation. Adv Funct Mater 26:3612–3620

Massoumi B, Entezami A (2002) Electrochemically controlled binding and release of dexamethasone from conducting polymer bilayer films. J Bioact Compat Polym 17:51–62

Melianas A, Quill TJ, LeCroy G et al (2024) Temperature-resilient solid-state organic artificial synapses for neuromorphic computing. Sci Adv 6:eabb2958. https://doi.org/10.1126/sciadv.abb2958

Mohammadi A, Hasan M-A, Liedberg B et al (1986) Chemical vapour deposition (CVD) of conducting polymers: polypyrrole. Synth Met 14:189–197. https://doi.org/10.1016/0379-6779(86)90183-9

Moussa M, El-Kady MF, Dubal D et al (2019) Self-assembly and cross-linking of conducting polymers into 3D hydrogel electrodes for supercapacitor applications. ACS Appl Energy Mater 3:923–932

Mushtaq F, Raza ZA, Batool SR et al (2022) Preparation, properties, and applications of gelatin-based hydrogels (GHs) in the environmental, technological, and biomedical sectors. Int J Biol Macromol 218:601–633

Nakabayashi K (2018) Direct arylation polycondensation as conjugated polymer synthesis methodology. Polym J 50:475–483. https://doi.org/10.1038/s41428-018-0039-5

Namkoong M, Guo H, Rahman MS et al (2022) Moldable and transferrable conductive nanocomposites for epidermal electronics. Npj Flex Electron 6:41. https://doi.org/10.1038/s41528-022-00170-y

Nezakati T, Seifalian A, Tan A, Seifalian AM (2018) Conductive polymers: opportunities and challenges in biomedical applications. Chem Rev 118:6766–6843

Ng WL, Goh MH, Yeong WY, Naing MW (2018) Applying macromolecular crowding to 3D bioprinting: fabrication of 3D hierarchical porous collagen-based hydrogel constructs. Biomater Sci 6:562–574

Nickels P, Dittmer WU, Beyer S et al (2004) Polyaniline nanowire synthesis templated by DNA. Nanotechnology 15:1524–1529. https://doi.org/10.1088/0957-4484/15/11/026

Nie S, Li Z, Yao Y, Jin Y (2021) Progress in synthesis of conductive polymer poly (3, 4-ethylenedioxythiophene). Front Chem 9:1137

Oh S, Cho J-I, Lee BH et al (2024) Flexible artificial Si-In-Zn-O/ion gel synapse and its application to sensory-neuromorphic system for sign language translation. Sci Adv 7:eabg9450. https://doi.org/10.1126/sciadv.abg9450

Ohayon D, Inal S (2020) Organic bioelectronics: from functional materials to next-generation devices and power sources. Adv Mater 32:2001439. https://doi.org/10.1002/adma.202001439

Otrokhov GV, Morozova OV, Vasil'Eva IS et al (2013) Biocatalytic synthesis of conducting polymers and prospects for its application. Biochemist 78:1539–1553

Papagiannopoulos A, Nikolakis S-P, Pamvouxoglou A, Koutsopoulou E (2023) Physicochemical properties of electrostatically crosslinked carrageenan/chitosan hydrogels and carrageenan/chitosan/Laponite nanocomposite hydrogels. Int J Biol Macromol 225:565–573

Park K-H, Jo EA, Na K (2007) Heparin/polypyrrole (PPy) composite on gold-coated matrix for the neurite outgrowth of PC12 cells by electrical stimulation. Biotechnol Bioprocess Eng 12:463–469

Peak CW, Stein J, Gold KA, Gaharwar AK (2018) Nanoengineered colloidal inks for 3D bioprinting. Langmuir 34:917–925

Pernaut J-M, Reynolds JR (2000) Use of conducting electroactive polymers for drug delivery and sensing of bioactive molecules. A redox chemistry approach. J Phys Chem B 104:4080–4090

Petty AJ, Keate RL, Jiang B et al (2020) Conducting polymers for tissue regeneration in vivo. Chem Mater 32:4095–4115

Pimentel R, Ko SK, Caviglia C et al (2018) Three-dimensional fabrication of thick and densely populated soft constructs with complex and actively perfused channel network. Acta Biomater 65:174–184

Poddar AK, Patel SS, Patel HD (2021) Synthesis, characterization and applications of conductive polymers: a brief review. Polym Adv Technol 32:4616–4641

Polikov VS, Tresco PA, Reichert WM (2005) Response of brain tissue to chronically implanted neural electrodes. J Neurosci Methods 148:1–18

Poverenov E, Li M, Bitler A, Bendikov M (2010) Major effect of electropolymerization solvent on morphology and electrochromic properties of PEDOT films. Chem Mater 22:4019–4025

Pranti AS, Schander A, Bödecker A, Lang W (2018) PEDOT: PSS coating on gold microelectrodes with excellent stability and high charge injection capacity for chronic neural interfaces. Sensors Actuators B Chem 275:382–393. https://doi.org/10.1016/j.snb.2018.08.007

Puiggalí-Jou A, del Valle LJ, Alemán C (2019) Drug delivery systems based on intrinsically conducting polymers. J Control Release 309:244–264. https://doi.org/10.1016/j.jconrel.2019.07.035

Qin J, Huang Z, Yin G et al (2016) Fabrication and protein conjugation of aligned polypyrrole-poly (L-lactic acid) fibers film with the conductivity and stability. J Nanosci Nanotechnol 16:2371–2377

Raia NR, Partlow BP, McGill M et al (2017) Enzymatically crosslinked silk-hyaluronic acid hydrogels. Biomaterials 131:58–67

Raj GK, Singh E, Hani U et al (2023) Conductive polymers and composites-based systems: An incipient stride in drug delivery and therapeutics realm. J Control Release 355:709–729

Rani S, Gupta T, Garg V et al (2022) Microstructural and electrochemical investigations of conductive bio-nanocomposite hydrogel based biosensing device. Mater Today Proc 62:638–643

Rebers L, Reichsöllner R, Regett S et al (2021) Differentiation of physical and chemical crosslinking in gelatin methacryloyl hydrogels. Sci Rep 11:3256

Ren X, Yang M, Yang T et al (2021) Highly conductive PPy–PEDOT: PSS hybrid hydrogel with superior biocompatibility for bioelectronics application. ACS Appl Mater Interfaces 13:25374–25382

Richardson RT, Wise AK, Thompson BC et al (2009) Polypyrrole-coated electrodes for the delivery of charge and neurotrophins to cochlear neurons. Biomaterials 30:2614–2624

Richardson JJ, Cui J, Björnmalm M et al (2016) Innovation in layer-by-layer assembly. Chem Rev 116:14828–14867. https://doi.org/10.1021/acs.chemrev.6b00627

Rivnay J, Owens RM, Malliaras GG (2014) The rise of organic bioelectronics. Chem Mater 26:679–685

Rocha I, Cerqueira G, Varella Penteado F, Córdoba de Torresi SI (2021) Electrical stimulation and conductive polymers as a powerful toolbox for tailoring cell behaviour in vitro. Front Med Technol 3:670274

Roncali J (1992) Conjugated poly (thiophenes): synthesis, functionalization, and applications. Chem Rev 92:711–738

Rumbau V, Pomposo JA, Eleta A et al (2007) First enzymatic synthesis of water-soluble conducting poly(3,4-ethylenedioxythiophene). Biomacromolecules 8:315–317. https://doi.org/10.1021/bm060949z

Ryu SB, Park KM, Park KD (2022) In situ graphene oxide-gelatin hydrogels with enhanced mechanical property for tissue adhesive and regeneration. Biochem Biophys Res Commun 592:24–30

Sagdic K, Fernández-Lavado E, Mariello M et al (2023) Hydrogels and conductive hydrogels for implantable bioelectronics. MRS Bull 48:495–505

Saha S, Mamun KA, Ahmed K et al (2021) Progress in brain computer interface: challenges and opportunities. Front Syst Neurosci 15:578875

Saxena S, Sen P, Soleymani L, Hoare T (2024) Anti-fouling polymer or peptide-modified electrochemical biosensors for improved biosensing in complex media. Adv Sens Res 3:2300170

Schmidt CE, Shastri VR, Vacanti JP, Langer R (1997) Stimulation of neurite outgrowth using an electrically conducting polymer. Proc Natl Acad Sci 94:8948–8953

Serruya MD, Harris JP, Adewole DO et al (2018) Engineered axonal tracts as "living electrodes" for synaptic-based modulation of neural circuitry. Adv Funct Mater 28:1701183. https://doi.org/10.1002/adfm.201701183

Shahab Ahmadi Seyedkhani (2024a) Biodegradable and Biohybrid Materials for Next-Generation Brain-Computer Interfaces. IntechOpen, Rijeka., p Ch. 4

Shahab Ahmadi Seyedkhani (2024b) Carbon Nanomaterials for Neural Interfaces: Synthesis, Properties and Applications. *Recent Progress and Development on Nanostructures*. IntechOpen, Rijeka., p Ch. 1 https://doi.org/10.5772/intechopen.1006603

Sheng H, Wang X, Kong N et al (2019) Neural interfaces by hydrogels. Extrem Mech Lett 30:100510

Shi G, Rouabhia M, Wang Z et al (2004) A novel electrically conductive and biodegradable composite made of polypyrrole nanoparticles and polylactide. Biomaterials 25:2477–2488. https://doi.org/10.1016/j.biomaterials.2003.09.032

Shi W, Yao Q, Donghui W et al (2022) Vapor phase polymerization of Ag QD-embedded PEDOT film with enhanced thermoelectric and antibacterial properties. NPG Asia Mater 14:47. https://doi.org/10.1038/s41427-022-00391-7

Shin Y, Lee HS, Hong YJ et al (2024) Low-impedance tissue-device interface using homogeneously conductive hydrogels chemically bonded to stretchable bioelectronics. Sci Adv 10:eadi7724. https://doi.org/10.1126/sciadv.adi7724

Shirakawa H (2001) The discovery of polyacetylene film: the dawning of an era of conducting polymers (Nobel lecture). Angew Chem Int Ed 40:2574–2580

Shirakawa H, Louis EJ, MacDiarmid AG et al (1977) Synthesis of electrically conducting organic polymers: halogen derivatives of polyacetylene, (CH). J Chem Soc Chem Commun 16:578–580. https://doi.org/10.1039/C39770000578

Simon DT, Kurup S, Larsson KC et al (2009) Organic electronics for precise delivery of neurotransmitters to modulate mammalian sensory function. Nat Mater 8:742–746

Stevenson G, Moulton SE, Innis PC, Wallace GG (2010) Polyterthiophene as an electrostimulated controlled drug release material of therapeutic levels of dexamethasone. Synth Met 160:1107–1114

Su D, Di Feng XJ, Che J, Xiao Y (2014) Application of conducting polymers in controlled drug delivery system. Prog Chem 26:1962

Sun J-Y, Keplinger C, Whitesides GM, Suo Z (2014) Ionic skin. Adv Mater 26:7608–7614. https://doi.org/10.1002/adma.201403441

Sun X, Zhong W, Zhang Z et al (2022) Stretchable, self-healable and anti-freezing conductive hydrogel based on double network for strain sensors and arrays. J Mater Sci 57:12511–12521

Suo H, Zhang D, Yin J et al (2018) Interpenetrating polymer network hydrogels composed of chitosan and photocrosslinkable gelatin with enhanced mechanical properties for tissue engineering. Mater Sci Eng C 92:612–620
Svirskis D, Travas-Sejdic J, Rodgers A, Garg S (2010) Electrochemically controlled drug delivery based on intrinsically conducting polymers. J Control Release 146:6–15
Tang X, Shen H, Zhao S et al (2023) Flexible brain–computer interfaces. Nat Electron 6:109–118. https://doi.org/10.1038/s41928-022-00913-9
Tang S, Yan Y, Lu X et al (2024) Nanocomposite magnetic hydrogel with dual anisotropic properties induces osteogenesis through the NOTCH-dependent pathways. NPG Asia Mater 16:16. https://doi.org/10.1038/s41427-024-00535-x
Thompson BC, Moulton SE, Ding J et al (2006) Optimising the incorporation and release of a neurotrophic factor using conducting polypyrrole. J Control Release 116:285–294
Thompson BC, Richardson RT, Moulton SE et al (2010) Conducting polymers, dual neurotrophins and pulsed electrical stimulation—dramatic effects on neurite outgrowth. J Control Release 141:161–167
Thoniyot P, Tan MJ, Karim AA et al (2015) Nanoparticle–hydrogel composites: Concept, design, and applications of these promising, multi-functional materials. Adv Sci 2:1400010
Tsuchida E, Abe K (2005) Interactions between macromolecules in solution and intermacromolecular complexes. In: Interact between Macromol Solut intermacromolecular complexes, pp 1–119
Umashankar PR, Mohanan PV, Kumari TV (2012) Glutaraldehyde treatment elicits toxic response compared to decellularization in bovine pericardium. Toxicol Int 19:51–58
Uz M, Mallapragada SK (2019) Conductive polymers and hydrogels for neural tissue engineering. J Indian Inst Sci 99:489–510
Wadhwa R, Lagenaur CF, Cui XT (2006) Electrochemically controlled release of dexamethasone from conducting polymer polypyrrole coated electrode. J Control Release 110:531–541
Wahab AHA, Saad APM, Harun MN et al (2019) Developing functionally graded PVA hydrogel using simple freeze-thaw method for artificial glenoid labrum. J Mech Behav Biomed Mater 91:406–415
Walde P, Kashima K, Ćirić-Marjanović G (2019) Synthesizing polyaniline with laccase/O2 as catalyst. Front Bioeng Biotechnol 7:165
Wang Y, Feng W (2022a) Preparation of conductive polymers BT – conductive polymers and their composites. In: Wang Y, Feng W (eds). Springer, Singapore, pp. 33–69
Wang Y, Feng W (2022b) Conductive polymers and their composites for biological application BT – conductive polymers and their composites. In: Wang Y, Feng W (eds) . Springer, Singapore, pp 153–203
Wang Y, Shi Y, Pan L et al (2015) Dopant-enabled supramolecular approach for controlled synthesis of nanostructured conductive polymer hydrogels. Nano Lett 15:7736–7741
Wang D, Noël V, Piro B (2016) Electrolytic gated organic field-effect transistors for application in biosensors—a review. Electronics 5:9
Wang GY, Lian K, Chu T-Y (2021a) Electrolyte-gated field effect transistors in biological sensing: A survey of electrolytes. IEEE J Electron Devices Soc 9:939–950
Wang J, Lin Y, Mohamed A et al (2021b) High strength and flexible aramid nanofiber conductive hydrogels for wearable strain sensors. J Mater Chem C 9:575–583
Ware T, Simon D, Rennaker RL, Voit W (2013) Smart polymers for neural interfaces. Polym Rev 53:108–129
Watanabe T, Ohtsuka A, Murase N et al (1996) NMR studies on water and polymer diffusion in dextran gels. Influence of potassium ions on microstructure formation and gelation mechanism. Magn Reson Med 35:697–705
Wei H, Kong D, Li T et al (2021) Solution-processable conductive composite hydrogels with multiple synergetic networks toward wearable pressure/strain sensors. ACS Sensors 6:2938–2951
Woods EF, Berl AJ, Kalow JA (2021) Advances in the synthesis of π-conjugated polymers by photopolymerization. ChemPhotoChem 5:4–11

Wu Q, Wei J, Xu B et al (2017) A robust, highly stretchable supramolecular polymer conductive hydrogel with self-healability and thermo-processability. Sci Rep 7:41566. https://doi.org/10.1038/srep41566

Wu N, Wan S, Su S et al (2021) Electrode materials for brain–machine interface: a review. InfoMat 3:1174–1194

Wu M, Yao K, Huang N et al (2023) Ultrathin, soft, bioresorbable organic electrochemical transistors for transient spatiotemporal mapping of brain activity. Adv Sci 10:2300504. https://doi.org/10.1002/advs.202300504

Xiang J, Shen L, Hong Y (2020) Status and future scope of hydrogels in wound healing: synthesis, materials and evaluation. Eur Polym J 130:109609

Xiao Y, Ye X, He L, Che J (2012) New carbon nanotube–conducting polymer composite electrodes for drug delivery applications. Polym Int 61:190–196

Xu C, Lee W, Dai G, Hong Y (2018) Highly elastic biodegradable single-network hydrogel for cell printing. ACS Appl Mater Interfaces 10:9969–9979

Yadav D, Yadav S, Veer K (2020) A comprehensive assessment of brain computer interfaces: recent trends and challenges. J Neurosci Methods 346:108918

Yang C, Suo Z (2018) Hydrogel ionotronics. Nat Rev Mater 3:125–142

Ye X, Li X, Shen Y et al (2017) Self-healing pH-sensitive cytosine- and guanosine-modified hyaluronic acid hydrogels via hydrogen bonding. Polymer (Guildf) 108:348–360. https://doi.org/10.1016/j.polymer.2016.11.063

Yokoyama F, Masada I, Shimamura K et al (1986) Morphology and structure of highly elastic poly (vinyl alcohol) hydrogel prepared by repeated freezing-and-melting. Colloid Polym Sci 264:595–601

Yokozawa T, Yokoyama A (2009) Chain-growth condensation polymerization for the synthesis of well-defined condensation polymers and π-conjugated polymers. Chem Rev 109:5595–5619

Yuan N, Xu L, Xu B et al (2018) Chitosan derivative-based self-healable hydrogels with enhanced mechanical properties by high-density dynamic ionic interactions. Carbohydr Polym 193:259–267

Yuk H, Lu B, Zhao X (2019) Hydrogel bioelectronics. Chem Soc Rev 48:1642–1667

Zhang Y, Won C, Chu C (1999) Synthesis and characterization of biodegradable network hydrogels having both hydrophobic and hydrophilic components with controlled swelling behavior. J Polym Sci Part A Polym Chem 37:4554–4569

Zhang G, Lv L, Deng Y, Wang C (2017) Self-healing gelatin hydrogels cross-linked by combining multiple hydrogen bonding and ionic coordination. Macromol Rapid Commun 38:1700018

Zhang D, Qiao H, Fan W et al (2020a) Self-powered ionic sensors overcoming the limitation of ionic conductors as wearable sensing devices. Mater Today Phys 15:100246

Zhang Q, Liu X, Duan L, Gao G (2020b) Nucleotide-driven skin-attachable hydrogels toward visual human–machine interfaces. J Mater Chem A 8:4515–4523

Zhang D, Jian J, Xie Y et al (2022) Mimicking skin cellulose hydrogels for sensor applications. Chem Eng J 427:130921

Zhao Y, Zhang B, Yao B et al (2020) Hierarchically structured stretchable conductive hydrogels for high-performance wearable strain sensors and supercapacitors. Matter 3:1196–1210

Zhao C, Liu G, Tan Q et al (2023) Polysaccharide-based biopolymer hydrogels for heavy metal detection and adsorption. J Adv Res 44:53–70

Zhou L, Fan L, Yi X et al (2018a) Soft conducting polymer hydrogels cross-linked and doped by tannic acid for spinal cord injury repair. ACS Nano 12:10957–10967

Zhou Y, Hu Y, Sun W et al (2018b) Radiotherapy-sensitized tumor photothermal ablation using γ-polyglutamic acid nanogels loaded with polypyrrole. Biomacromolecules 19:2034–2042

Ziai Y, Zargarian SS, Rinoldi C et al (2023) Conducting polymer-based nanostructured materials for brain–machine interfaces. Wiley Interdiscip Rev Nanomed Nanobiotechnol 15:e1895

Zotti G, Zecchin S, Schiavon G et al (2003) Electrochemical and XPS studies toward the role of monomeric and polymeric sulfonate counterions in the synthesis, composition, and properties of poly (3, 4-ethylenedioxythiophene). Macromolecules 36:3337–3344

Shahab Ahmadi Seyedkhani is a distinguished researcher and PhD candidate in the field of nanotechnology at Sharif University of Technology. With over a decade of expertise in designing and fabricating nanomaterials for advanced biotechnology applications—including tissue engineering, drug delivery, wound healing and biosensors, his current research focuses on developing nanostructured brain-computer interfaces (BCIs). As an Iranian young scientist, he has published numerous peer-reviewed scientific papers and authored several academic books. His primary research interest lies in investigating nanoscale interactions between living systems and biomaterials, with a particular emphasis on enhancing tissue-electronic interfaces.

Professor Azam Irajizad obtained her B.Sc. degree in Physics from Sharif University of Technology, Tehran. She earned her Ph.D. in Surface Physics from Sussex University, Brighton, in 1990. She is currently a Distinguished Professor in the Department of Physics and the Institute for Convergence Science and Technology (ICST). Her research focuses on the fabrication and characterization of materials at the nanoscale. With over 250 refereed publications, she specializes in nanostructures, sensors, biomaterials, and human-machine interfaces.

Associate Professor Raheleh Mohammadpour earned her B.Sc., M.Sc., and Ph.D. degrees from Sharif University of Technology. She has conducted research visits at Uppsala University and National Chiao Tung University. With over 100 refereed publications, her expertise lies in the development of self-powered nanosensors, human-machine interfaces, and wearable electronic devices.

Green Biopolymers as Carriers for Innovative Drug Delivery

7

Daniel Hernandez-Patlan, Inkar Castellanos-Huerta, Raquel López-Arellano, Guillermo Tellez-Isaias, and Bruno Solis-Cruz

Abstract

Recent research and technological advances have accelerated the development of green biopolymers because their chemical, electrical, optical and biological properties can be improved. These properties allow these green biomaterials to be biocompatible, biodegradable, bioinherent, non-immunogenic and less toxic, as well as to design more effective and biomimetic systems compared to conventional biopolymers. In this sense, the employment of green biopolymers based on polysaccharides, proteins, oligonucleotides, lipids, or their modifications and combinations obtained from different sources and technologies has increased mainly due to their biomedical and pharmaceutical applications, specifically for the development of new drug delivery systems. Therefore, this book chapter describes the applications of green biopolymers in the pharmaceutical field for

D. Hernandez-Patlan · B. Solis-Cruz (✉)
Laboratory 5: LEDEFAR, Multidisciplinary Research Unit, National Autonomous University of Mexico-Superior Studies Faculty at Cuautitlan (UNAM-FESC), Cuautitlan Izcalli, Mexico, Mexico

Nanotechnology Engineering Division, Polytechnic University of the Valley of Mexico, Tultitlan, Mexico, Mexico
e-mail: danielpatlan@comunidad.unam.mx; daniel.hernandez@upvm.edu.mx; brunosc@comunidad.unam.mx; bruno.solis@upvm.edu.mx

I. Castellanos-Huerta · G. Tellez-Isaias
Department of Poultry Science, University of Arkansas, Fayetteville, AR, USA
e-mail: icastell@uark.edu; gtellez@uark.edu

R. López-Arellano
Laboratory 5: LEDEFAR, Multidisciplinary Research Unit, National Autonomous University of Mexico-Superior Studies Faculty at Cuautitlan (UNAM-FESC), Cuautitlan Izcalli, Mexico, Mexico
e-mail: lopezar@unam.mx

R. Malviya, S. Sundram (eds.), *Sustainable Nanocomposites with Green Biomaterials*, Biomaterials, Bioengineering and Sustainability 2,
https://doi.org/10.1007/978-3-031-79110-9_7

the design and development of innovative systems for drug delivery, as well as the sources of obtaining said biomaterials, which are characterized by being renewable, sustainable, biodegradable, biocompatible and low toxicity, mainly.

Keywords

Green biopolymers · Drug delivery systems · Mucoadhesion · Modified biopolymers · Hybrid systems · Nanosystems

Abbreviations

16:0 PG	1,2-dipalmitoyl-sn-glycero-3-phospho-(1′-rac-glycerol) (sodium salt)
18:0 PEG2000 PE	1,2-distearoyl-sn-glycero-3-phosphoethanolamine-N-[methoxy(polyethylene glycol)-2000] (ammonium salt)
DSPC	1,2-Distearoyl-sn-glycero-3-phosphorylcholine
Da.	Dalton
FDA	Food and Drugs Administration
Fe_3O_4	Iron tetroxide, black iron oxide or magnetite
GRAS	Generally Recognized as Safe
Lipid ALC-0315	6-((2-hexyldecanoyl)oxy)-N-(6-((2-hexyldecanoyl)oxy)hexyl)-N-(4-hydroxybutyl)hexan-1-aminium
Lipid SM-102	8-[(2-hydroxyethyl)[6-oxo-6-(undecyloxy)hexyl]amino]-octanoic acid, 1-octylnonyl ester
PEG	Polyethylene glycol
PLGA	Poly (lactic-co-glycolic acid)
Pt(II)	Tetraammineplatinum (II)
SEEDS	Self-emulsifying drug delivery system

7.1 Introduction

The use of naturally derived polymers, known as biopolymers, in the research and development of new products has become a viable alternative to petroleum-derived polymers. since they present interesting properties of biodegradability and biocompatibility, lower toxicity and immunogenicity, as well as better biomimetic capacity (Baranwal et al. 2022; Kundu et al. 2022; Liu et al. 2023). Furthermore, they can help mitigate the impact of climate change, the depletion of natural resources and water pollution, mainly (Mori 2023).

Although the term "Green Polymers" emerged in the 1990s, this term does not necessarily encompass "Natural Polymers" since the first refer to polymers synthesized through green chemistry, that is, by minimizing or eliminating the use of materials that can be harmful to living beings (humans, animals and plants) and the environment (Gnanasekaran 2019). In contrast, "Natural Polymers" are biopolymers that are found in nature and can be obtained from plants and animals such as proteins, cellulose, natural rubber, silk, wool, or starch (Caillol 2020; Amarakoon et al. 2022). In this context, another term applied to biopolymers is "Bio-based Polymers", which refer to biopolymers that are synthesized from renewable resources such as plants, animals or microorganisms, but that are not necessarily biodegradable (Caillol 2020). Therefore, although it is known that all biopolymers are considered green, the term "Green Biopolymers" not only involves those obtained from sources such as plants, animals, and microorganisms, as well as biocompatibility and biodegradability (Al Mahmud et al. 2023), but also integrates sustainability (renewable natural resources) and green chemistry for its synthesis (eco-friendly), which are aspects that have been developing moderately in recent years (Singh 2022). In this sense, some synonyms used for "Green Biopolymers"are "Sustainable Biomaterials"and "Environmentally Friendly Materials"(Elfawy et al. 2023; Salleh et al. 2023).

There are several ways to classify green biopolymers, starting with their chemical composition, source of origin and biodegradability, mainly (Fig. 7.1). According to the chemical composition of green biopolymers, they are mainly classified into two groups: polysaccharides and proteins. Within the group of polysaccharides there are non-ionic polysaccharides: dextran, starch and cellulose; anionic polysaccharides: alginate, heparin, pectin, carrageenan and hyaluronic acid; and cationic polysaccharides: chitosan and pullulan (Baranwal et al. 2022; Elfawy et al. 2023). In the case of green biopolymers of protein origin some examples are collagen, gelatin, silk, keratin, sericin, fibrin, fibroin, casein, zein, sericin, elastin milk protein, whey protein, and soy protein (Nagarajan et al. 2019; Salleh et al. 2023).

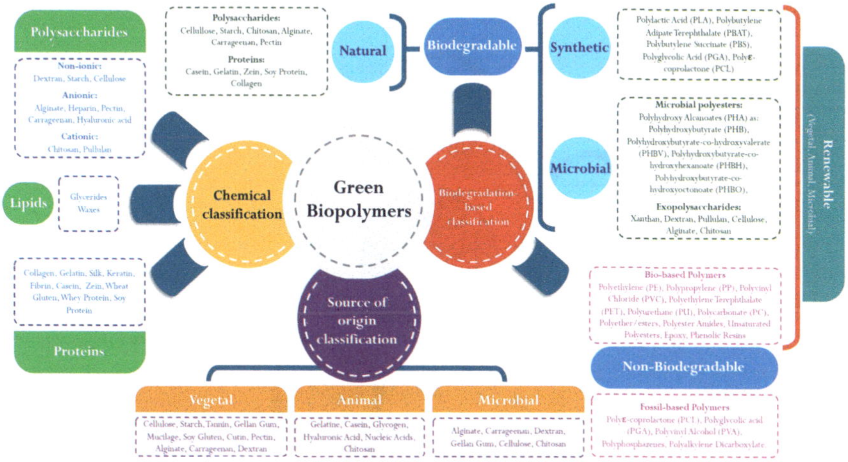

Fig. 7.1 Classification of green biopolymers

Although the two previous groups are the principal ones, another growing group is lipids such as triglycerides, phospholipids, and waxes (Gali et al. 2023). Another classification of green biopolymers considers the source of origin, which can come from plants/vegetables, animals and microorganisms, highlighting the last group derived from biotechnological advances (genetically modified organisms) and by which the production of biopolymers such as chitosan has been achieved (Pandey 2020; Huq et al. 2022). According to the nature of green biopolymers, they can be biodegradable or non-biodegradable, which makes them viable alternative materials as a replacement for petroleum-derived polymers (Pandey 2020). Biodegradable green biopolymers can be categorized into 3 groups considering their origin: natural, synthetic, and microbial. In the case of those of natural origin, they are obtained from renewable sources such as plants/vegetables and algae, and to a lesser extent from animals (Nazrun et al. 2023). Meanwhile, those of synthetic origin are obtained chemically (Samir et al. 2022) and the microbial ones are obtained under controlled cultivation conditions (biotechnology) and through a green approach that has become popular in the worldwide (Bhatia 2023). Although the term green biopolymer is related to renewable sources and biodegradation, there are polymers that are not necessarily biodegradable, such as all those polythioesters produced by microorganisms (Bio-based polymers) (Andreeßen and Steinbüchel 2019).

Regarding the use of green biopolymers, they have a wide field of application in the pharmaceutical, cosmetic, and food industries, and in the medical and agricultural sectors (Fig. 7.2). The applications of green biopolymers in the pharmaceutical industry mainly involve the development of controlled drug release systems, which can be both conventional (tablets and capsules, mainly) and non-conventional (micro and nanotechnological platforms) (Witika et al. 2020; Gheorghita et al. 2021; Opriş et al. 2024). In the medical field, green biopolymers have been used in

Fig. 7.2 Applications of green biopolymers in different fields of knowledge

tissue engineering and regeneration, implants and medical devices, as well as in the development of wound healing and biosensors for the detection of infectious diseases and the treatment of chronic diseases, as well as monitoring health status (Haleem et al. 2021; Das et al. 2023; RG et al. 2023). Likewise the applications of green biopolymers in the pharmaceutical industry, in the cosmetic industry the most important field of application is related to the development of cosmetic formulations for hair care, skin care, and oral care, as well as mesotherapy, which include conventional systems such as: emulsions and foams, and non-conventional systems such as micro/nanocapsules, liposomes, nanofibers and hydrogels (Tan et al. 2021; Mendoza-Muñoz et al. 2023). Another important field of application of green biopolymers lies in the food industry given that they have been used mainly for food preservation since they can encapsulate and release bioactive substances (nanocapsules) and prevent the contamination of food with microorganisms (edible films, coatings, and intelligent and active packaging), thus improving food safety (Samrat et al. 2022; Panda et al. 2024). Agriculture plays a fundamental role in sustaining human life by providing food, fiber and fuel, so the most important applications of green biopolymers include mulch film formation, seed coatings, superabsorbent polymers and agrochemical delivery systems (Verma et al. 2023). Considering this background, this book chapter aims to show the current events and trends of one of the principal uses of green biopolymers, as carriers for the innovative drug delivery systems.

7.2 Green Biopolymers in the Pharmaceutical Industry

Recently, biopolymers have positioned themselves as the most attractive sustainable alternative to replace petroleum-derived polymers (Balart et al. 2021). In fact, they were first used in the food industry as gelling, thickeners and/or stabilizers agents of food products, and in recent years they have also been used for the formation of film/coatings and packaging materials, as well as for the encapsulation of bioactive compounds in functional foods (Himashree et al. 2022; Bhaskar et al. 2023). However, the use of biopolymers has been significantly promoted in the biomedical and pharmaceutical sector (Wróblewska-Krepsztul et al. 2019) and although in the pharmaceutical industry they have been widely used as excipients for the development of different dosage forms such as tablets (immediate release and controlled release), capsules (hard and soft gelatin), suspensions, emulsions and gels for oral, parenteral, transdermal, buccal, ocular and nasal administration (Guru et al. 2023), currently their use have focused on the development of new non-conventional drug delivery systems for the treatment, control or prevention of diseases [36] because they present characteristics such as biodegradability, biocompatibility, sustainability, bioresorbability, flexibility, antibacterial activity, renewability and stability, as well as lower toxicity and immunogenicity, biomimetic properties, and are not carcinogenic or thrombogenic compared to petroleum-derived polymers (Gheorghita et al. 2021; Quinn et al. 2024) (Table 7.1). It is known that these characteristics are directly influenced by the properties that biopolymers have to form matrices (charge

Table 7.1 Main characteristics of green biopolymers, biopolymers and petroleum-based polymers

Characteristics	Green Biopolymers	Biopolymers	Petroleum-derived polymers	References
Source	Agro-resources	Agro-resources	Petroleum and gas	Abdel-Raouf et al. (2021), Gheorghita et al. (2021), Kučuk et al. (2023), Phutane et al. (2023) and Stuart-Fox et al. (2023)
Biodegradability	Yes	Yes	No	
Sustainability	High	Medium	Low	
Biocompatibility	Yes	Yes	Yes	
Toxicity	Low	Low	High	
Immunogenic	No	No	Yes	
Structure	Well defined	Well defined	Stochastic	
Biomimetic and bioinspired approaches	Yes	Yes	No	
Chemical backbone structure	Carbon, oxygen, and nitrogen	Carbon, oxygen, and nitrogen	Mostly carbon	
Batch to batch variation	Yes	Yes	No	
Physicochemical resistance	Low	Low	High	
Thermal stability	Low	Low	High	
Mechanical properties	Low	Low	High	
Availability	High	High	Decreasing	
Production costs	High (depends on the type)	High (depends on the type)	Low	
Highly resource-, eco-, and energy-effective mass production	High	Medium	Low	

distribution, molecular mass, and conformation), the product development conditions (concentration, pH, solvent, and temperature), and the type and concentration of the other additives/excipients in the formulation (preservatives, cross-linking agents, plasticizers, antioxidants, mainly) (Okolišan et al. 2022).

Despite biopolymers have many advantages over Petroleum-derived polymers, there are limitations in their obtention (Gheorghita et al. 2021):

1. As they are natural components, their properties depend mainly on the raw material used for their extraction. Therefore, their properties can vary according to their origin, climatic conditions, and obtaining methods.
2. Production costs are higher compared to petroleum-based polymers. However, the elimination of taxes on recycled raw materials and/or their waste through global legislation has lowered their production costs.
3. Although the production of biopolymers requires special equipment, today some existing equipment has been retrofitted to reduce production costs.
4. Lower performance in production, so its applications are limited.

Table 7.2 Bacteria and fungi used in the production of green biopolymers

Microorganisms	Species/Strain	Biopolymers
Bacteria	*Acetobacter polysaccharogenes, Acetobacter xylinum, Pseudomonas aeruginosa, Bacillus thuringiensis, Escherichia coli, Pasteurella multocida, Lactobacillus acidophilus, and Geobacillus thermonitrificans*	Polysaccharides
	Pseudomonas putida, Xanthomonas campestris, Corynebacterium glutamicum, *and* Escherichia coli	Polyamides
	Lactobacillus bulgaricus, lactobacillus delbrueckii, and lactobacillus leichmanni	Biopolyesters
	Pseudomonas putida, Aeromonas hydrophila, Ralstonia eutropha, Alcaligenes latus, Actinobacteria, Bacteroidetes, Cyanobacteria, Proteobacteria, Clostridia, and Bacilli	Polyhydroxylalkanoates
	Citrobacter freundii	Polyphosphates
Fungi	*Aspergillus Trichoderma, aspergillus penicillium, Agrocybe cylindracea, aspergillus terreus, and Fusarium equiseti*	Exopolysaccharides
	Absidia coerulea, Absidia glauca, Absidia blakesleeana, Mucor rouxii, aspergillus Niger, Phycomyces blakesleeanus, Trichoderma reesei, Colletotrichum lindemuthianum, Gongronella butleri, Pleurotus sajocaju, Rhizopus oryzae, and Lentinus edodes	Aminopolysaccharides
	Fomitopsis pinicola, Ganoderma lucidum, Pleurotus eryngii, Termitomyces albuminosus, Trametes versicolor Coprinus comatus, and Cordyceps kyushuensis	Intrapolysaccharides

Considering the above limitations, the use of microorganisms in conjunction with genetic engineering and biotechnology can be potential alternatives to optimize and improve the properties of biopolymers, reduce their production costs and establish a comprehensive reuse cycle. (Akinsemolu and Onyeaka 2023). However, only 1% of polymers worldwide are produced by microorganisms through their enzymatic system, which assembles fundamental units of sugars, hydroxylated fatty acids and amino acids to produce molecules with high molecular weight (Ghosh et al. 2021; Akinsemolu and Onyeaka 2023). Table 7.2 shows some types of biopolymers obtained by bacteria and fungi (Luft et al. 2020; Moradali and Rehm 2020; Akinsemolu and Onyeaka 2023).

7.3 Mucoadhesive Properties of Green Biopolymers

In addition to green biopolymers being of particular interest because of their properties such as biocompatibility, biodegradability, and versatility for use in pharmaceutical formulations, their mucoadhesive properties have allowed the development of new drug delivery systems that are capable of improving the effectiveness of treatments since they can prolong the contact time with the mucous membranes (nasal, oral, vaginal and even ocular) (Cazorla-Luna et al. 2021; Kimna et al. 2021).

The mucoadhesion mechanisms of green biopolymers are the same as Biopolymers and occur in two stages: (1) the contact stage and (2) the consolidation stage. The first stage begins with the wetting of the biopolymer to facilitate its interaction with the mucosa, followed by bioadhesion, which includes primary forces (ionic interactions and covalent bonds) and secondary forces (van der Waals, hydrophobic and hydrogen bonds) that favor the interaction between the biopolymer and mucosal glycoproteins (Fig. 7.3) (Tripathi et al. 2019). These interactions are mainly due to the chemical structure of the biopolymers and specifically to their functional groups, among which the alcohol (–OH), carboxyl (–COOH), amide (–NH_2), sulfate (–SO_4H) groups stand out (Surendranath et al. 2022).

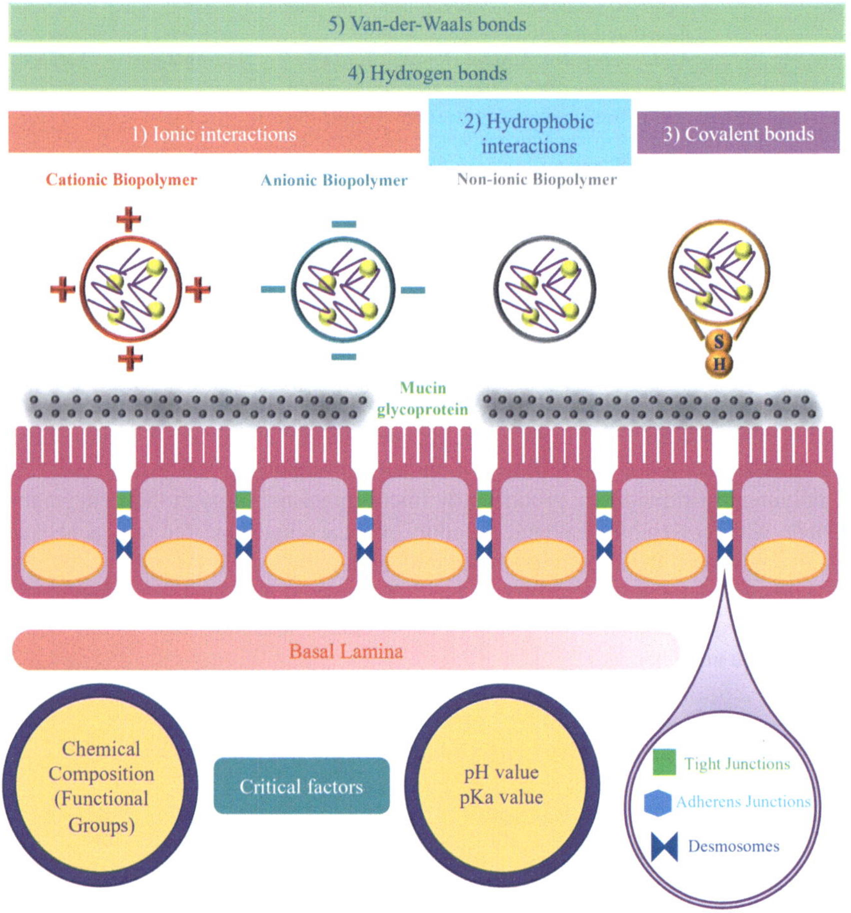

Fig. 7.3 Main interactions of biopolymers with the mucosal membrane

In this sense, to date four generations of mucoadhesive biopolymers are known. The first generation includes cationic biopolymers (chitosan, chitin and curdlan), which have great mucoadhesive properties given the strong electrostatic interactions that take place between their functional groups and the glycoproteins of the mucosa. Likewise, anionic biopolymers such as hyaluronic acid, alginate, heparin, xanthan gum, xylan, dextran, pectin, gellan, Mannan and inulin belong to the first generation. However, hydrophobic interactions, hydrogen bonds and van der Waals forces are predominant, so their mucoadhesive capacity is lower compared to cationic biopolymers. The last case of the first generation includes non-ionic biopolymers (hydroxypropyl methylcellulose, hydroxyethyl cellulose, methylcellulose, starch, agarose and pullulan), which are characterized by having the weakest mucoadhesion force (Surendranath et al. 2022; Fazal et al. 2023).

Meanwhile, in the case of second-generation mucoadhesive biopolymers, the bioadhesion mechanism entails the formation of covalent bonds between a specific functional group of the biopolymer and mucosal glycoproteins. Some examples of this type of biopolymers include thiolated biopolymers, acrylated biopolymers, and boronated biopolymers (Puri et al. 2020; Surendranath et al. 2022).

In the case of third generation biopolymers, these represent significant advances in the science of second generation biopolymers, with the main objective of improving biopolymer-mucosal interactions (Knoll et al. 2021).

Finally, fourth-generation mucoadhesive biopolymers include smart polymers, also known as stimulus-responsive or shape-memory polymers, that is, they can change their properties in response to environmental factors such as temperature, pH or light (Sharma et al. 2023).

The second stage of the mucoadhesion mechanism is critical and involves the maintenance of the biopolymer in the bioadhered zone and its penetration into the mucosal layer (Sharma et al. 2023), for the subsequent transport of the drug by transcellular transport (Active Transport), paracellular transport (Passive Transport) or trancytosis (Fig. 7.4) (Zhao et al. 2022). This mucoadhesion mechanism is applicable to both conventional pharmaceutical forms (Tablets, capsules, powders, pellets, patches and hydrogels) and non-conventional forms (Micro/nanoparticles, nanogels, nanofibers and films) developed from green biopolymers or biopolymers (Kumar et al. 2022).

There is no doubt that mucoadhesion is an interesting property that biopolymers have and that it is used during the development of new delivery systems because drug bioavailability can be improved by extending the contact time of the delivery system at the absorption site. However, problems may arise from the longer contact time of the drug delivery system with the mucosa, causing the manifestation of ulcers (Sharma et al. 2023). In addition, factors such as the pH in which the mucous membranes are located (ocular: 7.5–7.8, nasal: 6.3–6.7, oral: 6.8–7.4, pulmonary: 7.0, gastric: 1.0–3.0, intestinal: 5.9–7.5, colonic: 6.2–7.6, rectal: 6.8–7.9 and vaginal: 3.5–4.5), mucin concentration (ocular: 0.01%, nasal: 2–3%, buccal: 0.1–0.5%, pulmonary: 2–4%, gastric: 3%, intestinal: 1%, colonic: <5%, rectal: <5% and vaginal: 1–2%) and average layer thickness (ocular: 3–5 μm, nasal: 10–15 μm, buccal; 10–100 μm, pulmonary: 5–55 μm, gastric: 170 μm, intestinal: 0–37 μm, colonic:

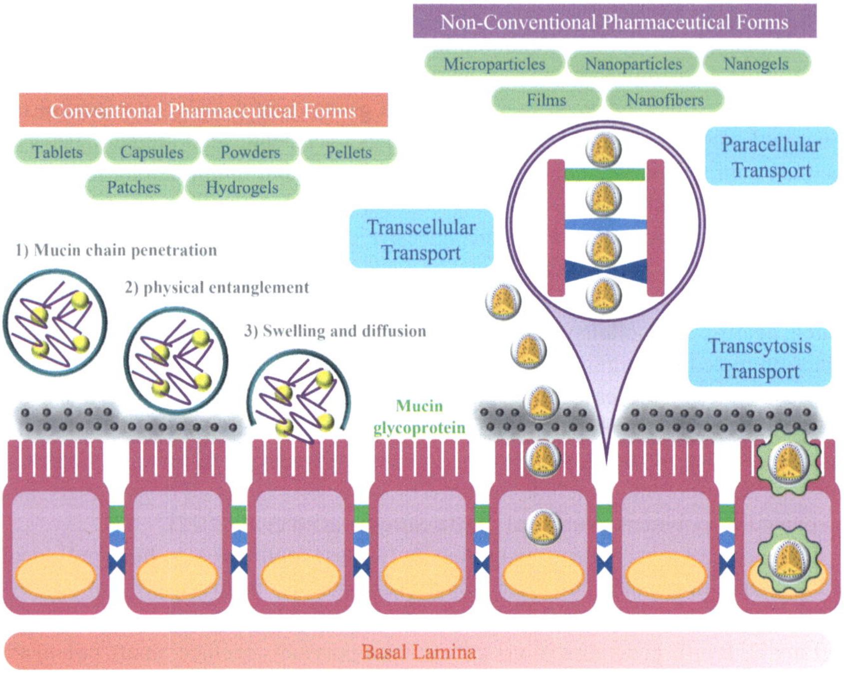

Fig. 7.4 Mucoadhesion mechanism and types of transport involved in conventional and non-conventional pharmaceutical forms developed with green biopolymers or Biopolymers

100 μm, rectal: 125 μm and vaginal: 20 μm), as well as hydrophobicity, surface charge and particle size of the pharmaceutical systems can compromise the mucoadhesion mechanism (Zhao et al. 2022).

7.4 Chemically Modified Green Biopolymers and Hybrid Systems

Many green biopolymers have been described to date, but due to some of their intrinsic properties that limit their application, chemical/physical methods have been developed to improve their properties (Ganie et al. 2021). These modifications, together with the development of hybrid systems (micro/nanosystems and micelles), heterogeneous mixtures of different materials, are strategies that have been used to functionalize green biopolymers into new materials with superior properties and applications (Das et al. 2023). According to the type of green biopolymer (ionic, non-ionic), modifications can be made in various functional groups, such as amines, hydroxyl, carboxylic acids, and aldehydes. Furthermore, modifications can be made to the above functional groups with thiol, azide or alkyne groups for additional chemical modifications (Muir and Burdick 2020).

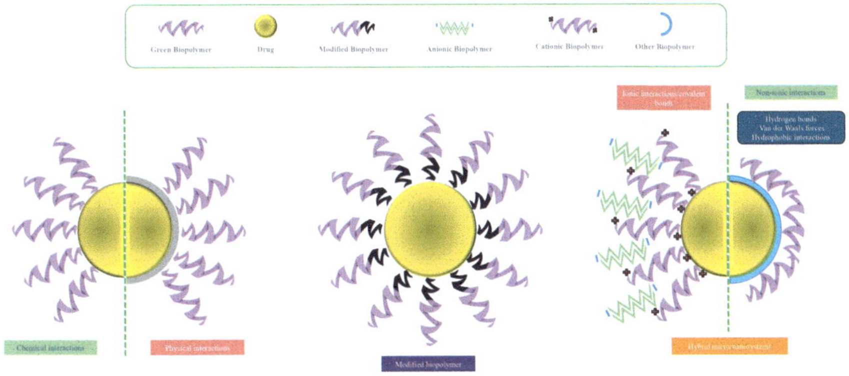

Fig. 7.5 Drug delivery systems considering green biopolymers, modified biopolymers, and hybrid micro/nanosystems

In the case of polysaccharides, most of them contain hydroxyl groups (-OH), which can be used to make modifications to the biopolymer (Muir and Burdick 2020). For their part, green biopolymers of a protein nature contain amino acids such as serine and tyrosine, which favor the formation of hydrogen bonds (Muir and Burdick 2020). Furthermore, polysaccharides such as hyaluronic acid, alginate, and chondroitin sulfate, as well as green protein biopolymers formed by amino acids such as aspartic acid and glutamic acid, have carboxyl groups (-COOH) at the end of the side chain (Baslé et al. 2010). Green biopolymers such as chitosan, and some lysine-conformed proteins and peptides contain amino groups (-NH_2 or -NR_2) (Baslé et al. 2010). Finally, amino acids such as cysteine and polysaccharides such as carrageenans that contain both hydroxyl (-OH) and sulfate (-OSO_3^-) groups can favor the formation of disulfide bonds due to their thiol groups (-SH) (Baslé et al. 2010). Figure 7.5 shows a representation of drug delivery systems with green biopolymers, modified biopolymers, and hybrid systems.

7.5 Polysaccharides

As previously described, green biopolymers have important applications in the development of conventional and non-conventional drug delivery systems since they can help improve the pharmacokinetic and pharmacodynamic properties of drugs and therefore their bioavailability, as well as improve therapeutic indices and reduce side effects, thus improving the overall effectiveness of the drug. (Okolišan et al. 2022).

Polysaccharides constitute, in general, one of the most important families of biopolymers because they are considered biocompatible, biodegradable, highly available and low toxicity, so they are extremely useful for the development of new drug delivery systems and even to protect highly sensitive drugs or molecules with biological activity (Kurczewska 2022).

7.5.1 Cellulose

Cellulose, the most abundant biopolymer in nature, is formed by chains of glucopyranose monosaccharides linked by 1,4-glycosidic bonds (Seremeta and Sosnik 2023). This biopolymer is a main component of plants and fibers such as cotton and linen (Basim et al. 2021) and can also be found in algae and bacteria (O'Brien et al. 2021). This biopolymer is insoluble in water, which limits its applications. However, its modification using esterification or etherification methods has allowed its derivatives to be soluble in aqueous media, increasing their viscosity power (thickening agent) and adhesiveness, as well as its film-forming capacity and emulsifying properties (Gupta et al. 2021). Therefore, these cellulosic derivatives have fields of application in the biomedical, pharmaceutical, food, cosmetics, chemical, textile, paper and biorefineries industries, mainly (Seremeta and Sosnik 2023). Cellulosic derivatives are divided into two large groups: (1) cellulose ethers, which are soluble in water and include carboxymethylcellulose, hydroxypropyl methylcellulose, methylcellulose, ethyl cellulose, hydroxypropyl cellulose, and hydroxyethyl cellulose; (2) cellulose esters as cellulose acetate (soluble in water) and cellulose acetate phthalate (insoluble in water) (Dong et al. 2016; Dai et al. 2019). These cellulosic derivatives are considered green biopolymers since their production costs are low, they can be obtained from renewable sources, they present excellent chemical stability, in addition to having good biological affinity (biocompatibility), moldability, porosity, biodegradability and being physiologically safe (non-toxic) (Gupta et al. 2021).

Although chemical modifications of cellulose have improved the properties of cellulosic derivatives, current global interest focuses on the integration of two or more different biopolymers to produce biopolymers with new characteristics and properties due to chemical modifications frequently require laborious procedures, special equipment, and greater amounts of energy and reagents, which can increase production costs (Shaghaleh et al. 2018). Therefore, the use of combinations of biopolymers or some other additives and/or the incorporation of nanoparticulate systems is interesting to improve the processability, mechanical resistance, thermal stability and other desired attributes of cellulosic materials such as flexibility, resistance or toughness, electrical conductivity, and antimicrobial and optical properties (Platnieks et al. 2023). Table 7.3 shows a summary of some applications of cellulosic biopolymer/hybrid materials as conventional and non-conventional drug delivery systems, as well as the main routes of administration.

7.5.2 Chitosan

Chitosan, the second most abundant biopolymer after cellulose, is a cationic green biopolymer that is biocompatible, adhesive, hemostatic, non-toxic and biodegradable, obtained from marine/food industry waste by the deacetylation of the N-acetyl glucosamine units of chitin (Visan and Cristescu 2023). Furthermore, it has been used in the development of different drug delivery systems for intravenous, nasal,

Table 7.3 Drug delivery applications of cellulosic biopolymer/hybrid materials

Green biopolymer	Type of biopolymer	Pharmaceutical dosage form	Drug	Routes of administration
Hydroxypropyl cellulose	Non-ionic	Capsular devices for pulsatile release. Extended-release tablets.	Acetaminophen (Fazal et al. 2023) Theophylline (Fazal et al. 2023)	Oral
Hydroxypropyl methylcellulose	Non-ionic	Controlled release tablets or capsules	Nitrofurantoin (Fazal et al. 2023)	
Carboxymethyl cellulose	Anionic	Floating tablets (Drug release for more than 8 hours). Hybrid aerogel Hydrogel films	Pantoprazole (Srebro et al. 2022) 5-Fluorouracil (Fazal et al. 2023) Doxorubicin (Fazal et al. 2023)	
Hydroxyethyl cellulose	Non-ionic	Iron oxide-coated nanoparticles for targeted drug delivery	Doxorubicin (Fazal et al. 2023)	Oral
Microcrystalline cellulose	Non-ionic	Nanocrystalline cellulose	Luteolin and luteoloside (Fazal et al. 2023)	
Ethyl cellulose	Non-ionic	Disk-shaped specimens	Acetaminophen and furosemide (Fazal et al. 2023)	
Alginate/carboxymethyl cellulose	Anionic	Patches	Atenolol (Al Mahmud et al. 2023)	Buccal
Carboxymethyl cellulose/graphene oxide bio-nanocomposite	Anionic	Hydrogel	Doxorubicin (Al Mahmud et al. 2023)	
Bacterial cellulose / methoxylated pectin	Anionic	Hydrogel	Levofloxacin (Cacicedo et al. 2018)	Oral
Sodium alginate/bacterial cellulose	Anionic	Hydrogel	Bovine serum albumin and lysozyme (Ji et al. 2021)	
N-trimethyl chitosan/Carboxymethyl cellulose	Anionic	Hydrogel	Ciprofloxacin (Lotfy and Basta 2020)	
Oxidized pectin/acyl hydrazide functionalized carboxymethyl cellulose	Anionic	Hydrogel	Limonin and doxorubicin	Parenteral

ocular, oral, and mucosal administration, the latter being important due to their mucoadhesive properties since the amino group of chitosan, which is positively charged under physiological conditions, can promote ionic interactions, formation of hydrogen bonds and hydrophobic interactions with the main constituents of the mucosa that are negatively charged (sialic acid and sulfonic acid) (Zhao et al. 2022; Fazal et al. 2023). Despite conventional drug delivery systems such as gels, tablets, biological membranes, powders, microemulsions and hydrogels, as well as micro and nanoparticulate systems as non-conventional systems, have been developed with this biopolymer. One of the main limitations of chitosan is its low solubility in aqueous solutions at pH higher than 6.3–6.5, so its mucoadhesive properties are limited since it only exists in acidic conditions (pH < 6) are its amino groups ionized, which favors this property (Mura et al. 2022). However, these limitations can be partially resolved by using chitosan salts or chemical modifications since their functional groups allow them to easily modify their backbone. After chemical modifications of chitosan, the most common resulting derivatives are trimethyl chitosan, carboxymethyl chitosan, and thiolated chitosan (Lang et al. 2020). Trimethyl chitosan and carboxymethyl chitosan are characterized by having a high aqueous solubility at neutral pH and in the case of thiolated chitosan its mucoadhesive properties are superior (Kumar et al. 2022). Some thiolated chitosans that have been developed are chitosan-cysteine, chitosan-glutathione, chitosanthoglycolic acid, chitosan-thiobutylamidine and chitosan-N-acetylcysteine (Kumar et al. 2022).

Unfortunately, the advantages offered by chemical modifications of chitosan are partial since the derivatives are required to be characterized not only physically-chemically, but also biopharmaceutically to guarantee their safety (Bellich et al. 2016). Likewise, their production costs are usually higher than the biopolymer itself, so they are less competitive in a market that has polymers/biopolymers with similar characteristics at a lower cost, with higher degrees of purity and reproducibility since they are available on the market chitosans that differ in molecular weight, degree of acetylation and acetylation pattern, which determines their physicochemical properties and biological activities (Mura et al. 2022).

Therefore, one of the trends in the development of new drug delivery systems is through the formation of biopolymer-biopolymer or biopolymer-drug conjugates since release characteristics, mucoadhesion and paracellular permeability can be improved (Das et al. 2022). Table 7.4 shows some applications of drug delivery systems developed with chitosan.

7.5.3 Alginate

Alginate is probably the most used anionic biopolymer in the medical-pharmaceutical terms because it is considered generally regarded as safe (GRAS) by the Food and Drugs Administration (FDA) and is even included in the European Pharmacopoeia (Cattelan et al. 2020; Fazal et al. 2023). The production costs of this biopolymer are low and it is characterized by being renewable, biodegradable, biocompatible, non-toxic, non-immunogenic and, most importantly, mucoadhesive (Gheorghita et al. 2021). The mucoadhesive properties of this biopolymer depend largely on its molecular weight, that is, the lower the molecular weight, the lower the mucoadhesive capacity (Kumar et al. 2022). Furthermore, given that it can form hydrogen

Table 7.4 Drug delivery applications of chitosan, modified chitosan, and hybrid biomaterials

Green biopolymer	Type of biopolymer	Pharmaceutical dosage form	Drug	Routes of administration
Chitosan	Cationic	Gels and films	Nystatin (Das et al. 2022)	Buccal
			Celecoxib (Parhi 2020)	
		Tablets	Chlorhexidine diacetate (Gheorghita et al. 2021)	
		Biological membrane	Mitomycin C and 5-fluorouracil (Parhi 2020)	Ocular
		Films	Lidocaine (Mahrous et al. 2020)	Oral
			Satranidazole (Gheorghita et al. 2021)	
		Powder (Binary mixtures)	Carvedilol (Sip et al. 2021)	Oral and topical
		Microemulsions	Clotrimazole (Parhi 2020)	Topical
		Hydrogel	Dopamine (Liu et al. 2022)	Spinal implant
		Nanoparticles	Sumatriptan succinate	Nasal
			Olanzapine	
			Ebastine (Parhi 2020)	
		Niosomal	Timolol maleate (Parhi 2020)	Ocular
		Nanoparticles	Cyclosporine A (Parhi 2020)	

(continued)

Table 7.4 (continued)

Green biopolymer	Type of biopolymer	Pharmaceutical dosage form	Drug	Routes of administration
		Nanoparticles	L-asparaginase (Vimal and Kumar 2021)	Oral
			Interferon-α (Gheorghita et al. 2021)	
			Amiodarone (Buyuk et al. 2020)	
		Microbeds and microgranules	Ampicillin (Das et al. 2022)	
		Microspheres	Acyclovir	
			Tetracycline (Gheorghita et al. 2021)	
		Microparticles	Salbutamol	Pulmonary
			Dapsone (Parhi 2020)	
		Nanoparticles	Itraconazole	
			Rifampicin	
			Prothionamide (Parhi 2020)	
		Sponges	Clotrimazole (Martins et al. 2023)	Vaginal
		Chitosomes	Metronidazole (Parhi 2020)	
		Nanoparticles	Metronidazole (Nayak et al. 2024)	
		Nanoparticles	Gadopentetic acid (Melo et al. 2018)	Rectal
		Nanoparticles	Doxorubicin (Verma et al. 2018)	Parenteral
Chitosanthoglycolic acid		Gels	Rhodamine (Federer et al. 2020)	Topical

Thiolated chitosan	Hydrogel	Ranibizumab and aflibercept (Federer et al. 2020)	Ocular
Thiolated chitosan	Nanoparticles	Galantamine (Singh and Mishra 2019)	Oral or intranasal
Thiolated chitosan	Nanoliposomes/liposomes	Docetaxel and dopamine (Federer et al. 2020)	Oral
Chitosan/polyvinylpyrrolidone	Films	Lidocaine (Samiraninezhad et al. 2023)	Buccal
Chitosan/polyvinylpyrrolidone	Mucoadhesive membranes	Lidocaine (Samiraninezhad et al. 2023)	Oral
		Betamethasone (Sizílio et al. 2018)	
Chitosan/gelatin	Gels	Nystatin	
Chitosan/alginate	Hydrogels	Insulin (Taokaew et al. 2023)	
Chitosan/acrylamide	Hydrogels	Insulin (Taokaew et al. 2023)	Parenteral
Thiolated chitosan/poly (malic acid)	Nanoparticles	Amoxicillin (Das et al. 2022)	Oral
Chitosan/cyclodextrin	Microcapsules	Na-L-thyroxine (Lakkakula et al. 2024)	
Chitosan/alginate	Microcapsules	Doxorubicin (Abasalizadeh et al. 2020)	Intratumoral

bonds between its carboxyl and hydroxyl functional groups and the glycoproteins of the mucosa, its mucoadhesive properties are good (Kumar et al. 2022). Alginate can be extracted from brown algae (*Macrocystis pyrifera, Laminaria hyperborea, Saccharina japonica* and *Ascophyllum nodosum*) in its sodium, calcium and magnesium form or can be produced by various species of bacteria such as Azotobacter vinelandii and some species of Pseudomonas, but the latter are not intended for commercialization (Gheorghita et al. 2021).

Its applications in the pharmaceutical sector are broad since the drug delivery systems developed with this biopolymer can be administered orally or parenterally. The drug delivery systems that have been developed are mainly based on immediate and controlled release tablets, gels, membranes, micro/nanoparticles, and micro/nanocapsules (Hariyadi and Islam 2020).

Despite the great variety of applications that alginate has in the pharmaceutical industry, it still has drawbacks that limit its use because it has poor properties of mechanical resistance, cellular affinity, gelation, wettability, and water sorption and diffusion (Sahoo and Biswal 2021). Therefore, these limitations have been mitigated through the use of strategies that involve its combination with other materials or it chemical and physical cross-linking, as well as chemical modifications and have provided the alginate-derivatives with new properties or improvement of existing ones (Sahoo and Biswal 2021; Farshidfar et al. 2023). In the case of chemical modifications of alginate, these are carried out in the secondary -OH groups (C-2 and C-3) or in the carboxyl group (C-6) (Rosiak et al. 2021). The reaction of hydroxyl Groups include acetylations, phosphorylations, and the Incorporation of Sulphate Residues, all in an acidic medium, as well as their oxidation, which favors to the formation of aldehyde groups and therefore greater flexibility to the biopolymer (Rosiak et al. 2021). Meanwhile, the carboxyl group can be converted to esters or amides, which increases the hydrophobicity in the initial part of the biopolymer (Rosiak et al. 2021). Below are some drug delivery systems developed with alginate (Table 7.5).

7.5.4 Hyaluronic Acid

Hyaluronic acid is another green biopolymer composed of repeating units of glucuronic acid and N-acetylglucosamine alternatively linked by β(1,4) and β(1,3) glycosidic bonds (Kłodzińska and Nielsen 2021). This biopolymer was traditionally obtained mainly from rooster combs, but currently it is obtained through fermentation processes using attenuated strains of *Streptococci* (Cerminati et al. 2021). Commercially it is available in different molecular weights, which range from 10^3 to 10^7 Da. This biopolymer is of utmost importance due to its numerous applications in the cosmetic and pharmaceutical industry and its market is estimated to exceed $15 billion by 2025 (Fallacara et al. 2018).

Furthermore, hyaluronic acid is a hydrophilic anionic biopolymer due to its structure containing hydroxyl and carboxyl groups (Seremeta and Sosnik 2023). Likewise, it can form viscous and elastic matrices since its chains adopt a spiral

Table 7.5 Drug delivery applications of alginate, modified alginate, and hybrid biomaterials

Green biopolymer	Type of biopolymer	Pharmaceutical dosage form	Drug	Routes of administration
	Anionic	Chewable tablets	(Jadach et al. 2022)	Oral
		Patches	Carvedilol (Sakhare Abhaykumar et al. 2019)	Topical
Alginate		Microspheres	Retinoic acid (Wang et al. 2018)	Ocular
		Nanoparticles	Zidovudine (Fazal et al. 2023)	Oral
			Rifampicin (Thomas et al. 2020)	
		Nanogels	Cisplatin/Gold (Mirrahimi et al. 2019)	
		Nanomicelles in spheroids	Dasatinib/zein-lactoferrin (Ragab et al. 2019)	
		Microspheres	Mesalamine (Patole and Pandit 2018)	
		Microcapsules	Mesenchymal stem cells (Choi et al. 2018)	Parenteral
		Microspheres	Tenofovir disoproxil fumarate (Avlani et al. 2024)	Vaginal
Amide functionalized alginate		Nanoparticles	Zidovudine (Raghav et al. 2023)	Oral
Alginate/carbopol		Hydrogel	Ezetimibe (Raghav et al. 2023)	
Alginate/agar		Hydrogel	Indomethacin (Raghav et al. 2023)	
Alginate/brushite		Hydrogel	Ibuprofen (Raghav et al. 2023)	
Alginate/ethylene glycol tetra acetic acid		Pastes/liposomes	Doxorubicin (Raghav et al. 2023)	Oral
Alginate/graphene oxide		Nanoparticles	Ciprofloxacin (Raghav et al. 2023)	
Alginate/β-cyclodextrin		Nanoparticles	5-fluorouracil (Raghav et al. 2023)	
Alginate/boronated chitosan		Nanoparticles	Paclitaxel (Kolawole et al. 2023)	Vaginal

configuration (Seremeta and Sosnik 2023). Therefore, it has excellent physicochemical properties such as biocompatibility, biodegradability, cell adhesion, water solubility, as well as non-inflammatory, non-toxic and non-immunogenic properties (Bayer 2020; Cerminati et al. 2021).

Hyaluronic acid plays an essential role as a carrier for the delivery and controlled release of different biological and chemical molecules, such as proteins, nucleic acids and antitumor drugs, through various routes of administration (Maghsoudnia et al. 2022). However, given that the rapid degradation of the biopolymer limits its use for the development of pharmaceutical products, chemical modifications of hyaluronic acid have established an important strategy to expand its applications, which consider at all times maintaining its properties of biocompatibility, biodegradability and mucoadhesiveness, as well as increasing its life time since it is easily hydrolyzable at pH below 4 or above 11 (Fallacara et al. 2018). Chemical modifications are mainly made in the hydroxyl and carboxyl groups of its chemical structure, allowing the formation of conjugated and cross-linked derivatives. However, there are marked differences since the conjugates include monofunctional compounds through covalent bonds with the purpose of obtaining carrier systems for the administration of drugs with better characteristics. On the other hand, cross-linking, polyfunctional compounds are joined by two or more covalent bonds to improve the mechanical, rheological and swelling properties of the biopolymer, as well as reduce their degradation rate, and thus obtain matrices that remain longer at the application site and with better release properties (Fallacara et al. 2018).

Among the chemical modifications of the hydroxyl group, esterifications with acylating agents, etherification with alkylating agents and oxidation to aldehydes (-CHO) stand out. For its part, modifications in the carboxyl group occur from its esterification with hydroxyl groups (-OH) and amidations with amino groups (-NH_2). Finally, modifications can also be carried out on the N-acetyl group such as deacetylations to recover the amino group (-NH_2) and amidations with carboxyl groups (-COOH) (Kłodzińska and Nielsen 2021). Table 7.6 shows both conventional and non-conventional drug delivery systems that are designed with hyaluronic acid.

7.5.5 Dextran

Dextran is a non-ionic biopolymer characterized by being biocompatible, non-toxic, non-immunogenic, and water-soluble. Its production process is considered green because this exopolysaccharide can be biosynthesized by lactic acid bacteria (Petrovici et al. 2023). It has been used in the development of different drug administration systems because it is widely safe in the pharmaceutical field. It is known that this biopolymer has been used commercially as a plasma substitute, iron carrier, anticoagulant and antithrombotic, as well as for the development of nanoparticulate systems for the administration of antioxidant agents and inducers of biomolecules such as interferon (Petrovici et al. 2023).

Table 7.6 Drug delivery applications of hyaluronic acid, modified hyaluronic acid, and hybrid biomaterials

Green biopolymer	Type of biopolymer	Pharmaceutical dosage form	Drug	Routes of administration
Hyaluronic acid	Anionic	Multilayers	Ciprofloxacin/Eosin (Bayer 2020)	Topical
		Conjugates	Taxol (Bayer 2020)	–
			Paclitaxel (Kłodzińska and Nielsen 2021)	
Hyaluronic acid		Film	Nisin (Bayer 2020)	Ocular
		Hydrogel - nanoparticles	Gambogic acid (Jing et al. 2024)	Intratumoral
			Naproxen (Espinosa-Cano et al. 2021)	—
		Inorganic nanoparticles with surface modification	Doxorubicin (Ashrafizadeh et al. 2021)	Intracellular
		Nanogel	AMP LLKKK18 antimicrobial peptide (Bayer 2020)	Oral
		Nanoparticles	Tacrolimus (How et al. 2020)	Topical
		Microparticles	Ibuprofen (Obaidat et al. 2018)	Oral
Thiol-functionalized hyaluronic acid		Hydrogel	Mesenchymal stem cells (Lee et al. 2022)	Parenteral
Hexanediamine-modified hyaluronic acid		Conjugate	Paclitaxel (Chen et al. 2018; Tiwari and Bahadur 2019)	–
Oxidized hyaluronic acid/carboxymethyl chitosan		Hydrogel	Acetylsalicylic acid (Zhang et al. 2024)	Parenteral
Hyaluronic acid/ methylcellulose		Hydrogel	Nimodipine (Huang and Huang 2018)	
Hyaluronic acid/ collagen/chitosan		Hydrogel	Alendronate (Klara et al. 2024)	
Hyaluronic acid/ methylcellulose		Hydrogel	α-Chymotrypsin (Huang and Huang 2018)	Rectal
Hyaluronic acid/ polylactic acid		Nanoparticles	Doxorubicin (Jia et al. 2022)	Intravenous
Hyaluronic acid/ chitosan/eudragit S100		Nanoparticles	Dexamethasone (Zhang et al. 2023)	Oral
Hyaluronic acid/ polyethylene glycol		Nanogels	Chlorhexidine diacetate (Huang and Huang 2018)	Topical

Like other biopolymers, dextran-based drug delivery systems exhibit poor mechanical properties, low drug association/encapsulation capacity, and poor cell adhesion. In this regard, chemically modified dextran derivatives, conjugates, nano-emulsions and micelles have been used to improve the intrinsic properties of dextran for potential applications in nanomedicine (Hu et al. 2021). This biopolymer can be easily modified because its chemical structure is abundant in hydroxyl groups, so reactions that involve esterifications, etherifications and oxidations can be carried out, which directly impacts its physicochemical properties such as solubility, ionic strength, balance, hydrophilic-lipophilic and resistance against enzymatic degradation, although the latter is not the most important because it is quite stable in the presence of enzymes such as amylase (Hu et al. 2021). In the case of dextran ethers, having an amphiphilic character, they can form micelles in water, which are useful for encapsulating and protecting highly lipophilic drugs or molecules with biological activity (Hu et al. 2021; Petrovici et al. 2023). The applications of conventional and unconventional dextran-based drug delivery systems are shown in Table 7.7.

7.5.6 Carrageenan

Carrageenans are anionic sulfated biopolymer of high molecular weight (100–1000 kDa) and considered as green biopolymers given that they are obtained from renewable natural resources through ecological extraction technology (red algae such as Chondrus crispus and Kappaphycus alvarezii) (Manna and Jana 2021; Flórez-Fernández et al. 2022).

This biopolymer is made up of repetitive sequence of galactose and 3,6-anhydrogalactose alternately linked by α-(1,3) bond and β-(1,4)-glycosidic bond. In this sense, carrageenans can be classified into six basic forms: kappa (κ), iota (τ), nu (η), mu (μ), lambda (λ), and theta (θ), which have differences in solubility, number and position of sulfate groups (22 to 35%), mode and source of extraction (Zia et al. 2017; Raghav et al. 2023). From the different carrageenans; kappa (κ), lambda (λ), and iota (τ) are those that have the greatest field of application in the pharmaceutical industry but it is known that factors such as temperature and pH affect their solubility (Raghav et al. 2023). Recently, the use of carrageenan has increased significantly in the pharmaceutical sector, specifically in the development of drug delivery systems because this biopolymer has good characteristics of biocompatibility, biodegradability, and non-toxicity, which is why it is considered GRAS by the FDA since 1973, and its monograph is even present in the European Pharmacopoeia, the British Pharmacopoeia and the United States Pharmacopoeia (Manna and Jana 2021; Raghav et al. 2023; Qureshi et al. 2024).

The chemical structure of carrageenan favors it to develop applications for drug administration through the use of systems based on hydrogel beads, films, microspheres, microcapsules, nanoparticles, emulsions, aerogels, nanospheres and microparticles, among others (Raghav et al. 2023) (Table 7.8). However, although carrageenans are approved by the FDA and recognized as a safe food additive, and it is known that the use of these biopolymers affects the immune system and blood coagulation due to their sulfate groups, as well as cause necrosis in the gastrointestinal tract when exposure is prolonged (1–8 days) (Oliyaei et al. 2022). Therefore, to resolve these limitations, it has been decided to make chemical modifications to

Table 7.7 Drug delivery applications of dextran, modified dextran, and hybrid biomaterials

Green biopolymer	Type of biopolymer	Pharmaceutical dosage form	Drug	Routes of administration
Dextran	Non-ionic	Dry powder inhaler	Rifampicin (Kadota et al. 2019)	Inhalation
		Hydrogels	Aniline trimer (Qu et al. 2019)	Topical
Dextran		Micelles	Doxorubicin (Delorme et al. 2020)	Intravenous
			Doxorubicin and bortezomib (Li et al. 2020)	
		Nanoparticles	Paclitaxel and silibinin (Huo et al. 2020)	
			Doxorubicin (Son et al. 2018)	
		Nanogel	siRNA/paclitaxel (Wang et al. 2020)	
		Nanogel	Doxorubicin (Yu et al. 2020)	–
		Nanomicelles	Rapamycin (Shaki et al. 2018)	
		Nanoparticles	Insulin (Jamwal et al. 2019)	Oral
Succinate dextran	Anionic	Micelles	Doxorubicin (Hu et al. 2021)	Intravenous
Oxidized dextran		Micelles	Doxorubicin, and bortezomib (Hu et al. 2021)	
Benzimidazole dextran	Cationic	Micelles	Doxorubicin (Hu et al. 2021)	–
Succinic anhydride dextran	Anionic	Nanoparticles	Doxorubicin (Hu et al. 2021)	Parenteral
Dextran/PLGA		Nanopolymersomes	Insulin (Hu et al. 2021)	Oral
Dextran, stearic acid, and poly (ethylene glycol)		Hybrid nanoparticle	Zidovudine (Petrovici et al. 2023)	Intravenous
Dextran/ dextran-folate		Nanoconjugates	Cabazitaxel (Petrovici et al. 2023)	

these biopolymers such as cationizations, carboxymethylations, acetylations, oxidations, phosphorylations, thiolations, which have improved the physicochemical and biological properties of this biopolymer (Raghav et al. 2023) (Table 7.8). Likewise, cross-linking with polymers such as polyvinyl alcohol, agar, and carboxymethylcellulose has been established as strategies (Manna and Jana 2021) (Table 7.8).

Table 7.8 Drug delivery applications of carrageenan, modified carrageenan, and hybrid biomaterials

Green biopolymer	Type of biopolymer	Pharmaceutical dosage form	Drug	Routes of administration
Carrageenan	Anionic	Films	Echinochrome (Volod'ko et al. 2024)	Ocular
		Tablets	Doxazosin (Raghav et al. 2023)	Oral
		Hydrogel	Tramadol (Javanbakht et al. 2019)	
			β-galactosidase (Silva et al. 2020)	
		Aerogels	Tetracycline (Agostinho et al. 2020)	
		Hydrogel	Coliphage vB_Eco2571-YU1 (Narayanan et al. 2024)	Topical
		Cryogels	Metformin (Saramas et al. 2022)	Transdermal
		Fast dissolving insert	Griffithsin (Derby et al. 2018)	Vaginal
Carrageenan		Nanocarrier	Doxorubicin (Sayed et al. 2023)	Intravenous
		Nanogel	Luteolin (Omran et al. 2023)	Ocular
		Nanoparticles	Insulin (Manna and Jana 2021)	Oral
		Microparticles	Ibuprofen (Obaidat et al. 2018)	
Thiolated carrageenan		Powder	Benzydamine (Kali et al. 2023)	Oral
Carrageenan/sodium alginate/PEG		Hydrogel	Cephradine (Qureshi et al. 2024)	Parenteral
Carrageenan-crosslinked chitosan/hydroxyapatite		Hydrogel	Ciprofloxacin (Manna and Jana 2021)	Oral
Uio-66/carrageenan		Hydrogel beads	Tramadol (Manna and Jana 2021)	
Carboxymethylated carraggeenan/chitosan		Nanoparticles	Insulin (Manna and Jana 2021)	
Carrageenan/polyacrylic acid		Nanogels	Levodopa (Manna and Jana 2021)	
Carrageenan/collagen		Films	Allopurinol	
Carraggeenan/Fe_3O_4 magnetic		Nanoparticles	Doxorubicin (Manna and Jana 2021)	–
Carrageenan/chitosan		Nanoparticles	Mercaptopurine (Manna and Jana 2021)	–
Carrageenan/chitosan		Nanoparticles	Bovine serum albumin (Qureshi et al. 2024)	Pulmonary and nasal

7.5.7 Proteins

After polysaccharides, proteins are considered important biopolymers in the development of drug delivery systems because of their abundance and availability, as well as their low toxicity and ease of modification to improve their properties (Hamid Akash et al. 2015). Some of the proteins most frequently used as drug transporters are silk fibroin, collagen, gelatin, and albumin, which are mainly of animal origin (Opriş et al. 2024), as well as zein and soy protein of plant origin (Jao et al. 2017). However, polyamino acids are biodegradable ionic polymers that are made up of a single type of amino acid and have been used more frequently recently given their properties. Poly γ-glutamic acid and poly l-lysine are some examples of this type of biopolymers (Fazal et al. 2023).

Although the stability (pH, salt concentration, temperature), immunogenicity, low permeation and poor solubility in water (in the case of proteins of plant origin) of protein-based polymers are some of their main limitations for the development of new drug delivery systems, modifications in their structure can improve their physicochemical properties to address these limitations (Nasrabadi et al. 2021; Lamp et al. 2022). The modifications of these biopolymers contemplate alterations at the structural level or functional groups to improve their functionality and bioactivity. The modifications can be physical, chemical or biological, mainly (Nasrabadi et al. 2021).

7.5.8 Collagen

Collagen is one of the most abundant proteins in the human body that is found mainly in connective tissues and that, due to its mechanical resistance, biocompatibility and degradability, has been used considerably for the development of drug delivery systems (Fazal et al. 2023). Collagen is a protein biomaterial of animal origin, which has been used in the pharmaceutical sector for its particular properties, is considered a green biopolymer because its bioproduction consists of bioengineering techniques that have improved its performance in response to market demand and the complexity of other extraction methods (Zhao et al. 2023; Guo et al. 2024). In addition, recently marine organisms (fish, jellyfish, sharks, starfish and sponges) have been used for the extraction of collagen, which is characterized by having a lower environmental impact and which is also sustainable since it supports Europe's zero waste strategy (Coppola et al. 2020).

Despite the excellent properties of collagen, some of the disadvantages of its use lie in the development of immunogenic reactions at its antigenic sites, its enzymatic degradation, its weak mechanical resistance and its low thermal stability (Mansour et al. 2023). To overcome these limitations, it has been decided to carry out modifications that include cross-linking, graft polymerization, and covalent mixing/conjugation with other biomaterials for the development of drug delivery systems such as films, sponges, hydrogels, microspheres, sheets, coatings, liposomes, discs, nanofibers, tablets, granules and nanoparticles (Arun et al. 2021). In fact, gelatin is a biopolymer produced from the acid/alkaline hydrolysis of collagen and is characterized by being easily cross-linkable for the formation of gels, which are useful for the oral administration of hydrophobic chemotherapeutics (Opriş et al. 2024). Some applications of collagen in the development of drug delivery systems are shown below (Table 7.9).

7.5.9 Albumin

Albumin is the most abundant plasma water-soluble protein in human blood, which is also characterized by being biocompatible, non-toxic, non-immunogenic, biodegradable and with the ability to easily bind cationic or anionic drugs since it contains different binding sites that favor the formation of amide bonds (Larsen et al. 2016; Fazal et al. 2023). Therefore, it is a biopolymer that has been used in the development of drug delivery systems given its availability, high affinity and stability (half-life of ~19 days). Albumin is marketed as ovalbumin, bovine serum albumin, or human serum albumin and was used for the development of albumin nanoparticles for the delivery of paclitaxel, a commercial product (Abraxane®) approved by the FDA for the treatment of metastatic breast cancer. Furthermore, there are also two products on the market (Victoza® and Levemir®) produced by Novo Nordisk based on albumin for the treatment of type 1 and type 2 diabetes (Larsen et al. 2016; Fazal et al. 2023). The use of albumin as a green biopolymer is because of its low cost, easy availability and purification, and more recently thanks to its green preparation (Hornok 2021).

However, its net negative charge under physiological conditions is a limitation that affects its effectiveness as a drug carrier; but modifications in its carboxyl, hydroxyl or amino functional groups can result in an improvement in its physicochemical properties (Hornok 2021). The main delivery systems for albumin-based drugs are conjugates and nanoparticulate systems such as nanoparticles and micro/nanoemulsions (Table 7.10).

7.5.10 Casein

Casein, the protein with the highest proportion in milk, is insoluble (isoelectric point: 4.6–4.8) and represents 80% of the total proteins in bovine milk. This biopolymer is economical, non-toxic, biodegradable, biocompatible and very stable, so it can be used for the development of drug delivery systems (Elzoghby et al. 2011). Furthermore, it has good surfactant and stabilizing properties, and can be used as an excellent emulsifier and gelling agent (Elzoghby et al. 2011). Likewise, it can form micellar structures in aqueous solution due to its amphiphilicity, film coatings, conjugates/complexes, hydrogels, floating beads, microparticles and nanoparticles. However, one of the limitations of this biopolymer is its immunogenicity/allergenicity, especially in pharmaceutical products that are administered intravenously since this protein is hydrolyzed orally in the gastrointestinal tract (Elzoghby et al.

Table 7.9 Drug delivery applications of collagen

Green biopolymer	Type of biopolymer	Pharmaceutical dosage form	Drug	Routes of administration
Gelatin	Cationic at physiological conditions	Film	Econazole nitrate (Dolci et al. 2020; Fazal et al. 2023)	Vaginal Buccal
Collagen		Sponges	Chloramphenicol Ibuprofen (Fazal et al. 2023)	Buccal
Collagen		Nanofilm	Doxorubicin (Fazal et al. 2023)	–
Gelatin/3-glycidoxypropyltrimethoxysilane		Porous material	5-fluorouracil (Fazal et al. 2023)	
Collagen/3-ethyl carbodiimide-hydrochloride/ malondialdehyde		Micellar nanoparticles	Silymarin (Arun et al. 2021)	Intraperitoneal
Collagen/glutaraldehyde		Hybrid matrix	Fludarabinen and epirubicin (Fazal et al. 2023)	Topical
Collagen/chitosan		Nanoparticles	Doxorubicin (Fazal et al. 2023)	Oral

Table 7.10 Drug delivery applications of albumin

Green biopolymer	Type of biopolymer	Pharmaceutical dosage form	Drug	Routes of administration
Albumin	Anionic at physiological conditions	Microparticles	Vancomycin (Fazal et al. 2023)	–
Albumin		Nanoparticles	Irinotecan (Fazal et al. 2023)	
Albumin		Nanoparticles	Adriamycin (Fazal et al. 2023)	Intravenous
Albumin		Tablets with micro/nanoparticles entrapping liposomes	Itraconazole (de Jesús Valle et al. 2017)	Vaginal
Bovine serum albumin/sodium alkyl sulfates bioconjugates		Bioconjugate	Ibuprofen (Fazal et al. 2023)	Oral
Albumin		Conjugate	Panobinostat (Fazal et al. 2023)	Parenteral

2011; Kianfar 2021). In order to mitigate the limitations of this biopolymer, casein conjugates with other biopolymers have been developed, which has made it possible to increase the stability of casein by delaying its hydrolysis by pepsin when drug delivery systems are administered via oral (Gandhi and Roy 2021). Table 7.11 shows some drug delivery systems based on casein and in combination with other polymers.

7.5.11 Whey Protein

Whey protein is another milk derivative made up mainly of β- lactoglobulin and α-lactalbumin (65%). This biopolymer is extremely versatile and ideal for the development of drug delivery systems since it is considered non-toxic, biocompatible and biodegradable (Farooq et al. 2019). Furthermore, this biopolymer is also considered GRAS, low cost, and great nutritional value (Ha et al. 2019).

Some of the whey protein-based drug delivery systems developed are: nanosuspensions, nanoemulsions, nanoparticles, nanosheets, microparticles, hydrogels, tablets and even solid dispersions (Farooq et al. 2019), which have the ability to bind hydrophobic drugs (Ha et al. 2019). However, because of they are susceptible to degradation by proteolytic enzymes, recent research has focused on promoting their physical or chemical cross-linking, as well as their combination with other biopolymers such as polysaccharides, lipids and/or other proteins to limit accessibility to the recognition motifs and thus increase their stability (Falsafi et al. 2022). In

Table 7.11 Drug delivery applications of casein

Green biopolymer	Type of biopolymer	Pharmaceutical dosage form	Drug	Routes of administration
Casein/glutamic acid/calcium ferrite	Anionic at physiological conditions	Nanohybrid carrier	Thymoquinone (Gandhi and Roy 2021)	Intravenous
Casein/sodium alginate		Nanoparticles	Doxorubicin (Gandhi and Roy 2021)	
Casein/chitosan		Nanoparticles	Bipyridine morpholine dithiocarbamate Pt(II) nitrate (Gandhi and Roy 2021)	Oral
Casein/sodium alginate		Nanoparticles	Caffeic acid phenethyl ester (Gandhi and Roy 2021)	
Casein/zein/lysine		Nanoparticles	Ferulic acid (Gandhi and Roy 2021)	
β-Casein		Micelles	Tipranavir, efavirenz, darunavir and ritonavir Naringenin (Gandhi and Roy 2021)	
Casein/genipin		Hydrogel	Bovine serum albumin (Elzoghby et al. 2011)	
Casein		Cross-linked micelles	Paclitaxel (Gandhi and Roy 2021)	Parenteral
Casein/iron oxide nanoparticles		Nanohybrid carrier	Cytarabine (Gandhi and Roy 2021)	–
Casein/PLGA		Nanoparticles	Paclitaxel (Gandhi and Roy 2021)	
β-Casein		Nanoparticles	Daunorubicin (Zahariev et al. 2023)	

general, the two ways to extract or obtain whey protein involve green technologies, which allows it to be considered a green biopolymer: 1) direct extraction of biomass (physical method) and 2) enzymatic process (Iñiguez-Moreno and Poonia 2023). Some drug delivery systems based on whey protein are shown in Table 7.12.

Table 7.12 Drug delivery applications of whey protein

Green biopolymer	Type of biopolymer	Pharmaceutical dosage form	Drug	Routes of administration
Denatured whey proteins	Anionic at physiological conditions	Powder	Theophylline (Farooq et al. 2019)	Oral
Whey protein		Hydrogel	Riboflavin (Farooq et al. 2019)	
Whey protein/alginate		Hydrogel microparticles	Insulin (Farooq et al. 2019)	
Whey protein		Microparticles	Vitamin E (Farooq et al. 2019)	
Whey protein		Nanoparticles	Coenzyme Q10 (Falsafi et al. 2022)	
Whey protein/soybean protein		Nanosuspension	Indomethacin Fenofibric acid (Farooq et al. 2019)	
Whey protein		Nanosuspension	Carvedilol (Farooq et al. 2019)	
Whey protein		Hydrogel	Chloramphenicol (Marzaman et al. 2022)	Topical
Whey protein		Microparticles	Amoxicillin and rifampicin (Deniz et al. 2022)	–
Whey protein		Nanocomplexes	Daidzein (Farooq et al. 2019)	
Whey protein/pectin		Nanocapsules	Folic acid (Farooq et al. 2019)	
Whey protein/chitosan		Nanoparticles	Doxorubicin (Yadollahi et al. 2023)	

7.5.12 Zein

Zein is considered a biopolymer easily obtained from corn and its yields are relatively high (44 to 79% depending on the extraction method). This biopolymer is characterized by being biocompatible, biodegradable and hydrophobic with wide applications in the pharmaceutical fields for the controlled and targeted administration of drugs with low water solubility from systems such as micro/nanoparticles, films, membranes and scaffolding (De Marco 2022). Like carrageenans, this biopolymer is approved by the FDA as a GRAS excipient and its production in 2019 was valued at USD 718.9 million with a growth projection of 6.3% by 2027 (Tortorella et al. 2021). Under physiological conditions, it is negatively charged, while at acidic pH it is positively charged (Isoelectric point: 6.2 to 6.8). Agricultural waste as plant derivatives are important sources for its extraction, which considers it as a green biopolymer due to its high availability, renewability, and low production cost (Tivano and Chiono 2023). Being a hydrophobic biopolymer, it is poorly soluble in water, so to solve this limitation, plasticizer or cross-linking agents have been included, as well as other biopolymers to obtain hybrid systems with better characteristics (Tortorella et al. 2021). Cross-linking of zein can be carried out with agents such as hydroxysuccinimide, formaldehyde, glutaraldehyde, succinic acid, citric acid and eugenol (Tivano and Chiono 2023). Some zein-based drug delivery systems are shown in Table 7.13.

Table 7.13 Zein-based drug delivery applications

Green biopolymer	Type of biopolymer	Pharmaceutical dosage form	Drug	Routes of administration
Zein	Anionic at physiological conditions	Nanoparticles	5-fluorouracil (De Marco 2022)	Intravenous
Zein		Nanoparticles	Artemether (De Marco 2022)	
Zein		Nanospheres	Atorvastatin (De Marco 2022)	Oral
Zein		Nanoparticles	Carvacrol (Shinde et al. 2020)	
Zein		Nanoparticles	Glibenclamide (De Marco 2022)	
Zein/phospholipid		Nanoparticles	Retinoic acid (De Marco 2022)	Oral, parenteral, depot, topical
Zein/sodium deoxycholate		Nanoparticles	Retinoic acid (De Marco 2022)	–
Zein/chondroitin sulfate		Nanoparticles	Docetaxel (Lee et al. 2021)	Intravenous
Zein/hydroxyapatite		Nanoparticles	Doxorubicin (Zha et al. 2020)	
Zein/tannic acid		Nanoparticles	Curcumin (Hu et al. 2016)	Oral
Zein/casein/pectin		Nanoparticles	Curcumin (Chang et al. 2017)	
Zein/chitosan		Nanoparticles	Resveratrol (Pauluk et al. 2019)	
Zein/chitosan		Nanoparticles	Docetaxel (De Marco 2022)	Parenteral
Pegylated zein		Nanospheres	Luteolin (El-Lakany et al. 2018)	Parenteral

7.5.13 Lipids

Lipids are non-polar organic molecules that can be obtained from plants, animals and insects, which are considered natural sources and are characterized by being insoluble in aqueous media such as water, but have been employed for the development of drug delivery systems since they can increase the stability, solubility and permeability of class 2 drugs (low solubility, high permeability) and 4 (low solubility, low permeability) of the biopharmaceutical classification system and therefore its bioavailability (Kesharwani et al. 2023). These materials are considered green hydrophobic biomaterials that are biocompatible, biodegradable, and have low toxicity and immunogenicity (Dymek and Sikora 2022; Jahangiri et al. 2024).

In general, lipids can be classified into 3 groups: simple lipids (fats, oils and waxes), derived lipids (steroids, terpenes and carotenoids) and compound lipids (phospholipids and glycolipids), but of these oils, triglycerides (glyceryl lipid), phospholipids and waxes are the most frequently used in the development of lipid-based delivery systems (Gali et al. 2023). In addition to the fact that this type of biopolymers can improve biopharmaceutical properties of drugs from group 2 and 4 of the biopharmaceutical classification system, also given its structure, it promotes lymphatic transport and limits the metabolization and degradation of drugs in the liver (minimization of first pass metabolism) (Gali et al. 2023).

The main drug delivery systems include emulsions [microemulsion, self-emulsifying drug delivery system (SEEDS), nanoemulsion, and pickering emulsion], vesicular systems [liposomes, niosomes, pharmacosomes, phytosomes, transfersomes, ethosomes, archaeosomes, vesosomes, colloidosomes, and herbosomes], y lipid particulate systems [lipospheres, solid lipid microparticles, solid lipid nanoparticles, nanostructure lipid carriers, and lipid drug conjugates] (Shrestha et al. 2014). These systems can be used for the administration of drugs parenterally (intravenous, intramuscular, subcutaneous and intradermal), local and nasal (Srivastav et al. 2023).

Although the development of lipid-based drug delivery systems is an interesting strategy for formulating poorly soluble drugs, their design must consider some points such as the use of surfactants since it is recommended that at lipid percentages greater than 60%, percentages less than 30% and 10% of surfactant and cosolvent should be used, respectively. Sometimes more than one surfactant must be used to stabilize the system. Likewise, it must be considered that medium chain lipids favor greater solubility and stability of the drug, but long chain lipids provide greater bioavailability of the drug (Prego et al. 2005). Table 7.14 shows some lipid-based drug delivery systems developed, as well as RNA-based vaccines.

Table 7.14 Lipid-based drug delivery systems

Green biopolymer	Type of biopolymer	Pharmaceutical dosage form	Drug	Routes of administration
16:0 PEG2000 PE/16:0 PG	Anionic	Liposomes	Bevacizumab (Karumanchi et al. 2018)	Ocular
Tristearin/tween 80/ sodium taurocholate		Solid lipid nanoparticles	Aripiprazole (Silki and Sinha 2018)	Oral
Precirol® ATO5			Rhein (Feng et al. 2017)	
Stearic acid/PEG2000			Doxorubicin (Yuan et al. 2013)	
Stearic acid/Eudragit L100			Linagliptin (Veni and Gupta 2019)	
Soybean Lecithin/tripalmitin/ stearic acid			Insulin (Xu et al. 2018)	
Sodium caprylate/ pluronic F127/ethyl butyrate		Nanomicelles	Carfilzomib (Rahdar et al. 2022)	Parenteral
Blend of vegetable oils		Nanoemulsion	Cefuroxime (Harun et al. 2018)	
Lipid ALC-0315/18:0 stearic acid (DSPC)		Lipid nanoparticles	mRNA molecule (Musielak et al. 2022)	
Lipid SM-102/1,2-distearyl-sn-glycero-3-phosphocholine			mRNA molecule (Musielak et al. 2022)	
Distearoyl phosphocholine		Liposomes	Glucosamine sulphate (Ji et al. 2019)	

7.6 FDA Approved Biopolymers for the Development of Drug Delivery Systems

Although biopolymers from different sources have been used for the development of new drug delivery systems, compliance with the corresponding regulations is important to guarantee their safety and effectiveness for medical use. Under this background, for the development of a new drug, safe and efficacious excipients (GRAS) must be used, as well as those that are within the recommended limits so as not to generate toxic or adverse events. Therefore, below is a list of some biopolymers approved by the FDA for use in drug development (Table 7.15). These biopolymers are characterized by being entirely biodegradable through metabolic and/or enzymatic processes. However, biopolymers such as polylactic-co-glycolic acid, polyglycolic acid and polylactic acid are also approved, which are metabolized in the body and eliminated in the form of carbon dioxide and which, by the way, have been used more frequently recently (Singh 2022).

Table 7.15 List of biopolymers approved by the FDA for the development of pharmaceutical drug dosage forms

Biopolymer	Routes of administration	Dosage form	Maximum potency per unit	Maximum daily exposure limit
Acacia	Oral	Powder and suspension	648–123 1 mg	7386 mg
Aspartame	Oral	Tablet and suspension	3.70–12 mg	14.8–18 mg
Carboxymethyl starch	Oral	Capsule and tablet	15 mg	–
Carboxymethyl cellulose	Oral and intramusculru:	Tablet, injection, capsule, drops, powder, suspension, jelly, film, dental enteral, ophthalmic, *etc.*	5 -241 mg	18–347 mg
Carrageenan	Dental, nasal, oral, inhaler, and topical	Paste, powder, capsule, suspension, tablet, lotion, and film	0.42–20.15 mg	1–15 mg
Cellulose microcrystalline cellulose	Nasal and oral	Spray, capsule, granule, powder, suspension, and tablet	13.1–1975 mg	60–800 mg
Collagen	Topical and oral	Gel, lotion, shampoo, liquid, and capsule	75–270 mg	–
Dextran	Ophthalmic and intravenous	Solution, injection, and powder	300 mg	–
Dextrose, dextrose monohydrate	Oral, intramuscular, intra-spinal, intravenous, and nasal	Lozenge, injection, powder, solution, spray, capsule, granule, pastille, syrup, and troche	50–1600 mg	200–30,144 mg
Galactose, galactose monohydrate	Oral and rectal	Powder, solution, and tablet	0.67–4.5 mg	–
Gelatin	Buccal, dental, intracavitary, intramuscular, intravenous, subcutaneous, nasal, oral, periodontal, and inhaler	Gum, paste, injection, powder, injection, solution, drops, capsule, elixir, and pastille	14–65 mg	288–4848 mg
Guar gum	Buccal, ophthalmic, oral, topical, and vaginal	Tablet, suspension, capsule, liquid, powder, lotion, and gum	0.12–35.4 mg	6–240 mg

(continued)

Table 7.15 (continued)

Biopolymer	Routes of administration	Dosage form	Maximum potency per unit	Maximum daily exposure limit
Hypromellose	Oral, nasal, ophthalmic, rectal, and topical	Suspension, tablet, troche, gel, cream, jelly, lotion, shampoo, solution, capsule, and granule	1.7–221 mg	10–1117 mg
Karaya gum	Buccal	Tablet	68.1 mg	–
Lactose, lactose monohydrate	Buccal, intramuscular, intravenous, and oral	Tablet, injection, powder, solution, capsule, and granule	120–1682 mg	150–4384 mg
Lanolin	Topical, nasal, ophthalmic, and rectal	Spray, ointment, cream, lotion, and emulsion	1.5–3.0 (% w/w)	–
Lecithin	Oral, rectal, topical, and vaginal	Injection, capsule, powder, suspension, tablet, suppository, gel, and cream	0.35–1(% w/w)	1 – 3900 mg
Low-substituted Hydroxypropyl cellulose	Oral	Tablet and capsule	25–94.29 mg	13–480 mg
Mannitol	Intravenous, ophthalmic, oral, and topical	Injection, capsule, granule, powder, lozenge, suspension, tablet, troche, wafer, cream, andemulsion	37.11–750 mg	14–6000 mg
Methylcellulose	Buccal, intraarticul<u, intramuscular, nasal, ophthalmic, oral, sublingual, and topical	Cream, tablet, injection, jelly, drops, solution, capsule, powder,suspension, syrup, aerosol, and sponge	0.1% w/v	–
Microcrystalline cellulose	Oral, buccal, intravitreal	Tablet, capsule, granule, pellet, powder, troche	1.66–789.6	144–29,520 mg
Pectin	Dental, nasal, and oral	Paste, spray, capsule, powder, and tablet	5.45–255 mg	–1 mg
Pullulan	Oral	Tablet	–	10 mg
Saccharin	Dental, oral, respiratory, sublingual, and topical	Paste, powder, drops, elixir, granule, solution, suspension, syrup, and tablet	0.9–20 mg	8 mg
Sodium alginate	Oral and topical	Capsule, lozenge, suspension, syrup, tablet, troche, and liquid	80–320 mg	700–955 mg
Sodium starch glycolate	Buccal, oral, and sublingual	Tablet, capsule, and powder	10–876 mg	5–540 mg

(continued)

Table 7.15 (continued)

Biopolymer	Routes of administration	Dosage form	Maximum potency per unit	Maximum daily exposure limit
Starch	Buccal and oral	Tablet, capsule, granule, powder, suspension, wafer, suppository, cream, and inserts	0.17–430 mg	–
Sucrose	Buccal, intravenous, and oral	Lozenge, tablet, troche, injectable liposome, injectable emulsion, suspension, powder, capsule, pellets, drops, elixir, granule, lozenge, syrup, and tablet	109–900 mg	10–12,000 mg
Tragacanth	Buccal, oral, and topical	Tablet, powder, suspension, wafer, and jelly	5–60 mg	12–70 mg
Xanthan gum	Ophthalmic, oral, rectal, and topical	Solution, capsule, film, granule, liquid, lozenge, paste, powder, suspension, tablet, troche, enema, suspension, aerosol foam, cream, gel, lotion, chewing gum, film, and spray	0.01–275 mg	–600 mg

7.7 Conclusion

Green biopolymers are materials obtained from sources such as plants, animals and microorganisms, which are considered biocompatible, biodegradable, non-toxic, non-immunogenic and with mucoadhesive properties in most cases. However, aspects such as sustainability, that is, obtaining them from renewable natural resources, and green extraction or synthesis methods (eco-friendly), have also gained great importance in recent years. In this sense, their use for the development of new conventional and non-conventional drug delivery systems has shown great advances in terms of the pharmacological efficacy of the treatment since physicochemical and biopharmaceutical properties can be improved.

Although there have been interesting and promising results with biopolymers based on polysaccharides, proteins and lipids, strategies have been developed to improve their intrinsic properties through physical and chemical modifications, as well as the development of hybrid systems, which consider combinations with other biopolymers, excipients, and/or other drug delivery systems such as nanotechnological platforms.

Finally, the trend in the development of new drug delivery systems is focused on the use of nanotechnology since it offers important advantages compared to conventional drug delivery systems. In fact, many dosage forms are based on nanoparticulate systems as can be seen in the tables corresponding to each biopolymer group.

References

Abasalizadeh F, Moghaddam SV, Alizadeh E et al (2020) Alginate-based hydrogels as drug delivery vehicles in cancer treatment and their applications in wound dressing and 3D bioprinting. J Biol Eng 14:1–22

Abdel-Raouf ME, Keshawy M, Hasan AM (2021) Green polymers and their uses in petroleum industry, current state and future perspectives. In: Crude oil – new technol recent approaches

Agostinho DAS, Paninho AI, Cordeiro T et al (2020) Properties of κ-carrageenan aerogels prepared by using different dissolution media and its application as drug delivery systems. Mater Chem Phys 253:123290

Akinsemolu A, Onyeaka H (2023) Exploring the role of green microbes in sustainable bioproduction of biodegradable polymers. Polymers (Basel) 15:4617

Al Mahmud MZ, Islam MD, Mobarak MH (2023) The development of eco-friendly biopolymers for use in tissue engineering and drug delivery. J Nanomater 2023:1–15

Amarakoon M, Alenezi H, Homer-Vanniasinkam S, Edirisinghe M (2022) Environmental impact of polymer fiber manufacture. Macromol Mater Eng 307:2200356

Andreeßen C, Steinbüchel A (2019) Recent developments in non-biodegradable biopolymers: precursors, production processes, and future perspectives. Appl Microbiol Biotechnol 103:143–157

Arun A, Malrautu P, Laha A et al (2021) Collagen nanoparticles in drug delivery systems and tissue engineering. Appl Sci 11:11369

Ashrafizadeh M, Mirzaei S, Gholami MH et al (2021) Hyaluronic acid-based nanoplatforms for doxorubicin: a review of stimuli-responsive carriers, co-delivery and resistance suppression. Carbohydr Polym 272:118491

Avlani D, Shivakumar HN, Kumar A et al (2024) Pre-exposure prophylactic mucoadhesive sodium alginate microsphere laden pessaries for intravaginal delivery of tenofovir disoproxil fumarate. Int J Biol Macromol 258:128816

Balart R, Garcia-Garcia D, Fombuena V et al (2021) Biopolymers from natural resources. Polymers (Basel) 13:1–9

Baranwal J, Barse B, Fais A et al (2022) Biopolymer: a sustainable material for food and medical applications. Polymers (Basel) 14:983

Basim P, Gorityala S, Kurakula M (2021) Advances in functionalized hybrid biopolymer augmented lipid-based systems: a spotlight on their role in design of gastro retentive delivery systems. Arch Gastroenterol Res 2:35–47

Baslé E, Joubert N, Pucheault M (2010) Protein chemical modification on endogenous amino acids. Chem Biol 17:213–227

Bayer IS (2020) Hyaluronic acid and controlled release: a review. Molecules 25:2649

Bellich B, D'Agostino I, Semeraro S et al (2016) "The good, the bad and the ugly" of chitosans. Mar Drugs 14:99

Bhaskar R, Zo SM, Narayanan KB et al (2023) Recent development of protein-based biopolymers in food packaging applications: a review. Polym Test 124:108097. https://doi.org/10.1016/j.polymertesting.2023.108097

Bhatia SK (2023) Microbial biopolymers: trends in synthesis, modification, and applications. Polymers (Basel). 15:1364

Buyuk NI, Arayici PP, Derman S et al (2020) Synthesis of chitosan nanoparticles for controlled release of amiodarone. Indian J Pharm Sci 82:10–36468

Cacicedo ML, Islan GA, Drachemberg MF et al (2018) Hybrid bacterial cellulose–pectin films for delivery of bioactive molecules. New J Chem 42:7457–7467
Caillol S (2020) Special issue "natural polymers and biopolymers II". Molecules 26:112
Cattelan G, Guerrero Gerbolés A, Foresti R et al (2020) Alginate formulations: current developments in the race for hydrogel-based cardiac regeneration. Front Bioeng Biotechnol 8:414
Cazorla-Luna R, Martín-Illana A, Notario-Pérez F et al (2021) Naturally occurring polyelectrolytes and their use for the development of complex-based mucoadhesive drug delivery systems: an overview. Polymers (Basel) 13:2241
Cerminati S, Leroux M, Anselmi P et al (2021) Low cost and sustainable hyaluronic acid production in a manufacturing platform based on Bacillus subtilis 3NA strain. Appl Microbiol Biotechnol 105:3075–3086
Chang C, Wang T, Hu Q et al (2017) Pectin coating improves physicochemical properties of caseinate/zein nanoparticles as oral delivery vehicles for curcumin. Food Hydrocoll 70:143–151
Chen Y, Peng F, Song X et al (2018) Conjugation of paclitaxel to C-6 hexanediamine-modified hyaluronic acid for targeted drug delivery to enhance antitumor efficacy. Carbohydr Polym 181:150–158
Choi S, Kim J-H, Ha J et al (2018) Intra-articular injection of alginate-microencapsulated adipose tissue-derived mesenchymal stem cells for the treatment of osteoarthritis in rabbits. Stem Cells Int 2018:2018
Coppola D, Oliviero M, Vitale GA et al (2020) Marine collagen from alternative and sustainable sources: extraction, processing and applications. Mar Drugs 18:214
Dai L, Cheng T, Duan C et al (2019) 3D printing using plant-derived cellulose and its derivatives: a review. Carbohydr Polym 203:71–86
Das SS, Kar S, Singh SK et al (2022) Chitosan-based systems for oral drug delivery applications. In: Chitosan in Drug Delivery, vol 12, pp 23–53
Das A, Ringu T, Ghosh S, Pramanik N (2023) A comprehensive review on recent advances in preparation, physicochemical characterization, and bioengineering applications of biopolymers. Polym Bull 80:7247–7312
de Jesús Valle MJ, Martín CM, Castañeda AZ, Navarro AS (2017) Albumin micro/nanoparticles entrapping liposomes for itraconazole green formulation. Eur J Pharm Sci 106:159–165
De Marco I (2022) Zein microparticles and nanoparticles as drug delivery systems. Polymers (Basel) 14:2172
Delorme V, Lichon L, Mahindad H et al (2020) Reverse poly (ε-caprolactone)-g-dextran graft copolymers. Nano-carriers for intracellular uptake of anticancer drugs. Carbohydr Polym 232:115764
Deniz KI, Ulag S, Gunduz O (2022) Investigation of the properties of encapsulated hydrophilic and hydrophobic drugs in whey protein microparticles. Mater Lett 324:132664
Derby N, Lal M, Aravantinou M et al (2018) Griffithsin carrageenan fast dissolving inserts prevent SHIV HSV-2 and HPV infections in vivo. Nat Commun 9:3881
Dolci LS, Albertini B, Di Filippo MF et al (2020) Development and in vitro evaluation of mucoadhesive gelatin films for the vaginal delivery of econazole. Int J Pharm 591:119979
Dong Y, Mosquera-Giraldo LI, Taylor LS, Edgar KJ (2016) Amphiphilic cellulose ethers designed for amorphous solid dispersion via olefin cross-metathesis. Biomacromolecules 17:454–465
Dymek M, Sikora E (2022) Liposomes as biocompatible and smart delivery systems–the current state. Adv Colloid Interf Sci 309:102757
Elfawy LA, Ng CY, Amirrah IN et al (2023) Sustainable approach of functional biomaterials–tissue engineering for skin burn treatment: a comprehensive review. Pharmaceuticals 16:701
El-Lakany SA, Elgindy NA, Helmy MW et al (2018) Lactoferrin-decorated vs PEGylated zein nanospheres for combined aromatase inhibitor and herbal therapy of breast cancer. Expert Opin Drug Deliv 15:835–850
Elzoghby AO, El-Fotoh WSA, Elgindy NA (2011) Casein-based formulations as promising controlled release drug delivery systems. J Control Release 153:206–216

Espinosa-Cano E, Huerta-Madronal M, Cámara-Sánchez P et al (2021) Hyaluronic acid (HA)-coated naproxen-nanoparticles selectively target breast cancer stem cells through COX-independent pathways. Mater Sci Eng C 124:112024
Fallacara A, Baldini E, Manfredini S, Vertuani S (2018) Hyaluronic acid in the third millennium. Polymers (Basel) 10:701
Falsafi SR, Karaca AC, Deng L et al (2022) Insights into whey protein-based carriers for targeted delivery and controlled release of bioactive components. Food Hydrocoll 133:108002
Farooq MA, Aquib M, Ghayas S et al (2019) Whey protein: a functional and promising material for drug delivery systems recent developments and future prospects. Polym Adv Technol 30:2183–2191
Farshidfar N, Iravani S, Varma RS (2023) Alginate-based biomaterials in tissue engineering and regenerative medicine. Mar Drugs 21:189
Fazal T, Murtaza BN, Shah M et al (2023) Recent developments in natural biopolymer based drug delivery systems. RSC Adv 13:23087–23121
Federer C, Kurpiers M, Bernkop-Schnürch A (2020) Thiolated chitosans: a multi-talented class of polymers for various applications. Biomacromolecules 22:24–56
Feng H, Zhu Y, Fu Z, Li D (2017) Preparation, characterization, and in vivo study of rhein solid lipid nanoparticles for oral delivery. Chem Biol Drug Des 90:867–872
Flórez-Fernández N, Falqué E, Domínguez H, Torres MD (2022) Green extraction of carrageenans from Mastocarpus stellatus. Polymers (Basel) 14:554
Gali L, Pirozzi A, Donsì F (2023) Biopolymer-and lipid-based carriers for the delivery of plant-based ingredients. Pharmaceutics 15:927
Gandhi S, Roy I (2021) Drug delivery applications of casein nanostructures: a minireview. J Drug Deliv Sci Technol 66:102843
Ganie SA, Ali A, Mir TA, Li Q (2021) Physical and chemical modification of biopolymers and biocomposites. In: Advanced green materials. Elsevier, pp 359–377
Gheorghita R, Anchidin-Norocel L, Filip R et al (2021) Applications of biopolymers for drugs and probiotics delivery. Polymers (Basel) 13:2729
Ghosh S, Lahiri D, Nag M et al (2021) Bacterial biopolymer: its role in pathogenesis to effective biomaterials. Polymers (Basel) 13:1242
Gnanasekaran D (2019) Green biopolymers and its nanocomposites in various applications: state of the art. In: Green Biopolym their Nanocomposites, pp 1–27
Guo X, Ma Y, Wang H et al (2024) Status and developmental trends in recombinant collagen preparation technology. Regener Biomater 11:rbad106
Gupta B, Mishra V, Gharat S et al (2021) Cellulosic polymers for enhancing drug bioavailability in ocular drug delivery systems. Pharmaceuticals 14:1201
Guru PR, Kar RK, Nayak AK, Mohapatra S (2023) A comprehensive review on pharmaceutical uses of plant-derived biopolysaccharides. Int J Biol Macromol 233:123454
Ha H-K, Rankin SA, Lee M-R, Lee W-J (2019) Development and characterization of whey protein-based nano-delivery systems: a review. Molecules 24:3254
Haleem A, Javaid M, Singh RP et al (2021) Biosensors applications in medical field: a brief review. Sensors Int 2:100100
Hamid Akash MS, Rehman K, Chen S (2015) Natural and synthetic polymers as drug carriers for delivery of therapeutic proteins. Polym Rev 55:371–406
Hariyadi DM, Islam N (2020) Current status of alginate in drug delivery. Adv Pharmacol Pharm Sci 2020:1–16
Harun SN, Nordin SA, Gani SSA et al (2018) Development of nanoemulsion for efficient brain parenteral delivery of cefuroxime: designs, characterizations, and pharmacokinetics. Int J Nanomedicine 13:2571–2584
Himashree P, Sengar AS, Sunil CK (2022) Food thickening agents: sources, chemistry, properties and applications – a review. Int J Gastron Food Sci 27:100468. https://doi.org/10.1016/j.ijgfs.2022.100468
Hornok V (2021) Serum albumin nanoparticles: problems and prospects. Polymers (Basel) 13:3759

How KN, Yap WH, Lim CLH et al (2020) Hyaluronic acid-mediated drug delivery system targeting for inflammatory skin diseases: a mini review. Front Pharmacol 11:549139
Hu S, Wang T, Fernandez ML, Luo Y (2016) Development of tannic acid cross-linked hollow zein nanoparticles as potential oral delivery vehicles for curcumin. Food Hydrocoll 61:821–831
Hu Q, Lu Y, Luo Y (2021) Recent advances in dextran-based drug delivery systems: from fabrication strategies to applications. Carbohydr Polym 264:117999
Huang G, Huang H (2018) Application of hyaluronic acid as carriers in drug delivery. Drug Deliv 25:766–772. https://doi.org/10.1080/10717544.2018.1450910
Huo M, Wang H, Zhang Y et al (2020) Co-delivery of silybin and paclitaxel by dextran-based nanoparticles for effective anti-tumor treatment through chemotherapy sensitization and microenvironment modulation. J Control Release 321:198–210
Huq T, Khan A, Brown D et al (2022) Sources, production and commercial applications of fungal chitosan: a review. J Bioresour Bioprod 7:85–98
Iñiguez-Moreno M, Poonia A (2023) Green technologies for treatment and utilization of whey towards sustainable exploitation. In: Whey valorization: innovations, technological advancements and sustainable exploitation. Springer, pp 63–81
Jadach B, Świetlik W, Froelich A (2022) Sodium alginate as a pharmaceutical excipient: novel applications of a well-known polymer. J Pharm Sci 111:1250–1261
Jahangiri F, Mohanty AK, Misra M (2024) Sustainable biodegradable coatings for food packaging: challenges and opportunities. Green Chem 26:4934–4974
Jamwal S, Ram B, Ranote S et al (2019) New glucose oxidase-immobilized stimuli-responsive dextran nanoparticles for insulin delivery. Int J Biol Macromol 123:968–978
Jao D, Xue Y, Medina J, Hu X (2017) Protein-based drug-delivery materials. Materials (Basel) 10:517
Javanbakht S, Shadi M, Mohammadian R et al (2019) Facile preparation of pH-responsive k-carrageenan/tramadol loaded UiO-66 bio-nanocomposite hydrogel beads as a nontoxic oral delivery vehicle. J Drug Deliv Sci Technol 54:101311
Ji X, Yan Y, Sun T et al (2019) Glucosamine sulphate-loaded distearoyl phosphocholine liposomes for osteoarthritis treatment: combination of sustained drug release and improved lubrication. Biomater Sci 7:2716–2728
Ji L, Zhang F, Zhu L, Jiang J (2021) An in-situ fabrication of bamboo bacterial cellulose/sodium alginate nanocomposite hydrogels as carrier materials for controlled protein drug delivery. Int J Biol Macromol 170:459–468
Jia Y, Chen S, Wang C et al (2022) Hyaluronic acid-based nano drug delivery systems for breast cancer treatment: recent advances. Front Bioeng Biotechnol 10:990145
Jing Y, Wang C, Li C et al (2024) Development of a manganese complex hyaluronic acid hydrogel encapsulating stimuli-responsive Gambogic acid nanoparticles for targeted Intratumoral delivery. Int J Biol Macromol 132348:132348
Kadota K, Yanagawa Y, Tachikawa T et al (2019) Development of porous particles using dextran as an excipient for enhanced deep lung delivery of rifampicin. Int J Pharm 555:280–290
Kali G, Fürst A, Efiana NA et al (2023) Intraoral drug delivery: highly thiolated κ-carrageenan as mucoadhesive excipient. Pharmaceutics 15:1993
Karumanchi DK, Skrypai Y, Thomas A, Gaillard ER (2018) Rational design of liposomes for sustained release drug delivery of bevacizumab to treat ocular angiogenesis. J Drug Deliv Sci Technol 47:275–282
Kesharwani R, Jaiswal P, Patel DK, Yadav PK (2023) Lipid-based drug delivery system (LBDDS): an emerging paradigm to enhance oral bioavailability of poorly soluble drugs. Biomed Mater Devices 1:648–663
Kianfar E (2021) Protein nanoparticles in drug delivery: animal protein, plant proteins and protein cages, albumin nanoparticles. J Nanobiotechnol 19:159
Kimna C, Winkeljann B, Hoffmeister J, Lieleg O (2021) Biopolymer-based nanoparticles with tunable mucoadhesivity efficiently deliver therapeutics across the corneal barrier. Mater Sci Eng C 121:111890

Klara J, Hinz A, Bzowska M et al (2024) In vitro/ex vivo evaluation of multifunctional collagen/chitosan/hyaluronic acid hydrogel-based alendronate delivery systems. Int J Biol Macromol 262:130142

Kłodzińska SN, Nielsen HM (2021) Hyaluronic acid-based nanosystems for drug delivery applications. In: Biopolymer-based nanomaterials in drug delivery and biomedical applications. Elsevier, pp 221–250

Knoll P, Le N-MN, Wibel R et al (2021) Thiolated pectins: in vitro and ex vivo evaluation of three generations of thiomers. Acta Biomater 135:139–149

Kolawole OM, Ifeanafor AR, Ifade WA et al (2023) Formulation and evaluation of paclitaxel-loaded boronated chitosan/alginate nanoparticles as a mucoadhesive system for localized cervical cancer drug delivery. J Drug Deliv Sci Technol 87:104810

Kučuk N, Primožič M, Knez Ž, Leitgeb M (2023) Sustainable biodegradable biopolymer-based nanoparticles for healthcare applications. Int J Mol Sci 24:3188

Kumar R, Islam T, Nurunnabi M (2022) Mucoadhesive carriers for oral drug delivery. J Control Release 351:504–559

Kundu R, Mahada P, Chhirang B, Das B (2022) Cellulose hydrogels: green and sustainable soft biomaterials. Curr Res Green Sustain Chem 5:100252

Kurczewska J (2022) Recent reports on polysaccharide-based materials for drug delivery. Polymers (Basel) 14:4189

Lakkakula J, Krause RWM, Barage S et al (2024) Exploring oral drug delivery: in vitro release and mathematical modeling of hydrophobic drug (Na-L-thyroxine) and its cyclodextrin inclusion complex in chitosan microparticles. Int J Biol Macromol 265:131019

Lamp A, Kaltschmitt M, Dethloff J (2022) Options to improve the mechanical properties of protein-based materials. Molecules 27:446

Lang X, Wang T, Sun M et al (2020) Advances and applications of chitosan-based nanomaterials as oral delivery carriers: a review. Int J Biol Macromol 154:433–445

Larsen MT, Kuhlmann M, Hvam ML, Howard KA (2016) Albumin-based drug delivery: harnessing nature to cure disease. Mol Cell Ther 4:1–12

Lee HS, Kang N-W, Kim H et al (2021) Chondroitin sulfate-hybridized zein nanoparticles for tumor-targeted delivery of docetaxel. Carbohydr Polym 253:117187

Lee H, Lee TW, Chandrasekharan A et al (2022) Injectable self-crosslinkable thiolated hyaluronic acid for stem cell therapy of atopic dermatitis. ACS Biomater Sci Eng 8:1613–1622

Li D, Su T, Ma L et al (2020) Dual-acidity-labile polysaccharide-di-drugs conjugate for targeted cancer chemotherapy. Eur J Med Chem 199:112367

Liu K, Dong X, Wang Y et al (2022) Dopamine-modified chitosan hydrogel for spinal cord injury. Carbohydr Polym 298:120047

Liu S, Yu J-M, Gan Y-C et al (2023) Biomimetic natural biomaterials for tissue engineering and regenerative medicine: new biosynthesis methods, recent advances, and emerging applications. Mil Med Res 10:16

Lotfy VF, Basta AH (2020) Optimizing the chitosan-cellulose based drug delivery system for controlling the ciprofloxacin release versus organic/inorganic crosslinker, characterization and kinetic study. Int J Biol Macromol 165:1496–1506

Luft L, Confortin TC, Todero I et al (2020) An overview of fungal biopolymers: bioemulsifiers and biosurfactants compounds production. Crit Rev Biotechnol 40:1059–1080

Maghsoudnia N, Eftekhari RB, Roshandel M et al (2022) Hyaluronic acid in drug delivery. In: Jana S, Jana S, Domb AJ (eds) Polysaccharide-based biomaterials: delivery of therapeutics and biomedical applications. The Royal Society of Chemistry

Mahrous GM, Shazly G, Zidan DE et al (2020) Formulation and evaluation of buccoadhesive films of lidocaine hydrochloride. J Adv Biomed Pharm Sci 3:53–59

Manna S, Jana S (2021) Carrageenan-based nanomaterials in drug delivery applications. In: Biopolymer-based nanomaterials in drug delivery and biomedical applications. Elsevier, pp 365–382

Mansour A, Romani M, Acharya AB et al (2023) Drug delivery systems in regenerative medicine: an updated review. Pharmaceutics 15:695

Martins F, Morgado DL, Sarmento B et al (2023) Chitosan-based sponges containing clotrimazole for the topical management of vulvovaginal candidiasis. Int J Pharm 647:123508

Marzaman ANF, Mudjahid M, Roska TP et al (2022) Development of chloramphenicol whey protein-based microparticles incorporated into thermoresponsive in situ hydrogels for improved wound healing treatment. Int J Pharm 628:122323

Melo M, Nunes R, Sarmento B, das Neves J (2018) Rectal administration of nanosystems: from drug delivery to diagnostics. Mater Today Chem 10:128–141

Mendoza-Muñoz N, Leyva-Gómez G, Piñón-Segundo E et al (2023) Trends in biopolymer science applied to cosmetics. Int J Cosmet Sci 45:699–724

Mirrahimi M, Abed Z, Beik J et al (2019) A thermo-responsive alginate nanogel platform co-loaded with gold nanoparticles and cisplatin for combined cancer chemo-photothermal therapy. Pharmacol Res 143:178–185

Moradali MF, Rehm BHA (2020) Bacterial biopolymers: from pathogenesis to advanced materials. Nat Rev Microbiol 18:195–210

Mori R (2023) Replacing all petroleum-based chemical products with natural biomass-based chemical products: a tutorial review. RSC Sustain 1:179–212

Muir VG, Burdick JA (2020) Chemically modified biopolymers for the formation of biomedical hydrogels. Chem Rev 121:10908–10949

Mura P, Maestrelli F, Cirri M, Mennini N (2022) Multiple roles of chitosan in mucosal drug delivery: an updated review. Mar Drugs 20:335

Musielak E, Feliczak-Guzik A, Nowak I (2022) Synthesis and potential applications of lipid nanoparticles in medicine. Materials (Basel) 15:682

Nagarajan S, Radhakrishnan S, Kalkura SN et al (2019) Overview of protein-based biopolymers for biomedical application. Macromol Chem Phys 220:1900126

Narayanan KB, Bhaskar R, Choi SM, Han SS (2024) Development of carrageenan-immobilized lytic coliphage vB_Eco2571-YU1 hydrogel for topical delivery of bacteriophages in wound dressing applications. Int J Biol Macromol 259:129349

Nasrabadi MN, Doost AS, Mezzenga R (2021) Modification approaches of plant-based proteins to improve their techno-functionality and use in food products. Food Hydrocoll 118:106789

Nayak R, Halder J, Rajwar TK et al (2024) Metronidazole loaded chitosan–phytic acid polyelectrolyte complex nanoparticles as mucoadhesive vaginal delivery system for bacterial vaginosis. Int J Biol Macromol 255:128212

Nazrun T, Hassan MK, Hossain MD et al (2023) Application of biopolymers as sustainable cladding materials: a review. Sustain For 16:27

O'Brien CT, Virtanen T, Donets S et al (2021) Control of the aqueous solubility of cellulose by hydroxyl group substitution and its effect on processing. Polymer (Guildf) 223:123681

Obaidat RM, Alnaief M, Mashaqbeh H (2018) Investigation of carrageenan aerogel microparticles as a potential drug carrier. AAPS PharmSciTech 19:2226–2236

Okolišan D, Vlase G, Vlase T, Avram C (2022) Preliminary study of κ-carrageenan based membranes for anti-inflammatory drug delivery. Polymers (Basel) 14. https://doi.org/10.3390/polym14204275

Oliyaei N, Moosavi-Nasab M, Mazloomi SM (2022) Therapeutic activity of fucoidan and carrageenan as marine algal polysaccharides against viruses. 3 Biotech 12:154

Omran S, Elnaggar YSR, Abdallah OY (2023) Carrageenan tethered ion sensitive smart nanogel containing oleophytocubosomes for improved ocular luteolin delivery. Int J Pharm 646:123482

Opriş O, Mormile C, Lung I et al (2024) An overview of biopolymers for drug delivery applications. Appl Sci 14:1383

Panda J, Mishra AK, Mohanta YK et al (2024) Exploring biopolymer for food and pharmaceuticals application in the circular bioeconomy: an agro-food waste-to-wealth approach. Waste Biomass Valoriz 15:1–31

Pandey J (2020) Biopolymers and their application in wastewater treatment. In: Emerging eco-friendly green technologies for wastewater treatment, pp 245–266

Parhi R (2020) Drug delivery applications of chitin and chitosan: a review. Environ Chem Lett 18:577–594

Patole VC, Pandit AP (2018) Mesalamine-loaded alginate microspheres filled in enteric coated HPMC capsules for local treatment of ulcerative colitis: in vitro and in vivo characterization. J Pharm Investig 48:257–267

Pauluk D, Padilha AK, Khalil NM, Mainardes RM (2019) Chitosan-coated zein nanoparticles for oral delivery of resveratrol: formation, characterization, stability, mucoadhesive properties and antioxidant activity. Food Hydrocoll 94:411–417

Petrovici AR, Pinteala M, Simionescu N (2023) Dextran formulations as effective delivery systems of therapeutic agents. Molecules 28:1086

Phutane P, Telange D, Agrawal S et al (2023) Biofunctionalization and applications of polymeric nanofibers in tissue engineering and regenerative medicine. Polymers (Basel) 15:1202

Platnieks O, Beluns S, Briede S et al (2023) Cellulose synergetic interactions with biopolymers: functionalization for sustainable and green material design. Ind Crop Prod 204:117310

Prego C, Garcia M, Torres D, Alonso MJ (2005) Transmucosal macromolecular drug delivery. J Control Release 101:151–162. https://doi.org/10.1016/j.jconrel.2004.07.030

Puri V, Sharma A, Kumar P, Singh I (2020) Thiolation of biopolymers for developing drug delivery systems with enhanced mechanical and mucoadhesive properties: a review. Polymers (Basel) 12:1803

Qu J, Liang Y, Shi M et al (2019) Biocompatible conductive hydrogels based on dextran and aniline trimer as electro-responsive drug delivery system for localized drug release. Int J Biol Macromol 140:255–264

Quinn M, Georgeadis S, Said M, et al (2024) Sustainable green biomaterials in biomimetic drug delivery. AAPS Newsmag

Qureshi MAUR, Arshad N, Rasool A et al (2024) Chitosan and carrageenan-based biocompatible hydrogel platforms for cosmeceutical, drug delivery, and biomedical applications. Starch-Stärke 76:2200052

Ragab D, Sabra S, Xia Y et al (2019) On-chip preparation of amphiphilic nanomicelles–in–sodium alginate spheroids as a novel platform against triple-negative human breast cancer cells: fabrication, study of microfluidics flow hydrodynamics and proof of concept for anticancer and drug deliver. J Pharm Sci 108:3528–3539

Raghav N, Vashisth C, Mor N et al (2023) Recent advances in cellulose, pectin, carrageenan and alginate-based oral drug delivery systems. Int J Biol Macromol 244:125357

Rahdar A, Hajinezhad MR, Sargazi S et al (2022) Pluronic F127/carfilzomib-based nanomicelles as promising nanocarriers: synthesis, characterization, biological, and in silico evaluations. J Mol Liq 346:118271

RG AP, Bajaj G, John AE et al (2023) A review on the recent applications of synthetic biopolymers in 3D printing for biomedical applications. J Mater Sci Mater Med 34:1–22

Rosiak P, Latanska I, Paul P et al (2021) Modification of alginates to modulate their physic-chemical properties and obtain biomaterials with different functional properties. Molecules 26:7264

Sahoo DR, Biswal T (2021) Alginate and its application to tissue engineering. SN Appl Sci 3:30

Sakhare Abhaykumar D, Biyani Kailash R, Sudke Suresh G (2019) Design and optimization of reservoir type transdermal patches of carvedilol. Skin (Los Angeles) 3:4

Salleh A, Maarof M, Mazlan Z et al (2023) The green approach-based biomaterials for tissue engineering application. In: Sustainable material for biomedical engineering application. Springer, pp 27–47

Samir A, Ashour FH, Hakim AAA, Bassyouni M (2022) Recent advances in biodegradable polymers for sustainable applications. Npj Mater Degrad 6:68

Samiraninezhad N, Asadi K, Rezazadeh H, Gholami A (2023) Using chitosan, hyaluronic acid, alginate, and gelatin-based smart biological hydrogels for drug delivery in oral mucosal lesions: a review. Int J Biol Macromol 252:126573

Samrat K, Sharath R, Chandraprabha MN et al (2022) Role of green nanocomposites in smart/active food packaging. In: Nanotechnology in intelligent food packaging, pp 59–77

Saramas T, Sakunpongpitiporn P, Rotjanasuworapong K et al (2022) Metformin delivery via iontophoresis based on κ-carrageenan cryogels. Int J Biol Macromol 223:702–712

Sayed A, Mahmoud AF, Aly AM et al (2023) Design of carrageenan based nanocarrier as a drug nanocarrier for tumor targeting: radiolabeling and biodistribution. J Drug Deliv Sci Technol 85:104573
Seremeta KP, Sosnik A (2023) Natural and semi-natural polymers. In: Biomaterials and biopolymers. Springer, pp 55–70
Shaghaleh H, Xu X, Wang S (2018) Current progress in production of biopolymeric materials based on cellulose, cellulose nanofibers, and cellulose derivatives. RSC Adv 8:825–842
Shaki H, Ganji F, Kempen PJ et al (2018) Self-assembled amphiphilic-dextran nanomicelles for delivery of rapamycin. J Drug Deliv Sci Technol 44:333–341
Sharma R, Kumar S, Malviya R et al (2023) Recent advances in biopolymer-based mucoadhesive drug delivery systems for oral application. J Drug Deliv Sci Technol 91:105227
Shinde P, Agraval H, Srivastav AK et al (2020) Physico-chemical characterization of carvacrol loaded zein nanoparticles for enhanced anticancer activity and investigation of molecular interactions between them by molecular docking. Int J Pharm 588:119795
Shrestha H, Bala R, Arora S (2014) Lipid-based drug delivery systems. J Pharm 2014:801820
Silki, Sinha VR (2018) Enhancement of in vivo efficacy and oral bioavailability of aripiprazole with solid lipid nanoparticles. AAPS PharmSciTech 19:1264–1273
Silva RC, Trevisan MG, Garcia JS (2020) β-galactosidase encapsulated in carrageenan, pectin and carrageenan/pectin: comparative study, stability and controlled release. An Acad Bras Cienc 92:e20180609
Singh S (2022) Biopolymers towards green and sustainable development. Bentham Science Publishers
Singh SK, Mishra DN (2019) Nose to brain delivery of galantamine loaded nanoparticles: in-vivo pharmacodynamic and biochemical study in mice. Curr Drug Deliv 16:51–58
Sip S, Paczkowska-Walendowska M, Rosiak N et al (2021) Chitosan as valuable excipient for oral and topical carvedilol delivery systems. Pharmaceuticals 14:712
Sizílio RH, Galvão JG, Trindade GGG et al (2018) Chitosan/pvp-based mucoadhesive membranes as a promising delivery system of betamethasone-17-valerate for aphthous stomatitis. Carbohydr Polym 190:339–345
Son S, Rao NV, Ko H et al (2018) Carboxymethyl dextran-based hypoxia-responsive nanoparticles for doxorubicin delivery. Int J Biol Macromol 110:399–405
Srebro J, Brniak W, Mendyk A (2022) Formulation of dosage forms with proton pump inhibitors: state of the art, challenges and future perspectives. Pharmaceutics 14:2043
Srivastav AK, Karpathak S, Rai MK et al (2023) Lipid based drug delivery systems for oral, transdermal and parenteral delivery: recent strategies for targeted delivery consistent with different clinical application. J Drug Deliv Sci Technol 15:104526
Stuart-Fox D, Ng L, Barner L et al (2023) Challenges and opportunities for innovation in bioinformed sustainable materials. Commun Mater 4:80
Surendranath M, Rekha MR, Parameswaran R (2022) Recent advances in functionally modified polymers for mucoadhesive drug delivery. J Mater Chem B 10:5913–5924
Tan C, Wang J, Sun B (2021) Biopolymer-liposome hybrid systems for controlled delivery of bioactive compounds: recent advances. Biotechnol Adv 48:107727
Taokaew S, Kaewkong W, Kriangkrai W (2023) Recent development of functional chitosan-based hydrogels for pharmaceutical and biomedical applications. Gels 9:277
Thomas D, KurienThomas K, Latha MS (2020) Preparation and evaluation of alginate nanoparticles prepared by green method for drug delivery applications. Int J Biol Macromol 154:888–895
Tivano F, Chiono V (2023) Zein as a renewable material for the preparation of green nanoparticles for drug delivery. Front Biomater Sci 2:1156403
Tiwari S, Bahadur P (2019) Modified hyaluronic acid based materials for biomedical applications. Int J Biol Macromol 121:556–571

Tortorella S, Maturi M, Buratti VV et al (2021) Zein as a versatile biopolymer: different shapes for different biomedical applications. RSC Adv 11:39004–39026
Tripathi J, Thapa P, Maharjan R, Jeong SH (2019) Current state and future perspectives on gastroretentive drug delivery systems. Pharmaceutics 11:193
Veni DK, Gupta NV (2019) Development and evaluation of Eudragit coated environmental sensitive solid lipid nanoparticles using central composite design module for enhancement of oral bioavailability of linagliptin. Int J Polym Mater Polym Biomater 69:407–418
Verma AK, Leekha A, Kumar V et al (2018) Biodistribution and in-vivo efficacy of doxorubicin loaded chitosan nanoparticles in Ehrlich Ascites Carcinoma (EAC) bearing Balb/C mice. J Nanomed Nanotechnol 9:510
Verma K, Sarkar C, Saha S (2023) Biodegradable polymers for agriculture. In: Biodegradable polymers and their emerging applications. Springer, pp 191–212
Vimal A, Kumar A (2021) Antimicrobial potency evaluation of free and immobilized L-asparaginase using chitosan nanoparticles. J Drug Deliv Sci Technol 61:102231
Visan AI, Cristescu R (2023) Polysaccharide-based coatings as drug delivery systems. Pharmaceutics 15:2227
Volod'ko AV, Son EY, Glazunov VP et al (2024) Carrageenan films as promising mucoadhesive ocular drug delivery systems. Colloids Surf B Biointerfaces 237:113854
Wang F, He S, Chen B (2018) Retinoic acid-loaded alginate microspheres as a slow release drug delivery carrier for intravitreal treatment. Biomed Pharmacother 97:722–728
Wang C, Guan W, Peng J et al (2020) Gene/paclitaxel co-delivering nanocarriers prepared by framework-induced self-assembly for the inhibition of highly drug-resistant tumors. Acta Biomater 103:247–258
Witika BA, Makoni PA, Matafwali SK et al (2020) Biocompatibility of biomaterials for nanoencapsulation: current approaches. Nano 10:1649
Wróblewska-Krepsztul J, Rydzkowski T, Michalska-Pożoga I, Thakur VK (2019) Biopolymers for biomedical and pharmaceutical applications: recent advances and overview of alginate electrospinning. Nanomater (Basel, Switzerland) 9. https://doi.org/10.3390/nano9030404
Xu Y, Zheng Y, Wu L et al (2018) Novel solid lipid nanoparticle with endosomal escape function for oral delivery of insulin. ACS Appl Mater Interfaces 10:9315–9324
Yadollahi Z, Motiei M, Kazantseva N et al (2023) Whey protein isolate-chitosan PolyElectrolyte nanoparticles as a drug delivery system. Molecules 28:1724
Yu K, Yang X, He L et al (2020) Facile preparation of pH/reduction dual-stimuli responsive dextran nanogel as environment-sensitive carrier of doxorubicin. Polymer (Guildf) 200:122585
Yuan H, Chen C-Y, Chai G et al (2013) Improved transport and absorption through gastrointestinal tract by PEGylated solid lipid nanoparticles. Mol Pharm 10:1865–1873
Zahariev N, Draganova M, Zagorchev P, Pilicheva B (2023) Casein-based nanoparticles: a potential tool for the delivery of daunorubicin in acute lymphocytic leukemia. Pharmaceutics 15:471
Zha L, Wang B, Qian J et al (2020) Preparation, characterization and preliminary pharmacokinetic study of pH-sensitive Hydroxyapatite/Zein nano-drug delivery system for doxorubicin hydrochloride. J Pharm Pharmacol 72:496–506
Zhang Y, Ma R, You C et al (2023) Hyaluronic acid modified oral drug delivery system with mucoadhesiveness and macrophage-targeting for colitis treatment. Carbohydr Polym 313:120884
Zhang F, Zhang S, Cui S et al (2024) Rapid self-healing carboxymethyl chitosan/hyaluronic acid hydrogels with injectable ability for drug delivery. Carbohydr Polym 328:121707
Zhao K, Xie Y, Lin X, Xu W (2022) The mucoadhesive nanoparticle-based delivery system in the development of mucosal vaccines. Int J Nanomedicine 17:4579–4598
Zhao Z, Deng J, Fan D (2023) Green biomanufacturing in recombinant collagen biosynthesis: trends and selection in various expression systems, vol 11. Biomater Sci, pp 5439–5461
Zia KM, Tabasum S, Nasif M et al (2017) A review on synthesis, properties and applications of natural polymer based carrageenan blends and composites. Int J Biol Macromol 96:282–301

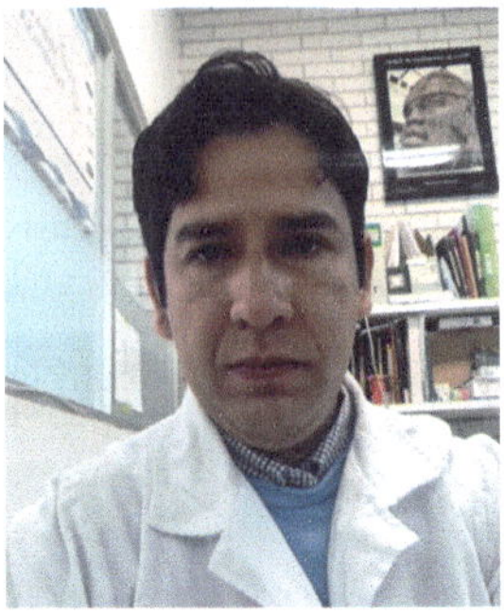

Dr. Daniel Hernandez-Patlan obtained the degree of biological pharmaceutical chemist he Faculty for Higher Studies (Facultad de Estudios Superiores, FES) Cuautitlán of the National Autonomous University of Mexico (UNAM) in 2011. Subsequently, in 2014 he obtained the degree of Master of Science, and in 2019 he graduated as Doctor of Science, in the Animal Production and Health program, UNAM. As part of your academic/professional training, he carried out research stays at the Department of Poultry Sciences at the University of Arkansas, United States, and at the Faculty of Pharmacy at the University of Clermont-Auvergne, France. Since 2010, Dr. Hernandez-Patlan has collaborated with the working group of the Laboratory of Tests of Pharmaceutical Development (LEDEFAR) at FES Cuautitlán, developing projects related to the development and validation of analytical methods, development of dissolution tests and dissolution studies, and in strategies for increasing the solubility and permeability of drugs through solid dispersions and micro/nanoparticulate systems. Furthermore, he has participated in various projects for the Secretary of Agriculture and Rural Development (SADER)/ the National Service for Agrifood Health, Safety and Quality (SENASICA), regarding the regulation of chemical-pharmaceutical and biotechnological products for veterinary use.

He is currently a full-time professor "A" at the Polytechnic University of the Valley of Mexico, belongs to the National System of Researchers (SNII) Level 1 of CONAHCYT, and has 4 years of experience in the pharmaceutical industry. Finally, his line of research includes the development of solid dispersions and micro/nanoparticle systems to encapsulate/associate phytobiotics, probiotics and enzymes as an alternative to the use of antibiotics to reduce bacterial resistance problems.

Inkar Castellanos-Huerta, a distinguished Veterinary Medicine and Animal Science professional, began his academic journey at the National Autonomous University of Mexico (UNAM) in 2002. He completed Clinical Pathology and Animal Production internships, focusing on poultry, and wrote a thesis on the Gumboro disease virus variants in Mexico. After graduation, he worked for 3 years in the food industry in quality control and health before transitioning to the veterinary industry as a sales representative in the bovine sector. He pursued postgraduate studies at UNAM in collaboration with CINVESTAV Irapuato and the University of Arkansas, focusing on expressing virus-like particles (VLPs) in plant tissue. He successfully demonstrated an alternative vaccine production model by expressing epitopes of the avian influenza virus, Campylobacter jejuni, and CD154. After earning his master's degree, Castellanos became the Director of Research for two Mexican companies producing veterinary vaccines. He led technology transfer initiatives, developed molecular diagnostic techniques, and directed CONACYT-supported projects. His work also included developing antigen expression systems in microalgae for poultry vaccines. He completed his Ph.D. at UNAM with honors, focusing on avian influenza virus proteins. Castellanos was invited to the University of Arkansas as a Postdoctoral Researcher, where he currently resides and works.

Raquel López-Arellano Full-time professor "C" at the Faculty for Higher Studies (Facultad de Estudios Superiores, FES) Cuautitlán of the National Autonomous University of Mexico (UNAM). She is specialist in industrial pharmacy, and doctor in Quality Systems and optimization of pharmaceutical processes from the Claude Bernard University Lyon 1, France, and Chemical Pharmaceutical Biology graduated from the Faculty of Chemical Sciences of the Benemérita Universidad Autónoma de Puebla, Mexico. Dr. López Arellano is a member of the National System of Researchers (SNII), level 2 and is Level "C" in the Full-Time Academic Staff Performance Bonus Program (PRIDE). Furthermore, she participates in the committee of experts in general methods of analysis of the Pharmacopoeia of the United Mexican States and is part of the National College of Pharmaceutical Biological Chemists of Mexico A.C. (CNQFBM A.C.). She has 7 years of experience in the pharmaceutical industry, has collaborated in 17 joint projects with the pharmaceutical industry and has directed 7 doctoral theses, 25 master's theses, 90 bachelor's theses in the pharmaceutical area. Likewise, she is a tutor for the postgraduate program in Animal Health and Production Science and Chemical Sciences at UNAM, as well as an accredited evaluator for CONAHCYT and in the Program to Support Research Projects and Technological Innovation (PAPIIT) at UNAM. Finally, Dr. López Arellano is responsible of the Laboratory of Tests of Pharmaceutical Development (LEDEFAR) at FES Cuautitlán and her research lines focus on the development of human and veterinary drugs, as well as the development and validation of analytical/bioanalytical methods.

Guillermo Tellez-Isaias Tellez was born in Mexico City, in 1963. He received his Doctor in Veterinary Medicine degree in 1986 and his Master in Science degree in Veterinary Sciences in 1989 from the National Autonomous University of Mexico (UNAM), and his Ph.D. from Texas A & M University. He worked as full Professor at UNAM for 16 years, 8 as head of the Avian Medicine Department at the College of Veterinary Medicine. Tellez was President of the National Poultry Science Association or Mexico during 1997–1998; is member of the Mexican Veterinary Academy and the Mexican National Research System. In 2001, Tellez worked during his sabbatical as Visiting Scientist at the Center of Excellence in Poultry Science of the University of Arkansas, where is still working as Research Professor. His research is focused on the advantages of the poultry gastrointestinal model to evaluate the beneficial effects of functional foods to enhance intestinal health and disease resistance.

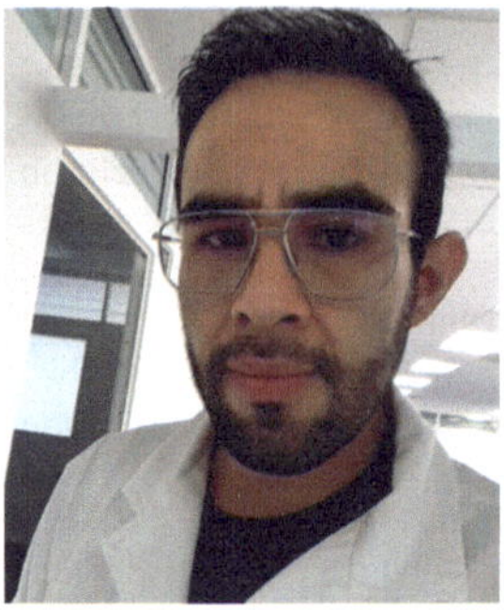

Bruno Solís-Cruz In 2011 he obtained the title of Biological Pharmaceutical Chemist from the National Autonomous University of Mexico, in 2014 he obtained the degree of Master's in Animal Production and Health Sciences, and in 2020 he graduated as a Doctor of Sciences from the same institution. Where he was recognized with 3 university merit medals. He has conducted and collaborated on research at the Department of Poultry Sciences at the University of Arkansas in the USA, as well as at the Faculty of Pharmacy at the University Clermont-Auvergne in France. He has published research articles in international peer-reviewed journals and international book chapters; and he has presented papers at national and international conferences.

Since 2010 he has been part of the working group of the Pharmaceutical Development Testing Laboratory (LEDEFAR) at the FESC, where he has carried out and collaborated in projects related to the development and validation of analytical methods, development of dissolution tests for the evaluation and characterization of drugs and medicines, permeability and toxicity tests in cell cultures, as well as in the search for new administration and release systems that allow improving the biopharmaceutical properties of active ingredients with low solubility, both in the human and veterinary areas. He has also participated in various projects for SAGARPA (now SADER)/SENASICA, regarding the regulation of chemical-pharmaceutical and biotechnological products for veterinary use. He is currently a subject professor at the Universidad Politécnica del Valle de México and an Analytical Development Chemist in the Research and Development laboratory of Bausch Health Companies.

Selective Bioadsorption of Gold and Platinum by Varying the Amount of Dead Biomass of *Aspergillus Niger* in Binary Aqueous Systems

8

Q. Gabriela Tapia, G. Ramiro Escudero, M. Raúl Cortes, and Rosa E. Pérez Sánchez

Abstract

Batch biosorption experiments were carried out with binary synthetic solutions and specific masses of the dead fungus *Aspergillus niger*. Solutions with known concentrations of gold or platinum and some of the metals copper, calcium, titanium, or aluminum were placed in contact with the fungus for 120 min at pH 4. The variables monitored were electrical conductivity, pH, and oxidation-reduction potential (ORP). At the end of the contact, aliquots of the residual aqueous medium were collected to analyze them using atomic absorption spectroscopy and UV-Vis techniques. The biomass was analyzed by FT-IR before and after contact with the solution.

The experimental results show that the bioadsorption is generally selective for gold and platinum. In the case of gold, almost 100% is adsorbed, while platinum observes a recovery limit of 71%. Iron recovery is less than 10%, up to 30% for calcium and aluminum, and 60% for copper. It is concluded that the extent of adsorption does not depend on the amount of biomass included in this work but rather on the active sites or functional groups related to metals in the solution.

The infrared spectra indicate that the functional groups of positive or neutral charge (N–H, C–H, and C–H2) on the surface of the biomass trap the ionic

Q. Gabriela Tapia · G. Ramiro Escudero (✉)
Institute of Research in Metallurgy and Materials, Universidad Michoacana de San Nicolas de Hidalgo, Morelia, Michoacán, Mexico
e-mail: 0618387A@umich.mx

M. Raúl Cortes · R. E. Pérez Sánchez
Pharmacobiology Faculty, Universidad Michoacana de San Nicolas de Hidalgo, Morelia, Michoacán, Mexico
e-mail: rosa.perez@umich.mx

R. Malviya, S. Sundram (eds.), *Sustainable Nanocomposites with Green Biomaterials*, Biomaterials, Bioengineering and Sustainability 2,
https://doi.org/10.1007/978-3-031-79110-9_8

compounds in the form of chlorides ($PtCl6^{2-}$, $AuCl4^{-}$), identified by the UV-Vis technique and the respective Pourbaix diagrams.

Keywords

Bioadsorption · Biomass · Gold · Platinum · *Aspergillus niger* · Aqueous solutions

Abbreviations

ORP	Oxidation-reduction potential
UV-Vis	Visible ultra violet technique
FT-IR	Fourier transform infrared analysis
PDA	Potato dextrose agar
CPD	Cultivated in a nutrient medium
mg/l	Milligrams per liter
mg/g	Milligrams per gram
EAA	Atomic absorption spectrometry
S/m	Siemens over meter
RmV	millivolts

8.1 Introduction

Currently, the use of precious metals has increased considerably (Golnaraghi et al. 2020). Gold and platinum are known as precious or noble metals due to their high stability due to the physical and chemical properties they present (Campos et al. 2005). These metals can be obtained from minerals or by processing some waste. From minerals, they are conventionally recovered by acid leaching, taking advantage of their chemical properties, such as the coordination of their complexes in a chloride medium and the alteration they present in their oxidation states (Doyle 1988; Suriya et al. 2013); subsequently, the application of conventional techniques allows their separation; however, they present certain limitations such as high cost, low recovery efficiency, and high sludge generation (Abdulaziz and Musayev 2017). Conversely, biosorption is a cost-effective technique where various biomaterials, including microorganisms, algae, and plant by-products, are used to capture and recover metals in complex aqueous solutions (Bahobil et al. 2017; Basu et al. 2022; Gouda and Taha 2023).

Microbial biomass comprises bacteria, fungi, yeasts, and microalgae. Due to their high availability, researchers consider these microbial species a feasible alternative to recovering metal ions (Güven et al. 2016). It has been shown that filamentous fungi have a prominent affinity for metal ions because they attract and capture them by physical-chemical mechanisms (Fawzy et al. 2017). Fungi are characterized by the negatively charged functional groups present in the cell wall, such as amines, hydroxyls, carboxyls, sulfhydryls, and phosphates, which are responsible for attracting metal ions (Güven et al. 2016); nevertheless, to increase the capture efficiency of metal ions, it is essential to study the factors that influence the biosorption process, such as the pH of the solution, the amount of biomass, and the contact time (Golnaraghi et al. 2020).

The pH influences the nature of the binding sites, the solubility of the metal, the chemical solution of metals, the activity of the functional groups of the biomass, and the competition between ions (Najam, and Andrabi 2016; Redha 2020); for its part, the concentration or amount of active sites or functional groups on the surface of the biomass influences the extent of adsorption of metal counter-ions; small amounts of biomass result in competition for active sites on the biosorbent, limiting the extent of biosorption. The increase in biomass then decreases the competition between the metal ions present to bind to the functional groups, specifically when more than one metal ion is present (Redha 2020; Moafi et al. 2016). Regarding the biomass-pregnant solution contact time, biosorption increases as the contact time progresses until reaching the equilibrium point where biosorption remains constant; when there is sufficient biomass, adsorption continues, and therefore, the extent of biosorption will depend on the contact time between the biosorbent and the aqueous medium (Gouda and Taha 2023).

The present work studies the recovery of precious metals from a solution loaded with gold and platinum through biosorption using dead biomass of *Aspergillus niger*. The importance of pH, biomass-solution contact time, and the relationship between the amount of biomass, and the percentage of metals recovered in the selectivity and efficiency of biosorption is established, all in the presence of metal ions copper, calcium, iron and aluminum.

8.2 Experimental Methodology

8.2.1 Materials and Methods

The *Aspergillus niger* strain was used as biomass; gold, platinum, calcium, magnesium, iron, titanium, and aluminum standards were used; all chemical reagents were acquired from Sigma-Aldrich.

8.2.2 Biomass Cultivation and Growth

An *Aspergillus niger* strain was plated on potato dextrose agar (PDA) and placed in an SHL incubator (LABV model 1525) for 72 h at 26 °C. The fungal strain was identified with macroscopic and microscopic monitoring; once the strain was isolated and purified, it was grown in a nutrient medium (CPD), incubated at 30 °C for 72 h at 150 rpm, and to obtain dead biomass it was dried in a FELISA oven between 60 and 70 °C for 12 h; finally, it was crushed in a porcelain mortar and stored in a glass container with a lid in the desiccator until it was required for the contact experiments.

8.2.3 The Optimal Amount of Biomass for Biosorption

Batch-type experiments were carried out from synthetic binary solutions with known concentrations of gold and platinum, separately, mixing each of them with known contents of one of the metals: calcium, iron, copper, or aluminum, according to the chemical analysis of a gold-bearing mineral from the central region of Mexico. The concentration of gold or platinum was set at 20 mg/l. In comparison, the other metals were added in concentrations of 50 mg/l so that it is possible to establish the biosorption selectivity and optimal biomass concentration to achieve maximum recovery of gold and platinum in the aforementioned binary systems. 25 ml of each of the synthetic binary solutions were used and contacted with 0.15, 0.30, 0.45, 0.60, 0.75, 1, 2, and 3 g of the dead biomass of *Aspergillus niger* in Falcon tubes separately for 120 min at pH 4. According to the bibliographic information, at pH 4, species of the metals included in this study do not precipitate.

The variables monitored were pH, electrical conductivity, and oxide-reduction potential (ORP). Aliquots were collected every 20 min during the first hour, and in the second hour, the aliquot was collected at 40 min. The aliquots and the residual solution were analyzed in an atomic absorption spectrometer (EAA—PinAacle 900H Perkin Elmer).

8.2.4 Biomass Characterization by FT-IR

The dead biomass was characterized by infrared spectroscopy (FT-IR) using a Bruker Tensor 27 model. The spectra were obtained with a resolution of 4 cm^{-1} in the spectral range of 4000 cm^{-1} 400 cm^{-1}. Previously dried samples were prepared with KBr pellets (Sigma Aldrich, spectroscopic grade). The characterization identified the main bands in the spectrum, and the vibration frequencies of the functional groups present in the biomass were observed after contact with the metal solution at pH 4.

8.2.5 Characterization of Aqueous Species by UV-Vis

The supernatant of the metal solution that was in contact with the dead biomass of *Aspergillus niger* was analyzed in a UV-Vis spectrophotometer at the 570 nm wavelength to identify the gold and platinum species formed in the synthetic aqueous solutions.

8.3 Results and Discussion

8.3.1 Optimal Biomass Amount and Bioadsorption Selectivity

Figure 8.1 shows the extent of adsorption by the dead biomass of *Aspergillus niger* for the case of the binary solution of gold and platinum (20 mg/l of each metal). The experimental results indicate that, in the case of platinum, the biosorption capacity reaches a maximum (up to 71.6%) from the first minutes of contact, regardless of the increase in biomass. Regarding gold, its adsorption increases with the amount of biomass until 96% of the initial metal is recovered in solution 120 min after the contact begins. In the latter case, it is established that the increase in the adsorbent surface offers more available or active sites for adsorption.

The information shown in Table 8.1 shows that the fungus in contact with the synthetic solution selectively adsorbs platinum; however, depleting the functional groups in the biomass related to the platinum ionic species reaches a maximum. On the other hand, ionic gold species find fewer active sites or functional groups available in the biomass, which increase with the amount of dead mass of *Aspergillus niger*, reaching higher recoveries concerning platinum.

Figures 8.2, 8.3, 8.4, and 8.5 show the adsorption capacity of dead biomass of *Aspergillus niger* for the cases of binary solutions containing gold or platinum mixed with known concentrations of any of the metals calcium, iron, copper, or aluminum.

As can be seen, gold is almost entirely adsorbed, selectively or preferentially, from the beginning of contact in the four binary solutions tested. It is observed that iron and calcium species are incipiently attracted to the functional groups of the biomass; aluminum species are barely attracted, contrary to copper species, which

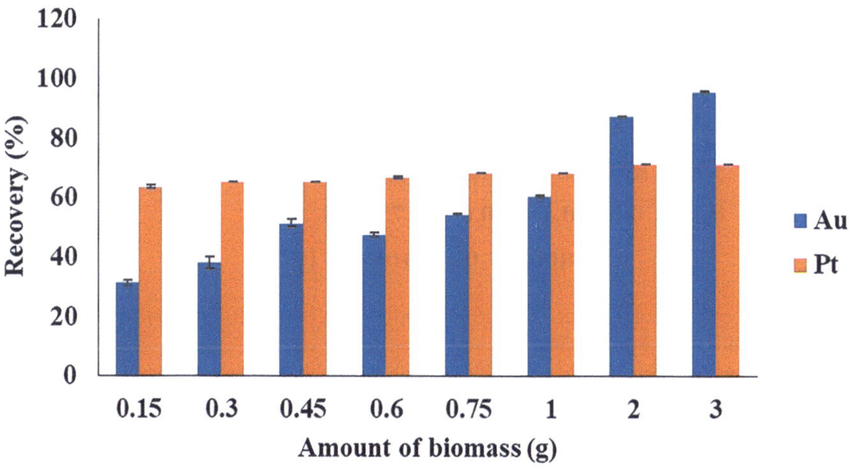

Fig. 8.1 The optimal concentration of dead biomass of *Aspergillus niger* for the adsorption of gold and platinum

Table 8.1 Percent recovery of gold and platinum as a function of the concentration of dead *Aspergillus niger* biomass

Amount of biomass (g)	Recovery of Au (%)	Recovery of Pt (%)
0.15	31.5	63.8
0.30	38.2	65.3
0.45	51.4	65.3
0.60	47.7	66.9
0.75	54.5	68.4
1.0	60.7	68.4
2.0	87.5	71.6
3.0	95.8	71.6

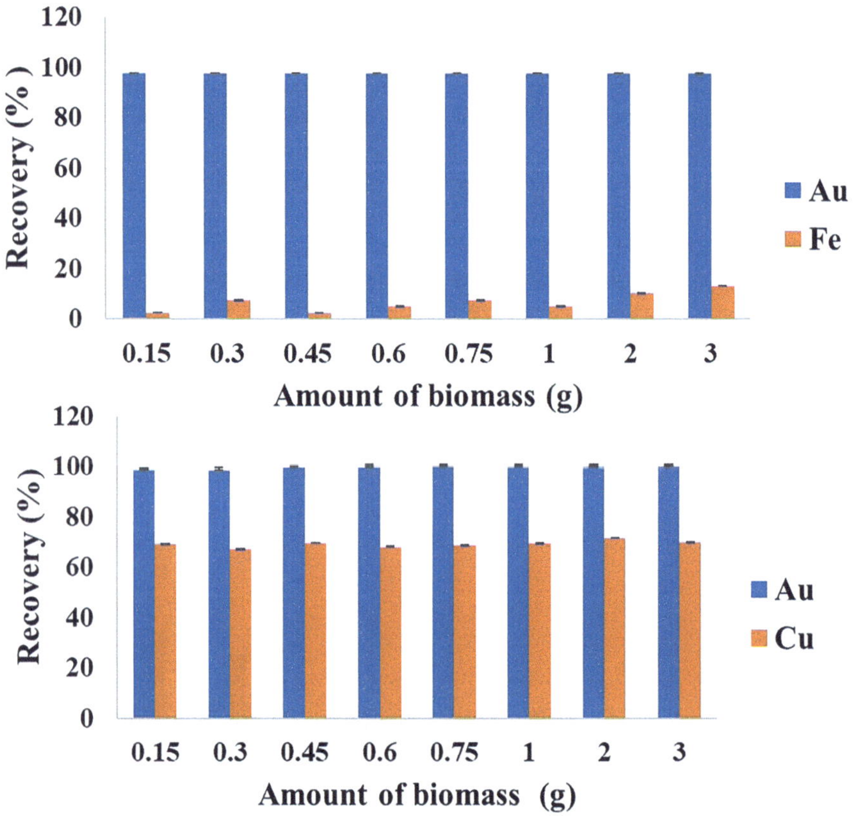

Fig. 8.2 The optimal concentration of *Aspergillus niger* dead biomass for maximum metal recoveries in the Au-Fe, and Au-Cu systems

have recovered up to 60% concerning the initial content. As in the case of platinum, the ionic species of the metals calcium, copper, aluminum, and iron, the ability to be adsorbed depends on the functional groups related to them on the surface of the biomass (Godlewska-Zylkiewicz et al. 2019).

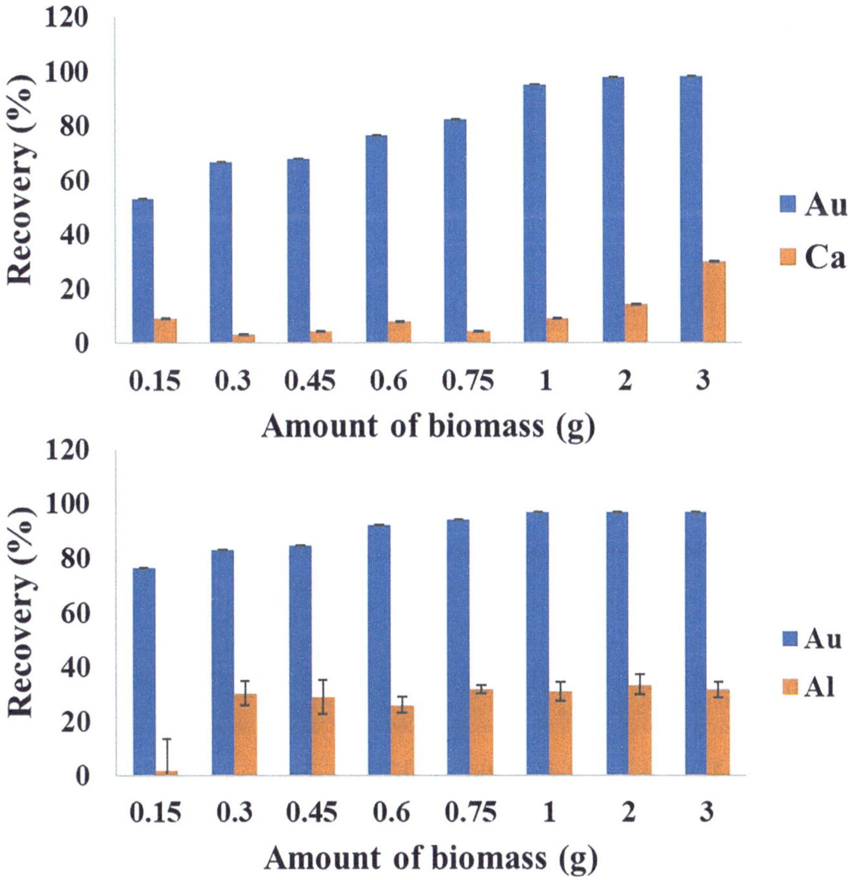

Fig. 8.3 The optimal concentration of *Aspergillus niger* dead biomass for maximum metal recoveries in the Au-Ca, and Au-Al systems

Figures 8.4 and 8.5 show the experimental bioadsorption results for binary solutions of platinum and some metals, such as calcium, iron, copper, or aluminum.

In all cases, platinum is adsorbed preferentially than the other four metals, although the recovery percentage ranges between 60% and 90%, depending on the functional groups available in the biomass or the selectivity of the biomass for the metal; for example, for the Pt-Fe system, the maximum recovery of platinum is 70% and does not depend significantly on the amount of biomass, instead on the functional groups related to the platinum species in solution. The biomass does not attract or capture iron metal species in this system. The Pt-Cu binary system observes a behaviour similar to the previous one. For the binary solutions Pt-Ca and Pt-Al, the adsorption of calcium and aluminum species does not depend on the amount of biomass, only on the functional groups related to these species, reaching recoveries that range from 30%.

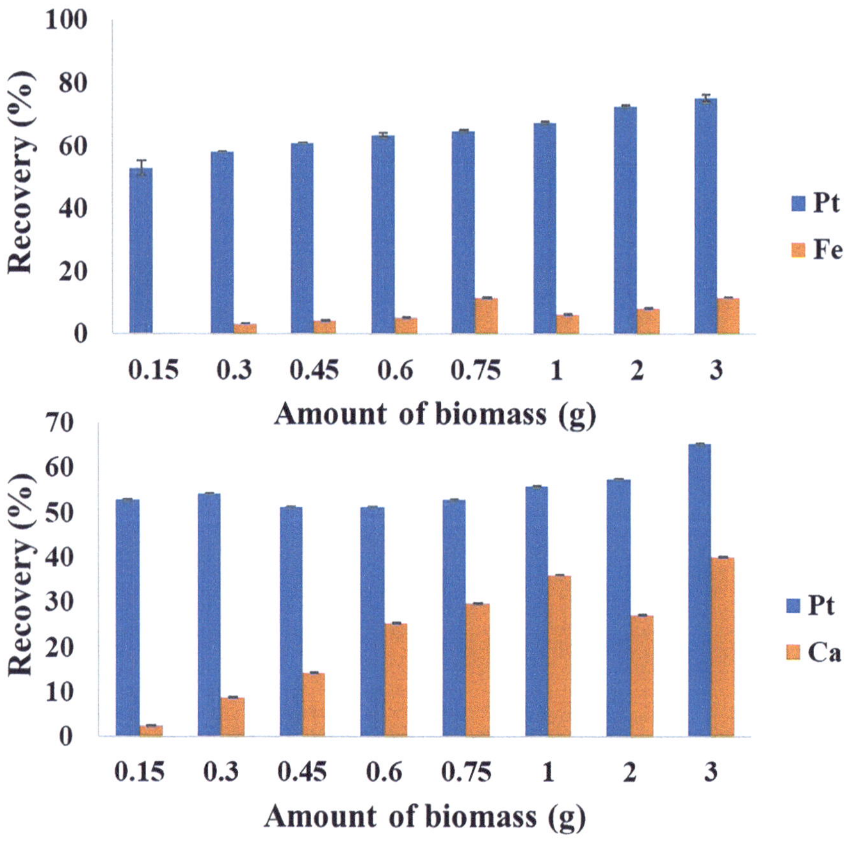

Fig. 8.4 The optimal concentration of dead biomass of *Aspergillus niger* for the Pt-Fe, Pt-Ca systems

Generally, it can be established that the amount of platinum species adsorbed by the biomass depends on the functional groups related to these species, without reaching recoveries close to 100% (Godlewska-Zylkiewicz et al. 2019).

8.3.2 Variations in pH, Electrical Conductivity, and ORP

During the biomass-pregnant solution contact experiments, the pH values, electrical conductivity, and oxide-reduction potential were recorded to establish any possible change due to the fungus's mass increase, in addition to identifying the species in equilibrium. According to the corresponding Pourbaix diagrams (Najam and Andrabi 2016; Torres Gómez (2024), under the corresponding experimental conditions. Table 8.2 shows these variations during the experimentation for the Pt-Au system. As can be seen, the pH varies slightly and remains in the range that rules out the formation of solid or solution phases that modify the bioadsorption kinetics.

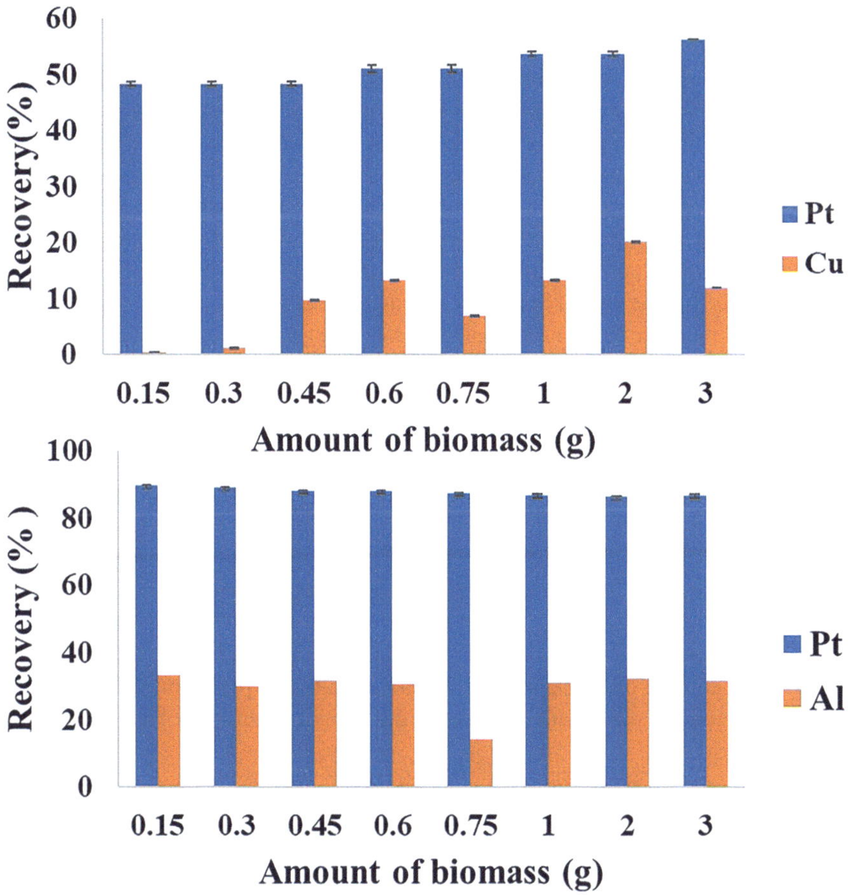

Fig. 8.5 The optimal concentration of dead biomass of *Aspergillus niger* for the Pt-Cu and Pt-Al systems

Table 8.2 Variations of pH, electrical conductivity, and oxidation-reduction potential (ORP) in the binary solutions during the adsorption experiments. Pt-Au binary system

Amount of biomass, g	pH	Electrical Conductivity, S/m	ORP (RmV)
0.15	3.65	3.8	993
0.3	3.64	4.8	1003
0.45	3.65	4	1001
0.6	3.45	4.1	1003
0.75	3.7	4	997
1	3.77	3.8	998
2	3.76	3.8	981
3	3.82	3.8	954

It is observed that for the biomass quantities of 0.15, 0.30, and 0.45 g, the pH remains constant at 3.64, while at 0.65 g of biomass, the pH decreases to 3.45, and subsequently increases gradually as the grams of biomass increase until it reaches 3.82; this is because the protonation or deprotonation of the functional groups takes place in the cell wall, a process required for biosorption to take place, so when the pH increases there is an intensification of the electrostatic forces involved in the adsorption process of the active sites, and consequently, the active sites on the cell wall are deprotonated, and the negative charge increases with an increase in the bonding of metals to the functional groups of the microbial biomass; On the other hand, at acidic pH the available adsorption sites are saturated by protons since there are more of them in the solution since metals exist as free ions (Gáplovská and Simonovicova 2018; Torres Gómez 2024).

Regarding the variation of electrical conductivity in the aqueous medium, it does not change. The variations are expected due to the electronic "noise" of the potentiometer. The minor variation in conductivity values is a product of the low concentration in parts per million of metal ions at the beginning and during the fungus-solution contact.

Changes in the oxide reduction potential (ORP) were also recorded; Fig. 8.6 compares these ORP values with the corresponding Pourbaix diagrams of the Pt-H_2O-HCl and Au-H_2O-HCl systems (Carriazo et al. 2007). It is verified that, according to the pH of the experiment, the aqueous species present are chlorides, and there is no type of precipitate or compound between platinum and gold, only chlorinated gold or platinum compounds; the above, regardless of the amount of biomass.

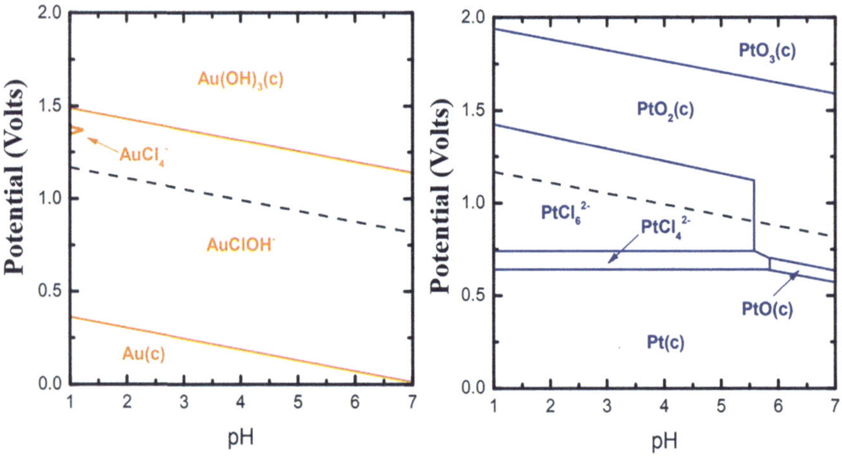

Fig. 8.6 Location of the aqueous species formed during the experiment, according to the Pourbaix diagrams for the Pt-H_2O-HCl and Au-H_2O-HCl systems [16]

8.3.3 Biomass Characterization by Infrared Spectroscopy (FT-IR)

The Fourier characterized the surface of the *Aspegillus niger* biomass transform infrared spectroscopy (FT-IR) technique. Figure 8.7 shows the infrared spectra of the dead biomass before and after contact with the platinum and gold metal solution at pH 4. The identified functional groups are presented in Table 8.3. In the infrared spectrum before contact with the metal solutions, bands are recorded at 3419 cm^{-1} corresponding to the stretching vibrations of the O–H (hydroxyl) groups of the polysaccharides and the vibrational tension of the N–H (amino) group corresponding to the proteins and the acetamido group of chitin (Gochev et al. 2010; Tran et al. 2013). The bands at 2925 cm^{-1} and 2855 cm^{-1} are associated with the asymmetric mode of CH_2 and with the stretching vibrations of the C–H (aliphatic groups) of fatty acids.

A 2144 cm^{-1} se presenta una banda atribuida a la vibración de tensión del grupo $C \equiv C$, bands at 1744 cm^{-1} and 1645 cm^{-1} typical of the stretching vibration of the C=O and the vibrational tension of the C=O bond (corresponding to the amides) were also recorded. In addition, bands appear at 1552 cm^{-1} due to the deformation of the amides (N–H), at 1451 cm^{-1} attributed to the asymmetric stretching of CH_3 (aliphatic group), and at 1375 cm^{-1} associated with the COO— carboxylate group

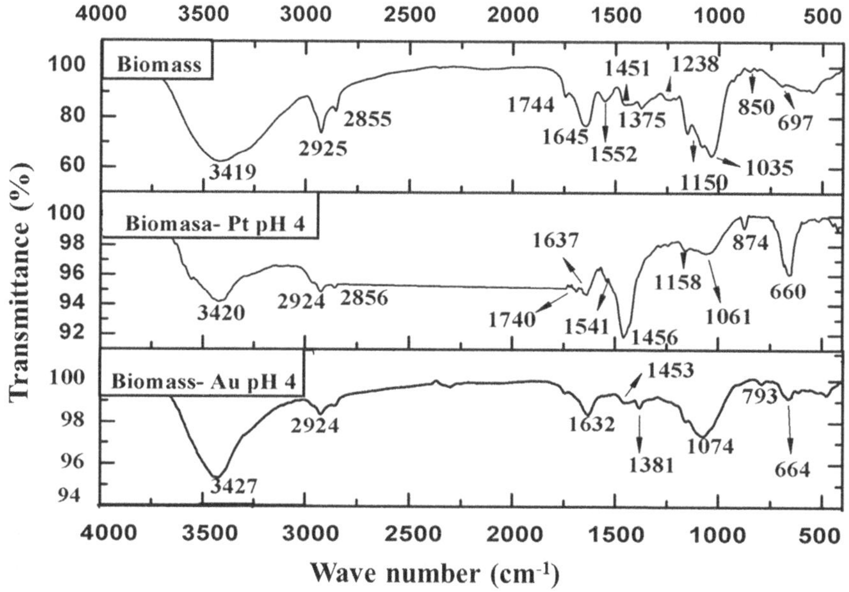

Fig. 8.7 FT-IR spectra of dead biomass before and after contact with platinum and gold solutions at pH 4

Table 8.3 Functional groups identified by FT-IR in the dead biomass of *Aspergillus niger* (Gochev et al. 2010; Tran et al. 2013)

Wave number	Bond	Charge	Functional group
3419.51	O–H, N–H	Anionic, cationic	Hidroxyl
2925.14	CH_2	Non-ionic	Methylene
2855.11	C–H	Non-ionic	Alkanes
1744.06	C=O	Anionic	Carboxylic acid
1645.5	C=O	Anionic	Carboxyl (corresponding to amide groupings)
1552.21	N–H	Cationic	Amide
1451.7	CH_3	Non-ionic	Methyl
1375.67	COO^-	Anionic	Carboxylate
1238.4	C–O, C–N	Anionic	Carboxylic acid, aliphatic amine
1150.97	C–N (alquil)	Anionic	Amine
1035.49	C–N (alquil)	Anionic	Amine
850.76	C–H (para)	Cationic	Aromatic ring
697.71	C–H (mono)	Anionic, cationic	Aromatic ring

(Gáplovská and Simonovicova 2018). The bands at 1150 cm^{-1} and 1035 cm^{-1} correspond to the stretching vibrations of chitin and amines typical of protein fractions, while the bands at 850 cm^{-1} and 697 cm^{-1} are associated with aromatic rings (=CH) (Gochev et al. 2010).

The band attributed to the stretching vibrations of the hydroxyl group (O–H, 3419 cm^{-1}) appears slightly out of phase due to the formation of intermolecular hydrogen bonds in the polysaccharides of the biomass and the metal (Pt^{2+} or Au^{3+}). The bands associated with the aliphatic groups (CH_2 and CH) of the fatty acids at 2925 cm^{-1} are slightly modified in both cases, while for biomass-Au, the band at 2855 cm^{-1} disappears completely. At 1451 cm^{-1}, it is modified in both cases, the most relevant modification being biomass-Pt. The above described is attributed to forming a covalent bond between the aliphatic groups and the metals in question.

From Fig. 8.7, the band associated with the carboxyl group (C=O) corresponding to the amide groups varies slightly at 1744 cm^{-1} for biomass-Pt and disappears in the case of biomass-Au, while at 1645 cm^{-1}; it is modified in both cases (Biomass-Pt and Biomass-Au), due to the intermolecular forces (Van der Waals forces) formed between C=O and the metal. The band of the amino group (N–H, 1552 cm^{-1}), due to the deformation of amides, disappears in biomass-Au and is modified in biomass-Pt, indicating the participation of this group in the biosorption of gold and platinum. One more band, associated with the carboxylate group (COO^-, 1375 cm^{-1}), disappears entirely in biomass-Pt and is modified in biomass-Au, while the band at 1238 cm^{-1} corresponding to the C–O (carboxyl) group disappears in both cases, due to the union of platinum and gold to the carboxyl group (COO^-, C–O).

The bands at 1150 cm^{-1} and 1035 cm^{-1} corresponding to the stretching vibrations of chitin and the amines of the protein fractions (CN^-) are modified for biomass-Pt, while they disappear for biomass-Au. These changes are attributed to the adsorption of platinum and gold. Bands are present at 850 cm^{-1} and 697 cm^{-1}

associated with the amino group (N–H); these bands are modified in both cases due to the adsorption of platinum and gold in the CH group (aromatic ring).

8.3.4 Characterization of Solutions by UV-Vis Spectroscopy

Using UV-Vis spectroscopy, the absorbance of the platinum and gold solutions (separately) at pH four was obtained, and the formation of chlorine complexes was verified. The spectra of Figs. 8.8 and 8.9 for Pt and Au, respectively, show these complexes. In the case of platinum, an absorption band is exhibited at 260 nm corresponding to $PtC16^{2-}$ (Georgieva and Andonovski 2003; Liang et al. 2011).

Figure 8.9 shows an absorption band at 310 nm corresponding to the $AuCl^{4}$— compound (Chan et al. 2020); therefore, the identified complex chlorines ($PtCl_6^{2-}$ and $AuCl^{4-}$) are adsorbing on the biomass surface. The information from the FT-IR spectra supports the above, where it is established that platinum and gold compounds are attracted to the functional groups present in the positively or neutrally charged biomass (NH, CH, and CH_2).

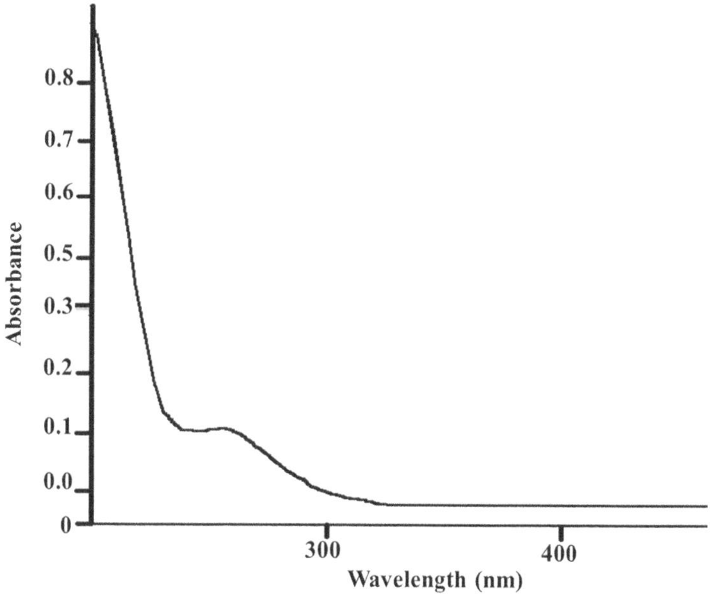

Fig. 8.8 UV-Vis spectrum of the platinum solution with a peak at 260 nm corresponding to the $PtCl_6^{2-}$ compound

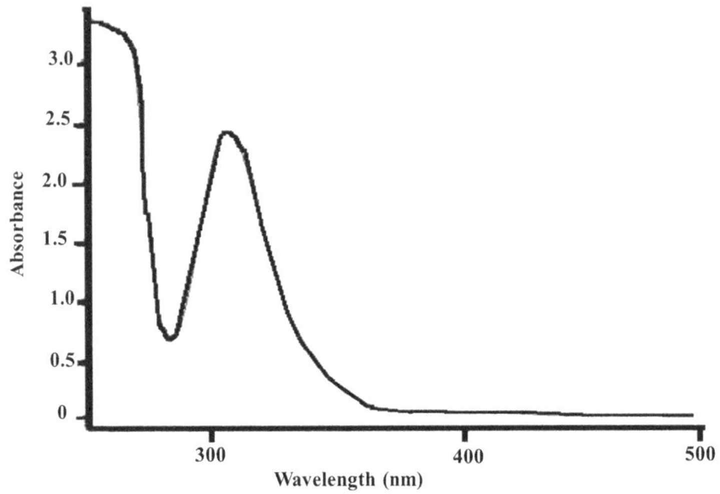

Fig. 8.9 The UV-Vis spectrum of the gold solution shows a peak at 310 nm, corresponding to the $AuCl^4$ compound

8.4 Conclusions

From the biosorption experiments of gold, platinum, and other metals at pH four and varying the mass of the fungus *Aspergillus niger*, the following conclusions are derived:

In general, bioadsorption is selective for gold and then for platinum. In the case of gold, almost 100% is adsorbed, while for platinum, the recovery limit is 71%.

Iron recovery is less than 10%, up to 30% for calcium and aluminum, and 60% for copper.

The extent of adsorption depends not on the biomass included in this work but on the active sites or functional groups related to metals in the solution.

The infrared spectra indicate that the functional groups of positive or neutral charge (N–H, C–H, and $C–H_2$) on the surface of the biomass trap the ionic compounds in the form of chlorides ($PtCl_6^{2-}$, $AuCl^{4-}$), identified by the UV-Vis technique and the respective Pourbaix diagrams.

Acknowledgments Gabriela Tapia-Quiroz deeply thanks the UMSNH for supporting her doctoral studies and for funding this research.

Interest Conflict The authors declare that there is no conflict of interest.

References

Abdulaziz M, Musayev S (2017) Multicomponent biosorption of heavy metals from aqueous solutions: a review. Pol J Environ Stud 26(4):1433–1441. https://doi.org/10.15244/pjoes/67975

Ali Redha A (2020) Removal of heavy metals from aqueous media by biosorption. Arab J Basic Appl Sci 27(1):183–193. https://doi.org/10.1080/25765299.2020.1756177

Bahobil A, Bayoumi RA, Atta HM, Sehrawey E (2017) Fungal biosorption for cadmium and mercury heavy metal ions isolated from some polluted localities in KSA. Int J Curr Microbiol App Sci 6(6):2138–2154

Basu A, Ali SS, Hossain SS, Asif M (2022) A review of the dynamic mathematical modeling of heavy metal removal with the biosorption process. PRO 10(6). https://doi.org/10.3390/pr10061154

Campos GC, Fernández AF, Aguirre MA, Candó MJ, Figueredo YP (2005) Metodologías Aplicables en la Geoquímica, Minería y Tecnología para la Determinación de Elementos del Grupo del Platino. Revista Internacional de Contaminación Ambiental 21(3):107–113. ISSN: 0188-4999

Carriazo JG, Uribe-Pérez M, Hernández-Fandiñom O (2007) Diagramas de predominancia, de Frost y de Pourbaix: tres contextos para desarrollar competencias en procesos de óxido-reducción. Sci Tech XIII(34):569–574

Chan JZ, Ali R, Shameli K, Salleh MSN, Lee KX, Mohamed Isa ED (2020) Green synthesis of gold nanoparticles using aqueous extract of clitoria ternatea flower. IOP Conf Ser Mater Sci Eng 808(1):012036. https://doi.org/10.1088/1757-899X/808/1/012036

Doyle FM (1988) Developments in hydrometallurgy. JOM 40(4):32–38., abr. 1988. https://doi.org/10.1007/BF03259019

Fawzy EM, Abdel MF, Zayat SAE (2017) Biosorption of heavy metals onto different eco-friendly substrates. J Bioremed Biodegr 8(3):10.4172/2155-6199.1000394

Gáplovská K, Simonovicova A (2018) Study of the binding sites in the biomass of *Aspergillus Niger* wild-type strains by FTIR spectroscopy. Chem Pap 72(9):2283–2288. https://doi.org/10.1007/s11696-018-0487-6

Georgieva M, Andonovski B (2003) Determination of platinum(IV) by UV spectrophotometry. Anal Bioanal Chem 375(6):836–839. https://doi.org/10.1007/s00216-003-1783-7

Gochev V, Velkova Z, Stoytcheva M (2010) Hexavalent chromium removal by waste mycelium of Aspergillus awamori. J Serb Chem Soc 75(4):551–564. https://doi.org/10.2298/JSC090331008G

Godlewska-Żyłkiewicz B, Sawicka S, Karpińska J (2019) Removal of platinum and palladium from wastewater using biosorption on fungi Aspergillus sp. and yeast Saccharomyces sp. Water 11(7). https://doi.org/10.3390/w11071522

Golnaraghi GA, Asasian-Kolur N, Sharifian S, Golnaraghi A (2020) Biosorption for sustainable recovery of precious metals from wastewater. J Environ Chem Eng 8(4):103996. https://doi.org/10.1016/j.jece.2020.103996

Gouda SA, Taha A (2023) Biosorption of heavy metals as a new alternative method for wastewater treatment: a review. Egypt J Aquat Biol Fish 27(2):135–153. https://doi.org/10.21608/ejabf.2023.291671

Güven K, Korcan E, Malkoç S, Aydin KM, Güven A (2016) Lead biosorption by a moderately Halophile Penicillium sp. isolated from Çamalti Saltern in Turkey. ANADOLU Univ J Sci Technol C Life Sci Biotechnol 5(1):10.18036/btdc.01000

Liang Z, Liu Y, Ng SS, Li X, Lai L, Luo S (2011) The effect of pH value on the formation of gold nanoshells. J Nanopart Res 13(8):3301–3311. https://doi.org/10.1007/s11051-011-0244-z

Moafi HF, Ansari R, Ostovar F (2016) Ag_2O/sawdust nanocomposite as an efficient adsorbent for removal of hexavalent chromium ions from aqueous solutions. J Mater Envirom Science 7(6):2051–2068

Najam R, Andrabi SMA (2016) Removal of Cu(II), Zn(II) and Cd(II) ions from aqueous solutions by adsorption on walnut shell-equilibrium and thermodynamic studies: treatment of effluents from electroplating industry. Desalin Water Treat 57(56):27363–27373. https://doi.org/10.1080/19443994.2016.1166350

Suriya J, Bharathiraja S, Rajasekaran R (2013) Biosorption of heavy metals by biomass of enterobacter cloacae isolated from metal-polluted soils. Int J ChemTech Res 5(3):1229–1238

Torres Gómez RDJ (2024) Recuperación de metales del grupo del platino por métodos hidrometalúrgicos». Accedido: 22 de abril de 2024. [En línea]. Disponible en: https://doi.org/10.24275/uami.79407x16z

Tran MV, Doan TT, Duong PN, Le MLP, Nguyen TPT (2013) Electrochemical behavior and morphology of nano catalyst for fuel cell: the effect of ultrasonic and microwave techniques. ECS Trans 50(2):2001–2008. https://doi.org/10.1149/05002.2001ecst

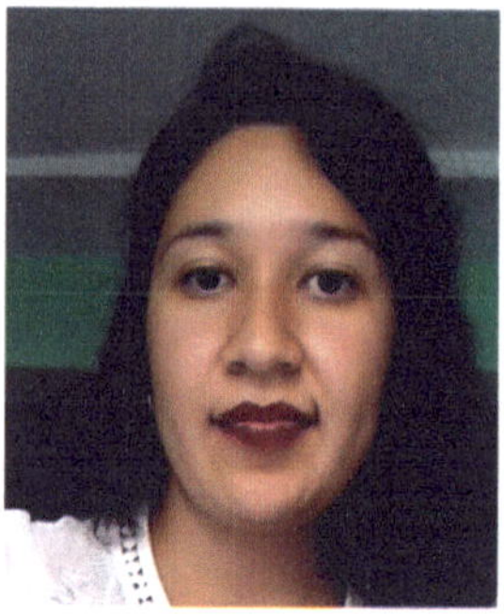

Q. Gabriela Tapia She is a full-time doctoral student in the Department of Extractive Metallurgy of the Metallurgy and Materials Research Institute of the Universidad Michoacana de San Nicolás de Hidalgo, Mexico. She has experience in food safety and quality control, human capital formation, and implementation of biotechnological techniques for the extraction of valuable metals.

G. Ramiro Escudero He is Chief Professor of the Extractive Metallurgy Department at Universidad Michoacana de San Nicolás de Hidalgo in México. Is professor and researcher at the Institute of Research in Metallurgy and Materials in the same University. He received his Ph.D. in Mineral Engineering from McGill University in Canada. He worked during 5 years in the blast furnace department of a steelmaking plant located in Mexico. He has about 150 publications, including two books, and book chapters in the areas of Mineral Processing, Phytomining, and Biomining, Wastewater Treatment, and Gas Disperser Design using Computer Simulators. He is a member of the National Research System in México. He is a member of editorial committees of several international journals. He has directed many master's and doctoral thesis. He is a Member of the TMS Minerals, Metals and Materials Society, a member of the Mexican Association for Research and Teaching in Chemical Engineering (AMIDIQ), a founding member of the Ibero-American Extractive Metallurgy Network and the Michoacán Academy of Sciences.

M. Raúl Cortes He is a professor and researcher at the Faculty of Chemistry and Pharmacobiology of the Universidad Michoacana de San Nicolás de Hidalgo in Mexico and chief of the Environmental Biotechnology Laboratory in the Faculty. He received his Doctorate in Engineering, emphasizing Water Sciences, from the Autonomous University of the State of Mexico. He has around 40 publications in specialized journals in the areas of Water Treatment and Environmental Biotechnology, particularly in searching for new adsorbent materials for removing heavy metals and arsenic from water. He is a member of the National Research System of Mexico. He has supervised several master's and doctoral theses.

Rosa E. Pérez Sánchez She is a professor and researcher at the Faculty of Chemistry and Pharmacobiology of the Universidad Michoacana de San Nicolás de Hidalgo in Mexico, is a member of the National Research System of Mexico.

She teaches the asignatures: Statistics, Biostatistics, and Experimental design subjets.

Hers research lines are: Impact of unconventional technologies for agricultural systems, generation of sustainable agricultural products through technological innovation, and Dynamics of structures, biodiversity, management, and utilization in forest and agricultural ecosystems.

She has supervised several master's and undergraduate theses.

Terpenes: Nature's Plasticizers for Sustainable Biopolymer Enhancement

9

Jaume Gomez-Caturla, Juan Ivorra-Martinez, Luis Quiles-Carrillo, Marina P. Arrieta, and Teodomiro Boronat

Abstract

This chapter provides a deep study of the use of terpenes, for enhancing the properties of biopolymers such as polylactic acid (PLA) and polyhydroxyalkanoates (PHAs). Terpenes and terpenoids are present in many plants as essential oils. With the growing environmental concern about developing sustainable alternatives to petroleum-based plastics, biobased polymers like PLA and PHAs are emerging as significant candidates to substitute the traditional materials due to their biodegradability. However, their applications, particularly in food packaging due to material limitations such as rigidity and brittleness.

Terpenes and terpenoids are natural solution to these challenges by serving as bio-based plasticizers into the polymer matrices of PLA and PHAs. Their incorporation leads to a modification of the polymers' structure that improves flexibility and enhancing mechanical properties. This chapter shows how terpenes lower the glass transition temperature (T_g) of these biopolymers and increase their ductility, making them more suitable for a broader range of applications, especially in the packaging industry where material flexibility and strength are crucial.

J. Gomez-Caturla · J. Ivorra-Martinez · L. Quiles-Carrillo · T. Boronat (✉)
Institute of Materials Technology (ITM), Universitat Politècnica de València (UPV), Plaza Ferrándiz y Carbonell 1, Alcoy, Alicante, Spain
e-mail: jaugoca@epsa.upv.es; juaivmar@doctor.upv.es; luiquic1@epsa.upv.es; tboronat@dimm.upv.es

M. P. Arrieta
Departamento de Ingeniería Química Industrial y del Medio Ambiente, Escuela Técnica Superior de Ingenieros Industriales, Universidad Politécnica de Madrid (ETSII-UPM), Madrid, Spain

Grupo de Investigación, Polímeros, Caracterización y Aplicaciones (POLCA), Madrid, Spain
e-mail: m.arrieta@upm.es

R. Malviya, S. Sundram (eds.), *Sustainable Nanocomposites with Green Biomaterials*, Biomaterials, Bioengineering and Sustainability 2,
https://doi.org/10.1007/978-3-031-79110-9_9

Moreover, the utilization of terpenes aligns with the increasing demand for eco-friendly polymer materials, improving the performance of PLA and PHAs but at the same time maintaining their environmentally sustainable profile. By enabling the production of more versatile and biodegradable polymers, terpenes are positioned as key contributors for the evolution of the biopolymer industry. This chapter highlights the potential of terpenes in creating sustainable, flexible materials for future applications, marking a significant step towards reducing the environmental footprint of industrial sectors such as the packaging sector.

Keywords

Sustainable material · Biodegradable · Plasticizer · Terpenes · Biomaterial · Essential oil · Terpenoid

Abbreviations

ATBC	Acetyl tributyl citrate
CED	Cohesive energy density
DSC	Differential scanning calorimetry
Et	Tensile modulus
GAc	Geranyl acetate
LAc	Lynalyl acetate
PEG	Polyethylene glycol
PHA	Polyhydroxyalkanoates
PHB	Polyhydroxybutyrate
PLA	Polylactic acid
RED	Relative energy difference
SFE	Supercritical fluid extraction
$T_{5\%}$	Temperature at 5% weight loss
T_{cc}	Cold crystallization temperature
TEC	Triethyl citrate
T_g	Glass transition temperature
TGA	Thermogravimetric analysis
T_m	Melting temperature
T_{max}	Maximum degradation temperature
X_c	Degree of crystallinity
δ	Solubility parameter
δ_d	Dispersive component of the solubility parameter
δ_h	Hydrogen bond component of the solubility parameter
δ_p	Polar component of the solubility parameter
$\varepsilon_b\%$	Elongation at break
σ_t	Tensile strength

9.1 Introduction

In today's society, polymers and environmental sustainability are big concerns due to the widespread use of polymers from petrochemical origin. These synthetic materials come from non-renewable resources like petroleum. They are used in many industries like packaging, construction, and electronics because of their versatility (Cerdá-Gandia 2023; Quiles-Carrillo et al. 2019). However, their common use has led to serious environmental issues. These materials don't break down easily in nature, which leads to them piling up in landfills and oceans. For all this reasons, they act as harmful materials for wildlife and marine life. In addition it strongly contributes to global warming and climate change due to their wide production that produces a lot of greenhouse gases (Torres-Giner 2021; Ramesh and Vinodh 2020).

As a result of the problems emerged, society is becoming aware and concerned about the need for sustainable materials. Governments are pushing for the implantation alternatives like biopolymers in the manufacturing of polymer made products. Those biopolymers are obtained from natural resources such as plants or byproducts by means of bacterial or fungi fermentation (Baranwal et al. 2022). Nowadays, polylactic acid (PLA) and polyhydroxyalkanoates (PHAs) are some of the most promising biopolymers with a good future perspective. PLA can be produced from plant-origin starch, so plants with a high starch content, like corn, are perfect candidates to be used as bio sources of this polymer. PHAs are made through natural processes by microbes that ferment a carbon source under proper conditions (Rezvani Ghomi et al. 2021; Saravanan et al. 2022). These materials could help reduce our dependency on oil-based resources and lessen our environmental impact, which is crucial for a sustainable future (Naser et al. 2021a, b; Ilyas et al. 2020).

However, replacing traditional plastics with biopolymers like PLA and PHAs isn't straightforward (Ivorra-Martinez et al. 2024). These biopolymers can vary a lot in how strong or flexible they are, which might make them unsuitable for certain uses. They also need specific conditions to be processed correctly and don't handle high temperatures well. Despite producing biopolymers is currently more expensive due to the cost of raw materials and sometimes the need for special equipment, the market for biopolymers is still growing, and not enough people know about their benefits, which slows down their acceptance (Naser et al. 2021a, b; Lee et al. 2020; Ivorra-Martinez et al. 2023a, b, c; Balart et al. 2021). Ongoing research is focused on making biopolymers better through genetic engineering for the improvement of the bacterial fermentation process, new processing techniques, and the creation of new material mixtures. Solving the technical and economic issues is key to making biopolymers more popular and taking big steps towards a sustainable and environmentally respectful approach to materials (Sharma et al. 2021; Wang et al. 2023).

In the realm of biopolymer development, the quest to enhance material properties such as flexibility, durability, and processability is paramount. This enhancement is often achieved through two primary strategies: the incorporation of plasticizers and the creation of polymer blends. Each approach allows to tune the main properties of biopolymers, making them more suitable for industrial applications such as single use packaging application and the development of new biomedical devices (Fredi and Dorigato 2023; Righetti et al. 2023).

Plasticizers are additives that improve the ductile properties and fluidity of a polymer material by blending them in the proper amount (generally up to 20% in weight of plasticizer). In biopolymer context, plasticizers can significantly improve ductile properties like elongation at break (ε_b%), stiffness and toughness. By integrating plasticizers, biopolymers can be made more malleable and easier to process, which is crucial for their use in industrial manufacturing processes like compression molding and extrusion processes (Vikhareva et al. 2022).

The selection of a proper plasticizers is crucial, many criteria can be employed like the environmental objectives and also the critical performance requirements. Researchers are focusing on getting a sustainable alternative to conventional petroleum-based plasticizers like phthalates, which are associated health and environmental risks. Furthermore, an effective plasticizer should exhibit good compatibility with the polymer matrix, ensuring at the same time an adequate flexibility and a low migration. To this effect, it should possess low volatility to minimize evaporation during thermal processing and use, thereby maintaining its efficacy over time. Additionally, an ideal plasticizer should be non-toxic, supporting broader health and safety standards, and it should have a minimal environmental footprint, ideally being recyclable or biodegradable. These criteria ensure that the plasticizers not only enhance the material's properties but also align with the overarching principles of environmental sustainability and health safety (Sree et al. 2023; Dragan et al. 2024).

In the study of polymer plasticization, various classical theories have been proposed to explain its effects. However, three predominant theories are highlighted: lubrication theory, gel theory, and free volume theory. All of them agree on a critical point: the efficacy of the plasticizer is intrinsically linked to its low molecular weight. This characteristic allows plasticizers to efficiently position themselves among the polymer chains, facilitating significant structural and dynamic changes in the material (Ivorra-Martinez et al. 2022).

According to lubrication theory, a plasticizer acts as a lubricant between polymer chains, diminishing intermolecular interactions and easing molecular movement under stress. This theory proposes that macromolecules flow past one another when the plasticized material undergoes an effort. Due to the plasticization, the internal friction of the polymer chains is reduces allowing to tune the stiffness of the material at the same time that the ductility is optimized. In this context, a low molecular weight plasticizer utilization is vital, as it allows effective insertion between the chains, minimizing friction and modifying the three-dimensional structure of the polymer. This alteration leads to a reduction in the energy necessary for molecular motion (Zhang and Geng 2020).

The gel theory proposes that polymer chains work as a three-dimensional network in which plasticizer are attached to the polymer molecules via weak secondary forces. The plasticizer can disrupt the interaction between the different polymer chains involved, avoiding their function as cross-linking points with adjacent chains. This process enhances the mobility of the chains, providing a more ductile material.

As proposed by this theory, the rigidity materials are formed by the existing polymer-polymer bonds. With the plasticizer introduction, a decrease in the number of possible sites that can form bonds between the chains is obtained (Muobom et al. 2020).

The free volume theory suggests that the free space between polymer chains allows their movement during the effort application. Polymers with a high stiffness have a low free volume structure. However, by the incorporation of a plasticizer, free volume increases providing more space that facilitates the free movement of the polymer chains. This enhancement in space significantly improves the ductility of the material and reduces its T_g. In this context, is important to select a low molecular weight plasticizer, as it helps in its proper distribution between the polymer molecules, maximizing the increase in terms of free volume (Hill et al. 2020).

The obtention of blends by a solving method involves mixing a polymer and plasticizer in a common solvent followed by solvent removal through evaporation, this method is frequently used as laboratory scale level as it cannot be properly scaled up in the industry. Melt blending is the industrial process which is employed in extrusion or injection molding methods. In this case the plasticizer is added to the molten polymer and stirred mechanically. Another interesting approach to incorporating plasticizer to a polymeric matrix involves the in-situ polymerization where the plasticizer is introduced while the polymerization process of the polymer is produced, integrating the plasticizer molecules as the polymer chains form. Each method is chosen based on the specific properties required in the final product and the selection of a suitable plasticizer and its concentration in the polymer matrix are critical factors that determine the success of the mixing process, with the ultimate goal of tailoring the properties of polymers to meet the demands of diverse applications, ranging from everyday consumer products to specialized industrial materials (Râpă et al. 2015; YousefniaPasha et al. 2021; Vijayakumar et al. 2021).

Another approach is the development of blends involving biopolymers, which represents an alternative way of enhancing polymer properties. By blending two or more polymers, it is possible to obtain materials with synergistic combination of properties that cannot be achieved by a single polymer. For instance, blending a PHA or PLA with other polymers it is possible to obtain materials with a balanced mechanical outcome, improved heat resistance, and enhanced biodegradability. Blending process of polymers can also involve the use of additives. To enhance the blend performance, is necessary to improve the interaction of both parts with a proper compatibilization strategy. These composites can provide improved strength, impact resistance or thermal stability, expanding the application potential of biopolymers in other industrial sectors (Di Lorenzo 2021; Acharjee et al. 2024; García-Campo et al. 2017; Montava-Jordà 2019).

In the exploration of biopolymer optimization, the employment of traditional plasticizers like phthalates has been widely studied along the years. These plasticizers were widely employed in the past to improve the flexibility, processability, and general physical properties of biopolymers in many industrial sectors. However, as phthalates are not derived from bio-based sources, they are associated with significant environmental and health concerns. The main issues from their potential application in packaging sector are related with endocrine effects and persistence in the environment, which have led to many regulatory restrictions that have reduced their utilization nowadays (Ivorra-Martinez et al. 2023a, b, c).

As a result of these concerns, there is a new trend for investigating alternative plasticizers that match the sustainability goals inherent in the use of biopolymers. Bio-based plasticizers, such as terpenes and terpenoids, present a promising solution. Terpenes are naturally occurring organic compounds, primarily found in form of essential oils of plants, and offer a renewable, less toxic option compared to traditional plasticizers that nowadays are widely employed in other industries like the cosmetic one (Tejada-Oliveros et al. 2024).

Terpenes are a group of organic compounds that can be found in many natural species like plants and insects as the main components of essential oils. Composed of simple hydrocarbons, each terpene molecule is built from isoprene units (C_5H_8) and additional functional groups. They are the fundamental building blocks of this biochemical family, with over 20,000 different types identified, each contributing with a unique scent and taste to its source (Inamdar et al. 2020; Fan et al. 2023; Feng 2021). Examples of well-known terpenes are limonene and pinene, both are characterized for their strong aromatic properties. Those compounds can be found in the oils of citrus fruits and pine trees, respectively. For example, these products have played a crucial role in medicine and aromatherapy due to their properties (Perveen and Al-Taweel 2018).

On the other hand, terpenoids are chemically modified terpenes that include additional functional groups such as alcohols, ethers, and aldehydes. These modifications, typically achieved through oxidation or rearrangement of the molecular structure. The proposed changes alter their chemical properties and also their biological activity making them more suitable for some applications. For example, menthol is a terpenoid that possesses an alcohol group, distinguishing it from its original terpene. By this modification, the range of functions and applications can be extended. It has to be highlighted that terpenes and terpenoids are commonly grouped under the broad term "terpenes"in everyday discourse (Perveen and Al-Taweel 2018).

Terpenes give plants their aromatic properties but at the same time have important ecological roles. They can help plants by repelling herbivores and attracting pollinators, and they help plants fight off insects and diseases. Because terpenes are found everywhere in nature and have so many different roles, they are very interesting for scientific research and practical uses, especially in creating sustainable materials (Ninkuu et al. 2021; Gad et al. 2022).

Terpenes are being studied for their potential as natural softeners that enhance the properties of biopolymers, making them better alternatives to less eco-friendly materials. When added to polymers, they fit between the polymer chains, reducing the forces that hold these chains tightly together as proposed by the plasticizing theories proposed above. This action makes the material more flexible, which is great for manufacturing and final use (Della Monica and Kleij 2020; Zeinali et al. 2021).

Chemically, terpenes are made up of isoprene units, which allows them to be modified easily to form terpenoids allowing them to interact well with various organic materials (Agus 2021). Terpenes and terpenoids can change the physical properties of polymers considerably by lowering characteristic temperatures of the mixture like the T_g, the temperature at which a polymer changes from being a hard and glass-like material to a soft and flexible one. This makes the biopolymers easier to shape and use at lower temperatures (Sessini et al. 2020; Charon et al. 2022). At the molecular level, terpenes and terpenoids reduce the forces that normally make polymer chains rigid and brittle acting as a plasticizer of the polymer matrix. By lowering the temperature at which polymers become flexible, terpenes make it easier to process biopolymers into various shapes, improving manufacturing and the quality of the end products. In addition, this T_g helps to achieve good ductile properties of the manufactured products at room temperature (Muobom et al. 2020; Zeinali et al. 2021).

Terpenes are naturally occurring compounds extensively utilized in sustainable practices due to their biodegradable nature. Nonetheless, their extraction methods, such as steam distillation or solvent extraction, can be energy-intensive and may involve harmful chemicals, which necessitates careful management. Utilizing industrial by-products as terpene sources not only mitigates these environmental effects but also embodies the principles of the circular economy by reducing the demand for virgin raw materials. The future of terpenes in promoting sustainability is promising, particularly with their increasing incorporation into biopolymers. Technological advancements that minimize the need for raw plant materials and more efficient extraction techniques will enhance their sustainability profile. To fully realize the environmental advantages of terpenes, it is crucial to scrutinize their production and usage while continuously advancing sustainable practices in their applications, thereby contributing to a more circular and responsible economy (Ben Salha et al. 2021; Rodríguez-Llorente et al. 2020; Sánchez-Martínez et al. 2021).

This book chapter delves into the striking role of terpenes and terpenoids natural compounds in enhancing the properties of biopolymers such as PLA and PHAs. Given the urgency to develop sustainable alternatives to petroleum-based polymers, the focus here is particularly on optimizing the mechanical behaviour of these biodegradable materials through the use of terpenes as bio-based plasticizers. The integration of terpenes into the polymer matrices of PLA and PHAs modifies their structural characteristics, thereby improving flexibility and overall mechanical resilience. It will detail the mechanisms by which terpenes lower the T_g of these biopolymers, which enhances their ductility and broadens their application scope, especially in industries like packaging where flexibility and durability are crucial.

9.2 Terpenes and Terpenoids as Natural Plasticizers of Biopolymers

9.2.1 Terpenes and Terpenoids Classification

Terpenes are a large and varied groups of natural organic compounds. They are fundamentally composed of isoprene units, each containing five carbon atoms with the molecular formula C_5H_8. These isoprene units repetitively link together, forming the basic structure of terpenes. Such linking can result in a spectrum of structural complexities, from the simpler monoterpenes to the more intricate diterpenes, sesquiterpenes, triterpenes, and larger polymeric forms (Agus 2021; Ben Salha et al. 2021).

The versatility in the structure of terpenes and terpenoids comes from the way the isoprene units are joined, either in head-to-tail or occasionally in head-to-head configurations. The classification of terpenes and terpenoid types based on their structure can be seen in Table 9.1.

The versatility of terpenes and terpenoid structures arises from the various ways isoprene units bond, either in head-to-tail or sometimes head-to-head configurations, allowing for diverse chemical behaviors and biological activities. Such diversity makes terpenes and terpenoids highly valuable in numerous applications across industries such as pharmaceutical, food, cosmetic, and agriculture. A deeper understanding of terpenoid structures aids in appreciating their functionalities and their extensive use in both natural and applied contexts.

9.2.2 Terpenes Extraction Process

Extraction methods for terpenes are employed at industrial level due to the high interest in many industries (Ben Salha et al. 2021). Common methods include:

Steam distillation: used extensively for extracting essential oils containing terpenes, this method involves passing steam through the raw material to vaporize volatile compounds that contain the terpene. The vapors are then condensed back into a liquid form, allowing for the separation and collection of oils. This method suits heat-stable compounds and is environmentally benign, primarily using water as a solvent (El Kharraf 2021).

Solvent extraction: this technique utilizes solvents such as hexane, ethanol, or methanol to dissolve essential oils from plant materials. After solvent evaporation, the essential oils are recovered. The efficiency and selectivity of this method depend significantly on the choice of solvent and extraction conditions, making it adaptable for extracting specific terpenoids (Ben Salha et al. 2021).

Supercritical fluid extraction (SFE): employing supercritical fluids like carbon dioxide, which at critical temperature and pressure exhibits properties of both gases and liquids, SFE allows for efficient terpenoid extraction without high temperatures. This method is environmentally friendly and leaves no toxic residues,

Table 9.1 Terpene and terpenoids types and their main characteristics

Name	Structure	Examples	Sources	Uses	References
Hemiterpenes	One isoprene unit (C5)	Isoprene	Biological processes of plants	Growth and defense mechanism of plants	Tetali (2019)
Monoterpenes and monoterpenoids	Two isoprene units (C10)	Limonene, pinene, geraniol esters and myrcene	Essential oils of citrus fruits, conifers and eucalyptus	Perfumes, flavorings and aromatherapy	Wojtunik-Kulesza (2019), Paulino (2022)
Sesquiterpenes and sesquiterpenoids	Three isoprene units (C15)	Farnesene, humulene and bisabolol	Essential oils of ginger, cedar, and hops	Anti-inflammatory, antibacterial, and antifungal	Li (2022), de Moura (2021)
Diterpenes and diterpenoids	Four isoprene units (C20)	Taxol, gibberellins and cafestol	Coffee beans, Pacific yew tree, and some flowering plants	Medicine, particularly taxol in cancer treatment	Georg (2021), Moeenfard and Alves (2020)
Triterpenes and triterpenoids	Six isoprene units (C30)	Squalene, oleanolic acid and beta-amyrin	Plants and animals, including human sebum	Precursors for steroids and in skincare	Noack (2019), Sun (2020)
Tetraterpenes and tetraterpenoids	Eight isoprene units (C40)	Beta-carotene, astaxanthin and lycopene	Colorful fruits and vegetables like carrots	Act as antioxidants	Baccouri and Rajhi (2021)
Polyterpenes and polyterpenoids	More than eight isoprene units	Natural rubbers, polyprenols	Latex from rubber trees	Produce rubber	Sahu (2020)

ideal for producing high-purity extracts, though it requires expensive equipment and precise control (Uwineza and Waskiewicz 2020).

These extraction methods are chosen based on specific needs, including the terpene type, source form, and intended extract application.

9.2.3 Terpenes Chemical Modification

Terpenes and terpenoids are widely used across various industries due to their extensive applications, and they often undergo chemical modifications to boost their compatibility and functionality, particularly when used as plasticizers in different polymers. These modifications help adjust the physical and chemical attributes of terpenoids to meet specific needs. Here are some common methods for chemically altering terpenoids:

Hydrogenation: is the process of adding hydrogen to a terpenoid molecule, often using catalysts like nickel or palladium. Unsaturated bonds are transformed to saturated ones allowing to modify the chemical structure and main properties of the terpenoid. Allows to improve the stability and also to decrease the reactivity of terpenoids, which is beneficial in environments that require less reactive substances. An example of a hydrogenating process is the limonene to obtain p-menthane, lowering volatility and improving thermal stability, making it suitable for high temperature applications (Hartati et al. 2021; Valencia et al. 2024).

Isomerization: changes the structure of terpenoids without modifying their molecular formula as it only involves rearranging the chemical structure. This can significantly affect the terpenoid's aroma and chemical properties, making it particularly interesting in the fragrance industry or in applications that need specific molecular interactions. An example is the isomerization of pinene, which yields camphene, a compound with distinct scent notes and chemical properties that are ideal for particular industrial uses (Wu et al. 2023).

Esterification and etherification: by this method it is possible to introduce ester or ether groups into terpenoids. Esterification involves an acid that reacts with alcohols to form esters. Etherification involves forming ethers from alcohols. An enhancement of the solubility of terpenoids in specific solvents or matrices is achieved at the same time that volatility is reduced, improving their effectiveness as plasticizers. An example of this modification is geraniol alcohols, which are converted into geranyl acetate through esterification, enhancing its solvent properties and compatibility with biopolymers (Rosyda et al. 2020; da Silva Corrêa et al. 2020).

Hydroxylation: hydroxyl groups are introduced to the terpenoid backbone, increasing the hydrophilicity allowing to enhance the compatibility with certain polymers and improves the plasticizing effect. Modifying sesquiterpenes like farnesene to produce farnesol, this process is being increasingly used in cosmetic and pharmaceutical industry as it allows to enhance functional properties (Zhang et al. 2023).

These chemical modifications not only improve the physical and chemical properties of terpenoids but also broaden their applications, particularly in enhancing the performance of biopolymers as plasticizers. Through precise selection and application of these techniques, manufacturers can finely tune terpenoid properties to meet specific industrial requirements, optimizing both performance and sustainability of the final products.

9.2.4 Examples of Terpenes and Terpenoids Employed as Plasticizers

As proposed, terpenes and terpenoids can be employed in many industries with different applications. In addition, these compounds have shown the ability to enhance the mechanical behaviour of biopolymers, making a more versatile product for commercial and environmental applications. Here are some examples of them that have been successfully integrated into polymers:

Limonene: This prominent cyclic monoterpene, often derived from citrus fruit peels like oranges and lemons, serves as an eco-friendly plasticizer due to its renewable source and ability to utilize waste from the citrus industry. Limonene's inclusion in polymers such as PLA and polyhydroxybutyrate (PHB), the simplest PHA (with only 3-hydroxybutyrate units) (Garcia-Garcia et al. 2022)], significantly enhances their flexibility and ductility, vital for packaging applications. It works by lowering the intermolecular forces that appear between the involved polymer chains, thus moving the T_g to temperatures close to room temperature and rendering the materials more pliable at lower temperatures. This property is particularly beneficial in the food packaging sector, where biodegradability and non-toxicity are paramount. Additionally, limonene may impart a mild, pleasant citrus scent, enhancing the sensory attributes of packaging materials (Brüster et al. 2019; Arrieta et al. 2013, 2014a, b, 2015). Limonene can be found in two isomer forms as it can be observed in Fig. 9.1, in which both structures are presented.

Geraniol: Found in the essential oils of roses, citronella, and lemongrass, geraniol is an acyclic monoterpene alcohol that serves as a renewable and versatile plasticizer. When transformed into geraniol esters, its ability to integrate into polymer matrices is enhanced, improving the flexibility and reducing the brittleness of biopolymers at lower temperatures. These properties make geraniol and its esters suitable for food packaging and other applications requiring direct contact with sensitive products. The inherent fragrance of geraniol may also add a subtle aroma to the packaging, improving the consumer experience without artificial additives (Ivorra-Martinez et al. 2023a, b, c; da Silva Corrêa 2020; Gomez-Caturla et al. 2024a, b). In Fig. 9.2, different geraniol esters are presented.

Carvone: Known for its distinctive scent, carvone is found in caraway seed, dill seed oil, and spearmint oil. As a plasticizer in biopolymers, carvone improves the ε_b% of the polymer, enhancing its ductility and flexibility which is advantageous

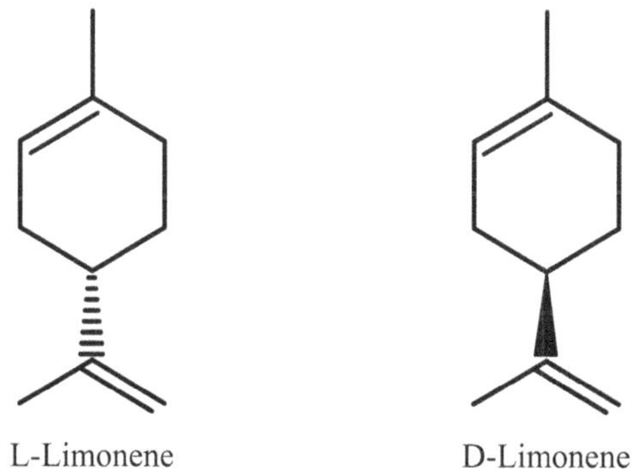

Fig. 9.1 Chemical structure of both isomer forms of limonene terpene

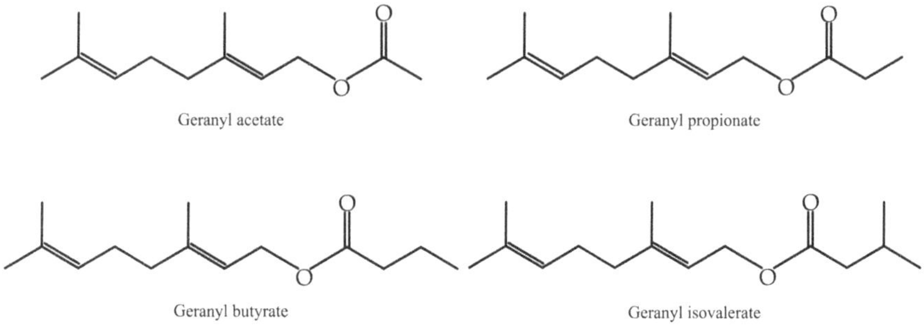

Fig. 9.2 Chemical structure of the geranyl ester terpenoids

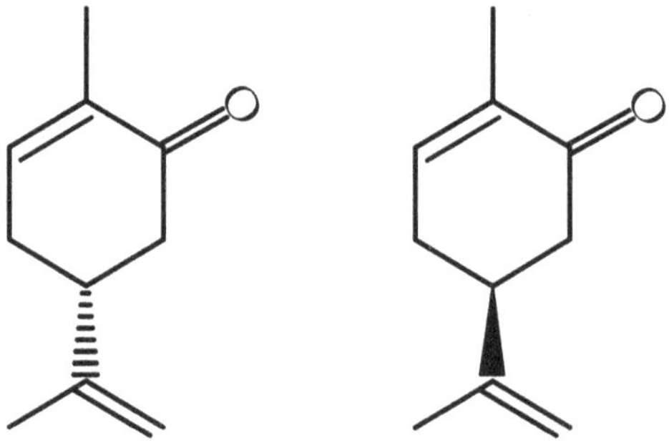

Fig. 9.3 Chemical structure of both isomer forms of carvone terpenoid

for applications requiring more pliable materials (Tejada-Oliveros et al. 2024). In Fig. 9.3, both isomer forms present in carvone are represented.

Citral: This acyclic monoterpenoid, a major component of lemongrass oil, is effective as a plasticizer, significantly increasing $\varepsilon_b\%$ and enhancing the material's flexibility by improving its miscibility and reducing tensile strength (σ_t) and tensile modulus (E_t) (Tejada-Oliveros et al. 2024). Figure 9.4 represents the chemical structure of citral.

Citronellal: Derived from plants like citronella and lemongrass, citronellal enhances the flexibility of PLA by lowering both the T_g and crystallization temperatures, aiding in material processing and utility (Tejada-Oliveros et al. 2024). Citronellal structure is presented in Fig. 9.5.

Eucalyptol: A major component of eucalyptus oil, eucalyptol serves as a plasticizer but shows the lowest miscibility with PLA among studied terpenoids. It still contributes to reducing the T_g, though less effectively than other monoterpenoids (Tejada-Oliveros et al. 2024). Figure 9.6 shows the representation of the eucalyptol structure.

Linalool: Commonly found in essential oils such as those of lavender, coriander, and laurel, linalool has been evaluated for its potential as a plasticizer in PHB. Incorporating linalool into PHB significantly reduces the T_g enhancing the material's flexibility and suitability for various applications requiring improved

Fig. 9.4 Chemical structure of citral terpenoid

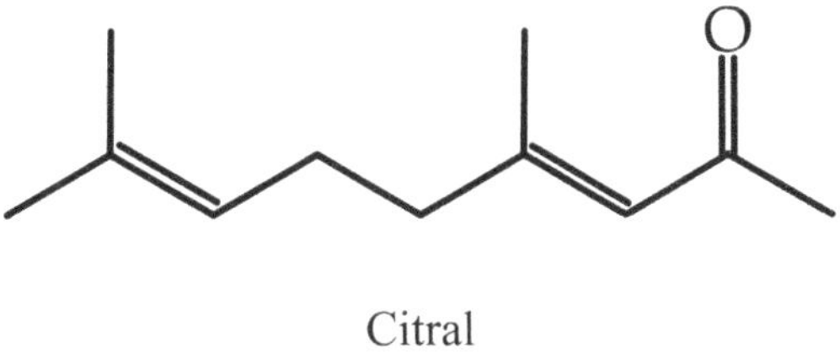

Fig. 9.5 Chemical structure of citronellal terpenoid

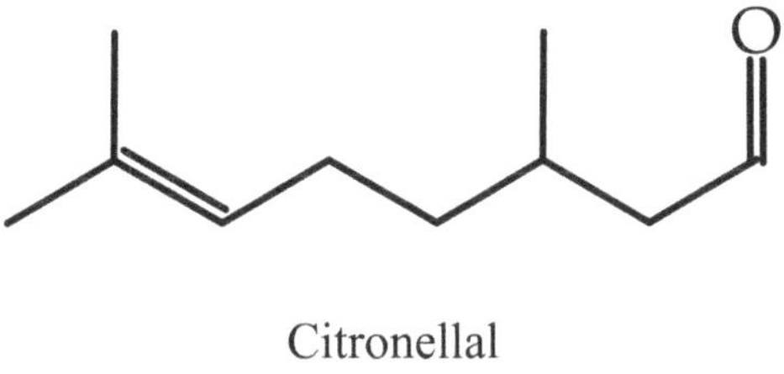

Fig. 9.6 Chemical structure of eucalyptol terpenoid

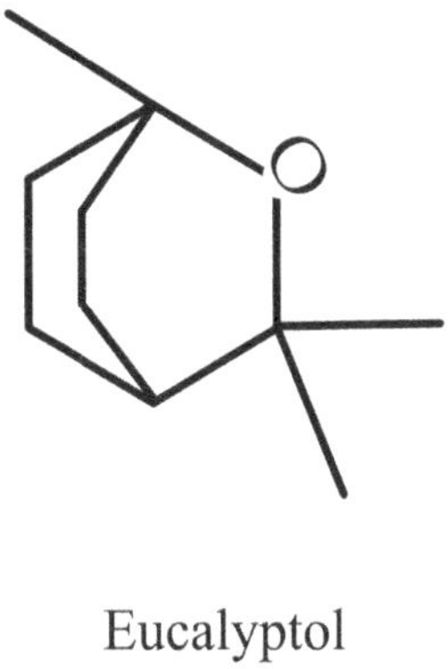

$\varepsilon_b\%$ and reduced brittleness (Mangeon et al. 2018). Linalool chemical structure is presented in Fig. 9.7.

Linalyl acetate: Derived from linalool and acetic acid and commonly found in lavender, linalyl is utilized as a monomeric plasticizer for PLA, significantly enhancing the polymer's ductility. Its addition to PLA formulations greatly improves the $\varepsilon_b\%$, making the polymer more flexible and less prone to cracking, particularly valuable in packaging applications (Gomez-Caturla et al. 2024a, b). Figure 9.8 shows the linalyl structure.

Myrcene: Is present in some essential oils obtained from plants like hops and cannabis, myrcene has been used as a bio-based plasticizer showing the ability to effectively lower the T_g and enhance the ductile properties such as the $\varepsilon_b\%$ impact strength of PLA, allowing to make it more flexible and durable for applications requiring robust, impact-resistant materials (Brüster et al. 2019). In Fig. 9.9, the chemical structure of myrcene is presented.

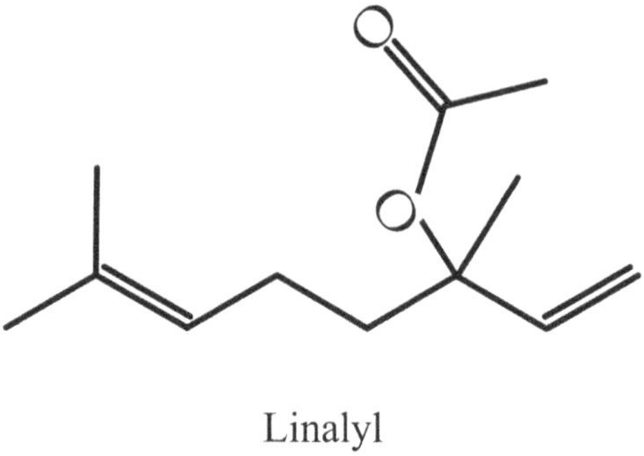

Fig. 9.7 Chemical structure of linalool terpenoid

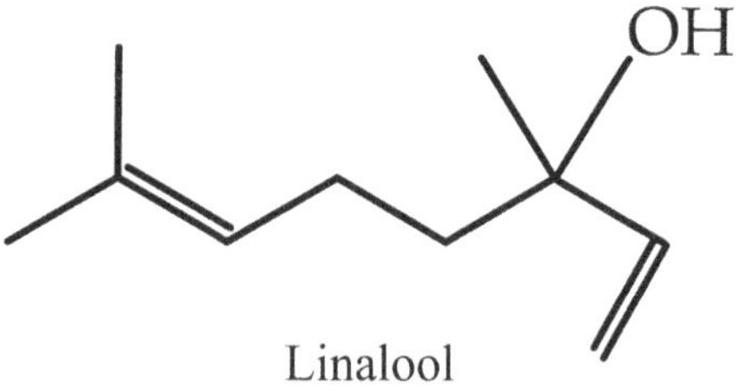

Fig. 9.8 Chemical structure of linalyl acetate terpenoid

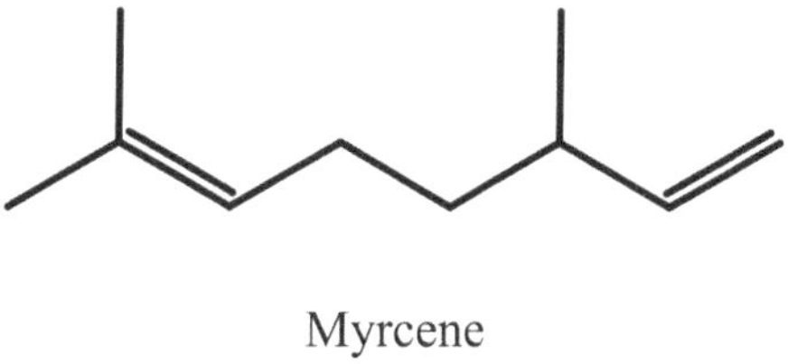

Fig. 9.9 Chemical structure of myrcene terpene

In the forthcoming Table 9.2, several pertinent physical properties of the terpenes previously discussed are summarized.

Table 9.2 of terpene and terpenoid-based plasticizers provides insightful data that highlights their potential in biopolymer applications. These natural compounds are not only derived from renewable sources but also offer beneficial properties that might reduce environmental impact when compared to traditional plasticizers.

Focusing on the specifics, terpenes such as limonene and myrcene have relatively low molecular weights at 136.2 g/mol and boiling points around 166 °C to 177 °C. This contrasts with traditional plasticizers for biobased and biodegradable polymers like ATBC (Acetyl Tributyl Citrate) and TEC (Triethyl Citrate), which have higher molecular weights and boiling points of 402.4 g/mol and 327 °C for ATBC, and 276.2 g/mol and 294 °C for TEC. The higher molecular weight and boiling point of ATBC and TEC suggest better thermal stability and could potentially support applications that require higher processing temperatures, a feature less prominent in the lighter terpenes. Despite the higher thermal stability of the traditional plasticizers, as could be observed later, terpenes are capable to properly plasticize biopolymers like PLA and PHA's.

Table 9.2 Main physical properties of the terpenes and terpenoids employed as plasticizers in biopolymers (Gomez-Caturla et al. 2024a, b)

Name	Molecular weight (g/mol)	Boiling point at 760 mm Hg (°C)	Appearance
Limonene	136.2	176–177	Colorless clear liquid
Geranyl acetate	196.3	240–245	Colorless clear liquid
Geranyl propionate	210.3	252–254	Colorless clear liquid
Geranyl butyrate	224.3	242–243	Colorless to pale yellow clear liquid
Geranyl isovalerate	238.3	276–278	Colorless to pale yellow clear oily liquid
Carvone	150.2	231	Colorless to pale yellow clear liquid
Citronellal	154.2	206	Colorless to pale yellow clear liquid
Eucalyptol	154.2	176–177	Colorless clear liquid
Linalool	154.2	194–197	Colorless clear liquid
Linalyl acetate	196.2	220	Colorless clear liquid
Myrcene	136.2	166–167	Colorless clear liquid

Regarding the appearance, most terpenes are colorless and clear, indicative of high purity, which is crucial in several applications for maintaining the transparency and original color of the biopolymer products. Some, like Geranyl butyrate and Geranyl isovalerate, present a pale-yellow hue, but still retain a clear quality. This is significant because even slight color variations can be critical in certain applications where product appearance is a key factor.

Phthalates are known to be good plasticizers but have a potential health and environmental risks. On the contrary, terpenes provide a sustainable alternative to the traditionally employed plasticizes. Their renewable nature and lower toxicity are highly valued in the current market that favors environmentally friendly materials. Additionally, in comparison with other biobased additives such as ATBC and TEC, terpenes generally offer lower molecular weights and boiling points. This might limit their application in high temperatures manufacturing process unless derived products with higher boiling points, are utilized.

9.2.5 Solubility of Terpenes and Terpenoids with Biopolymers

The solubility of terpenes and terpenoids with biopolymers such as PLA and various PHAs should be analyzed using the study of the solubility parameters to assess the theoretical effectiveness. This studied with the Hoftyzer and Van Krevelen model, which utilizes the cohesive energy density from functional group contributions of the polymer chains and also the plasticizer employed (Barmpalexis et al. 2019).

The principle "like dissolves like" suggests that plasticizer with solubility parameters similar to a polymer could effectively mix allowing to plasticize the matrix.

Applied to biopolymers like PLA or PHA, it is necessary to mix terpenes and terpenoids with similar solubility parameters than the selected biopolymer. Solubility parameters (δ) derive from the cohesive energy density (CED) and can be dived into dispersive solubility parameter (δ_d), polar solubility parameter (δ_p), and hydrogen-bonding solubility parameter (δ_h).

With the Hoftyzer and Van Krevelen method, the calculation the solubility parameter contributions from different functional groups can be conducted. For example, limonene mainly impacts the δ_d component due to its C-H bonds, whereas terpenoids with additional groups like alcohols or esters also affect the δ_p and δ_h components. By analyzing all the involved functional group contributions the solubility parameters can be managed to enhance properties such as flexibility and processability ensuring the proper durability of the material.

The compatibility and effectiveness of terpenes and terpenoids as plasticizers for biopolymers like PLA and PHAs are related with their solubility parameters. Strategic adjustment of the solubility by means of molecular design and functional group contributions can improved biopolymer properties in short and long term by reducing the migration phenomena. Making them suitable for industrial applications including packaging and biomedical sector. This theoretical approach aids to design the terpenoid modification by predicting the behaviour of the plasticized biopolymer.

Table 9.3 summarizes all the solubility parameters and the Relative Energy Difference (RED) for the employed biopolymers along with some terpenes and terpenoids. These values provide valuable information to assess the compatibility of these plasticizers with the specified biopolymers.

Regarding the dispersive component (δ_d) is related to non-polar interactions, a close match in this value between the plasticizer and the polymer should provide a good miscibility in the non-polar terms. Linalool, linalyl acetate, and geranyl acetate are good expels of how the dispersive component similar to PHB or PLA will ensure good compatibility.

The polar component (δ_p) and Hydrogen-Bonding component (δ_h) indicate the ability of molecules to interact through polar groups or hydrogen bonds. Linalool and geranyl acetate have higher δ_h values, suggesting better interaction potential with PHB through hydrogen bonding. Linalyl acetate and geranyl acetate show lower δ_p and δ_h values compared to PLA, suggesting some differences in terms of polar and hydrogen-bonding interactions but still within an acceptable range for effective plasticization.

The RED values provide a quantitative measure of how close the solubility parameters of a plasticizer are to the polymer. Values closer to zero indicate better compatibility. For PHB, both linalool and geranyl acetate show excellent compatibility with RED values significantly less than 1, indicating strong solubility potential. For PLA, both linalyl acetate and geranyl acetate show RED values under 1, confirming good but not perfect miscibility.

These values confirm the suitability of these terpenes and terpenoids as plasticizers for PHB and PLA, supporting their use in enhancing the properties of these

Table 9.3 Solubility parameters for different polymers and terpenoids (Tejada-Oliveros et al. 2024; Brüster et al. 2019; Gomez-Caturla et al. 2024a, b; Arrieta et al. 2014a, b; Hayat 2014)

Compound	δ (MPa$^{1/2}$)	δ_d (MPa$^{1/2}$)	δ_p (MPa$^{1/2}$)	δ_h (MPa$^{1/2}$)	RED
Polyhydroxybutyrate	18.9	15.0	6.5	9.5	–
Linalool	19.7	16.2	2.9	10.8	0.42
Geranyl acetate	18.4	14.7	2.9	10.8	0.40
Geraniol	16.3	14.7	2.9	10.8	–
Polylactic acid	22.2	15.3	8.4	10.9	–
Linalyl acetate	16.8	15.7	2.3	5.7	0.76
Geranyl formate	16.5	15.2	2.5	6.0	
Geranyl acetate	17.1	16.0	2.3	5.7	0.76
Geranyl propionate	16.9	15.9	2.1	5.5	0.80
Geranyl butyrate	16.8	15.8	2.0	5.3	0.81
Geranyl isovalerate	16.6	15.6	1.8	5.1	0.83
Geranyl caproate	16.8	15.9	1.7	5.0	0.85
Limonene	14.3	16.7	1.9	3.2	0.93
Myrcene	7.4	16.1	1.9	2.8	0.99
Carvone	17.0	15.9	5.0	3.6	0.77
Citral	17.9	16.5	4.7	5.1	0.68
Citronellal	17.4	16.0	4.4	5.0	0.69
Eucalyptol	15.0	14.2	2.4	4.2	0.87

biopolymers for various applications, including in areas requiring biodegradable materials with improved flexibility and processing characteristics.

9.3 Effect of Terpenes and Terpenoids in Terms of Mechanical Properties of Biopolymers

As it has been stated in the previous section, terpenes and terpenoids possess a great potential to be used as environmentally friendly additive for biopolymers such as PLA and PHAs. A vast amount of those components have great miscibility with those polymers and thus, they can be combined with them in order to enhance their ductility. Taking into consideration that PLA and PHAs as characteristic for possessing a very great fragility, this is of great interest, especially since those compounds do not compromise the environmentally friendly nature of PLA and PHAs, namely their biodegradability and their natural origin.

A wide range of terpenes and terpenoids have exhibited very positive results towards improving the mechanical performance of PLA and some polyhydroxyalkanoates, such as polyhydroxybutyrate (PHB). Table 9.4 gathers a fair amount of different terpenes such as geraniol or linalool, and terpenoids, such as geranyl acetate of linalyl acetate, added in different amounts to PLA and PHB; and their σ_t and impact strength mechanical properties.

With relation to the combination of non-ester terpenoids with PLA, Tejada-Oliveros et al. (2024) studied the effect of introducing carvone, citral, citronellal and eucalyptol at 10 and 20 wt.% on the mechanical behaviour of this polymer. It was clearly observed that these non-ester monoterpenoids decreased the σ_t of neat

Table 9.4 Tensile properties and Charpy impact properties for terpenoids in different proportions added to PLA and PHB biopolymers

Polymer matrix	Additive in wt. (amount in weight)	E_t (MPa)	σ_t (MPa)	ε_b% (%)	Impact strength (kJ/m^2)	References
PLA	Neat	3198	64.0	2.5	–	Gálvez (2020)
	20 ATBC	1678	31.8	138.0	–	
	Neat	3750	62.0	3.0	–	Brdlík (2022)
	15 PEG 400	750	20.0	57	–	
	Neat	3984	57.0	4.7	–	Gomez-Caturla et al. (2024a, b)
	10 Linalyl acetate	1347	15.9	298.4	–	
	20 Linalyl acetate	104	14.2	236.0	–	
	10 Geranyl acetate	1579	16.2	239.8	–	
	20 Geranyl acetate	193	14.5	215.0	–	
	Neat	3763	65.6	8.1	1.1	Gomez-Caturla et al. (2024a, b)
	10 Geranyl formate	1755	26.8	262.4	5.2	
	10 Geranyl acetate	1509	30.4	301.9	6.1	
	10 Geranyl propionate	1865	37.2	264.3	2.2	
	10 Geranyl butyrate	2239	32.2	250.0	2.3	
	10 Geranyl isovalerate	2698	43.2	237.6	3.0	
	10 Geranyl caproate	2341	41.6	235.3	2.4	
	Neat	3671	67.5	5.2	7.2	Tejada-Oliveros et al. (2024)
	10 Carvone	2973	47.3	7.8	7.4	
	20 Carvone	473	19.8	224.6	No break	
	10 Citral	2483	45.4	200.7	7.7	
	20 Citral	466	18.9	342.4	No break	
	10 Citronellal	2338	48.9	210.3	7.2	
	20 Citronellal	595	20.1	301.8	No break	
	10 Eucalyptol	1462	62.2	7.2	7.3	
	20 Eucalyptol	443	25.8	247.3	No break	
	Neat	3761	56.0	5.6	1.8	Gomez-Caturla et al. (2023a, b)
	20 α-terpinyl acetate	1861	15.9	327.6	5.1	
	Neat	1350	38	1.6	–	Arrieta et al. (2013)
	15 Limonene	840	24	152	–	
	20 Limonene	770	23	166	–	
PHB	Neat	1475	27.0	2.1	–	Mangeon et al. (2018)
	10 Linalool	860	21.7	5.0	–	
	20 Linalool	560	20.0	5.2	–	
	10 Geraniol	820	23.0	5.1	–	
	20 Geraniol	760	20.8	5.9	–	
	10 Geranyl acetate	710	22.3	7.4	–	
	20 Geranyl acetate	395	15.0	13.8	–	
	Neat	1640	26.3	7.4	2.2	Ivorra-Martinez et al. (2023a, b, c)
	20 Geranyl acetate	668	14.6	8.5	3.4	
	20 Geranyl butyrate	651	13.5	7.4	2.8	
	20 Geranyl propionate	606	13.9	8.4	3.1	
	20 Geranyl isovalerate	706	13.5	9.7	2.5	

PLA and increased its ε_b% and impact strength, especially at higher concentrations. Citral, which is highly miscible with PLA, causes the greatest decrease in σ_t from 67.5 MPa down to 18.9 MPa (around 74% reduction) when added at 20 wt.%. Conversely, eucalyptol has minimal effect on σ_t, maintaining values similar to neat PLA at 10 wt.%. The addition of these monoterpenoids reduces the stiffness of PLA

(E_t) considerably, with 20 wt.% carvone and citral exhibiting the lowest stiffness (around 89% decrease compared to pure PLA). However, they significantly increase the ε_b%, particularly at 20 wt.% concentration, with citral and citronellal leading to the highest ε_b% (over 300% increase compared to PLA). Impact strength was specially increased at 20 wt.% of each plasticizer, where samples could not be broken. The effectiveness of these monoterpenoids as plasticizers for PLA is attributed to their solubility parameters that ensure a proper miscibility with the polymer. Citral and citronellal demonstrate the highest plasticizing effect due to their high miscibility with PLA, while eucalyptol shows lower efficiency due to its lower miscibility. The results indicate that the plasticizing efficiency is directly related to miscibility, with higher miscibility leading to greater plasticizing effects.

Other studies have examined the efficacy of ester-based terpenoids, such as linalyl acetate and geranyl acetate, when used in PLA at concentrations of 10 and 20 wt.% (Gomez-Caturla et al. 2024a, b). Geranyl acetate (GAc) and linalyl acetate (LAc) notably reduce the E_t, demonstrating classic plasticization effects. For example, while pure PLA exhibits a high E_t up to 4000 MPa, this value significantly reduces to 104 MPa with 20 wt.% LAc. These terpenoids have been reported to facilitate chain mobility by reducing Van der Waals forces and also the polymer-polymer interactions, thereby decreasing both the E_t and strength. The σ_t of unmodified PLA typically is close to 60 MPa, by this modification it drops to 14–16 MPa in PLA formulations that incorporate these plasticizers. Moreover, the reduced interaction between polymer chains enhances the ε_b%, with values soaring to as high as 298.4% for PLA formulations containing 10 wt.% LAc, a stark contrast to the ε_b% of pure PLA, which is only 4.7%. Consequently, both terpenoids substantially enhance PLA's plasticity, offering increases in ε_b% that are comparable to, or even exceed, those of other commonly used PLA plasticizers.

Similar results were reported in a work that studied the effect of the chain length of different ester-based geraniol terpenoids in terms of mechanical behaviour of PLA (Gomez-Caturla et al. 2024a, b). In this study, geranyl formate, geranyl acetate, geranyl propionate, geranyl butyrate, geranyl isovalerate and geranyl caproate were manufactured at 10 wt.%. These plasticizers possessed an increasing molecular weight and boiling point as their carbon-carbon backbone became longer. All terpenoids drastically increased ε_b% of PLA, achieving values between 237% and 300%. Terpenoids of higher molecular weight such as geranyl caproate and isovalerate allow to obtain higher mechanical properties but lower ε_b% (237 and 235%, respectively) compared to plasticizers with lower molecular weight such as geranyl formate or acetate (262% and 300% ε_b%, respectively). This is related to the fact that smaller plasticizer molecules can be introduced more easily between the polymeric chains, thus exerting a more pronounced lubricity and reducing to a higher extent the intermolecular forces between PLA chains. Additionally, it was observed that smaller plasticizers possessed higher miscibility with PLA, according to the van Krevelen and Hofytzer theoretical method. Impact strength was also drastically increased in the case of geranyl formate, which improved it five-fold. This study demonstrated the importance of the molecular weight and size of the terpenoid molecule when combining them with polymers.

An additional study tested the effectiveness of a derivative of α-terpineol, a tertiary monoterpenoid alcohol that can be obtained from limonene or pinenes that is widely used in the flavors and fragrances industries (Sales et al. 2020). An ester of this monoterpenoid, α-terpinyl acetate, was used and combined both with neat PLA and a wood plastic composite made of PLA and tangerine peel flour (Gomez-Caturla et al. 2023a, b). This study fairly demonstrated the great effectiveness of this monoterpenoid as a plasticizer, which allowed to obtain an ε_b% of 327% at a 20 wt.% of composition, which is a value greater than the ones reported for the other monoterpenoids in this book. Regarding σ_t and E_t were decreased down to 16 and 1800 MPa. Which are similar results to those of geranyl acetate or linalyl acetate. This plasticizer allowed to obtain more than 190% ε_b% for samples with 10 wt.% a filler of tangerine peel flour which is a lignocellulosic byproduct that helps to increase the rigidity of the plasticized PLA formulations. Therefore, further corroborating its great plasticizer potential.

These findings suggest that natural-based plasticizers like monoterpenoids can compete with traditional PLA blends with plasticizers such as ATBC or polyethylene glycol (PEG), as it is illustrated in Table 9.4, where all the terpenoids presented possess equal or far superior mechanical properties than those of these traditional plasticizers. Such is the example of ATBC at 20 wt.%, which exhibits an ε_b% of 138%, while some terpenoids like linalyl acetate at the same proportion present ε_b% of 236%.

Regarding the addition of terpenoids to PHB, one of the most popular PHAs, there is little literature that reports studies about mixtures of these kind of polymers with terpenoids. However, Table 9.4 also reports the mechanical properties of different terpenoids that have been added to this polymer.

Mangeon et al. (2018) documented the combination of two alcohol-based terpenoids (geraniol and linalool) and one ester-based terpenoid (geranyl acetate), all of them at 10 and 20 wt.%. PHB is a very fragile polymer, with lower ε_b% even than PLA and lower σ_t and E_t. In this case, the addition of geraniol and linalool offered a little increase in ε_b% from a value of 2.1% up to 5–6%. On the other hand, the use of geranyl acetate allowed to obtain higher ε_b%, with a value of 13.8% for the sample with 20 wt.% geranyl acetate. This fact seems to indicate that, in general, ester-based terpenoids offer better results in terms of ductility than alcohol-based terpenoids, which was ascribed to an increase in free volume and molecular mobility by the ester group. In tensile test, σ_t and E_t were decreased from a value of 27 down to values between 15 and 23 MPa.

Another study reported the use of different terpenoids based on geraniol to plasticize PHB. In this study, geranyl acetate, geranyl butyrate, geranyl propionate and geranyl isovalerate at 20 wt.% were employed as natural plasticizers of PHB [29]. All of them are miscible components with PHB according to the theoretical solubility methodology developed by van Krevelen. In this case, neat PHB presented a σ_t of 26.5 MPa and a E_t of 1640 MPa. Those parameters were clearly decreased by the addition of the plasticizers down to values of σ_t in the 13–15 MPa range and E_t in the 600–700 MPa range. On the contrary, ε_b% of neat PHB was 7.4%, which is ascribed to its high crystallinity as a result of the accelerated physical aging that this polymer suffers at room temperature (McAdam et al. 2020), while the ε_b% of the plasticized samples increased up to certain point, reaching a maximum ε_b% of 9.7% for the sample with geranyl isovalerate.

The results observed for PHB imply that the effect that terpenoids have on PHAs is not as noticeable as in other aliphatic polymers such as PLA, where the difference in ε_b% is enormous. This is mainly ascribed to the large spherulites that grow in PHB during the cooling stage as a result of a low crystallization rate and nucleation density. Additionally, it suffers from physical aging, which promotes secondary crystallization that provokes internal shrinkage. Therefore, there is plenty of more terpenoids to be tested with several PHAs to try and improve their mechanical performance, or even inhibit its physical aging.

9.4 Effect of Terpenes and Terpenoids in Terms of Thermal Properties of Biopolymers

Thermal properties are also a very important property to be taken into account when it comes to plasticization, especially in the case of fragile polymers such as PLA of PHAs. In fact, one of the most limiting drawbacks of PLA is its great T_g, which is a lot higher than room temperature (normally around 60 °C) (Quiles-Carrillo et al. 2018). As the T_g indicates the point from which the material ceases to behave like a rigid polymer and becomes more rubber-like, this great T_g makes PLA to be quite rigid and fragile at room temperature, limiting its applications. Terpenoids, as plasticizers, are capable of modifying this rather limiting thermal transition.

In this context, Fig. 9.10 shows the evolution of the T_g for PLA and PHB with the composition of different terpenoids. As it was studied by Tejada-Oliveros et al. (2024), Carvone, Citral, Citronellal and Eucalyptol (non-ester terpenoids) are capable of considerably reducing the T_g of PLA. Carvone reduces the T_g from a value of 61 °C to approximately 52 °C for a composition of 20 wt.%; citral reduces it down

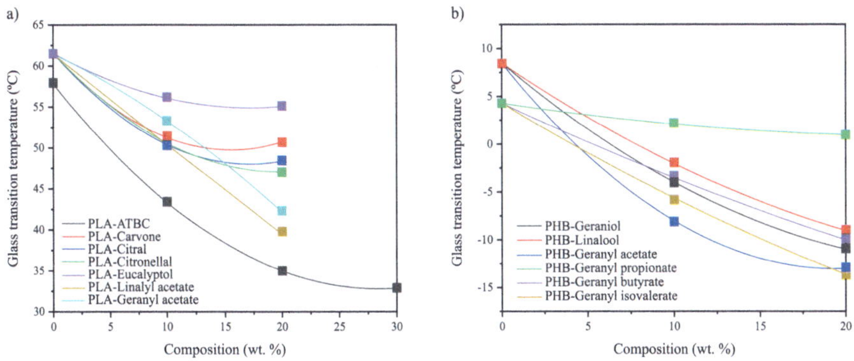

Fig. 9.10 Evolution of the glass transition temperature for: (**a**) PLA with ATBC and different terpenoids; (**b**) PHB and different terpenoids (Tejada-Oliveros et al. 2024; Gomez-Caturla et al. 2024a, b; Gálvez et al. 2020)

to approximately 49 °C for a 20 wt.% plasticizer composition; citronellal diminishes the T_g of PLA down to 48 °C; and eucalyptol did not reduce it to such a great extent, as a T_g of 56 °C was registered at 20 wt.% of eucalyptol. This reduction of the T_g is connected to changes in the internal structure of PLA, as these monoterpenoids are capable of decreasing the intensity of forces (van der Waals) between PLA chains, boosting their mobility, as well as increasing the free volume and exerting a lubricant effect that also enhances the mobility of PLA chains. This effect is less pronounced in the case of eucalyptol, which is the plasticizer that possesses the lowest miscibility with PLA. As it was expected, the higher the terpenoid composition, the lower the T_g. Another study with ester-based terpenoids studied the T_g of PLA with the composition of geranyl acetate and linalyl acetate (Gomez-Caturla et al. 2024a, b). In this case, T_g was reduced down to values of 42.5 and 39 °C for geranyl and linalyl acetate, respectively. This seems to point out the fact that ester-based terpenoids allow for a more efficient plasticization than non-ester based terpenoids, as it was also reported above in the mechanical behavior section. This is probably ascribed to a better miscibility (lower RED value) of the ester terpenoids with PLA than the non-ester ones.

Finally, these values have been compared with the work of Gálvez et al. (2020), who studied the effect of a more traditional plasticizer on PLA, ATBC. It can be seen that the reduction of the T_g is more acute in this case, which obtains a T_g of approximately 35 °C for a 20 wt. % ATBC composition. This fact implies that there is more research to be done in relation to find other terpenoids that allow for a further reduction of the T_g. Nonetheless, the presented terpenoids already exhibited quite a positive behavior towards reducing the brittleness of PLA and it should be taken into account that the PLA used in the work of Gálvez et al. was of a grade with lower T_g than the other studies.

With regard to the study of the T_g of PHB, Mangeon et al. (2018) observed that geraniol, linalool and geranyl acetate effectively reduced the T_g of PHB below 0 °C, moving it from 8.5 °C down to −11, −9 and −13 °C, respectively. This was ascribed to an increase in the mobility of the amorphous regions of the PHB polymer chains, and these results suggest that the three terpenoids are miscible with PHB. Additionally, this seems to again point out to the fact that ester-based terpenoids possess greater compatibility with PHB than non-ester terpenoids and thus, they offer a greater reduction in the T_g. Geranyl propionate, butyrate and isovalerate also promote a positive reduction of the T_g of PHB, reaching values of 1, −10 and −13.6 °C, respectively (Ivorra-Martinez et al. 2023a, b, c), being geranyl isovalerate the most effective one. This results perfectly match the ε_b% values shown in the previous section, where geranyl isovalerate was the one with the highest ε_b% of them. On the contrary, geranyl propionate seems to possess a lower miscibility with PHB, as the reduction of T_g was the lowest of them all. Anyway, this suggests that the three ester-based terpenoids are perfect candidates as environmentally friendly plasticizers for PHB.

In addition to the analysis of the T_g with different terpenoids made according to Fig. 9.10, Table 9.5 gathers several thermal transitions and parameters related to differential scanning calorimetry (DSC) and thermogravimetric analysis (TGA)

Table 9.5 Thermal properties obtained by DSC and TGA for different terpenoids in different proportions added to PLA and PHB biopolymers: Cold crystallization temperature (T_{cc}), melting temperature (T_m), degree of crystallinity (X_c), temperature at 5% weight loss ($T_{5\%}$), maximum degradation temperature (T_{max})

Polymer matrix	Additive in wt. (amount in weight)	DSC			TGA		References
		T_{cc} (°C)	T_m (°C)	X_c (%)	$T_{5\%}$ (°C)	T_{max} (°C)	
PLA	Neat	103.6	151.3	3.5	339.3[a]	367.6	Gálvez (2020)
	10 ATBC	93.4	145.4	3.1	302.1[a]	365.1	
	20 ATBC	90.4	143.5	3.3	280.9[a]	363.8	
	30 ATBC	76.7	137.6	4.1	262.4[a]	362.6	
	Neat	109.2	174.1	6.6	–	–	Brdlík (2022)
	15 PEG 400	84.1	167.9	29.5	–	–	
	Neat	139.9	160.1	2.7	333.2	373.8	Gomez-Caturla et al. (2024a, b)
	10 Linalyl acetate	101.2	161.0	6.2	292.2	379.2	
	20 Linalyl acetate	90.4	160.8	11.3	232.5	378.7	
	10 Geranyl acetate	103.6	164.1	8.7	292.5	378.3	
	20 Geranyl acetate	90.7	163.5	10.1	232.4	377.8	
	Neat	102.8	173.3	29.4	357.9	380.0	Gomez-Caturla et al. (2024a, b)
	10 Geranyl formate	104.6	172.4	14.5	233.3	375.9	
	10 Geranyl acetate	99.8	172.6	21.7	260.6	370.8	
	10 Geranyl propionate	99.8	173.4	22.8	260.7	378.5	
	10 Geranyl butyrate	91.2	172.2	39.2	261.1	373.3	
	10 Geranyl isovalerate	95.1	171.6	29.0	266.5	377.5	
	10 Geranyl caproate	95.3	171.5	26.5	279.8	378.1	
	Neat	102.8	173.5	12.4	354.3	383.6	Tejada-Oliveros et al. (2024)
	10 Carvone	98.9	172.3	11.1	256.4	382.5	
	20 Carvone	95.9	171.9	11.4	195.6	382.2	
	10 Citral	95.3	171.7	10.2	249.1	382.5	
	20 Citral	96.9	170.8	16.2	221.2	381.8	
	10 Citronellal	97.2	171.3	13.3	240.0	382.3	
	20 Citronellal	94.2	169.8	12.3	202.6	385.1	
	10 Eucalyptol	102.1	172.7	9.4	223.1	384.4	
	20 Eucalyptol	100.6	171.0	10.6	172.4	383.6	
	Neat	–	173.2	5.2	305.0	337.5	Gomez-Caturla et al. (2023a, b)
	20 a-terpinyl acetate	88.5	168.9	17.3	232.5	356.0	
PHB	Neat	–	182.0	57.0	275.0	–	Mangeon et al. (2018)
	10 Linalool	–	176.0	53.0	127.0	–	
	20 Linalool	–	179.0	60.0	–	–	
	10 Geraniol	–	175.0	53.0	137.5	–	
	20 Geraniol	–	174.0	59.0	–	–	
	10 Geranyl acetate	–	176.0	40.0	130.0	–	
	20 Geranyl acetate	–	170.0	36.0	–	–	
	Neat	–	171.0	47.4	–	–	Ivorra-Martinez et al. (2023a, b, c)
	20 Geranyl acetate	–	164.6	47.0	–	–	
	20 Geranyl butyrate	–	164.4	47.3	–	–	
	20 Geranyl propionate	–	163.9	44.1	–	–	
	20 Geranyl isovalerate	–	162.6	45.2	–	–	

[a]Values related to the temperature at which the material has lost 10 wt.% of its weight

results for PLA and PHB with some traditional plasticizer such as ATBC and a wide variety of terpenoids.

Regarding the DSC results for PLA, the research conducted by Gálvez et al. (2020), documented a notable reduction in the cold crystallization temperature (T_{cc}) from 103 °C in pure PLA to 76.7 °C in a sample containing 30 wt.% of ATBC. This temperature decrease becomes more pronounced as the proportion of the plasticizer in the mixture increases. The decline in this parameter is attributed to the effect of plasticizers on increasing the mobility of PLA chains. This enhanced mobility allows the chains to reorganize more readily when the material is reheated (during crystallization), thus facilitating a reduction in the temperature at which this crystallization occurs (Gomez-Caturla et al. 2023a, b). The melting temperature (T_m) of PLA does not change substantially; however, it is observed to decrease to 137 °C in a sample with 30 wt.% ATBC, compared to 151 °C in pure PLA. Another significant observation is the increase in the crystallinity of the polymer as the concentration of plasticizer in the blends increases. This increase in crystallinity is attributed to a nucleating effect caused by ATBC, which raises the number of crystal growth initiation points, thus facilitating the catalysis of crystallization. Specifically, crystallinity rises from 3.5% in neat PLA to 4.1% in the sample with 30 wt.% ATBC (Ivorra-Martinez et al. 2023a, b, c). A similar trend is noted with another common plasticizer, PEG 400, which leads to a decrease in the T_{cc} from 109 °C to 84 °C, a reduction in the T_m from 174 °C to 167 °C, and a significant enhancement in crystallinity, which drastically increases from 6% to 29.5%. These changes highlight the impactful role of plasticizers in modifying the thermal and crystalline properties of PLA, enhancing its processability and applicability in various industrial contexts (Brdlík et al. 2022).

Regarding the great variety of terpenoids studied, all seem to follow the same trend as traditional plasticizers. Linalyl acetate and geranyl acetate at 10 and 20 wt.% allow to decrease the T_{cc} thermal transition from 139.9 °C down to values between 90 and 102 °C. This decrease is very similar to the one reported for ATBC, which indicates a similar behavior between terpenoids and this effective common-used plasticizer. The T_m was practically not affected in this case, maintaining it in a range between 160 and 163 °C, which is not a significant variability. Finally, crystallinity was increased from 2.7% up to 10 and 11% for the addition of 20 wt.% of linalyl and geranyl acetate, respectively, as a result of an heterogeneous nucleating effect (Gomez-Caturla et al. 2024a, b). The different chain length geranyl ester terpenoids did not decrease the T_{cc} down to a great extent due to the low amount of plasticizer used (10 wt.%) achieving a minimum value at 91 °C for geranyl butyrate. These terpenoids did decrease the T_m down to a certain point, probably due to the fact that these components make the blend to be less thermally stable than neat PLA (Gomez-Caturla et al. 2024a, b). In relation to crystallinity, the terpenoids with higher molecular weight allowed to obtain higher crystallinities, especially geranyl butyrate, which obtained a crystallinity of 39% in comparison to neat PLA (29%). The last ester-based terpenoid shown in Table 9.5, α-terpinyl acetate, registered the highest decrease in the T_{cc} transition of all the plasticizers shown, achieving 88.5 °C of T_{cc}, as well as a drastic increase in crystallinity, from 5.2 up to 17.3%, which implies an increase of almost 250% (Gomez-Caturla et al. 2023a, b).

The study related to the non-ester terpenoids proved that these compounds lead to a less pronounced reduction in the T_{cc} than the ester-based terpenoids. This is in accordance with what was observed in the T_g analysis, which again points out to the fact that ester-based terpenoids are more desirable as plasticizers than non-ester terpenoids. Carvona, citral, citronellal and eucalyptol reduce the T_{cc} point from 102.8 °C to values between 95 and 102 °C (Tejada-Oliveros et al. 2024). As with the rest of terpenoids, the T_m is not greatly affected, as it stands between 169 and 173 °C for all the samples. Crystallinity is not much affected either, which could be related to a less pronounced plasticizer effect of the non-ester terpenoids.

The analysis of the thermal transitions of terpenoids in PHB is less significant than the one of PLA, as the T_{cc} is not observed in this case. As it is remarked by the studies shown, geranyl acetate is capable of reducing to a greater extent the T_m of PHB than non-ester terpenoids, as it decreases it from 182 down to 170 °C, while geraniol and linalool decrease it only to 175–176 °C, which indicates greater mobility of the polymer chains thanks to the addition of the acetate (Mangeon et al. 2018). Geranyl propionate, butyrate and isovalerate also decrease the T_m to values between 162 and 164 °C. The variation of crystallinity in the case of PHB is not so noticeable as in the case of PLA, as the plasticized samples maintain a similar degree of crystallinity as neat PHB, although it is observed that the higher the plasticizer content, the higher the crystallinity, which again corroborates the heterogeneous nucleating effect exerted by these terpenoids, which catalyzes crystallization (Ivorra-Martinez et al. 2023a, b, c).

TGA is a valuable tool for assessing the thermodegradation of biopolymers, providing significant information regarding the volatility of plasticizers when mixed with these materials. In Table 9.5, two main parameters are analyzed: the onset degradation temperature ($T_{5\%}$) and the temperature at maximum degradation rate occurs (T_{max}). The $T_{5\%}$ is defined as the temperature stars to loss weigh, takin as a criterion the loss of 5% of its mass. The maximum degradation rate temperature is denoted as T_{max} and represents the temperature at which the degradation rate is the highest. These parameters are crucial for understanding the thermal stability and degradation characteristics of biopolymers when plasticizers are incorporated, informing their suitability and performance under various thermal conditions.

The thermal stability of PLA when with terpenes and terpenoids, including ATBC and the onset degradation temperature is lowered. This decrease is primarily attributed to the higher volatility of the plasticizers compared to PLA, resulting in these additives decomposing at a lower temperature than the neat polymer. For example, the $T_{5\%}$ for neat PLA is approximately 339 °C. However, this temperature is reduced to 280 °C for a sample that contains a 20 wt.% of ATBC and further reduction to 262.4 °C for a sample with 30 wt.% ATBC. This behavior underscores the impact of plasticizer volatility on the thermal stability of PLA, indicating a critical consideration in the formulation and application of plasticized biopolymers. (Gálvez et al. 2020). As anticipated, the presence of a higher concentration of plasticizer in the material correlates with a decrease in the onset degradation temperature. This trend is further evidenced by the thermogravimetric results for linalyl acetate and geranyl acetate used as plasticizers in PLA. Specifically, the $T_{5\%}$ diminishes from 292 °C with a 10 wt.% composition of linalyl acetate to 232 °C with a 20 wt.% composition of geranyl acetate. Interestingly, a notable observation from the study is the effect of the boiling point and molecular weight of the terpenoids on the onset degradation temperatures. For example, geranyl formate, which is a lighter geranyl ester, shows a $T_{5\%}$ of 233 °C, whereas geranyl caproate, a geranyl ester with a significantly higher molecular weight and boiling point, exhibits a $T_{5\%}$ of 279 °C. This indicates that terpenoids with lower volatility and hence higher boiling points and molecular weights tend to confer better thermal stability to the plasticized PLA. This differentiation highlights the importance of selecting appropriate plasticizers based on their physical properties to optimize the thermal performance of PLA in various applications (Gomez-Caturla et al. 2024a, b). Additionally, non-ester terpenoids induce a greater decrease in the onset degradation temperature, as it is denoted by a value of 172.4 °C for PLA with 20 wt.% of eucalyptol (Tejada-Oliveros et al. 2024). This is clearly ascribed to the higher volatility of non-ester terpenoids (especially alcohol based terpenoids), which possess a far lower boiling point than ester-based terpenoids, as it has been stated in the first section of this chapter. Similar results are reported for PHB regarding the onset degradation temperature, as $T_{5\%}$ for neat PHB is 275 °C and the addition of alcohol based terpenoids such as linalool or geraniol drastically decrease it down to 127 and 137 °C, respectively (Mangeon et al. 2018).

Totally in contrast with $T_{5\%}$, the T_{max} does not vary significantly with the addition of ATBC or terpenoid-based plasticizers, as it can be observed by all the results shown in Table 9.5, where all T_{max} values are comprised between 370 and 382 °C for the terpenoid plasticizers, and between 362 and 367 °C for PLA plasticized with ATBC. This fact is ascribed to an early volatilization of the plasticizer in the mixture, which leaves practically only PLA left to be thermally degraded, thus not varying this parameter to a greater extent. All in all, it can be concluded that the addition of plasticizers is especially determinant in the temperature range that comprises the boiling point of the plasticizer used.

9.5 Conclusions

Terpenes and terpenoids, as natural plasticizers, play a crucial role in enhancing the flexibility and processability of biopolymers, particularly in polymer matrices such as polylactic acid (PLA) and polyhydroxyalkanoates (PHAs). Their integration into these biopolymers not only improves mechanical properties by enhancing the flexibility but also highlights their potential as sustainable alternatives to conventional petroleum-based plasticizers. This aligns with the increasing environmental urgency to shift towards sustainable material development. Utilizing renewable resources to derive terpenes supports environmental objectives, reduces reliance on non-renewable resources, and mitigates pollution in synthetic material-heavy industries.

The necessity for ongoing research is clear as current biopolymer limitations must be addressed to enhance their application scope. The enhancements that terpenes provide could lead to wider acceptance of biopolymers in competitive markets, such as packaging, which requires materials to be both flexible and strong. Continued innovation and technological advances are essential for developing more efficient methods for the extraction and integration of terpenes, which will support the scalability and economic feasibility of natural plasticizers in biopolymer production.

Acknowledgements This research is a part of the grant PID2020-116496RB-C22 and PID2021-123753NA-C32, funded by MCIN/AEI/10.13039/501100011033 and the grant TED2021-131762A-I00, funded by MCIN/AEI/10.13039/501100011033 and by the European Union "NextGenerationEU"/PRTR. Authors also thank Generalitat Valenciana-GVA for funding this research through the grant numbers AICO/2021/025 and CIGE/2021/094. J. G.-C. wants to thank FPU20/01732 grant funded by MCIN/AEI/10.13039/ 501100011033 and by ESF Investing in your future.

References

Acharjee S, Gogoi B, Bharali P, Sorhie V, Alemtoshi B (2024) Recent trends in the development of Polyhydroxyalkanoates (PHAs) based biocomposites by blending with different bio-based polymers. J Polym Res 31(4):98. https://doi.org/10.1007/s10965-024-03947-z

Agus H (2021) Terpene toxicity and oxidative stress. In: Toxicology, vol 1. Elsevier, pp 33–42

Arrieta M, López J, Ferrándiz S, Peltzer M (2013) Characterization of PLA-limonene blends for food packaging applications. Polym Test 32(4):760–768. https://doi.org/10.1016/j.polymertesting.2013.03.016

Arrieta M, López J, Hernández A, Rayón E (2014a) Ternary PLA–PHB–limonene blends intended for biodegradable food packaging applications. Eur Polym J 50(1):255–270. https://doi.org/10.1016/j.eurpolymj.2013.11.009

Arrieta M, López J, Hernández A, Rayón E (eds) (2014b) The potential of d (+)-limonene to improve pla-phb blends properties. In: Citrus: molecular phylogeny, antioxidant properties and medicinal uses. Nova Science

Arrieta M, López J, Ferrándiz S, Peltzer M (2015) Effect of d-limonene on the stabilization of poly (lactic acid). Acta Hortic 1065(1):719–725. https://doi.org/10.17660/ActaHortic.2015.1065.90

Baccouri B, Rajhi I (2021) Potential antioxidant activity of terpenes. Terpenes Terpenoids Recent Adv 1:53–62

Balart R, Garcia-Garcia D, Fombuena V, Quiles-Carrillo L, Arrieta M (2021) Biopolymers from natural resources. Polymers 13(15):2532. https://doi.org/10.3390/polym13152532

Baranwal J, Barse B, Fais A, Delogu G, Kumar A (2022) Biopolymer: a sustainable material for food and medical applications. Polymers 14(5):983. https://doi.org/10.3390/polym14050983

Barmpalexis P, Karagianni A, Katopodis K, Vardaka E, Kachrimanis K (2019) Molecular modelling and simulation of fusion-based amorphous drug dispersions in polymer/plasticizer blends. Eur J Pharm Sci 130(1):260–268. https://doi.org/10.1016/j.ejps.2019.02.004

Ben Salha G, Abderrabba M, Labidi J (2021) A status review of terpenes and their separation methods. Rev Chem Eng 37(3):433–447. https://doi.org/10.1515/revce-2018-0066

Brdlík P, Novák J, Borůvka M, Běhálek L, Lenfeld P (2022) The influence of plasticizers and accelerated ageing on biodegradation of PLA under controlled composting conditions. Polymers 15(1):140. https://doi.org/10.3390/polym15010140

Brüster B, Adjoua Y, Dieden R, Grysan P, Federico C, Berthé V et al (2019) Plasticization of polylactide with myrcene and limonene as bio-based plasticizers: conventional vs. reactive extrusion. Polymers 11(8):1363. https://doi.org/10.3390/polym11081363

Cerdá-Gandia R, Ivorra-Martinez J, Agüero Á, Quiles-Carrillo L, Gomez-Caturla J, Fenollar O et al (2023) Processing of biopolymer loaded with porous inorganic fillers encapsulating active substance for active food packaging applications. In: Biopolymers, vol 1. Elsevier, pp 207–231

Charon G, Peixinho J, Michely L, Guinault A, Langlois V (2022) Rosin natural terpenes as processing aid for polyhydroxyalkanoate: thermal, mechanical, and viscoelastic properties. J Appl Polym Sci 139(43):1–9. https://doi.org/10.1002/app.53052ff.ffhal-03816621f

da Silva CL, Henriques R, Rios J, Lerin L, de Oliveira D, Furigo A (2020) Lipase-catalyzed esterification of geraniol and citronellol for the synthesis of terpenic esters. Appl Biochem Biotechnol 190(1):574–583. https://doi.org/10.1007/s12010-019-03102-1

de Moura D, Rocha T, de Melo BD, da Silva M, dos Santos SM, Neta B et al (2021) Evaluation of the antioxidant, antibacterial, and antibiofilm activity of the sesquiterpene nerolidol. Arch Microbiol 203(7):4303–4311. https://doi.org/10.1007/s00203-021-02377-5

Della Monica F, Kleij A (2020) From terpenes to sustainable and functional polymers. Polym Chem 11(32):5109–5127. https://doi.org/10.1039/D0PY00817F

Di Lorenzo M (2021) Poly(l-lactic acid)/poly(butylene succinate) biobased biodegradable blends. Polym Rev 61(3):457–492. https://doi.org/10.1080/15583724.2020.1850475

Dragan V, Petrovics N, Kirchkeszner C, Tábi T, Szabó B, Eke Z (2024) The kinetics of swelling and migration: a case study of plasticized polylactic acid food contact plastics tested with ethanolic food simulants. Express Polym Lett 18(4):391–405. https://doi.org/10.3144/expresspolymlett.2024.29

El Kharraf S, Faleiro M, Abdellah F, El-Guendouz S, El Hadrami E, Miguel M (2021) Simultaneous hydrodistillation-steam distillation of Rosmarinus officinalis, Lavandula angustifolia and Citrus aurantium from Morocco, major terpenes: impact on biological activities. Molecules 26(18):5452. https://doi.org/10.3390/molecules26185452

Fan M, Yuan S, Li L, Zheng J, Zhao D, Wang C et al (2023) Application of terpenoid compounds in food and pharmaceutical products. Fermentation 9(2):119. https://doi.org/10.3390/fermentation9020119

Feng S, Gmitter F Jr, Grosser J, Wang Y (2021) Identification of key flavor compounds in citrus fruits: a flavoromics approach. ACS Food Sci Technol 1(11):2076–2085. https://doi.org/10.1021/acsfoodscitech.1c00304

Fredi G, Dorigato A (2023) Compatibilization of biopolymer blends: a review. Adv Ind Eng Polym Res. https://doi.org/10.1016/j.aiepr.2023.11.002

Gad H, Ramadan G, El-Bakry A, El-Sabrout A, Abdelgaleil S (2022) Monoterpenes: promising natural products for public health insect control-a review. Int J Trop Insect Sci 42(2):1059–1075. https://doi.org/10.1007/s42690-021-00692-4

Gálvez J, Correa Aguirre J, Hidalgo Salazar M, Vera Mondragón B, Wagner E, Caicedo C (2020) Effect of extrusion screw speed and plasticizer proportions on the rheological, thermal, mechanical, morphological and superficial properties of PLA. Polymers 12(9):2111. https://doi.org/10.3390/polym12092111

García-Campo M, Boronat T, Quiles-Carrillo L, Balart R, Montanes N (2017) Manufacturing and characterization of toughened poly (lactic acid)(PLA) formulations by ternary blends with biopolyesters. Polymers 10(1):3. https://doi.org/10.3390/polym10010003

Garcia-Garcia D, Quiles-Carrillo L, Balart R, Torres-Giner S, Arrieta M (2022) Innovative solutions and challenges to increase the use of poly(3-hydroxybutyrate) in food packaging and disposables. Eur Polym J 178(1):111505. https://doi.org/10.1016/j.eurpolymj.2022.111505

Georg G, Boge T, Cheruvallath Z, Clowers J, Harriman GC, Hepperle M et al (2021) The medicinal chemistry of taxol. In: Taxol, vol 1. CRC Press, pp 317–375

Gomez-Caturla J, Dominguez-Candela I, Medina-Casas M, Ivorra-Martinez J, Moreno V, Balart R et al (2023a) Improvement of poly (lactide) ductile properties by plasticization with bio-based tartaric acid ester. Macromol Mater Eng 308(7):2200694. https://doi.org/10.1002/mame.202200694

Gomez-Caturla J, Montanes N, Quiles-Carrillo L, Balart R, Garcia-Garcia D, Dominici F et al (2023b) Development of biodegradable PLA composites and tangerine peel flour with improved toughness containing a natural-based terpenoid. Express Polym Lett 17(8):789–805. https://doi.org/10.3144/expresspolymlett.2023.59

Gomez-Caturla J, Ivorra-Martinez J, Tejada-Oliveros R, Moreno V, Garcia-Garcia D, Balart R (2024a) Effect of the chain length of geraniol esters on the plasticization efficiency with poly (lactide). Polymer 290(1):126522

Gomez-Caturla J, Tejada-Oliveros R, Ivorra-Martinez J, Garcia-Sanoguera D, Balart R, Garcia-Garcia D (2024b) Development and characterization of new environmentally friendly polylactide formulations with terpenoid-based plasticizers with improved ductility. J Polym Environ 32(2):749–762. https://doi.org/10.1007/s10924-023-03000-y

Hartati FP, Bahruji H, Bakar M (2021) Review on heterogeneous catalysts for the synthesis of perfumery chemicals via isomerization, acetalization and hydrogenation. Flavour Fragr J 36(5):509–525. https://doi.org/10.1002/ffj.3671

Hayat K (2014) Citrus: molecular phylogeny, antioxidant properties and medicinal uses. Nova Science

Hill A, Thornton A, Hannink R, Moon J, Freeman B (2020) Role of free volume in molecular mobility and performance of glassy polymers for corrosion-protective coatings. Corros Eng Sci Technol 55(2):145–158. https://doi.org/10.1080/1478422X.2019.1701237

Ilyas R, Sapuan S, Kadier A, Kalil MS, Ibrahim R, Atikah M et al (2020) Properties and characterization of PLA, PHA, and other types of biopolymer composites. In: Advanced processing, properties, and applications of starch and other bio-based polymers, vol 1. Elsevier, pp 111–138

Inamdar A, Morath S, Bennett J (2020) Fungal volatile organic compounds: more than just a funky smell? Ann Rev Microbiol 74(1):101–116. https://doi.org/10.1146/annurev-micro-012420-080428

Ivorra-Martinez J, Peydro MA, Gomez-Caturla J, Boronat T, Balart R (2022) The potential of an itaconic acid diester as environmentally friendly plasticizer for injection-molded polylactide parts. Macromol Mater Eng 307(12):2200360. https://doi.org/10.1002/mame.202200360

Ivorra-Martinez J, Gomez-Caturla J, Montanes N, Quiles-Carrillo L, Dominici F, Puglia D et al (2023a) Effect of dibutyl itaconate on plasticization efficiency of a REX processed polylactide with peroxides. Polym Test 124:108059. https://doi.org/10.1016/j.polymertesting.2023.108059

Ivorra-Martinez J, Peydro MÁ, Gomez-Caturla J, Sanchez-Nacher L, Boronat T, Balart R (2023b) The effects of processing parameters on mechanical properties of 3D-printed polyhydroxyalkanoates parts. Virtual Phys Prototyp 18(1):e2164734. https://doi.org/10.1080/17452759.2022.2164734

Ivorra-Martinez J, Valencia Y, Gomez-Caturla J, Agüero A, Arrieta M, Boronat T et al (2023c) Plasticization of poly (3-hydroxybutyrate) with biobased terpenoid esters of geraniol. Express Polym Lett 17(8):773–788. https://doi.org/10.3144/expresspolymlett.2023.58

Ivorra-Martinez J, Gomez-Caturla J, Peydro M, Selles M, Boronat T (2024) Commercial grade of P (3HB-co-3HHX) final properties adjustment by the modification of cooling conditions in the injection molding and ageing process. Macromol Mater Eng:2400033. https://doi.org/10.1002/mame.202400033

Lee C, Sapuan S, Ilyas R, Lee S, Khalina A (2020) Development and processing of PLA, PHA, and other biopolymers. In: Advanced processing, properties, and applications of starch and other bio-based polymers, vol 1. Elsevier, pp 47–63

Li H-Y, Yang W-Q, Zhou X-Z, Shao F, Shen T, Guan H-Y et al (2022) Antibacterial and antifungal sesquiterpenoids: chemistry, resource, and activity. Biomol Ther 12(9):1271. https://doi.org/10.3390/biom12091271

Mangeon C, Michely L, Rios de Anda A, Thevenieau F, Renard E, Langlois V (2018) Natural terpenes used as plasticizers for poly(3-hydroxybutyrate). ACS Sustain Chem Eng 6(12):16160–16168. https://doi.org/10.1021/acssuschemeng.8b02896

McAdam B, Brennan Fournet M, McDonald P, Mojicevic M (2020) Production of polyhydroxybutyrate (PHB) and factors impacting its chemical and mechanical characteristics. Polymers 12(12):2908. https://doi.org/10.3390/polym12122908

Moeenfard M, Alves A (2020) New trends in coffee diterpenes research from technological to health aspects. Food Res Int 134(1):109207. https://doi.org/10.1016/j.foodres.2020.109207

Montava-Jordà S, Quiles-Carrillo L, Richart N, Torres-Giner S, Montanes N (2019) Enhanced interfacial adhesion of polylactide/poly (ε-caprolactone)/walnut shell flour composites by reactive extrusion with maleinized linseed oil. Polymers 11(5):758. https://doi.org/10.3390/polym11050758

Muobom S, Umar A-M, Soongseok Y, Brolin A-P (2020) A review on plasticizers and eco-friendly bioplasticizers: biomass sources and market. Int J Eng Res 9(5):1138–1144. https://doi.org/10.17577/IJERTV9IS050788

Naser AZ, Deiab I, Darras B (2021a) Poly (lactic acid)(PLA) and polyhydroxyalkanoates (PHAs), green alternatives to petroleum-based plastics: a review. RSC Adv 11(28):17151–17196. https://doi.org/10.1039/D1RA02390J

Naser AZ, Deiab I, Defersha F, Yang S (2021b) Expanding poly (lactic acid)(PLA) and polyhydroxyalkanoates (PHAs) applications: a review on modifications and effects. Polymers 13(23):4271. https://doi.org/10.3390/polym13234271

Ninkuu V, Zhang L, Yan J, Fu Z, Yang T, Zeng H (2021) Biochemistry of terpenes and recent advances in plant protection. Int J Mol Sci 22(11):5710. https://doi.org/10.3390/ijms22115710

Noack F, Heinze R, Heretsch P (2019) Contemporary synthetic strategies towards secosteroids, abeo-steroids, and related triterpenes. Synthesis 51(10):2039–2057. https://doi.org/10.1055/s-0037-1611576

Paulino B, Silva G, Araújo F, Néri-Numa I, Pastore G, Bicas J et al (2022) Beyond natural aromas: the bioactive and technological potential of monoterpenes. Trends Food Sci Technol 128(1):188–201. https://doi.org/10.1016/j.tifs.2022.08.006

Perveen S, Al-Taweel A (2018) Terpenes and terpenoids. Intechopen, Place

Quiles-Carrillo L, Blanes-Martínez M, Montanes N, Fenollar O, Torres-Giner S, Balart R (2018) Reactive toughening of injection-molded polylactide pieces using maleinized hemp seed oil. Eur Polym J 98(1):402–410. https://doi.org/10.1016/j.eurpolymj.2017.11.039

Quiles-Carrillo L, Boronat T, Montanes N, Balart R, Torres-Giner S (2019) Injection-molded parts of fully bio-based polyamide 1010 strengthened with waste derived slate fibers pretreated with glycidyl-and amino-silane coupling agents. Polym Test 77(1):105875. https://doi.org/10.1016/j.polymertesting.2019.04.022

Ramesh P, Vinodh S (2020) State of art review on life cycle assessment of polymers. Int J Sustain Eng 13(6):411–422. https://doi.org/10.1080/19397038.2020.1802623

Râpă M, Darie-Niţă R, Grosu E, Tănase E, Trifoi A, Papa T et al (2015) Effect of plasticizers on melt processability and properties of PHB. J Optoelectron Adv Mater 17(11–12):1778–1784

Rezvani Ghomi E, Khosravi F, Saedi Ardahaei A, Dai Y, Neisiany R, Foroughi F et al (2021) The life cycle assessment for polylactic acid (PLA) to make it a low-carbon material. Polymers 13(11):1854. https://doi.org/10.3390/polym13111854

Righetti G, Nasti R, Beretta G, Levi M, Turri S, Suriano R (2023) Unveiling the hidden properties of tomato peels: cutin ester derivatives as bio-based plasticizers for polylactic acid. Polymers 15(8):1848. https://doi.org/10.3390/polym15081848

Rodríguez-Llorente D, Cañada-Barcala A, Álvarez-Torrellas S, Águeda V, García J, Larriba M (2020) A review of the use of eutectic solvents, terpenes and terpenoids in liquid–liquid extraction processes. PRO 8(10):1220. https://doi.org/10.3390/pr8101220

Rosyda M, Aminah N, Kristanti A (2020) Various ester derivatives from esterification reaction of secondary metabolite compounds: a review. MOJ Ecol Environ Sci 5(3):141–151. https://doi.org/10.15406/mojes.2020.05.00187

Sahu P, Bhowmick A, Kali G (2020) Terpene based elastomers: synthesis, properties, and applications. PRO 8(5):553. https://doi.org/10.3390/pr8050553

Sales A, Felipe L, Bicas J (2020) Production, properties, and applications of α-terpineol. Food Bioprocess Technol 13(8):1261–1279. https://doi.org/10.1007/s11947-020-02461-6

Sánchez-Martínez J, Bueno M, Alvarez-Rivera G, Tudela J, Ibañez E, Cifuentes A (2021) In vitro neuroprotective potential of terpenes from industrial orange juice by-products. Food Anal Methods 12(1):302–314. https://doi.org/10.1039/d0fo02809f

Saravanan K, Umesh M, Kathirvel P (2022) Microbial polyhydroxyalkanoates (PHAs): a review on biosynthesis, properties, fermentation strategies and its prospective applications for sustainable future. J Polym Environ 30(12):4903–4935. https://doi.org/10.1007/s10924-022-02562-7
Sessini V, Palenzuela M, Damián J, Mosquera M (2020) Bio-based polyether from limonene oxide catalytic ROP as green polymeric plasticizer for PLA. Polymer 210(1):123003. https://doi.org/10.1016/j.polymer.2020.123003
Sharma V, Sehgal R, Gupta R (2021) Polyhydroxyalkanoate (PHA): properties and modifications. Polymer 212(1):123161. https://doi.org/10.1016/j.polymer.2020.123161
Sree C, Buddolla V, Lakshmi B, Kim Y-J (2023) Phthalate toxicity mechanisms: an update. Comp Biochem Physiol C 263(1):109498. https://doi.org/10.1016/j.cbpc.2022.109498
Sun B, Wu L, Wu Y, Zhang C, Qin L, Hayashi M et al (2020) Therapeutic potential of Centella asiatica and its triterpenes: a review. Front Pharmacol 11(1):568032. https://doi.org/10.3389/fphar.2020.568032
Tejada-Oliveros R, Gomez-Caturla J, Fenollar O, Montanes N, Ivorra-Martinez J, Garcia-Garcia D (2024) Assessment of non-ester monoterpenoids as biobased plasticizers for polylactide with improved ductile behaviour. Polymer 290(1):126572. https://doi.org/10.1016/j.polymer.2023.126572
Tetali S (2019) Terpenes and isoprenoids: a wealth of compounds for global use. Planta 249(1):1–8. https://doi.org/10.1007/s00425-018-3056-x
Torres-Giner S, Figueroa-Lopez KJ, Melendez-Rodriguez B, Prieto C, Pardo-Figuerez M, Lagaron J (2021) Emerging trends in biopolymers for food packaging. Sust Foood Packag Technol 1:1–33
Uwineza P, Waśkiewicz A (2020) Recent advances in supercritical fluid extraction of natural bioactive compounds from natural plant materials. Molecules 25(17):3847. https://doi.org/10.3390/molecules25173847
Valencia D, Martinez-Hernandez E, García A, Aburto J (2024) Sustainable hydrogenation of limonene to value-added products using Cu–Ni catalysts supported on KIT-5. J Clean Prod 434(1):140356
Vijayakumar V, Anothumakkool B, Kurungot S, Winter M, Nair J (2021) In situ polymerization process: an essential design tool for lithium polymer batteries. Energy Environ Sci 14(5):2708–2788. https://doi.org/10.1039/D0EE03527K
Vikhareva I, Aminova G, Mazitova A (2022) Development of a highly efficient environmentally friendly plasticizer. Polymers 14(9):1888. https://doi.org/10.3390/polym14091888
Wang J, Liu S, Huang J, Cui R, Xu Y, Song Z (2023) Genetic engineering strategies for sustainable polyhydroxyalkanoate (PHA) production from carbon-rich wastes. Environ Technol 30(1):103069. https://doi.org/10.1016/j.eti.2023.103069
Wojtunik-Kulesza K, Kasprzak K, Oniszczuk T, Oniszczuk A (2019) Natural monoterpenes: much more than only a scent. Chem Biodivers 16(12):e1900434. https://doi.org/10.1002/cbdv.201900434
Wu W, Chen Y, Liu N, Shi L, Meng X (2023) Synthesis of camphene by α-pinene isomerization using acid nano titanium dioxide catalysts. New J Chem 47(14):6822–6832. https://doi.org/10.1039/D3NJ00229B
YousefniaPasha H, Mohtasebi S, Tabatabaeekoloor R, Taherimehr M, Javadi A, Soltani Firouz M (2021) Preparation and characterization of the plasticized polylactic acid films produced by the solvent-casting method for food packaging applications. J Food Process Preserv 45(12):e16089. https://doi.org/10.1111/jfpp.16089
Zeinali R, Del Valle L, Franco L, Yousef I, Rintjema J, Alemán C et al (2021) Biobased terpene derivatives: stiff and biocompatible compounds to tune biodegradability and properties of poly (butylene succinate). Polymers 14(1):161. https://doi.org/10.3390/polym14010161
Zhang Y, Geng X (2020) Principle of biopolymer plasticization. In: Processing and development of polysaccharide-based biopolymers for packaging applications, vol 1. Elsevier, pp 1–19
Zhang Z, Wu Q, Ge Y, Huang Z, Hong R, Li A et al (2023) Hydroxylases involved in terpenoid biosynthesis: a review. Bioresour Bioprocess 10(1):39. https://doi.org/10.1186/s40643-023-00656-1

Dr. Jaume Gomez-Caturla successfully defended his doctoral thesis in 2024 at the Universitat Politècnica de València (UPV) in the department of Ingeniería Mecánica y de Materiales. Dr. Gomez-Caturla earned a degree in Chemical Engineering from the UPV in 2019, and he completed his Master's degree in Engineering, Processing, and Characterization of Materials. In 2021, he was granted from the Spanish Ministry. Under this last grant, he currently works at the UPV Alcoy campus as a member of the research staff and engages in teaching collaboration tasks. His research focuses on polymeric materials, particularly on integrating polymers with byproducts from different industries to align with the principles of the circular economy. Additionally, he has authored over 38 JCR indexed scientific articles and participated in 24 research congresses.

Dr. Juan Ivorra-Martinez successfully defended his doctoral thesis in 2023 at the Universitat Politècnica de València (UPV), where he was a student in the Department of Ingeniería Mecánica y de Materiales obtaining the Best Thesis Award from Polymers Journal. Now, he is a researcher in the Instituto Universitario de Investigación de Tecnología de Materiales (IUITM). His research primarily focused on developing polymeric materials with reduced environmental impacts, emphasizing their processing through various manufacturing techniques such as injection molding and additive manufacturing. He is author of more than 35 JCR manuscripts, 27 congresses and supervisor of 2 doctoral thesis.

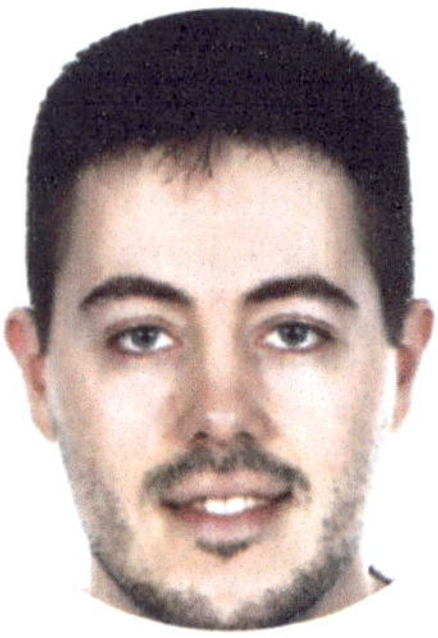

Dr. Luis Quiles-Carrillo was Senior Research Technician at the Universitat Politècnica de València in Spain between 2016 and 2021, and Assistant Professor Doctor from 2021 to 2023. Since December 2023 he is Full Professor at the UPV, within the department of Ingeniería Mecánica y de Materiales. He is currently working on the creation of new polymeric materials with high environmental efficiency following the circular economy principes. During his research carrier he has authored more than 68 JCR articles in the field of polymeric materials. He has worked as guest editor of four special issues of the JCR journals. He is currently in the Stanford Top 2% Scientists Rankings of most cited authors with an h-index of 29. He has also participated in more than 8 research projects, 34 conferences, 1 book and 2 doctoral theses.

Marina P. Arrieta is a faculty member and researcher at the Departamento de Ingeniería Química Industrial y del Medio Ambiente of the Universidad Politécnica de Madrid (UPM). She holds a degree in Biochemistry from the Universidad Nacional de Córdoba (Argentina, 2004). In 2006, she secured a permanent position at the Laboratorio de Seguridad Alimentaria y Control de Materiales of CEPROCOR (Center for Studies and Processes Córdoba, Argentina). She completed her PhD at the Instituto de Tecnología de Materiales of the Universitat Politècnica de València (ITM-UPV), where she was awarded the Extraordinary Doctoral Thesis Prize (UPV 2015). She has been a predoctoral fellow with MAEC-AECID and Santiago Grisolía. Later, she joined the Instituto de Ciencia y Tecnología de Polímeros (ICTP-CSIC), where she held a Juan de la Cierva-Training contract. From 2018 to 2020, she worked at the Universidad Complutense de Madrid, initially with a postdoctoral contract from UCM and later with a Juan de la Cierva-Incorporation contract. Currently, she is an I3 Contracted Doctor Professor at the Universidad Politécnica de Madrid and also is the Principal Investigator of POLCA (polymer, characterization and applications) research group. Her main research focus is on the valorization of agro-industrial waste for the development of sustainable polymeric materials for application in the agri-food sector, within a circular economy.

Dr. Teodomiro Boronat, a Full Professor in the department of Ingeniería Mecánica y de Materiales at the Alcoy Campus of Universitat Politècnica de València (UPV), brings over two decades of expertise in manufacturing processes, particularly in polymeric materials and polymer matrix composites. He leads the Formulations and Processing of Polymeric Materials Research Group at the Instituto Universitario de Investigación de Tecnología de Materiales (IUITM). A Fellow of the Sociedad de Ingeniería de Fabricación, Dr. Boronat is the former Editor-in-Chief of the Journal of Applied Research in Technology & Engineering and has guest edited special issues on environmentally friendly polymers and advanced materials in additive manufacturing. His research focuses on recycling polymeric material wastes, surface modification of polymers, and developing biodegradable and environmentally friendly composites. He has published extensively in high-impact journals, continuing to explore innovative solutions to the challenges facing modern manufacturing.

Tailore-Made Sustainable Nanocomposites for Injection Moulding and Fused Granulate Fabrication

10

Blaž Nardin and Silvester Bolka

Abstract

Thermoplastic composites with a PLA matrix are of commercial interest due to their bio-based nature and biodegradability. The main obstacle to their use in more demanding applications is their fragility and limited exploitation temperature. The present work describes the use of PLA in combination with polycarbonate (PC) and the addition of nanocrystalline cellulose (NCC) without prior surface treatment. Reactive compatibilizers were used in the production process to simultaneously produce a homogeneous thermoplastic mixture of PC and PLA and to enhance the surface interactions among NCC and the PC/PLA blend and additives. We performed a comparison between injection moulding and 3D printing using Fused Granule Fabrication (FGF) technology. The best results were obtained with the reinforcement of 1 wt% NCC in the PC/PLA matrix with compatibilizer and additives. Reactive extrusion with a thermoplastic mixture of PC and PLA reinforced with 1 wt% NCC with additives enhanced impact strength by more than 160% and toughness evaluated by tensile tests by more than 90% regarding to neat PLA in injection moulded parts. The comparison between injection-moulded and FGF parts showed that the stiffness was higher and the strength and toughness were lower in FGF parts than in injection-moulded parts.

Keywords

Nanocomposites · Polylactic acid · Nanocrystalline cellulose · Injection moulding · Fused granulate fabrication

B. Nardin (✉) · S. Bolka
Faculty of Polymer Technology, Slovenj Gradec, Slovenia
e-mail: blaz.nardin@ftpo.eu

R. Malviya, S. Sundram (eds.), *Sustainable Nanocomposites with Green Biomaterials*, Biomaterials, Bioengineering and Sustainability 2,
https://doi.org/10.1007/978-3-031-79110-9_10

Abbreviations

ABS	Acrylonitrile butadiene styrene
$CaCO_3$	Calcium carbonate
C_p	Specific heat
DMA	Dynamic mechanical analyser
DSC	Differential scanning calorimeter
EGMA	Ethylene-glycidyl methacrylate
E'	Storage modulus
E_t	Tensile stiffness
FDM	Fused deposition modelling
FGF	Fused granular fabrication
H_c	Crystallization enthalpy
H_m	Melting enthalpy
MEX	Material extrusion
mTPU	Modified thermoplastic polyurethane
NCC	Nanocrystalline cellulose
OIT	Oxidation induction time
PC	Polycarbonate
PLA	Polylactic acid
SEBS	Styrene-ethylene-butylene-styrene
SEBS-*g*-MA	Styrene-ethylene -butylene-styrene grafted with maleic anhydride
tan δ	Loss factor
T_c	Crystallization temperature
T_g	Glass transition temperature
TGA	Thermogravimetric analyser
T_m	Melting temperature
ε_m	Elongation at yield
ε_{tb}	Elongation at break
σ_m	Tensile strength

10.1 Introduction

For thermoplastic composites with added natural fibers, there must be close contact between the reinforcement and the thermoplastic matrix in the compounding phase, which leads to good interfacial interactions of natural fibers and the thermoplastic resin. The stiffness and strength of composites result from these interfacial interactions. Therefore, the choice of compatibilizer and reinforcement is very important (Krishnasamy et al. 2024). Thermoplastic composites with the addition of natural fibers are interesting for researchers due to their low density. However, they have other advantages over composites with thermoplastic matrix reinforced with glass or carbon fibers, e.g. they are bio-based, less expensive, and less harmful to the environment. However, the development of the topic of bio-based reinforcement of thermoplastic resins needs to continue to improve reinforcement-matrix interactions, production technologies, prolonged lifetime, and overall performance. The factors listed are key issues to meet the increasingly rigorous requirements of materials for industrial applications (Musa and Onwualu 2024; Syduzzaman et al. 2020; Chandgude and Salunkhe 2021; Li et al. 2020). To obtain cellulose fibers from lignocellulosic biomass, researchers are developing environmentally friendly methods that require fewer chemicals and are therefore more environmentally friendly. The development also aims to find green technologies for NCC (Abolore et al. 2024). The resulting NCC contains many hydroxyl groups that can be functionalized to achieve appropriate surface properties (Phanthong et al. 2018).

Polylactic acid (PLA) is a very interesting biomaterial from renewable sources that decomposes in industrial composting and is also commercially very attractive. It has an interesting combination of strength, stiffness, processing, and melting temperature. The major weakness of PLA is its brittleness. Researchers are investigating increasing the ductility of PLA mixtures with the help of compatibilized polymer blends. They are using petroleum-based elastomers, bio-based elastomers, rubber, tough thermoplastics, and biodegradable thermoplastics (Zhao et al. 2020). The disadvantages of PLA are poor barrier properties and a low glass transition temperature. These disadvantages can be avoided by mixing PLA with rubber and cellulose. The processing properties are also improved Mavai et al. (2024). PLA was modified with elastomers by melt blending, whereby a very high toughness was achieved (Zhang et al. 2024; Lu et al. 2014; Liu et al. 2014, 2016; Si et al. 2017; Xiu et al., 2014; Feng and Ye 2011; Lebarbe et al. 2015; Ho et al. 2008; Likittanaprasong et al. 2015; Han et al. 2013a, b; Han and Huang 2011; Shi et al. 2013; He et al. 2014; Dai et al. 2016). To simultaneously increase PLA's toughness and temperature stability, PLA/polycarbonate (PC) blends were produced by reactive compatibilization. Styrene glycidyl methacrylate was used as the compatibilizing agent. To achieve even higher toughness, an impact modifier was added. The toughness of the mixture produced in this way is further improved by annealing after processing at a temperature above the T_g of PLA. Annealing also changes the morphological structure of the PC/PLA mixture (Lin et al. 2015). For the PC/PLA mixture, hydrogenated SEBS was used to enhance ductility, and EGMA was used as a compatibilizer. The blending with PC increased the strength, elongation, and heat deflection temperature of the prepared mixture (Hashima et al. 2010). Two different compatibilizers were added to the PC/PLA/nanosilica nanocomposite. The nanocomposites were produced by melt extrusion and formed into test specimens using the hot press technique. Compared to the PC/PLA/nanosilica sample, the samples with added compatibilizers exhibit lower stiffness, lower tensile strength, and higher toughness.

Recently the report on the extraction of nanoparticles from waste biomass has become increasingly interesting for researchers. A major challenge in the field of optimization is technology and price (Irani-Kolash et al. 2020). With its unique properties, nanocellulose is the material of the future. In combination with thermoplastics, it contributes to new applications in many different industries. Thanks to its biodegradability, it is particularly interesting for use in combination with bio-based and biodegradable thermoplastics (Le Gars et al. 2020). Nanocellulose has a large surface area, and excellent mechanical characteristics, and its surface can be functionalized (Khorsandi et al. 2024; Thomas et al. 2018). The surface treatment of nanocellulose has received much attention with the introduction of a new biomimetic method. The challenges in surface treatment lie in improving the mechanical characteristics of thermoplastic nanocomposites with added nanocellulose as well as for other specific applications such as catalysis and biomedicine. Surface treatment of nanocellulose by biomimetics is used in both industrial and biomedical applications (Kanomata et al. 2020; Qian et al. 2023; Deng et al. 2022). Nanocellulose is obtained from bacteria, plants, algae, and animals. For large-scale production, nanocellulose is mixed with thermoplastics for packaging, thermal insulation, flame retardancy, and energy storage (Bacakova et al. 2019). NCC is usually produced by acid hydrolysis of cellulose. In this process, the amorphous areas are destroyed while the aligned areas, where the cellulose crystallizes remain intact. NCC is also referred to in research publications as cellulose nanowhiskers, nanocrystals and nanorods. Cellulose nanocrystals can also be spherical. In the case of NCC, the minimum distance must be less than 100 nm (Souza et al. 2018; Chavez-Guerreo et al. 2018). To successfully improve the properties of bionanocomposites with nanocellulose, it is important to know the characterization methods that allow high-quality characterization of the nanocellulose in terms of both shape and arrangement (Garavand et al. 2023). The surface interactions of NCC with the PLA resin were increased by surface treatment of NCC with poly(glutamic acid). The addition of NCC (5 wt%) increased the stiffness and strength of the PLA nanocomposite according to the PLA nanocomposite without chemical modification of NCC. In contrast to the neat PLA matrix, the stiffness and stability of the PLA nanocomposite with chemically modified NCC were lower, and the toughness was drastically reduced (Averianov et al. 2019).

Reactive extrusion is a cost-effective and eco-friendly production technology for new mixtures of thermoplastic materials or composites with advanced properties. Reactive extrusion enables the mixing of two immiscible phases, the modification and functionalization of polymers, and processing on existing compounding equipment (Formela et al. 2018). Polymer blends are used to improve the properties of existing polymers by compatibilization without the need to synthesize a new polymer. In-situ compatibilization is a new method for the production of nanocomposites. The immiscibility of most polymers manifests itself in phase separation in the morphology of the blends. This is reflected in uneven physical properties throughout the volume. Block or graft polymers can form an intermediate phase by reacting with a reactive functional group. This helps to improve interfacial adhesion between otherwise immiscible polymers. Graft polymers are added to the polymer mixture before compounding. Graft polymers are concentrated in the interphase areas and thus also act as emulsifiers, enabling better dispersion. One of the methods of polymer blending with compatibilization is the addition of a third, functionalized polymer to the immiscible polymer mixture. The functionalized polymer dissolves in one of the polymers and reacts with the other polymer to form chemical bonds. For the compatibilization to be successful, the functionalized polymer must have reactive groups (Chanda 2024). The approach of polymerization and compatibilization during compounding was developed for the production of nanocomposites. In this procedure, the monomer is polymerized either at the end of the chain or along the chain to form a different type of polymer. To obtain nanocomposites, the polymerization rate and phase separation must be sufficiently high. This is achieved by efficient mixing and a controlled residence time during pre-cooking (Hu et al. 2002; Cartier and Hu; Zhang et al. 2006a, b, 2011; Hu et al. 2007).

Thermoplastic nanocomposites are interesting for use in industrial applications because they both enhance strength and stiffness change the surface properties of the products and modify the other characteristics of the thermoplastic nanocomposite. According to the nanoparticle surface modification, these other properties can be controlled according to the requirements of the end products (Rashid et al. 2024).

PLA is a very commonly applied material for additive manufacturing, mainly thanks to its environmental friendliness. It can be produced on traditional plastic processing machines, and composites with a PLA matrix are also widely used. PLA filaments are often used for 3D printing. Material extrusion (MEX) is a 3D printing technology in which the material in the form of a cord is melted at processing temperature and applied through an extruder die on a flat plane. The nozzle or base moves during this process. This layer-by-layer additive manufacturing enables the production of any three-dimensional object (Wang et al. 2024). The problem with natural fibers mixed with PLA matrix for 3D printing is their unstable properties. It is precisely this property of natural fibers that prompts researchers to conduct intensive research in the field of biocomposites with a PLA matrix with natural fibers for 3D printing (Tuli et al. 2024). To avoid air entrapment during 3D printing, the researchers used sizing techniques based on silan for the modification of the flax fibers to ensure good bonding between the flax and the PLA resin. The appearance of the surface of the 3D-printed parts was improved. The stiffness and strength of the additively manufactured objects were comparable to the mechanical properties of parts produced by compression molding (Long et al. 2021). If nanocomposites intended for additive manufacturing are reinforced with more than 3 nanoparticles, agglomeration and thus a deterioration of the properties can occur (Heller et al. 2016). Nanocomposites with ABS matrix and organically modified montmorillonite (5 by weight) significantly improved the mechanical characteristics of additively manufactured objects (Weng et al. 2016). PLA nanocomposites with graphene and multi-walled carbon nanotubes were mixed in a compounder. The filament was extruded with a single-screw extruder. The mechanical properties of the additive manufactured objects increase by reinforcing with 0.2 wt% graphene and 0.1 wt% multi-walled carbon nanotubes (Plymill et al. 1955). The mechanical characteristics of the PLA nanocomposite with graphene nanoplatelets increased, while the toughness of the additively manufactured objects decreased (Caminero et al. 2019).

The manufactured 3D printing filament achieves good interlayer adhesion between PLA and cellulose. Nanocellulose fibrils were surface-modified by chemically attached L-lactide, which was also polymerized during the reaction. The surface change of the nanocellulose also enables homogeneous distribution in the thermoplastic resin. To enhance the PLA ductility, rubber and cellulose were added, which simultaneously lowered the T_g of the ternary blends. PLA nanocomposite crystallinity with added cellulose increases, as the cellulose functions as a nucleation nucleus in the PLA matrix. The supplement of nanofibrillated cellulose (20 by weight) to the PLA resin enhanced the tensile strength of the 3D-printed parts. The dipole-dipole attraction between molecules in the PLA resin (ester groups) and cellulose (hydroxyl groups) improved the interactions at the interface, which resulted in the enhancement of stiffness and strength. PLA nanocomposites with nanocellulose drastically increase the stiffness and toughness of the additive-manufactured objects, as the stiffness of the individual layer is greatly increased by the stiffness of the cellulose in the individual slice versus the neat PLA matrix. The interlayer bonding of the individual 3D-printed slices also increases. A further enhancement in the strength, stiffness, and temperature stability was achieved by annealing 3D-printed parts made of PLA nanocomposites (Gauss and Pickering 2023; Matsumoto et al. 2024; Tadi et al. 2024; Le Gars et al. 2020). The strengthening of PLA with cellulose nanofibers in the attendance of lignin increases the flexural stability of PLA, as lignin acts as a compatibility factor among the PLA resin and the cellulose nanofibers. 3D printing of PLA nanocomposites went smoothly thanks to the enhanced temperature resistance obtained. The mechanical properties correspond to those of the pure PLA matrix (Zhang et al. 2012).

In the additive manufacturing technology Fused Granular Fabrication (FGF), layer thickness and infill density influence the strength and stiffness of additive-manufactured products made from recycled polyethylene terephthalate and glycol-modified polyethylene terephthalate. The tensile strength increases with increasing layer thickness and filling density (Nguyen et al. 2024). 3D FGF-printed ABS test samples have similar tensile strength (64.9 MPa ± 2.1 MPa) compared to 3D FDM-printed ABS test samples (66.3 MPa ± 2.9 MPa) (Reich et al. 2019).

10.2 Used Ingredients and Characterization Techniques

10.2.1 Samples

The main purpose of this research report was to prepare a tough PC/PLA bionanocomposite with the addition of NCC. We chose the PC/PLA mixture because it achieves a very high toughness compared to pure PLA. In the study, we determined the influence of different amounts of added NCC into the PC/PLA mixture, mainly in the direction of increasing the already very high toughness. The main objective was to determine whether the reinforcement of the PC/PLA mixture with NCC further increases the toughness of the prepared bionanocomposites. We used surface-unmodified NCC. A combination of two compatibilizers was used to enhance the bonding between NCC and the PC/PLA mixture and a homogeneous distribution of NCC in the PC/PLA mixtures. The combination of these two compatibilizers also enables the production of a PC/PLA blend with improved properties as it ensures good miscibility of PLA and PC. To enhance the crystallization of PLA during cooling, we added $CaCO_3$ as a nucleating agent. For processing, we chose a reaction compounding with high shear and a long retention time of the melt at the pre-processing temperature. This was simulated by multiple compounding at high screw speeds for runs with 1 wt% and 2 wt% added NCC. For samples with 5 wt% added NCC, one cycle of reaction compounding was performed. After reaction compounding, the material was granulated. One series of samples was injected and the other series of samples was 3D printed using an FGF printer. The thermal characteristics of the prepared PC/PLA/NCC mixtures were characterized after reaction compounding and granulation. Using both processing technologies, samples complying with ISO 527 type 1BA and ISO 178 standards were produced and applied for the characterization of the mechanical and dynamic characteristics. We characterized the impact of the added NCC in bionanocomposites and the impact of the processing on the characteristics of the produced pieces. A PLAPC blend with added compatibilizer and one cycle of reactive compounding was used as a reference.

Market-accessible PLA Ingeo 4043D was delivered free of charge by the firm Plastika Trček, Slovenia. Market-accessible PC Lexan 243 R was bought from Sabic, Austria. NCC was delivered free of charge by the firm Navitas, Slovenia. Market-accessible SEBS-*g*-MA FG 1901 GT was bought from Kraton, Germany. A market-accessible modified TPU Kuramiron U TU-S5265 was bought from Kuraray, Germany. Market-accessible $CaCO_3$ Calplex Extra was delivered free of charge by the company Calcit, Slovenia. Table 10.1 contains the recipes for the samples.

10.2.2 Reactive Extrusion

On the Labtech LTE 20-44 twin-screw extruder all the samples were compounded after being premixed separately. The screws had a diameter of 20 mm, an L/D ratio of 44:1, a screw turning of 600 rpm, and an increasing temperature profile with a lower temperature (160 °C) when the material enters the extruder and a higher temperature when the material exits the extruder (200 °C). After compounding, the two filaments produced were led through a water bath and were cut to the shape of approx. 5 mm × 3 mm. For the second and third processing cycles, the samples of nanocomposites were processed with the same settings as the first processing.

Table 10.1 Recipes for the PC/PLA samples

Sample	PLA (wt.%)	PC (wt.%)	SEBS-*g*-MA (wt.%)	mTPU (wt.%)	$CaCO_3$ (wt.%)	NCC (wt.%)
PLAPC	42	40	10	5	3	0
PLAPC NCC1	41	40	10	5	3	1
PLAPC NCC1	40	40	10	5	3	2
PLAPC NCC5	37	40	10	5	3	5

10.2.3 Injection Moulding

The injection molding was performed on a Krauss Maffei 50-180 CX with a screw diameter of 30 mm and rising temperature settings. With a lower temperature (185 °C) when the material enters the injection moulding machine and a higher temperature when the material exits the injection moudling machine (200 °C). The mold temperature has been adjusted to 30 °C and the cooling time to 10 s. The back pressure has been adjusted to 150 bar and the plasticizing to 50 rpm. The moulding velocity has been adjusted to 60 mm/s and 20 mm/s for the last 1 mm.

10.2.4 3D Printing

All samples were 3D FGF printed with the 3D printer Tumaker Bigfoot 500 Pro Dual. For 3D printing, we used the samples listed above in granular form. The printing parameters are summarised in Table 10.2.

10.2.5 Description of the Test Methods for the Evaluation of the Bio(nano)composites

The measurements of thermal characteristics were carried out with a Mettler Toledo differential scanning calorimeter DSC 2, in an N_2 ambiance (20 mL/min). The first temperature scan was done at 10 °C/min from 0 to 200 °C. In the second step, there was an isothermal segment at 200 °C for 5 min to completely melt the thermoplastic resin. In the third step with the cooling rate of −10 °C/min, the temperature was decreased from 200 °C to 0 °C. At the end of the third segment, segments one through three were repeated. The crystallization temperature (T_c) and crystallization enthalpy (ΔH_c) were characterized by cooling, melting temperature (T_m), and melting enthalpy (ΔH_m) were characterized in the second heating.

For the evaluation of OIT on DSC, the first step was isothermal heating for 5 min. at 50 °C. The second step was a temperature scan with 20 °C/min from 50 to 200 °C. The third step was again isothermal heating at 200 °C for 5 min. All three steps were under N_2 atmosphere (50 ml/min). The fourth step was isothermal

Table 10.2 Parameter settings of 3D FGF printing

Parameter	Value
Nozzle temperature (°C)	220
Chamber temperature (°C)	160
Bed temperature (°C)	80
Layer height (mm)	0.20
Filling—share (%)	100
Nozzle diameter (mm)	0.8
Fan speed (%)	0
Print speed (mm/s)	3600
Extruder speed (mm/s)	Varied for optimal fill (360–400)

heating at 200 °C in an O_2 atmosphere (50 ml/min). OIT was evaluated as the time in the fourth step, where the heat flow deviated from the baseline, namely as the onset beginning of the deviation of the measured heat flow from the baseline.

The cold crystallization behavior was evaluated on a Mettler Toledo Flash DSC 1 with Huber TC45 intercooler and N_2 purge gas (50 ml/min). The first heating cycle was carried out from 15 °C to 200 °C with 500 °C/min. The sample temperature was then decreased from 200 °C to 15 °C, heated with 500 °C/s to 90 °C to age the samples for 100 s, cooled with −500 °C/s to 15 °C, heated with 500 °C/s to 120 °C. The cold crystallization step was done at 120 °C with cold crystallization times ranging from 0.1 to 1200 s, followed by cooling to 15 °C at 500 °C/min and heating from 15 °C to 200 °C at 500 °C/min. For the evaluation of the last heating section, the heating scan after the cold crystallization segment was taken. The T_m and ΔH_m were characterized. To evaluate the mass of the samples, a step at Δc_p was taken, which was compared with the step at Δc_p measured by standard DSC.

The thermogravimetric analysis was performed using the Mettler Toledo TGA/DSC 3+. Approximately 5 mg of the sample was heated from 40 °C to 600 °C (N_2, 20 ml/min) in an aluminum oxide crucible at 10 °C/min. The second step was a temperature scan (O_2, 20 ml/min) at 600–900 °C at 10 °C/min. Degradation temperature (T_d) and step at degradation were characterized.

A Shimadzu AG-X universal testing machine with a 10 kN load cell was applied to assess the bending tests according to ISO 178. The moving speed of the jaw was 2 mm/min, and the distance between the supports was 16 times the thickness of the samples (64 mm). 5 parallels were measured for each sample. The pieces were 10 mm wide and 4 mm thick. The bending stiffness (E_f), bending strength (σ_{fM}), and elongation at yield (ε_{fM}) were evaluated.

A Shimadzu AG-X universal testing machine with a 10 kN load cell was used to evaluate the tensile tests according to ISO 527. The jaw travel speed was 1 mm/min (0–0.4%), followed by 50 mm/min until break. The distance between the jaws was 50 mm (20 mm to measure the elongation to determine the E modulus. 5 parallels were measured for each sample. The specimens are 5 mm wide and 2 mm thick. The tensile stiffness (Et), tensile strength (σm), elongation at yield (εm), and elongation at break (εtb) were determined.

The thermomechanical properties were examined using a dynamic mechanical analyzer DMA 8000 from Perkin Elmer. During the test storage modulus (E') and the loss factor (tan δ) in the temperature range of 25–180 °C were recorded. The samples were square-shaped (45 × 5 × 2 mm, taken from ISO 527 type 1BA samples). Temperature scans (2 °C/min) with dynamic loading (1 Hz and 0,02 mm) in dual cantilever mode were performed.

The impact test was performed using an LY-XJJDS device according to ISO 179 with a pendulum of 5 J (unnotched samples). The distance between the supports (span) was 60 mm. The tests were carried out at room temperature. The specimens were 10 mm wide and 4 mm thick. Ten measurements were taken for each sample.

10.3 Results and Discussion

10.3.1 Thermal Properties: DSC

The addition of NCC moves T_g toward higher temperatures (Table 10.3). The T_g for the PC/PLA blend is 59.8 °C. The largest shift is seen with the addition of 1 wt% NCC (61.2 °C), and the T_g decreases with increasing addition of NCC (60.4 °C and 60.0 °C for 2 wt% and 5 wt% NCC addition, respectively). Crystallization during heating occurs when the PC/PLA blend is heated and the ΔH_c is 5.35 J/g. Crystallization during heating is not observed in the bionanocomposite with 1 wt% NCC added, but occurs again at 2 wt% (0.48 J/g) and 5 wt% NCC (0.07 J/g), where the crystallization enthalpy is drastically lower. The melting temperature increases with increasing NCC content, and thus also the ΔH_m minus the ΔH_c. The melting point temperature for the PC/PLA blend is 152.0 °C, with 1 wt.% NCC added it rises to 152.4 °C, with 2 wt.% NCC to 152.6 °C and with 5 wt.% NCC to 153.2 °C. The difference between the H_m and the H_c in the PC/PLA blend is 0.02 J/g, at 1 wt.% added NCC it increases to 0.37 J/g, at 2 wt.% NCC it decreases again to 0.17 J/g and at 5 wt.% NCC it increases again to 0.52 J/g. During heating, the glass transition for PC is not obvious as it overlaps with the PLA melting signal. On cooling, since PLA does not crystallize, the glass transition is visible for both thermoplastics (Table 10.4). T_g of PC is slightly reduced by the addition of NCC. For the PC/PLA blend, the glass transition for PC is 137.4 °C, with 1 wt.% NCC added it drops to 135.6 °C, for 2 wt.% NCC to 136.6 °C and for 5 wt.% NCC to 136.4 °C. The change in specific heat is the same for all samples. The glass transition for PLA increases significantly with the addition of NCC. For the PC/PLA blend, the glass transition for PLA is 53.7 °C, with 1 wt.% NCC added it increases to 54.1 °C, for 2 wt.% NCC to 54.5 °C and for 5 wt.% NCC to 54.2 °C. The change in specific heat is highest for the PC/PLA blend and amounts to 0.08 J/gK. supplemented by 1 wt.% nanocrystalline cellulose, it drops to 0.06 J/gK, supplemented by 2 wt.% and 5 wt.% nanocrystalline cellulose to 0.07 J/gK.

10.3.2 Thermal Properties: DSC OIT

Oxidation induction time measurements (Fig. 10.1) for the PC/PLA blend and the bionanocomposite with added NCC show that three times processing samples with 1 wt% and 2 wt% added NCC lowers the OIT versus the OIT of the PC/PLA

Table 10.3 Differential dynamic calorimetry evaluated thermal transitions for heating

Sample	T_g (°C)	ΔC_p (J/gK)	T_{cc} (°C)	ΔH_{cc} (J/g)	T_m (°C)	ΔH_m (J/g)	Diff. ΔH_m (J/g)
PLAPC	59.8	0.106	123.6	5.35	152.0	5.37	0.02
PLAPC NCC1	61.2	0.100	–	–	152.4	0.37	0.37
PLAPC NCC2	60.4	0.110	129.3	0.48	152.6	0.65	0.17
PLAPC NCC5	60.0	0.090	130.0	0.07	153.2	0.59	0.52

Table 10.4 Differential dynamic calorimetry evaluated thermal transitions for cooling

Sample	T_{g1} (°C)	ΔC_{p1} (J/gK)	T_{g2} (°C)	ΔC_{p2} (J/gK)
PLAPC	53.7	0.08	137.4	0.02
PLAPC NCC1	54.1	0.06	135.6	0.02
PLAPC NCC2	54.5	0.07	136.6	0.02
PLAPC NCC5	54.2	0.07	136.4	0.02

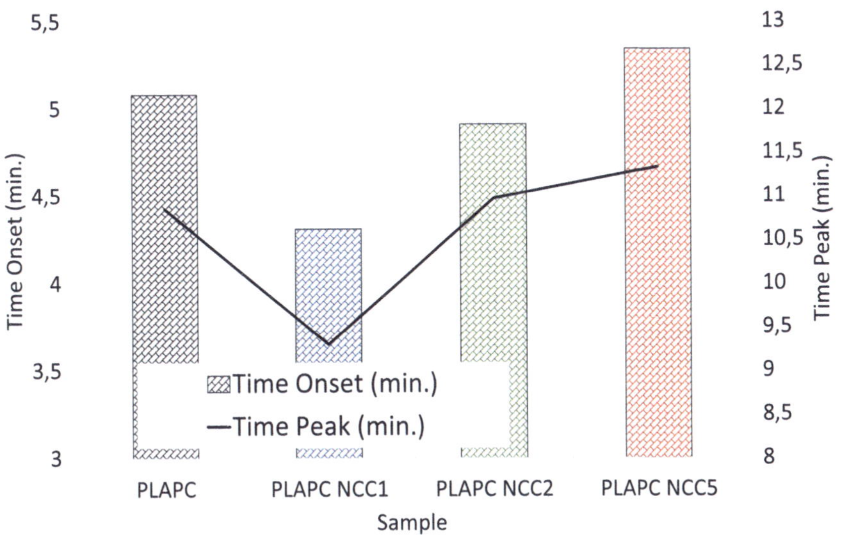

Fig. 10.1 Summarized results of oxidation induction time: onset time (bars) and peak time (line)

mixture. The OIT for the sample with 5% by weight added NCC is higher than for the PC/PLA mixture. We can therefore summarise from the results that increasing the addition of NCC to the PC/PLA blend increases the OIT, while repeated processing has a negative effect on the OIT and lowers it. This means that PLA is oxidatively stabilized for a single additional processing, but with multiple processing the oxidative stabilization of PLA is no longer sufficient and degradation occurs even with shorter retention times of the melt at the processing temperature.

10.3.3 Thermal Properties: Flash DSC

For the evaluation of the crystallization kinetics flash DSC results were taken (Figs. 10.2 and 10.3). The start of the creation of crystalline domains after aging at 90 °C 100 was at 100 s at 120 °C s for all samples. Lower residence periods at elevated temperatures are not satisfactory for the development of crystalline domains. The addition of NCC to the PC/PLA mixture reduces the formation of crystal moieties formation up to a residence time of 600 s at elevated temperatures, and thereafter the addition of NCC increases crystal moieties formation. Crystall moieties

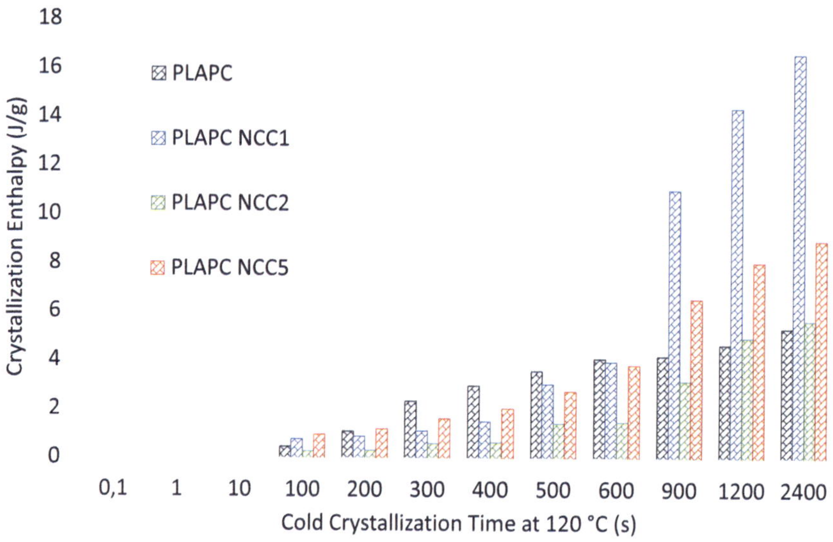

Fig. 10.2 Dependence of ΔHc of samples on aging time and cold crystallization at elevated temperature

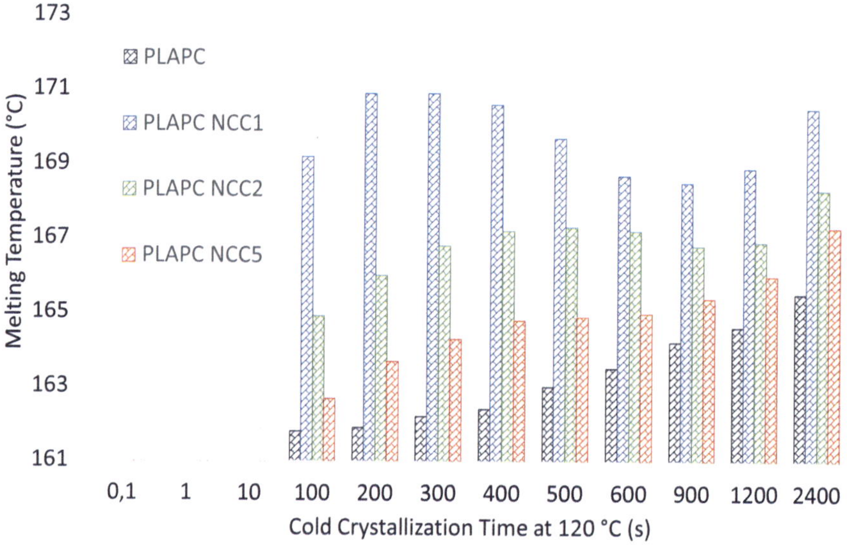

Fig. 10.3 Dependence of melting temperature of samples on aging time and cold crystallization at elevated temperature

formation for blend PC/PLA increases linearly with longer residence time at elevated temperatures. If 1 wt.% NCC is added to the bionanocomposite, the increase in crystal moieties formation is linear up to 400 s, after which the increase is exponential. If 2 wt.% NCC is added to the bionanocomposite, the increase in crystal moieties formation is linear up to 600 s, after which the increase is exponential. When 5% by weight NCC is added to the bionanocomposite, the increase in crystal moieties formation is linear up to 500 s, after which the increase is exponential. We can therefore summarise that the NCC suppresses crystallization at heating at reduced residence periods and promotes crystallization at heating at extended residence periods at elevated temperatures. The movement of PLA sequences at high temperatures achieves a breakpoint after 400 s at 1 wt. NCC loading and after 600 s at 2 wt.% NCC reinforcing the for the generation of crystalline domains. At the loading of 5 wt% NCC reinforcing, 100 s is appropriate because of the increased content of NCC pieces in the PC/PLA blend. It can also be concluded from the results that the conglomeration of NCC appears to a lesser extent grow with the proportion of NCC. The melting temperature grows with the proportion of NCC in the bionanocomposite. In the PC/PLA blend, the melting point increases linearly as the residence time at elevated temperature increases (from 162 at 100 s at 120 °C to 165 °C at 2400 s at 120 °C). At an NCC content of 1 wt.%, the melting temperature is practically independent of the residence time at elevated temperature and lies between 169 °C and 171 °C. With an NCC loading of 2 wt%, the melting point temperature increases up to a residence time of 400 s at elevated temperature (from 165 °C to 167 °C), after which it no longer changes with increasing residence time at elevated temperature. With an NCC loading of 3 wt%, the melting temperature increases linearly with increasing residence time at elevated temperatures (from 163 °C at 100 s at 120 °C to 167 °C at 2400 s at 120 °C). From the results, it can be summarized that 1 wt% NCC addition is sufficient for the heterogeneous crystallization of PLA around the nuclei represented by NCC particles. With a higher addition of NCC, steric hindrances already occur during crystallization, so the resulting crystalline areas become smaller because they interfere with each other while crystals are being formed and the development of the crystals is thus interrupted. The biggest difference in melting point temperature is at shorter residence times at elevated temperatures, which means that less time is needed for additional annealing of the produced pieces to reach the final degree of crystallinity of the PC/PLA/NCC bionanocomposites.

From the DSC results it is evident, that the combination of compatibilizers when NCC is added to the PC/PLA blend successfully compatibilizes the surface unmodified NCC with the PC/PLA matrix. The T_g for PLA shifts slightly to higher temperatures for the bionanocomposites and slightly to lower temperatures for PC, indicating that the materials are at least slightly miscible thanks to the added compatibilizers. The higher difference in ΔH_m and ΔH_c and not least the higher difference in specific heat at glass transition for PLA indicate that the addition of NCC in the PC/PLA blend acts as a nucleus for the crystallization of PLA upon cooling. This prediction is also confirmed by the fact that for the bionanocomposites, the crystallization during heating for PLA is much less pronounced (or not visible at all in the case of the reinforcement of 1 wt% NCC) than for the PC/PLA blend. The highest glass transition for PLA is achieved at 1 wt% added NCC, while the T_g for PC is the lowest. At 5 wt% added NCC, we obtain the lowest glass transition and the highest melting point for PLA as well as the largest discrepancy between ΔH_m and ΔH_c. The addition of NCC to a PC/PLA blend increases the oxidation induction time, but at the same time, the oxidation induction time is reduced by multiple reactive compounding. With multiple reactive compounding, antioxidants would have to be added prior to reactive compounding to increase the oxidation induction time of the bionanocomposites. The NCC loading in the PC/PLA blend functions as a nucleation promoter for the PLA matrix resulting in heterogeneous crystallization. For heterogeneous crystallization, 1 wt% NCC is sufficient to increase the melting temperature of PLA. With a higher NCC loading (2 or 5 wt%), steric hindrances occur during crystallization and the melting temperature of PLA decreases.

10.3.4 Thermal Properties: TGA

Table 10.5 presents the results of TGA evaluations. We obtain a two-stage decomposition. In the first decomposition around 360 °C, about 80 wt.% of the samples decompose, in the second decomposition around 455 °C between 12 and 15 wt.% of the samples decompose. About 5% by weight are soot and between 0.4 and 1.3% are inorganic residues. NCC loading reduces the extent of initial damage. At 1 wt.% NCC loading, the temperature of the first decomposition drops drastically and is only 338.6 °C. The decrease in T_d can be explained by the lowest oxidation induction time, which means that the matrix was degraded the most in this sample, affecting the lower first T_d. Despite the lower T_d, the amount of decomposition (80.4 wt%) was lower than in the PC/PLA blend, where the first decomposition occurred at 358.3 °C, with 81.9 wt% of the sample decomposed. With an NCC loading of 2 wt.%, the temperature of the first decomposition was already higher and was 363.6 °C, with 79.8 wt.% of the sample being decomposed. At an NCC loading of 5 wt%, the first decomposition temperature was 360.1 °C, and 79.0 wt% of the sample decomposed. The second decomposition took place at the lowest temperature for the PC/PLA blend and was 448.6 °C, at which 12.1 wt% of the sample decomposed. As the NCC loading increased, the temperature of the second decomposition and the amount of the second decomposition increased. Thus, the temperature of the second decomposition at 1 wt NCC loading was 454.8 °C, with 12.8 wt% of the sample decomposing. At an NCC loading of 2 wt.%, the temperature of the second decomposition was 456.1 °C, with 14.6 wt.% of the sample decomposing.

Table 10.5 Summarized results of the thermogravimetric analysis for all samples

Sample	T_{d1} (°C)	Decomposition 1 (%)	T_{d2} (°C)	Decomposition 2 (%)	Ashes (%)	Inorganic residue (%)
PLAPC	358.3	81.9	448.6	12.1	4.9	1.1
PLAPC NCC1	338.6	80.4	454.8	12.8	5.9	0.9
PLAPC NCC2	363.6	79.8	456.1	14.6	5.2	0.4
PLAPC NCC5	360.1	79.0	457.5	14.4	5.3	1.3

At an NCC loading of 5% by weight, the temperature of the second decomposition was 457.5 °C and 14.4% by weight of the sample decomposed.

From the results, it could be concluded that NCC loading reduces the extent of the first degradation, increases the extent of the second degradation, and causes more ash formation. A higher NCC loading increased the temperature of the second degradation and also the amount of degraded material in the second degradation. The inorganic residue after TGA is practically not affected by NCC loading, as the amount of NCC loading is low. NCC loading increases the thermal stability of PC/PLA/NCC bionanocomposites.

10.3.5 Mechanical Characteristics: Bending Tests

The reinforcement of PC/PLA with NCC considerably enhances the toughness of the bionanocomposites (Table 10.6). A clear difference between the mechanical characteristics of pieces manufactured using the technologies of injection moulding and additive-manufactured FGF pieces could be observed in that of the injection-molded and 3D-FGF-printed test samples can be seen in the flexural rigidity and flexural strength. With injection molding, we achieve a significantly higher bending stiffness and flexural strength. The elongation of the injection-molded test pieces is also above the elongation of 3D-FGF-printed parts. For the PC/PLA blend, the bending stiffness is 2 GPa, for the injection-molded parts it is higher for the sample with 1 wt% NCC loading and is 2.1 GPa. At 2 wt% NCC loading it drops to 2 GPa, and at 5 wt% NCC loading to 1.9 GPa. For the printed 3d-FGF samples, the flexural strength is 1.5 GPa at 1 wt% NCC loading, 1.7 GPa at 2 wt% NCC loading, and 1.4 GPa at 5 wt% NCC loading. The flexural strength for the PC/PLA blend is 48 MPa. It is higher for all injection molded parts with NCC loading, which means that we achieved good interphase interactions between the PC/PLA blend and the NCC surface during reaction mixing. At 1 wt% NCC loading, the flexural strength increased to 57 MPa, at 2 wt% NCC loading to 53 MPa, and 5 wt% NCC loading to 52 MPa. The flexural elongation was lowest with the PC/PLA blend and amounted to 3.1%. With NCC loading, it was higher for all samples. Injection moulded samples exhibited higher flexural elongation than 3D-FGF printed samples. At 1 wt% NCC loading, the flexural elongation was 4.6% for injection molded and 4.4% for 3D printed FGF samples. At an NCC loading of 2% by weight, the flexural

Table 10.6 Bending stiffness, strength, and elongation of the samples

Sample	E_f (GPa)	σ_{fM} (MPa)	ε_{fM} (%)
PLAPC	2.03 ± 0,02	48.2 ± 1.1	3.13 ± 0.28
PLAPC NCC1-IM	2.06 ± 0.01	57.4 ± 0.3	4.62 ± 0.14
PLAPC NCC2-IM	1.99 ± 0.01	53.4 ± 0.2	4.75 ± 0.08
PLAPC NCC5-IM	1.94 ± 0.01	51.6 ± 0.5	4.77 ± 0.13
PLAPC NCC1-FGF	1.54 ± 0.09	42.5 ± 2.0	4.40 ± 0.44
PLAPC NCC2-FGF	1.72 ± 0.11	47.7 ± 2.6	4.28 ± 0.26
PLAPC NCC5-FGF	1.42 ± 0.22	43.9 ± 4.9	4.54 ± 0.19

elongation was 4.8% for the injection molded and 4.3% for the printed 3D FGF samples. At an NCC loading of 5% by weight, the flexural elongation was 4.8% for the injection-molded and 4.5% for the printed 3D FGF samples.

The flexural stiffness of the pieces produced by injection moulding was above the flexural stiffness of 3D-FGF-produced pieces. It was highest with 1 wt% NCC loading and exceeded the flexural stiffness of the PC/PLA blend. For all other samples with NCC loading, the bending stiffness was lower than the bending stiffness of the PC/PLA mixture. The highest flexural strength of the 3D-FGF produced piece was the sample with 2 wt% NCC loading. The flexural strength of the pieces produced by injection moulding was above the flexural strength of the 3D FGF-produced pieces. For the injection moulded samples, the NCC loading was higher than the flexural strength of the PC/PLA blends in all cases, with the highest flexural strength achieved at 1 wt% NCC loading. For the 3D printed FGF samples, it was lower than the flexural strength of the PC/PLA blend at all NCC loadings, being highest at 2 wt% NCC loading. The flexural elongation was higher than the flexural elongation of the PC/PLA blend for all samples with NCC loading. The flexural elongation was also higher for the pieces produced by injection moulding than for the 3D FGF produced pieces. For the pieces produced by injection moulding and 3D FGF produced pieces, the flexural elongation was highest at an NCC loading of 5% by weight. For the samples with an NCC loading of 2 wt, the difference in flexural properties was the smallest depending on the manufacturing process. We can conclude that we successfully modified the NCC surface during reaction compounding with the combination of compatibilizers. For the injection molded samples, we achieved higher flexural strength, which is a good indicator that the samples had good interlayer adhesion bonding of the NCC and PC/PLA mixture.

10.3.6 Mechanical Characteristics: Tensile Tests

The tensile strength of the bionanocomposites is considerably enhanced by the addition of NCC to PC/PLA (Table 10.7). A clear difference between the strength of the pieces produced by injection moulding and the 3D-FGF produced pieces, as seen in the bending tests, is no longer visible in the tensile tests. Both the pieces produced by injection moulding and the 3D-FGF produced pieces achieve the best strength and stiffness at an NCC loading of 1 wt. With increasing NCC loading, the

Table 10.7 Tensile stiffness, strength, and elongation of the samples

Sample	E_t (GPa)	σ_m (MPa)	ε_m (%)	ε_{tb} (%)
PLAPC	2.17 ± 0.16	31.6 ± 0.3	3.5 ± 0.3	4.8 ± 0.6
PLAPC NCC1-IM	2.38 ± 0.26	41.3 ± 0.5	4.5 ± 0.2	8.9 ± 0.9
PLAPC NCC2-IM	2.25 ± 0.27	37.1 ± 0.1	4.2 ± 0.1	8.7 ± 0.4
PLAPC NCC5-IM	2.23 ± 0.07	36.6 ± 0.5	4.7 ± 0.2	8.6 ± 0.6
PLAPC NCC1-FGF	2.40 ± 0.28	38.6 ± 4.9	4.6 ± 0.2	6.7 ± 0.9
PLAPC NCC2-FGF	2.26 ± 0.20	35.5 ± 0.5	4.4 ± 0.3	8.8 ± 1.7
PLAPC NCC5-FGF	1.42 ± 0.15	34.2 ± 4.1	5.1 ± 0.4	8.0 ± 1.6

tensile stiffness and strength decrease, and the elongation at break also decreases in the injection-molded products. With an NCC reinforcing of 1 wt% and 2 wt%, the tensile stiffness of 3D FGF produced pieces and injection-molded samples are practically the same. With an NCC loading of 2 by weight, the fracture pattern is practically identical for injection-molded and 3D-FGF-printed samples.

For the PC/PLA blend, the tensile stiffness is 2.2 GPa; for the injection-molded pieces, it is higher for the sample with 1 wt.% NCC loading and is 2.4 GPa. At 2 wt% NCC loading it drops to 2.3 GPa, at 5 wt% NCC loading to 2.2 GPa. For the printed 3d-FGF samples, the tensile stiffness is 2.4 GPa at 1 wt% NCC loading, 2.3 GPa at 2 wt% NCC loading, and 1.4 GPa at 5 wt% NCC loading. The strength of the PC/PLA mixture at the tensile test is 32 MPa. It is higher for all injection molded parts with NCC loading, which means that we achieved good interphase interactions between the PC/PLA blend and the NCC surface during reaction mixing. At 1 wt% NCC loading, the tensile strength is enhanced to 41 MPa, at 2 wt% NCC loading it decreased to 37 MPa, and at 5 wt% NCC loading it decreased to 37 MPa. The elongation at break was lowest with the PC/PLA blend and amounted to 4.8%. It was higher for all samples with NCC loading. On average, injection moulded samples had a higher elongation at break than 3D-FGF-printed samples. At 1 wt% NCC loading, the elongation at break was 8.9% for injection molded and 6.7% for 3D printed FGF samples. With an NCC loading of 2% by weight, the elongation at break was 8.7% for the injection moulded and 8.8% for the printed 3D FGF samples. At an NCC loading of 5% by weight, the elongation at break was 8.6% for the injection-molded and 8.0% for the printed 3D FGF samples.

The reinforcement of the PC/PLA matrix with th*e* NCC increases the rigidity, resistance, and ductility of bionanocomposites. Favorable inter-layer synergies among NCC and PC/PLA and homogeneous distribution of NCC in the PC/PLA matrix are ensured by appropriate reaction extrusion parameters at NCC concentrations of 1 wt.% to 5 wt.%. A comparison of two technologies, injection moulding, and additive manufacturing FGF, demonstrates that at an NCC loading of 2 wt%, the tensile strength properties of pieces produced by injection moulding and the 3D-FGF produced pieces are very similar. Reaction extrusion of PC and PLA with proper compatibilizers and reinforced with NCC ensures the mixability of PLA and PC as well as providing excellent surface interaction and distribution of the NCC in the PC/PLA mixture at NCC levels less than 5 wt%. We can conclude that with the combination of compatibilizers, we successfully modified the NCC surface during reaction compounding. We achieved higher rigidity, resistance, and ductility for the injection-molded pieces and also for the 3D FGF-printed samples, which is a good indicator that the samples had good interfacial interactions of the NCC and the PC/PLA mixture. The tensile test measurement distribution showed that injection molding additionally homogenizes the bionanocomposite compared to 3D FGF printing. Also, holding the pressure by compressing the bionanocomposite improves the inter-layer synergies among NCC and PC/PLA mixture thanks to the combination of compatibilizers by reducing the free volume of the bionanocomposite.

10.3.7 Dynamic Mechanical Properties

The curves for the E´ (Table 10.8) have the first drop at the T_g for the PLA matrix at approx. 60 °C. In the cold crystallization region, the E´ of PLA increases and peaks at about 115 °C. After that, the storage modulus decreases. The storage modulus then decreases and has a second drop at the T_g for the PC matrix. The storage modulus curve is similar for all samples, regardless of NCC loading and processing technology, i.e. injection molding or 3D FGF printing. The results for the storage modulus agree very well with the bending test characteristics. For injection molded samples, the storage modulus is higher than for 3D FGF printed samples, and only at an NCC loading of 1 wt% is the E´ higher than for the PC/PLA blend. As the NCC loading increases, the storage modulus decreases for both the pieces produced by injection moulding and the 3D-FGF produced pieces. The E´ for the PC/PLA blend is 1.7 GPa. At 1 wt% NCC loading, the Storega modulus is 1.8 GPa and 1.3 GP for injection molded and 3D FGF printed samples, respectively. With an NCC loading of 2% by weight, the storage modulus is 1.7 GPa or 1.2 GP for the pieces produced by injection moulding and the 3D-FGF produced pieces. At 5 wt% NCC loading, the storage modulus is 1.7 GPa and 1.2 GP for injection molded and 3D FGF produced pieces, respectively.

The loss factor curves (Fig. 10.4) show expressed signals at 68 °C and 160 °C for PLA and PC respectively. As in the DSC measurements, the T_g for PLA is a slight upward move towards elevated temperatures and the glass transition for PC is a slight downward move towards reduced temperatures in the DMA measurements, which can be attributed to the use of the right combination of compatibilizers that allow at least partial miscibility of the PLA and PC matrix. The height of the expressed signal of the tan δ at the first peak, the T_g of the PLA matrix, is lower for all NCC-loaded samples than for the PC/PLA blend, which means that inter-layer synergies among NCC and PC/PLA mixture are achieved in all designs. For the second peak, the glass transition for PC, the peaks of pieces produced by injection moulding with NCC are above those of the PC/PLA matrix, which means that we obtain a slightly elastic deformation and that these samples dampen vibrations better. For the pieces produced by injection moulding, the peak height for the glass transition is highest for PLA at 2 wt% NCC addition and lowest at 5 wt% NCC addition, and the peak heights are very comparable to each other. For the 3D printed FGF samples, the peak height for the glass transition for PLA is highest at an NCC

Table 10.8 Summarized results from the DMA tests for PP

Sample	E′ at 30 °C (GPa)	Tan δ_1 (°C)	Tan δ_2 (°C)
PLAPC	1.72	68.0	160.2
PLAPC NCC1-IM	1.75	69.1	160.5
PLAPC NCC2-IM	1.66	68.5	160.0
PLAPC NCC5-IM	1.65	68.1	159.4
PLAPC NCC1-FGF	1.27	68.4	158.8
PLAPC NCC2-FGF	1.22	68.2	159.6
PLAPC NCC5-FGF	1.23	68.4	158.8

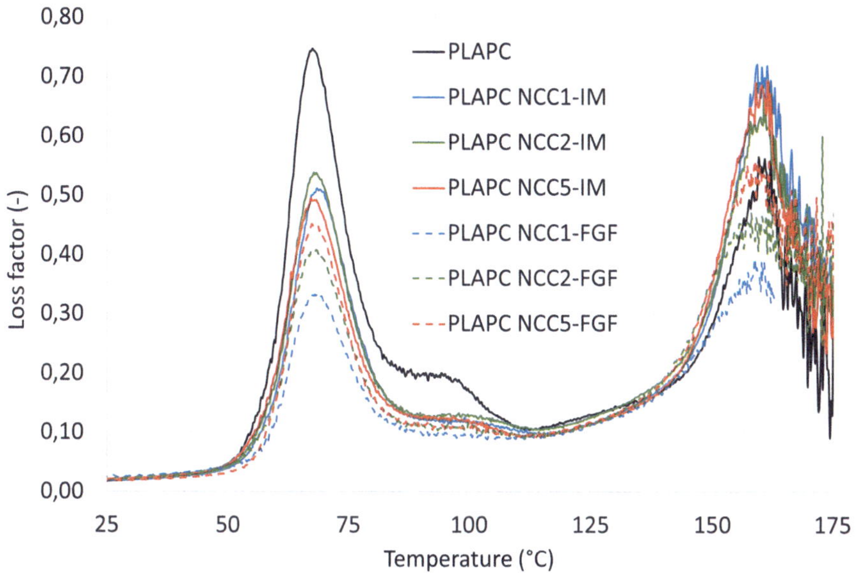

Fig. 10.4 Summarized results of loss factor measured with DMA for all samples

addition of 5 wt% and lowest at an NCC addition of 1 wt%, and the difference in peak heights is obvious. Based on the evaluations, it could be assumed partial NCC agglomeration is occurring in 3D FGF-printed samples with NCC loading of 1 wt.%, partial agglomeration of NCC is obtained, which is the reason for the lower elastic response of samples with NCC loading of 2 wt% and 5 wt%. For injection-molded samples, plasticization additionally homogenizes the bionanocomposite and ensures an excellent arrangement of the NCC nanoparticles in the PC/PLA mixture thanks to a properly selected combination of compatibilizers.

From the DMA tests we can conclude, that high-quality bionanocomposites can be produced by reactive compounding with appropriately selected reactive compounding parameters, even if the NCC is not surface pretreated by a suitable compatibilizer combination. The results agree well with the flexural and DSC tests and show the distinct contrast among injection molded and additive manufactured samples. If we want to obtain stiffer parts, we must use injection molding, if we want to obtain a more elastic response and consequently less damping, we must use 3D FGF printing.

10.3.8 Impact Properties

The reinforcement of PC/PLA with the NCC increased the impact strength at the injection molded bionanocomposites, at the 3D FGF produced pieces, the toughness of the samples with NCC loading is comparable to that of PC/PLA mixtures, considering not only the average value but also the dispersion of the measurements

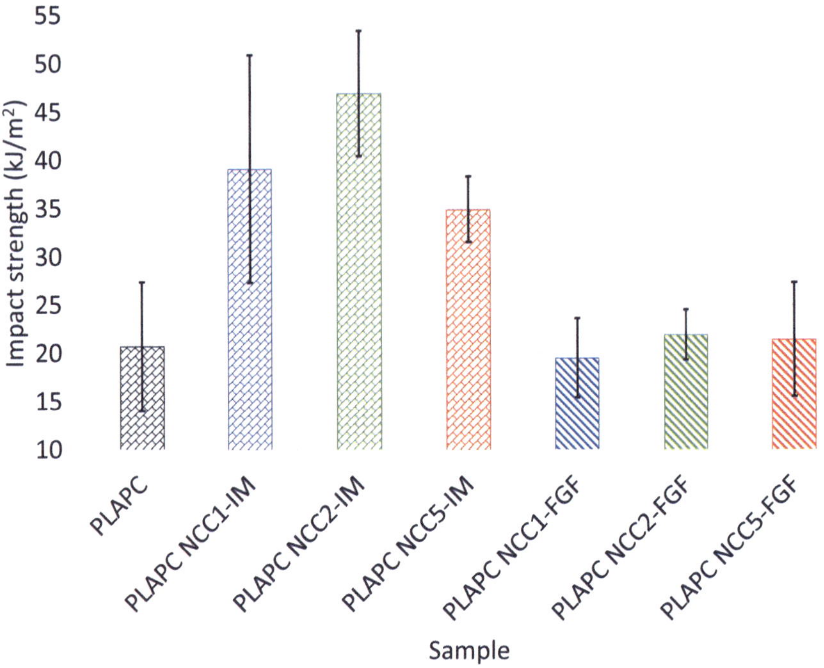

Fig. 10.5 Summarized results of impact strength for all samples

(Fig. 10.5). The highest toughness is achieved in both processing technologies at 2 wt% NCC loading. For injection-molded samples, the lowest toughness is at an NCC loading of 5 wt%, and for 3D-FGF-printed samples at an NCC loading of 1 wt%. For the PC/PLA blend, the toughness is 21 kJ/m^2, at 1 wt% NCC loading 39 kJ/m^2 and 20 kJ/M^2 for injection-molded and 3D-FGF-printed samples respectively. At 2% NCC loading by weight, it is 47 kJ/m^2 and 22 kJ/M^2 for injection-molded and 3D-printed FGF samples, respectively. At an NCC loading of 5 by weight, it is 35 kJ/m^2 and 21 kJ/M^2 for injection-molded and 3D-printed FGF samples respectively.

The reinforcement of the PC/PLA matrix with the NCC increases the toughness of bionanocomposites. Good surface interaction between NCC and PC/PLA and homogeneous distribution of NCC in the PC/PLA matrix are ensured by appropriate reaction extrusion parameters at NCC concentrations of 1 wt.% to 5 wt.% even without NCC surface treatment by combining suitable compatibilizers. To achieve higher toughness of the produced parts, injection molding is a better processing technology than 3D FGF printing, as the holding pressure during injection molding ensures a smaller free volume in the finished parts and thus enables better surface interactions among the NCC skin and the PC/PLA mixture. Better interphase interactions could withstand higher impact forces and achieve higher energies in impact tests.

10.4 Conclusion

We analyzed the thermal properties of reaction compounded PC/PLA /NCC bionanocomposites and compared the mechanical and DMA properties as well as the toughness of injection moulded and 3D-FGF parts. Surface-unmodified NCC was used for the production of bionanocomposites. To achieve good inter-layer interactions among the NCC surface and the PC/PLA mixture, a combination of two compatibilizers, modified TPU and SEBS-g-MA, was used.

It could be shown that the NCC loading of the PC/PLA matrix with the right combination of compatibilizers and reaction compounding raises the glass transition temperature, that the NCC loading acts as a nucleus and causes heterogeneous crystallization, which increases the melting enthalpy and shifts the Tg of PLA towards higher temperatures, while the Tg of PC occurs at lower temperatures since PLA and PC are at least partially miscible due to the added compatibilizers. The OIT increases with increasing NCC loading but simultaneously decreases with multiple reaction mixtures. The melting point temperature increases with longer residence time at higher temperatures and is increased by the NCC loading. The temperature of the first and second degradation increases with NCC loading, the proportion of the first decomposition decreases and the proportion of the second decomposition increases.

A comparison between the parts produced by injection moulding and those produced with 3D-FGF showed that the bending characteristics, dynamic mechanical properties and toughness of the injection moulded parts are better than those of the parts printed with 3D-FGF. The best bending properties and dynamic mechanical properties are achieved at an NCC loading of 1 wt%, with the highest toughness at an NCC loading of 2 wt%. The tensile strength characteristics depend only on the NCC loading and not on the manufacturing technology of the parts. At 1 and 2 wt% NCC loading, they are fully comparable between injection moulded and 3D-FGF printed parts.

Reaction compounding is also a suitable technology for the production of bionanocomposites with added NCC without prior surface bleaching if suitable compatibilizers are used and sufficient shear is ensured during reaction compounding for good homogenization and distribution of the nanoparticles in the PC/PLA matrix. In addition, it has been shown that such processing requires an excellent thermo-oxidatively stabilised matrix, otherwise partial degradation of the matrix will occur due to the high shear forces during processing and multiple processing, which means a long residence time at the processing temperature.

It has been shown that 3D-manufactured parts are still less stable and only comparable to injection-moulded parts to a limited extent. It is therefore still a major challenge to improve the technology of 3D FGF printing in order to make the products even more comparable with injection-moulded parts.

References

Abolore RS, Jaiswal S, Jaiswal AK (2024) Green and sustainable pretreatment methods for cellulose extraction from lignocellulosic biomass and its applications: a review. Carbohydr Polym Technol Appl 7

Averianov IV, Stepanova MA, Gofman IV, Nikolaeva AL, Korzhikov-Vlakh VA, Karttunen M et al (2019) Chemical modification of nanocrystalline cellulose for improved interfacial compatibility with poly(lactic acid). Mendeleev Commun 29(2):220–222

Bacakova L, Pajorova J, Bacakova M, Skogberg A, Kallio P, Kolarova K et al (2019) Versatile application of nanocellulose: from industry to skin tissue engineering and wound healing. Nano 9

Caminero MÁ, Chacón JM, García-Plaza E, Núñez PJ, Reverte JM, Becar JP (2019) Additive manufacturing of PLA-based composites using fused filament fabrication: effect of graphene nanoplatelet reinforcement on mechanical properties, dimensional accuracy and texture. Polymers (Basel) 11(5)

Cartier HÂ, Hu GH. A novel reactive extrusion process for compatibilizing immiscible polymer blends [Internet]. www.elsevier.com/locate/polymer

Chanda M (2024) Compatibilization phenomenon in polymer science and technology: chemical aspects. Adv Ind Eng Polym Res 7:363–372

Chandgude S, Salunkhe S (2021) In state of art: mechanical behavior of natural fiber-based hybrid polymeric composites for application of automobile components. Polym Compos 42:2678–2703

Chávez-Guerrero L, Sepúlveda-Guzmán S, Silva-Mendoza J, Aguilar-Flores C, Pérez-Camacho O (2018) Eco-friendly isolation of cellulose nanoplatelets through oxidation under mild conditions. Carbohydr Polym 181:642–649

Dai J, Bai H, Liu Z, Chen L, Zhang Q, Fu Q (2016) Stereocomplex crystallites induce simultaneous enhancement in impact toughness and heat resistance of injection-molded polylactide/polyurethane blends. RSC Adv 6(21):17008–17015

Deng J, Song Q, Liu S, Pei W, Wang P, Zheng L et al (2022) Advanced applications of cellulose-based composites in fighting bone diseases. Compos Part B Eng 245: 110221

Feng F, Ye L (2011) Morphologies and mechanical properties of polylactide/thermoplastic polyurethane elastomer blends. J Appl Polym Sci 119(5):2778–2783

Formela K, Zedler HA, Tercjak A (2018) Reactive extrusion of bio-based polymer blends and composites–current trends and future developments. Express Polym Lett 12(1):24–57

Garavand F, Nooshkam M, Khodaei D, Yousefi S, Cacciotti I, Ghasemlou M (2023) Recent advances in qualitative and quantitative characterization of nanocellulose-reinforced nanocomposites: a review. Adv Colloid Interf Sci 318:102961

Gauss C, Pickering KL (2023) A new method for producing polylactic acid biocomposites for 3D printing with improved tensile and thermo-mechanical performance using grafted nanofibrillated cellulose. Addit Manuf 61

Han JJ, Huang HX (2011) Preparation and characterization of biodegradable polylactide/thermoplastic polyurethane elastomer blends. J Appl Polym Sci 120(6):3217–3223

Han L, Han C, Dong L (2013a) Effect of crystallization on microstructure and mechanical properties of poly[(ethylene oxide)-block-(amide-12)]-toughened poly(lactic acid) blend. Polym Int 62(2):295–303

Han L, Han C, Dong L (2013b Jan) Morphology and properties of the biosourced poly(lactic acid)/poly(ethylene oxide-b-amide-12) blends. Polym Compos 34(1):122–130

Hashima K, Nishitsuji S, Inoue T (2010) Structure-properties of super-tough PLA alloy with excellent heat resistance. Polymer (Guildf) 51(17):3934–3939

He YS, Zeng JB, Liu GC, Li QT, Wang YZ (2014) Super-tough poly(L-lactide)/crosslinked polyurethane blends with tunable impact toughness. RSC Adv 4(25):12857–12866

Heller BP, Smith DE, Jack DA (2016) Effects of extrudate swell and nozzle geometry on fiber orientation in fused filament fabrication nozzle flow. Addit Manuf 12:252–264

Ho CH, Wang CH, Lin CI, Der LY (2008) Synthesis and characterization of TPO-PLA copolymer and its behavior as compatibilizer for PLA/TPO blends. Polymer (Guildf) 49(18):3902–3910

Hu GH, Li H, Feng LF (2002) Communications to the editor a two-step reactive extrusion process for the synthesis of graft copolymers with polyamides as grafts

Hu GH, Hoppe S, Feng LF, Fonteix C (2007) Nano-scale phenomena and applications in polymer processing. Chem Eng Sci 62(13):3528–3537

Irani-Kolash E, Moshiri-Gomchi N, Talebi-Liasi A, Sabahi S, Bahri-Laleh N, Mehdipour-Ataei S et al (2020) Preparation of an enhanced nanohybrid alloy based on polylactic acid/polycarbonate/nanosilica. Plast Rubber Compos 49(6):263–270

Kanomata K, Fukuda N, Miyata T, Lam PY, Takano T, Tobimatsu Y et al (2020) Lignin-inspired surface modification of nanocellulose by enzyme-catalyzed radical coupling of coniferyl alcohol in Pickering emulsion. ACS Sustain Chem Eng 8(2):1185–1194

Khorsandi D, Jenson S, Zarepour A, Khosravi A, Rabiee N, Iravani S et al (2024) Catalytic and biomedical applications of nanocelluloses: a review of recent developments. Int J Biol Macromol 268

Krishnasamy S, Mohit, Kumar H, Parameswaranpillai J, Sanjay, Rangappa M et al (2024) Interfacial bonding characteristics in natural fiber reinforced polymer composites fiber-matrix interface in biocomposites. (eds. Krishnasamy S, Kumar MH, Parameswaranpillai J, Ramgappa SM, Singchin S). Springer, Singapore

Le Gars M, Dhuiège B, Delvart A, Belgacem MN, Missoum K, Bras J (2020) High-barrier and antioxidant poly(lactic acid)/nanocellulose multilayered materials for packaging. ACS Omega 5(36):22816–22826

Lebarbé T, Grau E, Alfos C, Cramail H (2015) Fatty acid-based thermoplastic poly(ester-amide) as toughening and crystallization improver of poly(l-lactide). Eur Polym J 65:276–285

Li M, Pu Y, Thomas VM, Yoo CG, Ozcan S, Deng Y et al (2020) Recent advancements of plant-based natural fiber–reinforced composites and their applications. Compos Part B Eng 200

Likittanaprasong N, Seadan M, Suttiruengwong S (2015) Impact property enhancement of poly (lactic acid) with different flexible copolymers. IOP Conf Ser Mater Sci Eng 87:012069

Lin L, Deng C, Lin GP, Wang YZ (2015) Super toughened and high heat-resistant poly(lactic acid) (pla)-based blends by enhancing interfacial bonding and pla phase crystallization. Ind Eng Chem Res 54(21):5643–5655

Liu GC, He YS, Zeng JB, Xu Y, Wang YZ (2014) In situ formed crosslinked polyurethane toughened polylactide. Polym Chem 5(7):2530–2539

Liu Z, Luo Y, Bai H, Zhang Q, Fu Q (2016) Remarkably enhanced impact toughness and heat resistance of poly(l -Lactide)/thermoplastic polyurethane blends by constructing stereocomplex crystallites in the matrix. ACS Sustain Chem Eng 4(1):111–120

Long Y, Zhang Z, Fu K, Li Y (2021) Efficient plant fibre yarn pre-treatment for 3D printed continuous flax fibre/poly(lactic) acid composites. Compos Part B Eng 227:109389

Lu X, Wei X, Huang J, Yang L, Zhang G, He G et al (2014) Supertoughened poly(lactic acid)/polyurethane blend material by in situ reactive interfacial compatibilization via dynamic vulcanization. Ind Eng Chem Res 53(44):17386–17393

Matsumoto K, Takemura K, Kitamura R, Katogi H, Tanaka T, Takagi H (2024) Cellulose nanofiber-introduced continuous-ramie yarn-reinforced polylactic acid filament for 3D printing: novel fabrication process and mechanical properties. Compos Part A Appl Sci Manuf 176:107836

Mavai S, Bains A, Sridhar K, Rashid S, Elossaily GM, Ali N et al (2024) Formulation and application of poly lactic acid, gum, and cellulose-based ternary bioplastic for smart food packaging: a review. Int J Biol Macromol 268:131687

Musa AA, Onwualu AP (2024) Potential of lignocellulosic fiber reinforced polymer composites for automobile parts production: current knowledge, research needs, and future direction. Heliyon 10

Nguyen PQK, Panta J, Famakinwa T, Yang RC, Ahmed A, Stapleton M et al (2024) Influences of printing parameters on mechanical properties of recycled PET and PETG using fused granular fabrication technique. Polym Test 132

Phanthong P, Reubroycharoen P, Hao X, Xu G, Abudula A, Guan G (2018) Nanocellulose: extraction and application. Carbon Resour Convers 1:32–43

Plymill A, Minneci R, Greeley DA, Gritton J, Greeley D (1955) Graphene and carbon nanotube PLA composite feedstock graphene and carbon nanotube PLA composite feedstock development for fused deposition modeling development for fused deposition modeling graphene and carbon nanotube PLA composite feedstock development for fused deposition modeling [Internet]. https://trace.tennessee.edu/utk_chanhonoproj/1955

Qian H, Liu J, Wang X, Pei W, Fu C, Ma M et al (2023) The state-of-the-art application of functional bacterial cellulose-based materials in biomedical fields. Carbohydr Polym 300

Rashid AB, Haque M, Islam SMM, Uddin Labib KMR (2024) Nanotechnology-enhanced fiber-reinforced polymer composites: recent advancements on processing techniques and applications. Heliyon 10:e24692

Reich MJ, Woern AL, Tanikella NG, Pearce JM (2019) Mechanical properties and applications of recycled polycarbonate particle material extrusion-based additive manufacturing. Materials 12(10):1642

Shi YY, Zhang WB, Yang JH, Huang T, Zhang N, Wang Y et al (2013) Super toughening of the poly(L-lactide)/thermoplastic polyurethane blends by carbon nanotubes. RSC Adv 3(48):26271–26282

Si WJ, An XP, Zeng JB, Chen YK, Wang YZ (2017) Fully bio-based, highly toughened and heat-resistant poly(L-lactide) ternary blends via dynamic vulcanization with poly(D-lactide) and unsaturated bioelastomer. Sci China Mater 60(10):1008–1022

Souza SF, Mariano M, Reis D, Lombello CB, Ferreira M, Sain M (2018) Cell interactions and cytotoxic studies of cellulose nanofibers from Curauá natural fibers. Carbohydr Polym 201:87–95

Syduzzaman M, Faruque MAA, Bilisik K, Naebe M (2020) Plant-based natural fibre reinforced composites: a review on fabrication, properties and applications. Coatings 10:1–34

Tadi SP, Maddula SS, Mamilla RS (2024) Sustainability aspects of composite filament fabrication for 3D printing applications. Renew Sust Energ Rev 189

Thomas B, Raj MC, Athira BK, Rubiyah HM, Joy J, Moores A et al (2018) Nanocellulose, a versatile green platform: from biosources to materials and their applications. Chem Rev 118:11575–11625

Tuli NT, Khatun S, Rashid AB (2024) Unlocking the future of precision manufacturing: a comprehensive exploration of 3D printing with fiber-reinforced composites in aerospace, automotive, medical, and consumer industries. Heliyon 10

Wang X, Huang L, Li Y, Wang Y, Lu X, Wei Z et al (2024) Research progress in polylactic acid processing for 3D printing. J Manuf Process 112:161–178

Weng Z, Wang J, Senthil T, Wu L (2016) Mechanical and thermal properties of ABS/montmorillonite nanocomposites for fused deposition modeling 3D printing. Mater Des 102:276–283

Xiu H, Huang C, Bai H, Jiang J, Chen F, Deng H et al (2014) Improving impact toughness of polylactide/poly(ether)urethane blends via designing the phase morphology assisted by hydrophilic silica nanoparticles. Polymer (Guildf) 55(6):1593–1600

Zhang CL, Feng LF, Hu GH (2006a) Anionic polymerization of lactams: a comparative study on various methods of measuring the conversion of εcaprolactam to polyamide 6. J Appl Polym Sci 101(3):1972–1981

Zhang XM, Bin XZ, Feng LF, Song XB, Hu GH (2006b) Assessing local residence time distributions in screw extruders through a new in-line measurement instrument. Polym Eng Sci 46(4):510–519

Zhang CL, Feng LF, Gu XP, Hoppe S, Hu GH (2011) Kinetics of the anionic polymerization of ε-caprolactam from an isocyanate bearing polystyrene. Polym Eng Sci:2261–2272

Zhang CL, Feng LF, Hoppe S, Hu GH (2012) Compatibilizer-tracer: a powerful concept for polymer-blending processes. AICHE J 58(6):1921–1928

Zhang C, Zhou T, Gu G, Cai C, Hao D, Zou G et al (2024) Super-tough poly(lactic acid)/silicone rubber thermoplastic vulcanizates: the organic and inorganic synergistic interfacial compatibilization. Int J Biol Macromol 258:129110

Zhao X, Hu H, Wang X, Yu X, Zhou W, Peng S (2020) Super tough poly(lactic acid) blends: a comprehensive review. RSC Adv 10:13316–13368

Blaž Nardin is associated professor at the Faculty of Polymer Technology, Slovenj Gradec, Slovenia. He graduated at the University of Ljubljana, Faculty of Mechanical Engineering, where he obtained also his Master degree and PhD. After the graduation he started to work at Slovenian Tool and Die development Centre, where he was a researcher, head of CAE Department and director. In 2006, he was invited to the industry, where he took over the position of General manager of company Gorenje Orodjarna, d.o.o., where he styed on this position for full 10 years. After that he became vice president for Dishwashing and Cooling in Gorenje Group. In 2020 he joined FTPO as a dean, researcher and teacher. He is lecturing Polymer processing technologies, Production Technologies and Manufacturing of moulds. His research interests include injection moulding and additive manufacturing of thermoplastics, material characterisation and simulation of injection moulding.

Silvester Bolka is senior lecturer at the Faculty of Polymer Technology, Slovenj Gradec, Slovenia. He earned his Bachelor of Polymer Technology in 2010 and Master of Polymer Technology in 2013 from Faculty of Polymer Technology. He has 14 years of teaching and research experience and 20 years of industrial experience in the field of polymer processing research. His research interests include compounding, injection moulding and additive manufacturing of thermoplastics and thermoplastic (nano)composites, recycling and up-cycling of thermoplastics, characterization of polymers and additive manufacturing of tool inserts for injection moulding of recycled thermoplastics up-cycled with thermoset reinforcement.

Advancing Eco-friendly Biomedicine: Sustainable Nanocomposites as Platforms for Genetic Vaccines

11

Asra Hamidi [Jivam] and Masoumeh Poudineh

Abstract

The quest for sustainable solutions in biomedicine has driven the creation of cutting-edge platforms that use eco-friendly nanocomposites for delivering genetic vaccines. This book chapter explores the potential of sustainable nanocomposites as advanced platforms for genetic vaccines, emphasizing their eco-friendly nature and promising applications in modern medicine. Researchers are paving the way for efficient and sustainable vaccination strategies by utilizing nanocomposites, which offer benefits like improved stability, precise release mechanisms, and focused delivery. This chapter delves into the design, synthesis, and characterization of sustainable nanocomposites, highlighting their biocompatibility and biodegradability. Additionally, we explain how integrating genetic material into these nanocomposites enables precise modulation of immune responses, opening new avenues for personalized medicine and disease prevention. This chapter seeks to motivate biomedical researchers and practitioners by thoroughly examining the latest progress and upcoming opportunities, encouraging the use of sustainable nanocomposites in the creation of next-generation genetic vaccines, thereby promoting a greener and healthier future.

A. Hamidi [Jivam] (✉)
Quantum Immunogenix-A Third Eye Network INC – USA, Erbil, Kurdisran Region, Iraq
e-mail: Asra.Mahmood.Reza@gmail.com

M. Poudineh
Zahedan, Sistan and Baluchestan, Iran

R. Malviya, S. Sundram (eds.), *Sustainable Nanocomposites with Green Biomaterials*, Biomaterials, Bioengineering and Sustainability 2,
https://doi.org/10.1007/978-3-031-79110-9_11

Keywords

Sustainable nanocomposites · Eco-friendly biomedicine · Vaccine delivery systems · DNA vaccines · RNA vaccines · Biocompatibility · Controlled release · Targeted delivery · Modulation of immune responses · Cancer immunotherapy

Abbreviations

Ag_2O	Silver oxide
$AgO\text{-}TiO_2$	Silver oxide-titanium dioxide
AP-NPs	Antigen-presenting nanoparticles
CD4+	Cluster of differentiation 4+
CD8+	Cluster of differentiation 8+
CD28	Cluster of differentiation 28
CD80	Cluster of differentiation 80
CO_2	Carbon dioxide
$CoFe_2O_4$	Cobalt ferrite
COVID-19	Coronavirus disease 2019
CpG	Cytosine-phosphate-guanine
CpG oligonucleotides	Cytosine-phosphate-guanine oligonucleotides
CRISPR-Cas	Clustered regularly interspaced short palindromic repeats
DNA	Deoxyribonucleic acid
FeCo	Iron-cobalt
FePt	Iron-platinum
FT-IR	Fourier transform infrared spectroscopy
GFP	Green fluorescent protein
GO	Graphene oxide
GQDs	Graphene quantum dots
LCSS particles	Liquid core–solid shell particles
LNPs	Lipid nanoparticles
MERS-CoV	Middle east respiratory syndrome coronavirus
MHC	Major histocompatibility complex
MHC I	Major histocompatibility complex Class I
mRNA	Messenger RNA
MWs	Molecular weights
NCHs	Nanocomposite hydrogels
NGQDs	Nitrogen-doped graphene quantum dots
NPs	Nanoparticles
NY-ESO antigen	New York esophageal squamous cell carcinoma antigen
pDNA	Plasmid DNA
PEG	Polyethylene glycol
PEGylation	Polyethylene glycolylation (polyethylene glycol coating)
PEI	Poly ethylenimine

pH	Potential of hydrogen
PLA	Poly(lactic acid)
PLGA	Poly(lactic-co-glycolic acid)
RNA	Ribonucleic acid
saRNA	Self-amplifying RNA
srRNA	Self-replicating RNA
TCR	T cell receptor
TGA	Thermogravimetric analysis
TiO_2	Titanium dioxide
Tregs	Regulatory T cells
WHO	World Health Organization

11.1 Introduction

The landscape of modern medicine is undergoing a profound transformation, driven by a collective urgency to embrace sustainable solutions without compromising on efficacy and innovation. Amidst this paradigm shift, the realm of vaccine development stands as a pivotal frontier, where researchers and practitioners are relentlessly pursuing eco-friendly alternatives to traditional vaccine delivery systems. At the forefront of this movement are sustainable nanocomposites, heralding a new era of advanced platforms for genetic vaccines (Anwer et al. 2023).

This book chapter embarks on a journey into the realm of sustainable nanocomposites, unraveling their immense potential in revolutionizing vaccine delivery strategies. Our exploration delves deep into the intricacies of designing, synthesizing, and characterizing these remarkable nanomaterials, while also shedding light on their eco-friendly attributes and promising applications in genetic vaccine delivery.

Traditional vaccine formulations often face challenges related to stability, controlled release, and targeted delivery, necessitating innovative approaches to address these limitations (Wang et al. 2023a). Sustainable nanocomposites emerge as a beacon of hope, offering a multifaceted solution that not only overcomes existing hurdles but also opens doors to unprecedented levels of precision and efficacy in vaccination (Akbarian et al. 2024). Researchers are tapping into the distinctive features of nanocomposites, including increased resistance to degradation, precise release patterns, and targeted delivery to specific cells or tissues, to fully realize the potential of genetic vaccines (Hatae et al. 2023). These nanomaterials serve as versatile carriers, capable of encapsulating genetic material, be it DNA or RNA, and orchestrating a finely tuned immune response for optimal disease prevention and treatment (Wang et al. 2023b).

One of the defining features of sustainable nanocomposites is their intrinsic biocompatibility and biodegradability, aligning seamlessly with the principles of eco-friendly biomedicine. Unlike conventional materials that may pose environmental risks or induce immune reactions, these nanomaterials offer a harmonious synergy with biological systems, ensuring minimal adverse effects while maximizing therapeutic outcomes (Jamalipour Soufi and Iravani 2020). Moreover, the controlled release kinetics afforded by sustainable nanocomposites enable precise modulation of immune responses, a critical aspect in designing tailored vaccination strategies (Hao et al. 2020). Whether it's enhancing antigen presentation, promoting immunogenicity, or fine-tuning immunomodulatory pathways, these nanocarriers empower researchers to sculpt immune responses with unparalleled precision (Cai et al. 2023).

The marriage of sustainable nanocomposites and genetic vaccines holds immense promise across a spectrum of medical domains (Meher et al. 2024). From personalized cancer immunotherapy to prophylactic measures against infectious diseases, the versatility of these platforms knows no bounds (Qamar 2021). By encapsulating genetic material encoding specific antigens or immunomodulatory factors, nanocomposites pave the way for bespoke vaccine formulations tailored to individual genetic profiles and disease susceptibilities (Miao et al. 2023).

As we navigate through the realm of green sustainable nanocomposites and genetic vaccines, it becomes evident that our journey is not only about scientific progress but also about embracing responsibility towards our planet and future generations. The convergence of eco-friendly biomedicine and cutting-edge technology heralds a greener and healthier future, where sustainable solutions form the cornerstone of medical innovation (Ahmad et al. 2024a).

Through a comprehensive review of recent advancements, future prospects, and real-world applications, this chapter aims to inspire researchers, practitioners, and policymakers alike to join hands in leveraging sustainable nanocomposites for developing next-generation genetic vaccines. Together, we can forge a path towards a world where healthcare is not just effective but also sustainable, paving the way for a brighter tomorrow.

11.2 Research Methodology

This study investigates the potential of sustainable nanocomposites as advanced platforms for genetic vaccines, highlighting their environmentally friendly characteristics and promising applications in contemporary medicine. A systematic review was conducted on 204 peer-reviewed scientific articles, books, and theses published between 2018 and 2024. The review focused on genetic vaccine platforms, applications, synthesis, and the benefits of eco-friendly sustainable nanocomposites, specifically as vaccine and drug platforms, and immune modulation. The gathered information was synthesized and analyzed through thematic categorization of challenges and proposed solutions. The study acknowledges potential biases and underscores the necessity for further experimental validation of proposed solutions. Nevertheless, it provides an in-depth and methodical perspective on utilizing sustainable nanocomposites to advance the creation of future genetic vaccines and cancer immunotherapies, thus promoting a greener and healthier world.

11.3 Overview of Genetic Vaccines and Their Significance

Genetic vaccines represent a cutting-edge approach in immunization, employing nucleic acids like DNA or RNA to trigger a targeted immune reaction against particular pathogens. Their distinguished characteristics encompass outstanding stability, robust resistance against denaturation, reproducibility, versatility, ease of purification, cost-effectiveness, a low-risk profile in application, and convenient storage capabilities. These attributes have garnered substantial attention and underscore the exceptional appeal of genetic vaccines in the field of immunization. This innovative class of products holds immense promise for transforming preventive and therapeutic approaches to non-congenital diseases, boasting advantages in efficacy, adaptability, and expedited development. This section delves into the foundational aspects of genetic vaccines, elucidates their diverse categories and mechanisms of action, and underscores their pivotal role in enhancing public health initiatives and disease management (Mondal et al. 2024).

11.3.1 Types of Genetic Vaccines

Genetic vaccines encompass several types, including DNA vaccines, RNA vaccines, and viral vector vaccines. DNA vaccines work by directly inserting plasmid DNA that codes for antigenic proteins into host cells, leading to their expression and an immune response. In contrast, RNA vaccines use messenger RNA to produce antigenic proteins, achieving a comparable immunogenic outcome. Viral vector vaccines, however, rely on altered viruses as carriers to transport genes encoding antigens into host cells, provoking strong immune reactions (Shafaati et al. 2021).

11.3.2 DNA Vaccines

DNA vaccines offer several intrinsic benefits compared to protein-based and cellular vaccines. They leverage transcription and/or translation to display antigens through both MHC I and II pathways. In contrast, protein-based vaccines depend on phagocytosis, while cellular vaccines present antigens immediately after administration (Kozak and Hu 2024). Compared to traditional vaccines, DNA vaccines provide a range of advantages, including cost-effective manufacturing, straightforward production methods, exceptional stability, and an excellent safety profile (Shafaati et al. 2021). Unlike traditional protein antigens that necessitate intricate purification processes, various DNA constructs possess stable physical and chemical characteristics, allowing for easy purification in a single step. This simplifies storage, transportation, and distribution, especially in remote regions. Moreover, DNA vaccines avoid the risk of virulence transformation seen in live attenuated vaccines and bypass the typical adverse effects associated with inactivated cell vaccines (Ledesma-Feliciano et al. 2023). DNA vaccines offer versatility in design, facilitated by the manipulation capabilities of molecular technology. Tailored DNA vaccines can precisely express desired antigens or incorporate genes encoding a variety of antigens into a plasmid, thereby providing immunity against multiple diseases. For example, the inclusion of chimeric cytokine genes as adjuvants in the antigen plasmid can enhance immune responses (Kozak and Hu 2024). Distinct patterns found within DNA constructs can trigger immune responses regardless of the presence of genes encoding antigens (Arakelian et al. 2022). DNA-based immunizations provoke a diverse range of immune reactions, encompassing both humoral and cellular responses. This diversity is particularly vital for combating diseases that demand strong cellular immunity, such as anti-tumor responses (Eskandari et al. 2024). They have proven effective in generating immunity that protects against viral infections (Ledesma-Feliciano et al. 2023). Questions have arisen about the possibility of genetic integration into the host genome associated with DNA-based therapies (Eskandari et al. 2024). However, DNA vaccines are meticulously designed to mitigate potential risks. Typically, the DNA employed is in circular plasmid form, distinct from the linear chromosomes within the host genome, thus diminishing the probability of integration. Moreover, the limited amount of DNA administered and its targeted delivery to specific cell types further diminish integration risks. Extensive research supports the assertion of minimal genetic integration risk (Lu et al. 2024). Progress in adjuvant technologies, such as nanomaterials with self-adjuvant properties, significantly bolster the effectiveness of DNA vaccines. These innovative materials inherently possess adjuvant properties, facilitating antigen delivery, stimulating innate immune responses, regulating immune signaling, and concurrently delivering multiple components to amplify immune responses (Akbarian et al. 2024). Additionally, optimizing delivery systems, including nanocarrier modifications, enhances DNA vaccines'stability, delivery, and overall immune response (Eusébio et al. 2024). In ongoing clinical trials, DNA vaccines show promise in diverse applications, supported by effective delivery systems that maximize immunogenicity and translational success (Ledesma-Feliciano et al. 2023).

There are several types of DNA vaccines, each with unique attributes and applications:

11.3.2.1 Plasmid DNA Vaccines

Plasmid DNA vaccines represent a genetic vaccination approach aimed at eliciting immune responses by introducing specific DNA sequences into host cells. These vaccines comprise circular DNA molecules known as plasmids, carrying genes that encode antigens or immunogenic proteins derived from pathogens. Following administration, host cells internalize the plasmid DNA, serving as templates for antigenic protein synthesis. This intracellular antigen production initiates immune responses, including T cell activation and antibody generation against the targeted pathogen. Plasmid DNA vaccines are recognized for their stability, simplicity in production, and capacity to provoke both cellular and humoral immune responses. They have been investigated for diverse infectious diseases and hold potential in vaccine development endeavors (Eusébio et al. 2021).

11.3.2.2 Minigene DNA Vaccines

Minigene DNA vaccines focus on delivering short DNA sequences encoding specific antigenic epitopes to stimulate targeted immune responses against pathogens. By using smaller, more focused minigenes, these vaccines enhance immunogenicity while reducing the risk of unwanted immune reactions. They show promise in combatting infectious diseases and are being explored as a targeted approach in cancer immunotherapy. Minigene DNA vaccines target tumor-specific epitopes, aiming to activate the immune system against cancer cells while sparing healthy tissues. This personalized strategy offers advantages like potent anti-tumor responses and minimal off-target effects, contributing to the quest for effective cancer therapies. Ongoing research shows promising results in inducing tumor-specific immune responses and improving patient outcomes (Pandya et al. 2023).

11.3.2.3 Heterologous Prime-Boost Vaccines (mRNA-DNA Hybrid Vaccines)

Combining mRNA and DNA technologies, these vaccines encode mRNA sequences within DNA plasmids. These vaccines use a combination of mRNA and DNA to elicit a robust immune reaction. The mRNA component typically primes the immune system, while the DNA component acts as a boost to further enhance and sustain immunity. This approach is being explored in various vaccine development efforts, especially for diseases like COVID-19 (Liu et al. 2024a).

11.3.2.4 Self-replicating DNA Vaccines

Self-replicating DNA vaccines are an innovative vaccination approach that delivers circular DNA containing antigen genetic information into cells. This DNA can replicate independently, ensuring sustained antigen production within the body. This continuous antigen presence triggers a robust and enduring immune response, offering advantages like easy production, stability, and potential for multi-antigen encoding. Self-replicating DNA vaccines show promise in combating various diseases and cancers, marking a significant stride in vaccine technology for preventive healthcare (de Moor 2020).

11.3.2.5 Electroporation-Assisted DNA Vaccines

Electroporation-assisted DNA vaccines use electrical pulses to boost the delivery of DNA vaccines, enhancing their effectiveness. This technique creates temporary pores in cell membranes, allowing for increased uptake of the DNA vaccine and leading to a strong immune reaction. These vaccines offer benefits such as improved immunogenicity, targeted delivery to specific cells, and potential for personalized vaccine development. Research in electroporation-assisted DNA vaccines aims to advance effective vaccination against diseases like infections and cancers (Kisakov et al. 2023).

Each variant of DNA vaccine presents unique benefits, such as simplified production processes, resilience, and the capacity to trigger strong immune reactions. Their adaptability renders them attractive options for addressing a wide array of challenges.

11.3.2.6 Key Takeaways

In summary, DNA vaccines present a transformative approach to immunization, offering advantages such as cost-effectiveness, stability, safety, and the ability to induce robust immune reactions. Their flexible construction allows for tailored protection against multiple diseases, and advancements in adjuvants and delivery systems further enhance their efficacy (Lu et al. 2024). Despite concerns about genetic integration, stringent safeguards ensure their safety, paving the way for their promising role in global health and disease prevention efforts (Pandya et al. 2023).

11.3.3 RNA Vaccines

RNA vaccines employ an innovative method by introducing synthetic RNA molecules, typically messenger RNA (mRNA), into host cells. These RNA molecules contain instructions for synthesizing specific antigenic proteins within the host cells. Upon production, the antigen is displayed on the cell surface, prompting the immune system to detect and initiate a response against the pathogen (Wallis et al. 2019).

The essential elements of RNA vaccines comprise the RNA sequence encoding the antigen, engineered vaccine vectors, and stabilizing agents crucial for maintaining mRNA stability and effectiveness. Following administration, the RNA is enclosed within vectors, aiding in cellular uptake and intracellular transport. Within the cell, ribosomes translate the mRNA into antigenic proteins, prompting an immune response marked by antibody generation and the activation of immune cells (Gebre et al. 2021).

RNA vaccines encompass various types based on their design, delivery methods, and mechanisms of action. The main types of RNA vaccines include:

11.3.3.1 mRNA Vaccines (Messenger RNA Vaccines)

mRNA vaccines utilize engineered mRNA molecules containing specific genetic instructions for encoding antigens or immunogenic proteins derived from pathogens. Upon administration, host cells internalize the mRNA, serving as templates for protein synthesis. Consequently, antigenic proteins are produced, eliciting an immune reaction. Notable examples of mRNA vaccines include COVID-19 vaccines developed by Pfizer-BioNTech and Moderna (Gebre et al. 2021).

11.3.3.2 Self-amplifying RNA Vaccines (saRNA Vaccines)

Self-Amplifying RNA vaccines represent a cutting-edge approach designed to leverage the innate machinery within host cells, facilitating self-replication and thereby augmenting antigen production to elicit robust immune responses. This sophisticated technology, akin to mRNA vaccines, incorporates genetic instructions for antigen production. However, what sets saRNA vaccines apart is their additional encoding of self-replication components, which enable a heightened level of antigen expression (Bloom et al. 2021). The primary benefit of saRNA vaccines is their capacity to achieve heightened levels of antigen expression, resulting in robust immune responses. This characteristic not only boosts vaccine efficacy but also holds the promise of reducing required doses, thereby offering a potential means to optimize vaccine distribution and utilization. Significantly, a multitude of experimental vaccines targeting various infectious diseases leverage the capabilities of self-amplifying RNA technology, highlighting its versatility and potential impact in the field of immunization (Tregoning et al. 2023).

11.3.3.3 Self-replicating RNA Vaccines (Replicon RNA Vaccines)

srRNA vaccines represent a cutting-edge approach in vaccine technology, leveraging RNA replicons capable of self-replication within host cells. This unique characteristic enables srRNA vaccines to amplify antigen expression, leading to robust and durable immune responses. The advantages of replicon RNA vaccines include enhanced antigen production, potential for dose sparing, and prolonged immune stimulation, making them promising candidates for effective immunization strategies. Experimental vaccines targeting viral diseases are increasingly utilizing replicon RNA technology to harness the innate self-replication machinery and enhance immunogenicity against infectious pathogens (Aliahmad et al. 2023).

RNA vaccines are crucial due to their distinct advantages in rapid development, flexibility, safety, induction of robust immune responses, and scalability in manufacturing (Bloom et al. 2021).

However, RNA vaccines, particularly mRNA vaccines, face challenges due to their interaction with lipid nanoparticles (LNPs) and inherent instability, affecting their effectiveness and accessibility. Enhancing stability requires optimizing mRNA structure, screening excipients, and improving manufacturing processes. Challenges include susceptibility to degradation, logistical hurdles in storage and transportation, short-lived immune responses, and optimizing delivery systems for efficient immune cell targeting. Interdisciplinary efforts are needed to overcome these challenges and maximize the potential of RNA vaccines in global vaccination efforts (Cheng et al. 2023).

RNA vaccines, including mRNA, self-amplifying RNA (saRNA), and replicon RNA vaccines, hold immense promise for revolutionizing global immunization strategies. However, they face challenges such as stability issues, delivery optimization, and the need for durable immune responses (Tregoning et al. 2023).

The future of RNA vaccines lies in enhancing stability through mRNA structure optimization, excipient screening, and improved manufacturing processes (Cheng et al. 2023). Advanced delivery systems offer targeted immune cell delivery, maximizing vaccine efficacy. Next-generation RNA vaccine platforms, like saRNA and replicon RNA, promise enhanced antigen expression and prolonged immune stimulation, addressing challenges and improving vaccine impact (Aliahmad et al. 2023).

11.3.3.4 Key Takeaways

Interdisciplinary efforts are crucial to overcoming RNA vaccine challenges and maximizing their potential in global vaccination efforts against diseases. Continued research and innovation will shape the future of RNA vaccines, ensuring their role in global health security and pandemic response (Malik and Waheed 2024).

11.4 The Significance of Sustainable Nanocomposites in Vaccine Delivery

Hybrid nanomaterials, which combine diverse materials at the nanoscale level, are pivotal in advancing biomedical capabilities. Their distinct properties and synergistic effects make them adaptable across numerous domains in biomedicine. These materials find active applications in various fields such as drug platforms, diagnostics, imaging, tissue engineering, cancer therapies, biosensing, gene modulation, antimicrobial strategies, and vaccine distribution. Their utility extends to nano-immunotherapies, organ simulations, biomimicry, neuro-regulation, precise treatments, phototherapy, exosome manipulation, and stem cell innovations. However, unlocking their complete potential requires intensive research, thorough safety assessments, and regulatory clearances for a seamless transition into clinical settings (Kumar and Kumar 2024; Priyadarshi et al. 2024). In the realm of modern medicine and biotechnology, sustainable nanocomposites have emerged as groundbreaking tools in vaccine delivery (Wang et al. 2023a). These nanocomposites, designed for efficient vaccine encapsulation and targeted delivery, play a pivotal role in enhancing vaccine efficacy, stability, and safety (Tan et al. 2022). In this exploration of the latest advancements, we delve into the importance of sustainable nanocomposites in vaccine delivery, considering their impact on immunization strategies, public health initiatives, and environmental sustainability.

11.4.1 Enhanced Vaccine Stability and Shelf Life

Sustainable nanocomposites contribute to minimizing vaccine wastage by safeguarding against environmental stressors (Fan et al. 2024). Temperature fluctuations, a common challenge in many regions, can compromise vaccine potency. Nanocomposite formulations act as protective barriers, mitigating the impact of temperature changes and preserving vaccine efficacy (Kumar et al. 2024). This attribute is particularly crucial for vaccines that require stringent temperature control, such as mRNA vaccines, ensuring their integrity until administration (Cheng et al. 2023).

Moreover, sustainable nanocomposites play a pivotal role in addressing the challenges associated with vaccine distribution in remote or resource-limited areas. The protective properties of these nanocomposites extend to transportation and storage conditions, reducing reliance on complex cold chain infrastructures. This advancement not only improves vaccine accessibility but also contributes to cost-effectiveness and sustainability in healthcare delivery systems worldwide (Malik and Waheed 2024).

Another significant advantage of sustainable nanocomposites is their ability to improve the targeted delivery of vaccines. By incorporating specific targeting ligands or functional groups onto the nanocomposite surface, researchers can achieve precise delivery to immune cells or target tissues. This targeted approach optimizes immune responses, potentially reducing vaccine doses and minimizing adverse effects (Jin et al. 2023).

Sustainable nanocomposites can provide controlled release kinetics, ensuring gradual antigen release at the target site and extending immune system stimulation. This method of controlled release aids in developing long-lasting immune responses and diminishes the need for frequent vaccine administrations, thereby optimizing immunization programs and healthcare resource utilization (Kadkhodaie Kashani et al. 2024).

Furthermore, sustainable nanocomposites facilitate the development of novel vaccine formulations, including next-generation mRNA and DNA vaccines. These advanced vaccines hold immense promise for combating emerging infectious diseases and personalized immunotherapies (Fonseca et al. 2024). Nanocomposite-based platforms play a pivotal role in enhancing the delivery of genetic material, ensuring efficient cellular uptake, and promoting sustained antigen expression for prolonged immune stimulation (Ding et al. 2024a).

11.4.2 Adjuvant Properties and Immune Modulation

In addition to serving as vaccine carriers, sustainable nanocomposites can possess inherent adjuvant properties or be engineered to incorporate immunostimulatory molecules. Adjuvants are essential for enhancing immune reactions to vaccines by activating innate immune pathways, promoting antigen presentation, and modulating immune cell activation and differentiation (Sharma and Otto 2024). By integrating adjuvant functionalities into nanocomposite vaccine formulations, researchers can fine-tune immune responses, achieve desired immune outcomes (e.g., targeting tumor-associated macrophages), and enhance vaccine efficacy against challenging pathogens or antigens (Jain et al. 2023; Miao et al. 2023). This immune modulation capability is essential for developing vaccines with broad protective coverage, including against emerging infectious diseases and antigenically diverse pathogens (Nordin et al. 2023).

11.4.3 Reduced Environmental Impact and Biodegradability

Sustainable nanocomposites hold promise in vaccine delivery not only due to their biomedical benefits but also their potential to align with environmental sustainability goals. While not all nanocomposites are inherently eco-friendly, their design and composition can be tailored to minimize environmental impact (Osman et al. 2024). By integrating biocompatible and biodegradable components sourced from renewable materials, including plant-based polymers, sustainable nanocomposites can significantly lower the environmental impact associated with vaccine delivery systems. These materials are meticulously chosen to ensure compatibility with biological systems while fostering environmental stewardship (Tilwari and Saxena 2022). Furthermore, eco-friendly synthesis methods and manufacturing processes are instrumental in increasing the sustainability of nanocomposites (Jamalipour Soufi and Iravani 2020). Techniques such as green chemistry principles, solvent-free synthesis, and recycling approaches contribute to reducing waste generation and energy consumption throughout the nanocomposite lifecycle (Saddique et al. 2024; Hadi and Ali 2024).

The intentional design of sustainable nanocomposites also extends to end-of-life considerations, where materials can be engineered for biodegradation or recyclability, minimizing long-term environmental impact (Maham et al. 2020). This approach fosters a circular economy model, where resources are utilized efficiently and waste is minimized (Coltelli et al. 2023). Moreover, the design of biodegradable nanocomposites ensures that they undergo degradation and clearance within the body without generating harmful byproducts or long-term environmental residues. This eco-friendly approach aligns with global initiatives for sustainable healthcare practices and green technology development, fostering a more environmentally conscious approach to vaccine delivery and biomedical innovations (Anju et al. 2020).

11.4.4 Obstacles and Strategic Directions

In spite of the promising benefits of sustainable nanocomposites in vaccine delivery, several challenges and avenues for further research exist. These include optimizing nanocomposite formulations for stability, safety, and scalability, addressing concerns regarding immunogenicity and potential adverse reactions, and advancing nanotechnology-based manufacturing processes (Hatae et al. 2023; Ravinayagam and Jermy 2022).

Additionally, ongoing research focuses on exploring novel nanocomposite materials, advanced nanoparticle engineering techniques, and integration with cutting-edge drug delivery systems (such as hydrogels or microneedle patches) to enhance vaccine delivery efficiency and therapeutic outcomes (Ahmad et al. 2021; Chen et al. 2023a). Collaboration between multidisciplinary fields, including nanotechnology, immunology, materials science, and pharmacology, is essential for driving innovation and translating research findings into practical, impactful solutions for global vaccination challenges (Chatterjee et al. 2021).

11.4.5 Key Takeaways

Sustainable nanocomposites represent a significant advancement in vaccine delivery strategies, providing a blend of biomedical efficacy, environmental sustainability, and targeted immunization capabilities (Miao et al. 2023). Their relevance spans various applications, including enhancing vaccine stability, immune modulation, personalized vaccine formulations, and advancing global immunization initiatives (Nordin et al. 2023).

As we continue to explore the potential of nanotechnology and sustainable materials in vaccine science, the adoption of nanocomposite-based vaccine delivery systems shows immense promise for addressing current and future healthcare challenges (Shahid et al. 2024). These challenges include infectious disease outbreaks, vaccine accessibility, and environmental stewardship (Kuperkar et al. 2024; Malik et al. 2023; Tilwari and Saxena 2022). Embracing innovation in sustainable nanocomposites sets the stage for a healthier and more resilient world through effective and sustainable vaccination strategies (Chatterjee et al. 2021).

11.5 Environmental Advantages of Eco-friendly Sustainable Nanocomposites

The nanocomposites, regarded as eco-friendly materials for being fully biodegradable, renewable, and non-hazardous, have sparked considerable enthusiasm for promoting sustainable development efforts (Sapuan et al. 2023). Polymer nanocomposites utilize nanomaterials for reinforcement, resulting in a substantial interfacial area per unit volume, with minimal spacing between the filler nanoparticles and the polymer matrix (Hiremath et al. 2021). The molecular interactions between the filler particles and the matrix endow polymer nanocomposites with distinctive properties that are absent in conventional polymers or traditional macro-composites (Priyadarshi et al. 2024; Ingole et al. 2024). Nevertheless, nanoparticles, nanotubes, nanofilms, nanofibers, nanoflakes, and similar materials incorporated into nanocomposites have the potential to pose significant health risks and cause substantial environmental pollution (Harun-Ur-Rashid et al. 2023). This section explores the environmental benefits of eco-friendly sustainable nanocomposites, highlighting their potential to address key challenges while meeting the demands of modern applications.

11.5.1 Biodegradability, Renewability, and Eco-toxicity

Polymer nanocomposites, known as environmentally friendly materials due to their complete biodegradability, renewability, and benign nature, have garnered significant interest in advancing sustainable development initiatives (Sapuan et al. 2023). Polymer nanocomposites that utilize nano-materials for reinforcement result in a substantial interfacial area relative to their volume, with minimal spacing between the filler nanoparticles and the polymer matrix (Ingole et al. 2024; Harun-Ur-Rashid et al. 2023). The intricate molecular interactions occurring between filler particles and the matrix impart polymer nanocomposites with distinctive characteristics not found in conventional polymers or traditional macro-composites (Priyadarshi et al. 2024). Conversely, nanoparticles, nanotubes, nanofilms, nanofibers, nanoflakes, and similar materials incorporated into non-green nanocomposites have the potential to pose significant health risks and cause substantial environmental pollution (Yashas et al. 2021). Therefore, there is a pressing need to prioritize the development of sustainable green nanocomposites, ensuring their performance while adhering to rigorous environmental and safety standards.

Green nanocomposites offer unique advantages for medical and industrial applications owing to their distinct chemical and physical characteristics. These polymers are produced by incorporating nanofillers into biopolymer matrices. Fillers like cellulose-based materials and carbon quantum dots enhance polymer properties (Anwer et al. 2023; Li et al. 2024a). Sustainable bio-nanocomposites serve as an eco-friendly substitute for conventional vaccine platforms, reducing reliance on toxic viral and non-viral vectors (Malik et al. 2023).

For illustration, clay minerals like palygorskite, halloysite, sepiolite, kaolinite, and montmorillonite have acquired gaining traction in scientific research and industries due to their naturally occurring properties, which offer non-toxic, eco-friendly, and cost-effective advantages (Biddeci et al. 2023). These minerals exhibit unique crystal structures and physicochemical properties, especially when reduced to nanometric dimensions, leading to distinctive chemical and physical characteristics compared to traditional materials. They find potential applications across various fields as alternatives to toxic or expensive synthetic materials (Alahakoon et al. 2023). In pharmaceutical settings, clay minerals find widespread use as excipients, serving as both inactive ingredients and active substances. They play multifaceted roles, enhancing organoleptic and physical properties, acting as lubricants, efficiently dispersing active compounds, and fulfilling functions as gelling, emulsifying, and thickening agents. Additionally, their distinctive properties, including specific surface area, high adsorption capacity, chemical inertness, cation exchange capacity, interlayer reactions, and low toxicity, render them invaluable for various pharmaceutical applications (Biddeci et al. 2023). With their abundance in nature, affordability, and eco-friendliness, clay minerals can be modified or transformed structurally to incorporate different substances and create functional materials, including drug carriers and vaccine delivery systems, thereby enhancing their utility in various healthcare solutions (Ahmadizadegan and Esmaielzadeh 2024).

11.5.2 Reduced Carbon Footprint

The synthesis and production of eco-friendly nanocomposites align with green chemistry principles, energy-efficient processes, and renewable resources, resulting in reduced carbon emissions compared to conventional materials (Barua et al. 2022). Their application as carriers for genetic vaccines not only improves vaccine efficacy and safety but also promotes sustainable healthcare practices by minimizing environmental impact (Salman et al. 2023).

Various crucial methodologies are utilized in crafting eco-friendly sustainable nanocomposites, encompassing solvent-free synthesis, microwave-assisted synthesis, and the utilization of bio-based precursors. These approaches not only reduce carbon emissions but also augment the biocompatibility and biodegradability of the nanocomposites, rendering them ideal contenders for genetic vaccine delivery systems (Zhang et al. 2024a; Malode et al. 2024). Energy-efficient processes such as microwave-assisted synthesis and electrospinning play a critical role in diminishing the carbon footprint associated with nanocomposite production. These processes demand less energy input and result in lower carbon emissions compared to traditional methods. Furthermore, incorporating renewable energy sources further bolsters the sustainability of nanocomposite synthesis (Saadh et al. 2024).

Eco-friendly sustainable nanocomposites exhibit exceptional characteristics as carriers for genetic vaccines, including enhanced stability, controlled release, and precise targeting (Malik et al. 2023). These nanocomposites enable fine-tuning of immune responses, ultimately leading to heightened vaccine efficacy and safety. Additionally, their eco-friendly attributes contribute significantly to sustainable healthcare practices by minimizing environmental impact throughout the vaccine lifecycle (Constantino et al. 2023).

The adoption of eco-friendly sustainable nanocomposites as genetic vaccine platforms signifies a significant advancement in healthcare solutions focused on reducing the carbon footprint (Silva et al. 2023). By synthesizing these nanocomposites in accordance with green chemistry principles and energy-efficient processes, the medical sector can progress towards achieving sustainable development goals while simultaneously improving vaccine efficacy and safety. Ongoing research and innovation in this area are vital for driving sustainable healthcare practices and fostering environmental stewardship (Raj et al. 2022; El-Sayed and Kamel 2020).

11.5.3 Key Takeaways

The environmental advantages of eco-friendly sustainable nanocomposites are substantial, stemming from their biodegradability, renewability, and non-toxic properties (Feng et al. 2021). These materials, renowned for their complete biodegradability, are reinforced with nanomaterials, resulting in a significant interfacial area per unit volume and distinctive molecular interactions. Despite potential hazards to health and the environment posed by nanoparticle integration, the emphasis on sustainable green nanocomposites remains imperative (Ahmad et al. 2024b). They offer distinct advantages in medical and industrial applications, exemplified by clay minerals like palygorskite and montmorillonite, which are eco-friendly and cost-effective (Alahakoon et al. 2023). Moreover, their minimized carbon footprint resulting from synthesis and production, which adhere to green chemistry principles and employ energy-efficient techniques, renders them ideal for genetic vaccine delivery systems, advocating for sustainable healthcare practices and environmental stewardship (Silva et al. 2023).

11.6 Development and Creation of Environmentally Friendly Sustainable Nanocomposites

Within the intricate realm of healthcare innovation, the development and creation of eco-friendly sustainable nanocomposites emerge as transformative tools, especially in vaccine platforms (Salman et al. 2023). This section delves into the precise engineering of environmentally conscious nanocomposites tailored to enhance vaccine delivery, stability, and efficacy. By harnessing renewable polymers and eco-friendly synthesis routes, researchers are navigating a new frontier where sustainability meets cutting-edge healthcare solutions, promising a greener, more effective approach to immunization and disease prevention (Ahmad et al. 2021).

11.6.1 Designing Sustainable Nanocomposites for Genetic Vaccine Platforms

Advancements in in-silico approaches and technology for designing and synthesizing stable nanocomposites have facilitated the creation of vaccine platforms with precise characteristics, including surface chemistry, molecular size requirements, and unique physicochemical functional properties (Chouhan and Rangi 2023; Hassanzadeh 2020; Rajput et al. 2023). Incorporating effective drug molecules like doxorubicin and CpG oligonucleotides into these vectors enhances their efficiency in drug delivery and cellular absorption, consequently amplifying therapeutic effects while reducing off-target effects. This advancement has not only stabilized genetic vaccines on nanocomposite-based platforms but also improved their ability to penetrate target cells, thereby enhancing immune efficiency (Ding et al. 2024a; Mehta et al. 2022). Vectors based on stable nanocomposites can also serve as biomarkers, thus contributing to the advancement of vaccine research (Qamar 2021; Huda et al. 2021).

For illustration, polymer-based nanocomposite hydrogels (NCHs) have emerged as effective vehicles for controlled drug release, drawing significant attention. These nanocomposites blend the characteristics of nanoparticles (NPs) and hydrogels, showcasing enhanced capabilities compared to traditional hydrogels (Shahid et al. 2024).

As a second example, carbon-based materials like graphene oxide (GO) and graphene quantum dots (GQDs) have attracted considerable interest as potential platforms (Ahmadi et al. 2023). GO, characterized as the oxidized form of graphene sheets, has demonstrated solubility in water and versatile interactions with drugs and nucleic acids, establishing its utility as a drug delivery and biosensor platform. However, despite its promising transfection efficiency, GO's significant toxicity has constrained its practical in vivo applications (Hassan et al. 2024; Akmal et al. 2024).

Conversely, GQDs with their distinct properties, such as a size range of 1–10 nm and functional groups like hydroxyl, carbonyl, alcohol, and carboxylic groups, present a compelling alternative (Ding et al. 2024b). These nanoparticles exhibit relatively low toxicity, facilitating their application in nanomedicine and bio-imaging (Akmal et al. 2024). GQDs have been explored for their ability to interact with α-synuclein fibrils and demonstrate low cytotoxicity, a crucial factor in biomedical applications (Hassan et al. 2024). Furthermore, their capacity to load various drug molecules and genetic vaccines through physicochemical conjugation underscores their potential in therapeutic interventions (Malode et al. 2024).

The synthesis of positively charged N-doped graphene quantum dots (NGQDs) for genetic vaccine delivery represents a significant advancement in this domain.

Utilizing microwave-assisted hydrothermal reactions between polyethyleneimine (PEI) and citric acid precursors, NGQDs have shown exceptional transfection efficiency at considerably lower concentrations compared to their cytotoxic levels (Luisetto et al. 2022).

NGQDs (N-doped graphene quantum dots) nanocomposites are recognized as eco-friendly and sustainable nanocomposites (Khan et al. 2024). Derived from graphene, a naturally abundant and environmentally friendly material consisting of carbon atoms arranged in a hexagonal lattice, NGQDs offer several advantages over other carbon-based materials including graphene oxide (GO), including lower toxicity levels (He et al. 2024a). This reduced toxicity makes NGQDs particularly promising for biomedical applications, especially in vaccine platforms (Wang et al. 2023b; Luisetto et al. 2022). Moreover, NGQDs demonstrate high biocompatibility, ensuring they are well-tolerated by biological systems and minimizing adverse reactions or immune responses (Dejpasand et al. 2024) (Fig. 11.1).

The versatility of NGQDs extends to their ease of functionalization with various biomolecules, including DNA, RNA, or proteins, enhancing their specificity and efficacy as vaccine delivery platforms (Dejpasand et al. 2024). These characteristics, combined with their ability to transfect genetic vaccines without requiring additional chemical modifications, position NGQDs as cost-efficient options for large-scale synthesis—a critical consideration for future clinical applications (Malode et al. 2024).

For example, in a study, researchers synthesized N-doped graphene quantum dots (NGQDs) for gene transfection of mRNA and plasmid DNA (pDNA). Created from PEI and citric acid via microwave-assisted hydrothermal reaction, NGQDs effectively complex with genes due to their positive charge. They exhibited blue fluorescence, stability at room temperature, and low toxicity. Gene transfection experiments with mRNA encoding GFP showed NGQDs' efficiency was comparable to the commercial agent Lipofectamine. Flow cytometry analysis revealed NGQDs achieved up to 50% transfection efficiency, surpassing Lipofectamine in some cases. These results suggest NGQDs could serve as superior gene delivery platforms. Figure 11.2 illustrates the flow cytometry analysis of HeLa cells transfected with various groups, highlighting the enhanced efficiency of NGQDs (Ahn et al. 2021).

In summary, NGQDs offer a promising path for the development of eco-friendly and sustainable vectors, particularly valuable in vaccine platforms, where safety, biocompatibility, and functionalization are paramount (Khan et al. 2024).

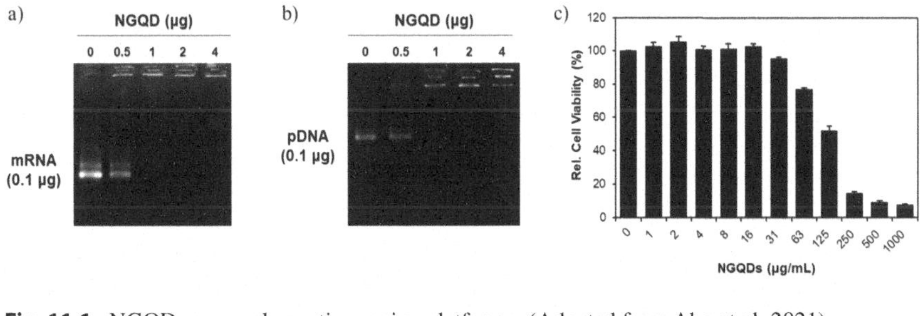

Fig. 11.1 NGQDs as novel genetic vaccine platforms. (Adapted from Ahn et al. 2021)

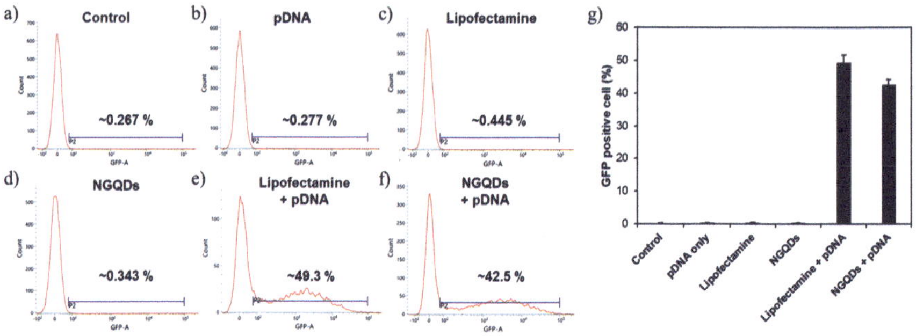

Fig. 11.2 Depicts the flow cytometry assessment of HeLa cells following a 24-h transfection with various groups: (**a**) control, (**b**) pDNA solely, (**c**) Lipofectamine alone, (**d**) NGQDs alone, (**e**) Lipofectamine + pDNA complex, (**f**) NGQDs + pDNA complex groups, and (**g**) the transfection efficacy of pDNA for each group. (Adapted from Ahn et al. 2021)

When considering the design of environmentally-friendly sustainable nanocomposites for genetic vaccine platforms, several critical principles come into play. These include:

11.6.1.1 Biocompatibility and Safety

Green sustainable nanocomposites prioritize biocompatibility, ensuring compatibility with biological systems, reduced cytotoxicity, and minimal immune reactions (Alshangiti et al. 2023). Surface modifications and polymer coatings further enhance biocompatibility, supporting safe vaccine delivery and modulation of immune responses (Rajput et al. 2023).

11.6.1.2 Controlled Release Kinetics

The design of nanocomposite platforms focuses on achieving controlled release kinetics for genetic vaccines, optimizing antigen presentation, immune cell activation, and maintaining sustained immune stimulation. Tunable release profiles enhance vaccine efficacy, reduce dosing frequency, and improve patient compliance (Nguyen Dang et al. 2024).

For example, In the realm of nanocomposite vaccine platforms, biodegradable core-shell microparticles stand out as promising biomaterials. They offer versatile applications that can significantly enhance vaccine efficacy and patient adherence, a critical consideration for areas with limited healthcare infrastructure (Kuperkar et al. 2024). These microparticles facilitate vaccine administration through a single injection, transdermal delivery of vaccines, cancer immunotherapy, and oral drug release triggered by pH changes, overcoming the constraints of conventional drug delivery methods (Xiao et al. 2023).

Central to their potential is the ability to control drug release, crucial for diseases requiring multiple treatments to improve patient adherence (Liu et al. 2024b). Studies reveal a global patient adherence rate of 50%, underscoring the need for advanced delivery systems (Xu et al. 2023a). Furthermore, these microparticles show promise in simplifying dosing regimens, reducing healthcare visits, and combating infectious disease outbreaks, especially in resource-constrained settings (Kuperkar et al. 2024).

Researchers delved into understanding the microstructural changes during hydrolysis-based degradation in core-shell microparticles, crucial for designing nanocomposite vaccine platforms with controlled release kinetics. They investigated a model cubic core-shell microparticle featuring a cap and base composed of varying molecular weights (MWs) of PLGA. The membrane, which was rapidly degrading, turned opaque and became distorted during hydrolysis, creating an area with pore formation that triggered a pulsatile release of the drug. Mechanical testing showed a decline in the material's properties as degradation occurred (Sarmadi et al. 2022).

Pore formation and cap deformation are key factors controlling release kinetics in vaccine platforms (Chen et al. 2023a). This study showed that a regionally controlled degradation pattern resulted in a delayed pulsatile release through interconnected pores (Sarmadi et al. 2022). The composition of particles, especially polymer properties like molecular weight and end groups, significantly influences release time rather than size or shape, which is crucial for customizing release kinetics (Graf et al. 2023). The incorporation of polyethylene glycol (PEG) as a pore-forming agent changes the release kinetics, leading to an earlier release caused by increased water absorption and hydrolysis (Sarmadi et al. 2022). Figure 11.3 illustrates how different design factors affect the release rate of biodegradable core-shell microparticles in this study.

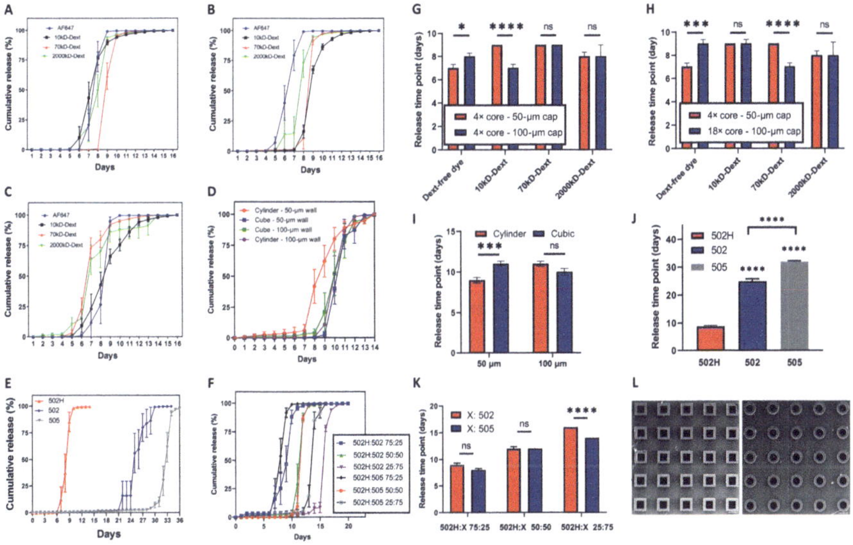

Fig. 11.3 Shows an examination of how various design factors impact the release rates in biodegradable core-shell microparticles. These include variations in core size, cap thickness, cap material, and blend of polymers used. (*), (***), and (****) denote statistically significant differences with P values less than 0.05, 0.001, and 0.0001, respectively. (Adapted from Sarmadi et al. 2022)

Overall, this work contributes significant insights into understanding and controlling pulsatile release kinetics in biodegradable microparticles, providing valuable insights for the development of advanced drug platforms, including nanocomposite vaccine platforms.

11.6.1.3 Targeted Delivery Strategies

Functionalized nanocomposites incorporate targeting ligands, antibodies, or peptides for precise vaccine delivery to immune cells, lymphoid tissues, or antigen-presenting cells (Cao et al. 2022). Targeted delivery enhances vaccine uptake, immune cell activation, and antigen-specific immune responses (Dowaidar 2024).

For illustration, a study on targeted delivery strategies using microfluidics to create uniform liquid core–solid shell particles (LCSS particles) reveals significant advancements in vaccine delivery. The method ensures enhanced reliability and uniformity, which are critical for consistent vaccine efficacy. Nanocomposite particles produced through this approach demonstrate delayed release capabilities, allowing for pulsatile release after a lag phase, essential for precise vaccine targeting. The research underscores the versatility of nanocomposites in customizing the timing of immune responses through particle formulation adjustments. Particularly noteworthy is the enhanced immunogenicity and efficacy achieved by incorporating a priming vaccine alongside a booster dose encapsulated within the particles, demonstrating the effectiveness of targeted delivery using sustainable nanocomposites within a single-injection approach (Guyon 2023).

11.6.2 Principles of Eco-friendly Sustainable Nanocomposite Design

The principles of eco-friendly sustainable nanocomposite design encompass a comprehensive approach aimed at optimizing material performance while minimizing environmental impact. Central to this concept lies the choice of environmentally friendly, renewable, and biodegradable materials as foundational components for nanocomposite structures. These materials not only diminish the environmental impact linked to production alongside enhance the overall sustainability of the final product (Coltelli et al. 2023). Sustainable nanocomposite design emphasizes efficient resource use, waste reduction, and recycling strategies throughout the manufacturing process to promote circularity and minimize the ecological footprint (Saadh et al. 2024).

Moreover, these design principles prioritize functionality and performance enhancement through precise control of nanostructure morphology, composition, and interfacial interactions. This entails harnessing nanoscale phenomena like quantum confinement effects, surface functionalization, and tailored interfaces to attain specific attributes, including enhanced mechanical strength, thermal stability, and precise control over release kinetics (Chouhan and Rangi 2023; Hassanzadeh 2020). The incorporation of renewable nanofillers, bio-derived polymers, and green synthesis methods strengthens the sustainability aspect of nanocomposite design, opening avenues for environmentally friendly solutions across various sectors such as biomedicine, energy storage, and environmental remediation (Barua et al. 2022; El-Sayed and Kamel 2020; Ruman et al. 2024).

These principles are integral to green sustainable nanocomposite design, encompassing a range of strategies and technologies aimed at achieving both environmental friendliness and enhanced material performance:

11.6.2.1 Material Selection

The foundation of green sustainable nanocomposite design lies in the careful selection of materials with minimal environmental impact and maximal functionality (Feng et al. 2021; Bhat et al. 2024). Biocompatible and biodegradable polymers, including chitosan, cellulose derivatives, and poly (lactic acid) (PLA), are frequently chosen due to their renewable properties and minimal toxicity (Anwer et al. 2023). Inorganic nanomaterials, such as clay minerals (e.g., montmorillonite), titanium dioxide (TiO_2), and carbon nanotubes, contribute to the structural integrity, mechanical strength, and functional properties of nanocomposites (Jackson et al. 2024).

11.6.2.2 Lifecycle Assessment

Sustainable nanocomposite design incorporates lifecycle assessment methodologies to comprehensively evaluate the environmental footprint of materials, processes, and end-of-life disposal strategies (Khosravi et al. 2024). Various factors, including energy consumption, greenhouse gas emissions, waste generation, and material recyclability, are thoughtfully evaluated to reduce environmental impact and advance circular economy principles (Shanmugam et al. 2021). Life cycle thinking ensures that sustainable nanocomposites are not only efficient during use but also sustainable throughout their production and disposal stages, contributing to a more environmentally conscious approach in material design and utilization (Saddique et al. 2024).

11.6.2.3 Green Chemistry Principles

The adoption of green chemistry principles is integral to sustainable nanocomposite design, emphasizing the use of benign solvents, renewable resources, and environmentally friendly synthesis routes (Hsu et al. 2024; Verma et al. 2024). Green solvents like water or supercritical CO_2 reduce chemical hazards and waste generation during nanocomposite fabrication (Pandiyarajan et al. 2024).

Biomimetic strategies, drawing inspiration from the design principles found in nature, including the emulation of efficient and resource-conserving processes observed in biological systems, contribute to the advancement of sustainability and innovation in nanocomposite development (Afshan and Bari 2023).

11.6.2.4 Functional Performance

Sustainability in nanocomposite design is achieved without compromising functional performance but rather aims to enhance it through synergistic material combinations and tailored properties (Bhat et al. 2024). Sustainable nanocomposites have showcased enhancements in mechanical strength, thermal stability, barrier properties, and functional attributes, including catalytic activity and optical characteristics, surpassing those of conventional materials (Hassan et al. 2021). These expanded capabilities enable diverse applications, spanning from lightweight and durable materials utilized in the automotive and aerospace industries, to advanced biomedical devices, and highly sensitive environmental sensors (Sharma et al. 2024a; Pengsomjit et al. 2024).

11.6.3 Synthesis Methods and Materials Used

Several synthesis methods and materials have been developed for the production of sustainable nanocomposites:

11.6.3.1 Green Synthesis Approaches

Green synthesis methods for sustainable nanocomposites encompass solvent-free processes, water-based formulations, and bioinspired assembly techniques (Raj et al. 2022). Utilizing green solvents, natural stabilizers, and enzymatic reactions helps minimize environmental impact and reduce chemical hazards (Osman et al. 2024).

11.6.3.2 Biomimetic Nanocomposites

Biomimetic nanocomposites mimic biological structures and functions by incorporating bioactive molecules, cell-targeting motifs, or immune-modulating agents. These biomimetic design principles enhance nanocomposite functionality and improve therapeutic outcomes (Chen et al. 2023b; Wang et al. 2023c).

11.6.3.3 Nanostructure Engineering

Nanostructure engineering techniques such as self-assembly, layer-by-layer deposition, or electrospinning allow for precise control over nanocomposite morphology, surface properties, and functionalization (Schneider et al. 2021; Sonowal and Gautam 2024). These nanostructured designs optimize vaccine delivery, immune cell targeting, and therapeutic performance (Ahmadi et al. 2023).

11.6.4 Polymer-Based Nanocomposites

Polymer-based nanocomposites represent sophisticated materials that integrate polymers with nanoscale fillers, amplifying their properties and performance (Li et al. 2024a). These nanocomposites typically consist of a polymer matrix, such as plastics, rubbers, or resins, reinforced with nanoparticles or nanofillers, such as nanoparticles, nanotubes, or nano-clays (Hassan et al. 2021; Wu et al. 2024).

There are multiple techniques available for producing polymer-based nanocomposites, with two of them being especially notable:

11.6.4.1 Solution Mixing

This technique involves blending polymer solutions with nanomaterial dispersions, which are then subjected to solvent evaporation or precipitation to create nanocomposite films or membranes (Salabat and Mirhoseini 2022). The selection of solvent is pivotal as it influences compatibility, uniform dispersion, and interfacial adhesion between polymer matrices and nanofillers (Hiremath et al. 2021; Li et al. 2024a).

11.6.4.2 In-situ Polymerization

Polymerization techniques like emulsion polymerization or melt blending allow for the direct synthesis of nanocomposites by integrating nanofillers during polymer formation. This method ensures thorough mixing at the molecular level, resulting in enhanced mechanical properties, thermal stability, and functional improvements (Wu et al. 2024; Salabat and Mirhoseini 2022).

Nanocomposites based on polymers are extensively utilized across various applications, particularly notable in the realm of vaccine delivery systems (Kuna et al. 2024). By integrating nanofillers into polymer matrices, researchers have the ability to craft nanocomposite materials endowed with controlled release properties, enhanced stability, and improved targeting for vaccine delivery. These systems provide benefits such as prolonged release kinetics, preservation of vaccine antigens, and precise delivery to targeted cells or tissues, thus catalyzing progress in biomedicine and healthcare (Graf et al. 2023).

11.6.5 Inorganic Nanocomposites

Inorganic nanocomposites are advanced materials that consist of inorganic nanoparticles or nanofillers embedded within an inorganic matrix (Hiremath et al. 2021). These materials merge the distinctive attributes of nanoscale inorganic materials, encompassing robust mechanical strength, thermal stability, electrical conductivity, and optical characteristics, with the structural benefits of inorganic matrices (Gomes Souza et al. 2024; Wang et al. 2020).

The synthesis of inorganic nanocomposites employs techniques like sol-gel synthesis and layer-by-layer assembly (Shange et al. 2024). Sol-gel processes create nanocomposite networks through precursor solutions, offering tunable porosity, surface chemistry, and mechanical properties (Zhang et al. 2024b). For instance, silica-based nanocomposites synthesized via sol-gel methods find applications in catalysis, adsorption, and biomedical scaffolds, including vaccine delivery systems due to their tailored properties (Gupta et al. 2024).

Layer-by-layer assembly entails depositing alternating layers of polymers and nanomaterials onto a substrate, allowing for meticulous control over both film thickness and functionality (Guzmán et al. 2022). Inorganic nanocomposites show immense potential across various sectors, from electronics to biomedical devices, owing to their customizable properties and versatile applications (Alshangiti et al. 2023; Al-Amiery et al. 2024). In addition to their technological advantages, inorganic nanocomposites also offer eco-friendly benefits (Hao et al. 2020). The precise control over composition and functionality in their synthesis processes, including sol-gel synthesis and layer-by-layer assembly, allows for the incorporation of environmentally friendly materials (Shange et al. 2024; Guzmán et al. 2022). For example, using sustainable nanomaterials like clay nanoparticles or metal oxides in layer-by-layer assembly can contribute to reducing environmental impact (Jackson et al. 2024). Moreover, the augmented characteristics of inorganic nanocomposites, including heightened durability and efficiency, have the potential to foster resource conservation and diminish waste in diverse applications. These environmentally friendly attributes position inorganic nanocomposites as a promising path for sustainable materials development across industries (Wang et al. 2021).

11.6.6 Hybrid Nanocomposites

Hybrid nanocomposites are advanced materials that combine two or more distinct types of nanomaterials or polymers to achieve unique and synergistic properties (Kumar and Kumar 2024). One common approach in creating hybrid nanocomposites is through core-shell structures, where different materials are layered to form a core coated with a shell. For example, a magnetic core coated with a biopolymer shell can create core-shell nanoparticles with applications in targeted drug delivery or environmental remediation (Zhang et al. 2024a; Liu et al. 2024b; Büyüktiryaki et al. 2024).

Surface functionalization is pivotal for optimizing the performance of hybrid nanocomposites. Strategies such as grafting, cross-linking, or chemical modification are utilized to alter the surface properties of these nanocomposites (Schneider et al. 2021). This leads to enhanced compatibility, stability, and reactivity, resulting in improved dispersion within matrices and controlled release characteristics. Functionalized nanocomposites are utilized across diverse fields, encompassing drug delivery systems and catalytic applications (Chiang et al. 2024; Shakil 2024; Mujtaba et al. 2023).

When considering vaccine delivery, hybrid nanocomposites present numerous advantages. They have the capacity to encapsulate and safeguard vaccine antigens, thereby enhancing both their stability and bioavailability (Sharma and Otto 2024; Ren et al. 2021). Additionally, functionalized nanocomposites can be tailored to achieve controlled release of antigens, ensuring a sustained immune response and potentially reducing the number of doses required for vaccination (Cao et al. 2022). Overall, hybrid nanocomposites hold promise as innovative materials for advancing vaccine delivery technologies (Creighton 2021).

11.6.7 Methods for Environmentally Friendly Synthesis

Methods for Environmentally Friendly Synthesis for green sustainable nanocomposites, particularly in the context of vaccine delivery, are becoming increasingly prominent due to their environmentally friendly nature and potential for biomedical applications. These approaches encompass a range of strategies aimed at reducing environmental impact while producing nanocomposites with desirable properties (Sivanesan et al. 2021).

11.6.7.1 Biogenic Synthesis

The utilization of biological entities, such as microorganisms, enzymes, or plant extracts, in the synthesis of nanocomposites represents a sustainable and eco-friendly approach (Kulal and Kodialbail 2021; Tripathi and Chung 2021; Hasan et al. 2024; Seth and Meena 2024; Shi et al. 2024). Biogenic synthesis routes frequently obviate the necessity for harsh chemicals and energy-intensive procedures, resulting in the creation of biocompatible nanomaterials endowed with distinctive properties (Shi et al. 2024; Madkour 2018). This method holds particular promise for vaccine delivery applications due to its ability to produce biocompatible nanomaterials suitable for this purpose (Salman et al. 2023).

For instance, researchers used a bacterial-based synthesis method to create Ag_2O/AgO-TiO_2 nanocomposites aimed at dye degradation in industrial waste. By employing the cell-free supernatant from Alcaligenes aquatilis culture broth, they produced nanocomposite particles with an average crystallite size of 38 nm. This technique highlights an environmentally sustainable, cost-effective, and efficient approach for producing nanocomposites with potential applications in solar light-mediated photocatalytic wastewater treatment (Kulal and Kodialbail 2021).

11.6.7.2 Microwave-Assisted Synthesis

Microwave irradiation accelerates reaction kinetics and facilitates the rapid synthesis of nanocomposites with controlled morphology, particle size, and crystallinity, thereby reducing energy consumption, reaction times, and enhancing material properties, making microwave-assisted methods viable for scalable and sustainable production. This technique has shown significant potential for the efficient production of nanocomposites tailored for vaccine delivery applications. Another promising technique is microwave-assisted synthesis, which utilizes microwave irradiation to expedite reaction kinetics and achieve controlled morphology and particle size in nanocomposites, further reducing energy consumption, shortening reaction times, and enhancing material properties. This method is suitable for scalable and sustainable production of nanocomposites specifically designed for vaccine delivery applications (Hassan 2024).

Overall, green synthesis approaches offer a pathway for developing environmentally friendly and biocompatible nanocomposites tailored for efficient vaccine delivery systems. These methods contribute not only to sustainability but also provide opportunities to enhance the effectiveness and safety of vaccines through innovative nanomaterial design (Salman et al. 2023; Sivanesan et al. 2021).

11.6.8 Advancements and Future Directions

Recent advancements in sustainable nanocomposite design and synthesis have paved the way for novel applications and technological breakthroughs. Hybrid nanocomposites combining biopolymers, nanoparticles, and functional additives exhibit enhanced performance in areas such as energy storage, water purification, biomedical implants, and sustainable packaging (Sharma et al. 2024a; Kalia et al. 2024; Vagena et al. 2024). Nanocomposites with stimuli-responsive properties, self-healing capabilities, or bioactive functionalities are actively explored for next-generation materials and devices (Saddique et al. 2024; Li et al. 2024b).

Future directions in sustainable nanocomposite research emphasize multifunctional materials, smart nanostructures, and integrated systems tailored for specific applications (Fan et al. 2024; Ingole et al. 2024). Leveraging emerging technologies, including artificial intelligence, additive manufacturing, and nanofabrication techniques, shows potential for speeding up materials discovery, optimization, and scale-up (Gomes Souzaet al. 2024; Colorado et al. 2024; De Moraes et al. 2024). For instance, advanced nanofabrication utilizing e-beam lithography has been employed to create nanocomposites with periodic arrays of soft magnetic FeCo-based nano-inclusions within a hard magnetic FePt matrix. High-resolution imaging and chemical analysis revealed the formation of Kirkendall voids due to annealing-induced diffusion. These nanocomposites act as model systems for micromagnetic simulations, demonstrating excellent correlation between measured and simulated magnetization reversal curves, and informing the design of new high-performance bulk magnets (De Moraes et al. 2024). This research could have significant applications in vaccine and medical biotechnology platforms. For example, magnetic nanostructures developed using this method are applicable in targeted drug platforms, enabling precise control and guidance to specific body locations using external magnetic fields, thereby improving vaccine efficiency and effectiveness. Additionally, these nanostructures can aid in the development of advanced diagnostic tools by creating sensitive biosensors capable of detecting biomarkers at very low concentrations, which enhances early disease diagnosis and monitoring. Furthermore, these magnetic nanostructures could be utilized in hyperthermia therapies, generating local heat to kill cancer cells without harming healthy tissues. Precise control over the nanoparticles' size, shape, and magnetic properties is essential for optimizing the effectiveness and safety of these treatments. Overall, the methods and principles established in this research can greatly improve vaccine delivery, disease diagnosis, and medical treatments, thereby contributing to scientific and educational progress in these areas (Selim et al. 2024). Partnerships among academia, industry, and regulatory bodies are crucial for fostering innovation, establishing sustainability standards, and ensuring the responsible use of nanocomposites across various sectors (Khosravi et al. 2024).

11.6.9 Key Takeaways

The development of green sustainable nanocomposites for vaccine platforms represents a significant step forward in healthcare innovation (Sivanesan et al. 2021). By combining renewable polymers with eco-friendly synthesis routes, researchers are creating materials that enhance vaccine delivery while minimizing environmental impact (Tripathi and Chung 2021). Key principles such as biocompatibility, controlled release kinetics, and targeted delivery strategies guide the design process, ensuring efficiency and sustainability (Constantino et al. 2023).

Advancements in synthesis methods and materials offer diverse avenues for creating these nanocomposites, promising a greener approach to immunization and disease prevention (Feng et al. 2021). Working together across academia, industry, and regulatory organizations is crucial in promoting innovation and guaranteeing the responsible use of these materials (Khosravi et al. 2024).

11.7 Properties and Benefits of Sustainable Nanocomposites

Nanocomposites are a versatile class of materials utilized in numerous industries such as healthcare, electronics, energy, and environmental remediation (Kumar et al. 2024; Barua et al. 2022; Al-Amiery et al. 2024; Gohar et al. 2024). Recently, there has been an increasing focus on developing sustainable nanocomposites that not only demonstrate superior performance but also adhere to eco-friendly principles throughout their lifecycle (Feng et al. 2021).

11.7.1 Enhanced Stability and Controlled Release

As we explained before, sustainable nanocomposites offer a range of properties and benefits (Hassan et al. 2021). One key advantage is their enhanced stability, which contributes to longer product lifespans and improved performance (Harun-Ur-Rashid et al. 2023). Additionally, these nanocomposites enable the controlled release of substances, allowing for precise delivery and utilization of active ingredients in various applications (Liu et al. 2024b).

11.7.2 Targeted delivery and Biocompatibility

Nanocomposites, particularly sustainable variants, have attracted considerable attention due to their distinctive properties and advantages across a range of fields, encompassing healthcare, environmental remediation, energy, and materials science (Chatterjee et al. 2021; Shanmugam et al. 2021; Büyüktiryaki et al. 2024; Sivanesan et al. 2021). Sustainable nanocomposites combine environmentally friendly materials with advanced nanotechnology, resulting in materials that offer enhanced stability, controlled release capabilities, targeted delivery mechanisms, and high biocompatibility (Cao et al. 2022; Chiang et al. 2024; Shakil 2024).

11.7.3 Enhanced Stability and Controlled Release

11.7.3.1 Improved Mechanical Properties

Sustainable nanocomposites often exhibit superior mechanical properties compared to conventional materials (Zhang et al. 2024b). Utilizing nanofillers like clay nanoparticles, carbon nanotubes, or graphene enhances polymer matrices, leading to improved strength, stiffness, and fracture resistance (He et al. 2024a; Jackson et al. 2024; Tripathy and Gupta 2024). The enhanced mechanical properties of sustainable nanocomposites position them as promising options for vaccine platforms. Their increased strength, stiffness, and fracture resistance, resulting from nanofiller integration, enhance the durability and reliability crucial for efficient vaccine delivery systems (Omidian et al. 2024).

11.7.3.2 Thermal Stability and Flame Retardancy

Nanocomposites provide improved thermal stability and fire resistance, rendering them appropriate for high-temperature uses and materials resistant to flames. Nanofillers, such as layered silicates or phosphorus-based compounds, function as obstacles to the transfer of heat and the spread of flames, thereby diminishing the likelihood of thermal degradation or combustion (Gupta et al. 2024; Huang et al. 2024).

While key factors such as biocompatibility, controlled release, and stability under physiological conditions remain top priorities for vaccine nanocomposites, ensuring that these materials maintain their structural integrity and safety, including thermal stability and flame retardancy, is crucial for their effective use in vaccine delivery (Sharma and Otto 2024; Vagena et al. 2024; Dukarski et al. 2024).

Furthermore, sustainable flame retardant nanocomposites provide added safety and reliability in construction materials, electronics, and protective clothing (Dukarski et al. 2024; Farkas et al. 2024).

11.7.3.3 Barrier Properties

Sustainable nanocomposites excel in providing barrier properties against gases, liquids, and contaminants (Tripathy and Gupta 2024). Nanofillers with high aspect ratios and strong interfacial interactions, such as montmorillonite clay or metal oxide nanoparticles, create tortuous pathways that impede molecule diffusion and permeation (Ahmadizadegan and Esmaielzadeh 2024; Pagliero et al. 2024). This barrier effect is beneficial for packaging materials, food preservation, gas separation membranes, and environmental barriers in coatings or construction materials (Hadi and Ali 2024; Cao et al. 2024; Anjum et al. 2024; Singh et al. 2024).

11.7.3.4 Controlled Release and Drug Delivery

Sustainable nanocomposites enable accurate modulation of drug release kinetics and therapeutic payloads, rendering them indispensable in drug delivery systems and biomedical applications (Ren et al. 2021; Pagliero et al. 2024). Nanocomposite carriers can encapsulate pharmaceutical compounds, proteins, or genetic materials and release them with precision in response to particular stimuli, including changes in pH, temperature, or the presence of enzymes (Kadkhodaie Kashani et al. 2024). This controlled release capability enhances drug efficacy, reduces side effects, and enables targeted delivery to diseased tissues or cells (Omidian and Chowdhury 2023).

11.7.4 Targeted Delivery and Biocompatibility

11.7.4.1 Surface Functionalization for Targeting

Sustainable nanocomposites have the capability to undergo surface functionalization by incorporating targeting ligands, peptides, or antibodies, facilitating site-specific delivery and improving cellular uptake (Meher et al. 2024; Manchanda et al. 2024; Prasad et al. 2024). Functionalized nanocarriers have the ability to identify and attach to particular receptors or biomarkers on target cells, thereby facilitating their internalization and payload delivery (Kadkhodaie Kashani et al. 2024). Targeted delivery strategies play a crucial role in cancer therapeutics, precision medicine, and tissue engineering applications (Chen et al. 2023b).

11.7.4.2 Mucosal and Intracellular Delivery

Nanocomposites enable the efficient delivery of bioactive molecules, vaccines, or therapeutic agents across mucosal barriers and into intracellular compartments (Haider et al. 2024). Nanoparticles with tailored surface properties and size distribution can penetrate mucosal surfaces such as respiratory, gastrointestinal, ocular mucosa, or the brain, as well as traverse cellular membranes, leading to enhanced bioavailability and efficacy. This ability is essential for advancing oral vaccines, inhalable therapies, and intracellular drug delivery systems (Amekyeh et al. 2024; Khalil et al. 2024; Sharma et al. 2024b).

11.7.4.3 Biocompatibility and Reduced Immunogenicity

Eco-friendly sustainable nanocomposites prioritize biocompatibility and reduced immunogenicity to ensure compatibility with biological systems (Oladipo et al. 2023). Biodegradable polymers, natural biomaterials, and surface modification techniques contribute to enhanced biocompatibility, reduced cytotoxicity, and improved biodegradation profiles, rendering them essential for vaccine platforms and various biomedical applications (Madkour 2018; Song et al. 2018). Moreover, biocompatible nanocomposites play a crucial role in biomedical implants, regenerative medicine, and tissue scaffolds, promoting minimal inflammatory responses and optimal tissue integration (Tripathy and Gupta 2024; Verstappen et al. 2024; Das et al. 2021).

11.7.4.4 Stimuli-responsive Delivery Systems

Sustainable nanocomposites can be engineered with stimuli-responsive properties, responding to external cues or physiological conditions for on-demand drug release. Stimuli-responsive nanocarriers can be triggered by changes in pH, temperature, enzymatic activity, or magnetic fields, enabling spatiotemporal control over therapeutic delivery. These responsive delivery systems enhance therapeutic efficacy, minimize systemic exposure, and mitigate off-target effects in personalized medicine approaches (Li et al. 2024b; Mishra et al. 2024).

11.7.5 Applications and Future Perspectives

The properties and benefits of sustainable nanocomposites have spurred advancements and innovations in numerous fields:

11.7.5.1 Biomedical Applications

Sustainable nanocomposites are integral in diverse biomedical applications, including drug delivery, tissue engineering, diagnostic imaging, and implantable devices (Jamalipour Soufi and Iravani 2020; Ahmad et al. 2024a; Al-Amiery et al. 2024). Tailored nanocomposite formulations provide targeted drug delivery, controlled release profiles, and improved biocompatibility, supporting applications in vaccine platforms, personalized therapies, and regenerative medicine strategies (Haider et al. 2024).

11.7.5.2 Environmental Remediation

Nanocomposites contribute to environmental sustainability through applications such as water purification, pollutant remediation, and green energy technologies (Yashas et al. 2021; Singh et al. 2024). Sustainable nanocomposite membranes, adsorbents, and photocatalysts enable the efficient removal of contaminants, energy-efficient processes, and renewable energy generation, aligning with global environmental conservation goals (Osman et al. 2024; Ruman et al. 2024).

11.7.5.3 Functional Materials

Sustainable nanocomposites are pivotal in advancing functional materials, imbuing them with improved properties such as conductivity in polymers, antimicrobial capabilities in coatings, and flexibility in electronics (Anjum et al. 2024; Farkas et al. 2024). Nanocomposite-based sensors, actuators, and energy storage devices demonstrate improved performance, durability, and eco-friendly attributes, thereby driving innovation in electronics, sensors, and smart materials (Pengsomjit et al. 2024; Colorado et al. 2024).

11.7.5.4 Industrial and Agricultural Applications

Nanocomposites find applications across various industrial sectors, including automotive, aerospace, packaging, and construction (Lu et al. 2024; Cao et al. 2024; Vafaeva et al. 2024). Sustainable nanocomposites offer lightweight materials, enhanced mechanical properties, and functional coatings that contribute to improved performance, energy efficiency, and reduced environmental impact in both industrial processes and consumer products (Yashas et al. 2021; Saadh et al. 2024; Silva et al. 2023; Wang et al. 2020; Omidian and Chowdhury 2023; Sandua et al. 2022).

11.7.6 Key Takeaways

Future perspectives in sustainable nanocomposites research focus on advancing materials design, synthesis methods, and functionalization strategies to address emerging challenges and opportunities (Gohar et al. 2024). Multidisciplinary collaborations, integration of computational modeling and simulation techniques, and adoption of green synthesis approaches are key drivers for accelerating sustainable nanocomposite innovations across industries (Verma et al. 2024; Afshan and Bari 2023; Vafaeva et al. 2024). Standardization, regulatory frameworks, and sustainability assessments also play critical roles in ensuring the responsible deployment and societal acceptance of nanocomposite technologies (Huang et al. 2024).

11.8 Applications of Sustainable Nanocomposites in Genetic Vaccines

11.8.1 DNA Vaccine Delivery

Sustainable nanocomposites serve as efficient carriers for DNA vaccines, protecting genetic material from degradation, enhancing cellular uptake, and promoting immune response activation (Nordin et al. 2023; Dowaidar 2024). DNA-loaded nanocomposites offer targeted gene delivery, antigen expression, and immunogenicity for infectious diseases, cancer, and genetic disorders (Sharma and Otto 2024; Qamar 2021).

To illustrate, a study investigated turmeric-coated $CoFe_2O_4$ nanoparticles, examining their structure, toxicity, DNA interaction, and potential for magnetic drug delivery. Various analyses, including FT-IR, TGA, and magnetic hysteresis loop simulation, confirmed the interaction between biomolecules and nanoparticles. In vivo toxicity tests revealed no significant effects, while DNA binding investigations indicated intercalative binding and favorable thermodynamic factors, highlighting the potential of this nanocomposite for biomedical applications such as magnetic drug delivery (Mirzaee et al. 2024).

11.8.2 RNA Vaccine Formulations

RNA-based nanocomposites facilitate the delivery of mRNA vaccines, enabling the rapid translation of antigenic proteins, activation of immune cells, and elicitation of adaptive immune responses (Rehan et al. 2024). RNA-loaded nanocomposites also support vaccine stability, antigen presentation, and efficacy against viral infections, emerging pathogens, and cancer (Zdrehus et al. 2024; Rinoldi et al. 2021).

For example, MERS-CoV, a deadly coronavirus with a 34% mortality rate, lacks an approved vaccine despite being listed as a WHO priority pathogen since 2012. Efforts are ongoing to develop prophylactic vaccines, with advancements in nucleic acid-based platforms showing promise. In a recent study, pDNA and mRNA vaccines targeting MERS-CoV's S.FL gene were developed and tested in mice. The optimized pDNA vaccine, without encapsulation, produced the strongest antibody response. Conversely, nano-porous aluminosilicate encapsulation enhanced the immune response to the mRNA vaccine. These results highlight the potential of nano-silica as an improved carrier for mRNA vaccines against MERS-CoV (Almansour and Jermy 2024).

11.8.3 Multifunctional Immunomodulation

Sustainable nanocomposites combine genetic vaccine delivery with immunomodulatory functionalities, which can include adjuvants, cytokines, or immune checkpoint inhibitors (He et al. 2024b). These multifunctional nanocomposites enhance immune cell activation, modulate cytokine levels, and promote antigen-specific immune responses, offering potential for personalized immunotherapy approaches (Nguyen Dang et al. 2024).

11.9 Incorporating Genetic Vaccines into Nanocomposites

The convergence of genetic vaccines and nanocomposites represents a paradigm shift in vaccine design and delivery, offering novel strategies for enhanced immunogenicity, targeted delivery, and personalized vaccination approaches (Nguyen Dang et al. 2024). Genetic vaccines, comprising DNA and RNA variants, leverage pathogen genetic material to trigger immune responses, while nanocomposites offer versatile platforms for encapsulating vaccines, regulating their release, and modulating the immune system (Constantino et al. 2023). In this exploration, we delve into the principles, techniques, and implications of incorporating genetic vaccines into nanocomposites, highlighting the latest advancements and potential applications in preventive medicine, infectious disease control, and public health.

11.9.1 Techniques for Incorporation

Genetic vaccines can be encapsulated within nanocomposites using techniques such as nanoprecipitation, emulsification, layer-by-layer assembly, or self-assembly. These methods ensure uniform distribution, protection from enzymatic degradation, and controlled release of genetic material (Sonowal and Gautam 2024; Chaudhari et al. 2024).

Surface Modification: Nanocomposites can be surface-modified with functional groups, polymers, or biomolecules to improve vaccine stability, cellular uptake, and interactions with immune cells. Surface engineering enhances nanocomposite biocompatibility, circulation time, and targeting capabilities (Meher et al. 2024).

Electrospinning and Nanofiber Assembly: Electrospinning techniques produce nanocomposite nanofibers that encapsulate drugs, offering high surface area, tunable release profiles, and potential for mucosal delivery or tissue targeting (Creighton 2021).

11.9.2 Applications and Future Directions

The future of incorporating genetic vaccines into nanocomposites promises groundbreaking advancements in vaccine technology. As research progresses, we anticipate the development of even more sophisticated techniques for the encapsulation of genetic material within nanocomposites. These techniques will ensure both the precise and efficient delivery of vaccines and the enhancement of their stability and bioavailability. This will lead to improved immunogenicity and therapeutic efficacy (Dehghani et al. 2024).

One exciting area of future development is the integration of smart materials and nanotechnology into genetic vaccine-nanocomposite platforms. This approach could facilitate the creation of vaccines that dynamically react to specific signals within the body, including the presence of pathogens or alterations in the microenvironment (Rehan et al. 2024). These "smart vaccines" could offer enhanced targeting, controlled release, and on-demand immune stimulation, revolutionizing the way we prevent and treat diseases (Mishra et al. 2024).

Furthermore, advances in nanocomposite surface engineering will enable the creation of vaccines with tailored properties for different applications (Sarasati et al. 2023). Functionalization with particular ligands, peptides, or antibodies could enable targeted delivery to distinct cell types or tissues, enhancing vaccine uptake and effectiveness while reducing off-target effects (He et al. 2024b).

Additionally, integrating nanoscale adjuvants and immunomodulators into genetic vaccine-nanocomposite formulations holds great promise for enhancing immune responses and promoting long-lasting immunity (Sarasati et al. 2023). These advancements could result in vaccines that require lower doses, cause fewer side effects, and enhance patient compliance (Chavda and Apostolopoulos 2024).

The future of genetic vaccines integrated into nanocomposites is characterized by ongoing innovation, collaboration across disciplines, and a relentless pursuit of safer, more effective, and personalized vaccination strategies (Sarasati et al. 2023; Hughes et al. 2024). These advancements have the potential to address current healthcare challenges and significantly impact global public health outcomes (Sharma et al. 2024b).

11.9.3 Key Takeaways

In summary, the fusion of genetic vaccines with nanocomposites represents a pivotal shift in vaccine technology (Almansour and Jermy 2024). This convergence offers novel approaches for improved immunogenicity, precise delivery, and personalized vaccination strategies (Fonseca et al. 2024). Techniques such as nano-precipitation, surface modification, and electrospinning ensure efficient encapsulation and controlled release of genetic material within nanocomposites, enhancing stability and efficacy (Tan et al. 2022; Mujtaba et al. 2023; Sandua et al. 2022; Chaudhari et al. 2024).

Looking ahead, advancements in smart materials and nanotechnology hold promise for creating responsive "smart vaccines" that adapt to body cues, revolutionizing disease prevention (Mishra et al. 2024). Integrating nanoscale adjuvants and immunomodulators into genetic vaccine-nanocomposite formulations also show potential for stronger immune responses and lasting immunity (Rehan et al. 2024).

Overall, ongoing innovation and interdisciplinary collaboration drive the quest for safer, more effective, and tailored vaccination approaches. These developments are poised to address healthcare challenges and significantly impact global public health in the future (Hughes et al. 2024).

11.10 Modulation of Immune Responses

Nanocomposites, consisting of nanostructured materials integrated into a matrix, provide versatile platforms for immune modulation, allowing precise control over immune responses, immunomodulation, and targeted therapeutic interventions (Alshangiti et al. 2023). The unique properties of nanocomposites, including their tunable physicochemical characteristics, surface functionalities, and interactions with immune cells, contribute to their advantages in immune modulation (Prasad et al. 2024; Oladipo et al. 2023). This exploration delves into the advantages, challenges, and solutions of using nanocomposites for immune modulation, with a special focus on their applications in vaccine development.

11.10.1 Advantages of Using Nanocomposites for Immune Modulation

11.10.1.1 Optimized Drug Delivery

Nanocomposites enhance the bioavailability and pharmacokinetics of immunomodulatory agents, such as cytokines, antibodies, or small molecules, by providing controlled release kinetics, prolonged circulation, and targeted delivery to immune cells or tissues (Kuna et al. 2024; Li et al. 2024b; Zdrehus et al. 2024; He et al. 2024b). Surface modifications, such as PEGylation (polyethylene glycol coating), enable stealth properties, reduce immunogenicity, and improve nanocomposite stability in biological environments, enhancing drug delivery efficiency (Oladipo et al. 2023).

11.10.1.2 Cell-Specific Targeting

Functionalized nanocomposites can be designed with ligands, antibodies, or peptides that identify specific receptors or biomarkers on immune cells, promoting targeted delivery and uptake (Manchanda et al. 2024). Delivering immunomodulators specifically to dendritic cells, macrophages, T cells, or B cells improves therapeutic efficacy, minimizes off-target effects, and reduces systemic toxicity (Jain et al. 2023).

11.10.1.3 Immune Cell Activation

Nanocomposites interact with immune cells, including dendritic cells, macrophages, and lymphocytes, modulating their activation, differentiation, cytokine secretion, and antigen presentation (Hao et al. 2020). Stimuli-responsive nanocomposites, activated by environmental cues or external triggers (e.g., pH, temperature, light, temperature, enzymatic activity, or magnetic fields), enable spatiotemporal control over immune cell activation and immune responses (Shi et al. 2024).

11.10.1.4 Cytokine Modulation

Nanocomposites regulate cytokine signaling pathways, modulate pro-inflammatory and anti-inflammatory cytokine profiles, and balance immune responses to achieve immune homeostasis. Encapsulation of cytokines, growth factors, or regulatory molecules within nanocomposites offers sustained release, prolonged bioactivity, and localized effects, supporting tissue repair and immunoregulation (Omidian et al. 2024; Nie et al. 2024).

11.10.1.5 Induction of Tolerance

Nanocomposites play a crucial role in DNA/RNA vaccine design by facilitating the induction of tolerance. They promote immune tolerance by inducing regulatory T cells (Tregs), suppressing pro-inflammatory responses, and dampening autoimmune reactions in autoimmune disorders or transplantation settings. Additionally, nanocomposites hold promise for creating a tolerogenic environment conducive to DNA/RNA vaccine efficacy (Koushki et al. 2021; Wang et al. 2024; Qin et al. 2023). Antigen-specific tolerance can be achieved using nanocomposites carrying antigenic peptides, antigens, or immunomodulatory agents. This fosters antigen-specific immune tolerance and immune regulation (Scotland et al. 2023).

11.10.1.6 Immunosuppression and Inflammatory Control

Nanocomposites loaded with immunosuppressive agents, such as corticosteroids, anti-inflammatory drugs, or immune checkpoint inhibitors, modulate immune cell activation, inhibit cytokine release, and mitigate inflammatory responses (Qin et al. 2023). Localized delivery of these immunosuppressive nanocomposites reduces systemic side effects, enhances therapeutic efficacy, and minimizes the risk of immune-related adverse events (Dehghani et al. 2024).

11.10.1.7 Biocompatible Materials

Eco-friendly nanocomposites utilize biocompatible materials, such as biodegradable polymers, natural biomolecules, or hydrogels, ensuring compatibility with biological systems, reduced cytotoxicity, and minimal inflammatory responses (Song et al. 2018; Das et al. 2021). Biodegradable nanocomposites degrade in a controlled manner, are metabolized, and cleared from the body, reducing long-term accumulation and toxicity risks (Wang et al. 2021).

11.10.1.8 Safety Profiles

Engineered nanocomposites undergo rigorous safety assessments, biocompatibility testing, and toxicity evaluations to ensure their safety profile, immunogenicity, and long-term biocompatibility (Thakur and Kumar 2024). Surface modifications, polymer coatings, and nanocomposite formulations optimize safety, reduce immunogenicity, and enhance cellular uptake without eliciting adverse immune reactions (Ravinayagam and Jermy 2022; Rajput et al. 2023).

11.10.1.9 Multifunctional Nanocomposites

Nanocomposites serve as multifunctional platforms that integrate multiple therapeutic modalities, combining immunomodulators, drugs, antigens, or biomolecules to achieve synergistic effects and therapeutic outcomes (Hughes et al. 2024). Combination therapies using nanocomposites offer personalized treatment strategies, address complex diseases, and overcome therapeutic resistance by targeting multiple pathways or immune checkpoints simultaneously (Sandbhor et al. 2024).

11.10.1.10 Therapeutic Combinations

Nanocomposites enable combination therapies by co-delivering immunomodulators with chemotherapy, radiotherapy, or targeted therapies, enhancing treatment efficacy, overcoming drug resistance, and minimizing systemic toxicity (Zhang et al. 2024c). Immunochemotherapy using nanocomposites combines immunomodulation with cytotoxic agents, enhancing anti-tumor immune responses, tumor cell killing, and long-term immune memory for durable remission (Xu et al. 2023b; Harshitha et al. 2023).

11.10.2 Recent Advances and Clinical Applications

11.10.2.1 Nanoparticle-Based Immunotherapy

Nanocomposite-based immunotherapy is advancing in cancer treatment, infectious diseases, autoimmune disorders, and chronic inflammatory conditions, demonstrating efficacy in preclinical models and clinical trials (Ding et al. 2024b; Koushki et al. 2021; Xu et al. 2023b; An et al. 2024). Leading innovations in immunotherapy include nanoparticle vaccines, immune checkpoint inhibitors, and nanocomposite drug delivery systems, providing precision medicine and tailored treatment options (Sandbhor et al. 2024).

11.10.2.2 Regenerative Immunology

Nanocomposites play a pivotal role in regenerative immunology, combining immunomodulation with tissue engineering, stem cell therapies, and organ transplantation to promote tissue repair, regeneration, and immune tolerance (Ding et al. 2024a;

Verstappen et al. 2024; Nie et al. 2024; Scotland et al. 2023; Xu et al. 2023b). Immuno-regenerative nanocomposites modulate immune responses, reduce graft rejection, and enhance tissue integration, supporting functional recovery and long-term graft survival in regenerative medicine applications (Qin et al. 2023).

11.10.3 Translation to Clinical Practice

Clinical translation of nanocomposite-based immunomodulation requires comprehensive preclinical studies, pharmacokinetic evaluations, immunogenicity assessments, and regulatory approval pathways (Wei et al. 2024).

Standardization, scalability, cost-effectiveness, and manufacturing considerations are essential for scaling up nanocomposite production, ensuring product quality, and facilitating widespread clinical adoption (Hadi and Ali 2024).

11.10.4 Key Takeaways

Nanocomposites are poised to revolutionize immune modulation and therapeutic interventions due to their versatile advantages. They offer optimized drug delivery, cell-specific targeting, immune cell activation, cytokine modulation, tolerance induction, and immunosuppression (Nguyen Dang et al. 2024; Manchanda et al. 2024; Wang et al. 2024; An et al. 2024). Recent advancements in nanoparticle-based immunotherapy and regenerative immunology show promising results in preclinical and clinical settings (Scotland et al. 2023).

For successful clinical adoption, rigorous preclinical studies, pharmacokinetic evaluations, and regulatory approvals are essential (Konwarh et al. 2024). Standardization, scalability, cost-effectiveness, and rigorous quality control are essential considerations for the widespread implementation of therapies using nanocomposites (Vasudevan and Senthil Kumar 2024). Overall, nanocomposites represent a transformative approach in healthcare, addressing complex medical challenges and improving patient outcomes (Anju et al. 2020).

11.11 Challenges and Considerations of Sustainable Nanocomposites in Vaccine Delivery

11.11.1 Safety and Immunogenicity

Ensuring the safety, immunogenicity, and long-term effects of genetic vaccine-nanocomposite formulations requires comprehensive preclinical and clinical evaluations (Konwarh et al. 2024). Considerations include immune responses, potential adverse effects, immunotoxicity, and long-term immune memory (Harshitha et al. 2023).

11.11.2 Regulatory Approval and Standardization

Regulatory approval pathways, manufacturing standards, and quality control measures for genetic vaccine-nanocomposite products need alignment with regulatory agencies and industry standards. Standardized protocols, validation assays, and clinical trial designs support regulatory compliance and product commercialization (Tibério and Reis 2023).

11.11.3 Cost and Scalability

Cost-effective production, scalability, and manufacturing scalability of genetic vaccine-nanocomposite platforms are crucial for widespread adoption and global accessibility. Addressing cost barriers, production efficiencies, and supply chain logistics optimizes vaccine affordability and availability (Tibério and Reis 2023).

11.11.4 Key Takeaways

The integration of sustainable nanocomposites in vaccine delivery presents both opportunities and challenges (Hatae et al. 2023). Ensuring safety, immunogenicity, and long-term effects of genetic vaccine-nanocomposite formulations through rigorous evaluations is paramount (Wang et al. 2023c). Regulatory alignment, standardization of manufacturing processes, and adherence to quality control measures are crucial for obtaining regulatory approvals and commercializing these products. Moreover, addressing cost barriers, optimizing production efficiencies, and ensuring scalability are essential for enhancing vaccine affordability and accessibility on a global scale (Tibério and Reis 2023). These considerations collectively contribute to the advancement and successful implementation of sustainable nanocomposites in vaccine delivery. This paves the way for improved public health outcomes worldwide (Rinoldi et al. 2021).

11.12 Applications and Future Directions

Nanocomposites are transforming biomedicine with genetic vaccines, CRISPR-Cas delivery, and biosensing applications, highlighting regulatory and safety concerns for clinical use (Wallis et al. 2019; Huda et al. 2021; Li et al. 2024c; Sheikhzadeh et al. 2021).

11.12.1 Infectious Disease Vaccines

Incorporating genetic vaccines into nanocomposites holds promise for combating infectious diseases such as viral infections, bacterial pathogens, and emerging zoonotic threats (Uyanga et al. 2024). Nanocomposite-based vaccines offer rapid response capabilities, adaptability to antigenic variability, and the potential for multivalent or combination vaccines (Dowaidar 2024).

11.12.2 Cancer Immunotherapy

Exploration of nanocomposites incorporating genetic vaccines in cancer immunotherapy involves targeting tumor-specific antigens, tumor-associated antigens, or immune checkpoint inhibitors. These formulations stimulate anti-tumor immune responses, enhance immune cell infiltration into tumors, and synergize with other immunotherapeutic modalities (Zhang et al. 2024c). Table 11.1 shows various nanocomposites designed to enhance the immune system's response in cancer treatment. These innovative materials are engineered to improve the delivery and efficacy of cancer vaccines, ultimately leading to more effective and sustained anti-tumor immune responses.

11.12.3 Allergen-Specific Immunotherapy

Investigations into nanocomposite-based genetic vaccines are underway for allergen-specific immunotherapy targeting allergic diseases, asthma, and autoimmune conditions (Cai et al. 2023; Fonseca et al. 2024; Rai et al. 2023). Targeted delivery of allergen-encoding genetic material modulates immune tolerance, reduces allergic responses, and holds potential for disease modification and long-term remission (Xu et al. 2023c).

Table 11.1 Nanocomposites that enhance the immune system's response in cancer treatment

Type of nanocomposite	Associations	Effects
Magnetic/iron oxide nanoparticles	Ovalbumin interferon-γ polycytidylic-acid imiquimod	The gathering of nanocomposites at the tumor location prompts an immune reaction against the tumor
Mesoporous silica nanoparticles	Doxorubicin ovalbumin CpG-oligodeoxynucleotides	Prompts robust immune reactions, including the activation of CD8+ T-cells, leading to improved tumor-fighting capabilities
Multi-walled carbon nanotubes	Ovalbumin CpG-Oligodeoxynucleotides NY-ESO antigen	Promotes a potent immune response involving CD4+ and CD8+ T cells
Nanogels	Ovalbumin and polyinosinic: polycytidylic acid	Efficiently transports the antigen to dendritic cells, resulting in robust and specific adaptive immune responses
Dendrimers	Ovalbumin combined with CpG-oligodeoxynucleotides	Triggers an elevated immune response mediated by T cells

Adapted from Zdrehus et al. (2024)

11.12.4 Emerging Technologies and Platforms

Advancements in nanotechnology, biomaterials, and genetic engineering are driving innovation in genetic vaccine-nanocomposite platforms (Xiao et al. 2023). Integrated approaches, such as mRNA-based nanocomposites, CRISPR-Cas9 delivery systems, and personalized vaccine formulations, are expanding the scope of genetic vaccination strategies (Rinoldi et al. 2021).

11.12.5 CRISPR-Cas Delivery Systems

Sustainable nanocomposites are being explored as carriers for CRISPR-Cas gene editing systems, enabling targeted genome modifications, gene therapies, and precision medicine applications. CRISPR-loaded nanocomposites offer precise gene delivery, cellular uptake, and therapeutic genome editing capabilities (Li et al. 2024c).

11.12.6 Environmental Monitoring and Biosensing

Nanocomposite-based biosensors and diagnostic platforms integrate genetic material, nanomaterials, and biomolecules, for environmental monitoring, pathogen detection, and disease diagnostics. Biosensing nanocomposites provide rapid, sensitive, and specific detection of biomarkers, pollutants, and infectious agents (Maham et al. 2020; Sheikhzadeh et al. 2021).

11.12.7 Regulatory Considerations and Safety Assessments

The translation of sustainable nanocomposites into clinical practice requires rigorous regulatory evaluations, safety assessments, and biocompatibility testing (Thakur and Kumar 2024). Standardized protocols, quality control measures, and ethical considerations ensure the safety, efficacy, and environmental sustainability of nanocomposite-based biomedical technologies (Das et al. 2021; Vasudevan and Senthil Kumar 2024).

11.12.8 Key Takeaways

Nanocomposites are revolutionizing biomedicine with their applications in genetic vaccines, CRISPR-Cas delivery, and biosensing technologies (Creighton 2021; Rinoldi et al. 2021; Sheikhzadeh et al. 2021). These advancements hold promise for combating infectious diseases, improving cancer immunotherapy, and offering innovative solutions for allergen-specific treatments (Hao et al. 2020; Ding et al. 2024b; Xu et al. 2023c). However, the successful integration of nanocomposites into clinical practice necessitates rigorous attention to regulatory and safety considerations (Thakur and Kumar 2024).

11.13 Conclusion

In the realm of biomedicine, the integration of sustainable nanocomposites marks a paradigm shift towards eco-friendly and efficient vaccination strategies (Ahmad et al. 2024a). This journey has been illuminated by the exploration of genetic vaccines and their significance, coupled with the innovative design and synthesis of sustainable nanocomposites tailored for vaccine delivery (Chavda and Apostolopoulos 2024). Throughout this endeavor, the overarching goal has been to create a greener, healthier future while addressing complex healthcare challenges (Hasan et al. 2024).

The environmental advantages offered by eco-friendly sustainable nanocomposites are profound. Their biodegradability, renewability, reduced carbon footprint, and minimized environmental impact make them indispensable in sustainable healthcare practices (Yashas et al. 2021; Hsu et al. 2024). By protecting genetic material, improving vaccine stability, facilitating targeted delivery, and regulating immune responses, these nanocomposites are crucial in enhancing the efficacy and accessibility of genetic vaccines (Tan et al. 2022; Jin et al. 2023).

Furthermore, the applications of sustainable nanocomposites in genetic vaccines are multifaceted and promising (Ahmad et al. 2024a). From DNA vaccine delivery to RNA vaccine formulations and multifunctional immunomodulation, nanocomposites offer versatile platforms for tailored vaccination strategies (Eusébio et al. 2024; Amekyeh et al. 2024; Mirzaee et al. 2024). Their integration into smart vaccines, coupled with surface engineering and immunomodulatory functions, heralds a new era of personalized immunotherapy and disease prevention (Cai et al. 2023; Prasad et al. 2024; Mishra et al. 2024).

However, challenges and considerations abound, from safety and regulatory approval to cost-effectiveness and scalability (Anju et al. 2020; Wei et al. 2024). Addressing these challenges requires collaborative efforts across disciplines, rigorous preclinical and clinical evaluations, standardized protocols, and regulatory compliance (Das et al. 2021). Only through such comprehensive approaches can we ensure the responsible deployment and widespread adoption of sustainable nanocomposites in vaccine and drug delivery systems (Khalil et al. 2024; Chaudhari et al. 2024).

Looking ahead, the future of biomedicine intertwined with sustainable nanocomposites holds immense promise (Mujtaba et al. 2023). Innovations in CRISPR-Cas delivery systems, cancer immunotherapy, allergen-specific treatments, biosensing technologies, and personalized medicine are on the horizon (Arakelian et al. 2022; Li et al. 2024c; Sheikhzadeh et al. 2021; Rai et al. 2023). These advancements, fueled by multidisciplinary collaborations and regulatory frameworks, pave the way for transformative solutions in global public health, environmental stewardship, and patient-centric healthcare (Sharma et al. 2024b).

In summary, the journey towards advancing eco-friendly biomedicine with sustainable nanocomposites is both a scientific endeavor and a moral imperative (Madkour 2018). By leveraging the capabilities of nanotechnology alongside genetic vaccines and sustainable practices, we set forth on a trajectory towards a planet that is not only environmentally friendly but also healthier and more robust for the generations to follow (Uyanga et al. 2024).

References

Afshan N, Bari A (2023) Advanced bionanomaterials for environmental clean recovery processes. In: Sustainable nanomaterials for biosystems engineering, 1st edn. Apple Academic Press, p 28. eBook ISBN 9781003333517

Ahmad NN, Ghazali NN, Wong YH (2021) Concept design of transdermal microneedles for diagnosis and drug delivery: a review. Adv Eng Mater 23(12):2100503. https://doi.org/10.1002/adem.202100503

Ahmad A, Haneef M, Ahmad N, Kamal A, Jaswani S, Khan F (2024a) Biological synthesis of silver nanoparticles and their medical applications (review). World Acad Sci J 22. https://doi.org/10.3892/wasj.2024.237

Ahmad I, Aftab MA, Fatima A, Mekkey SD, Melhi S, Ikram S (2024b) A comprehensive review on the advancement of transition metals incorporated on functional magnetic nanocomposites for the catalytic reduction and photocatalytic degradation of organic pollutants. Coord Chem Rev 514:215904. https://doi.org/10.1016/j.ccr.2024.215904. ISSN 0010-8545

Ahmadi S, Ebrahimi Warkiani M, Rabiee M, Iravani S, Rabiee N (2023) Carbon-based nanomaterials against SARS-CoV-2: therapeutic and diagnostic applications. OpenNano 10:100121. https://doi.org/10.1016/j.onano.2023.100121. ISSN: 2352-9520

Ahmadizadegan H, Esmaielzadeh S (2024) Organically modified clay as an enhancement filler in novel polyimide mixed matrix membranes for gas separation. Polym Bull 81:3273–3296. https://doi.org/10.1007/s00289-023-04882-7

Ahn M, Song J, Hong BH (2021) Facile synthesis of N-doped graphene quantum dots as novel transfection agents for mRNA and pDNA. Nanomaterials 11:2816. https://doi.org/10.3390/nano11112816

Akbarian M, Tayebi L, Redwan EM, Chavda VP, Uversky VN (2024) Vector-based vaccine delivery and associated immunity: current status and way forward. In: Chavda VP, Vora LK, Apostolopoulos V (eds) Developments in immunology: advanced vaccination technologies for infectious and chronic diseases. Academic Press, pp 73–98. https://doi.org/10.1016/B978-0-443-18564-9.00014-X

Akmal MH, Yari Kalashgrani M, Mousavi SM, Rahmanian V, Sharma N, Gholami A, Althomali RH, Rahman MM, Chiang WH (2024) Recent advances synergistic of GQD-based hydrogel for bioimaging and drug delivery in cancer treatment. J Mater Chem B. https://doi.org/10.1039/D4TB00024B

Al-Amiery AA, Fayad MA, Abdul Wahhab HA, Al-Azzawi WK, Mohammed JK, Majdi HS (2024) Interfacial engineering for advanced functional materials: surfaces, interfaces, and applications. Results Eng 22:102125. https://doi.org/10.1016/j.rineng.2024.102125. ISSN 2590-1230

Alahakoon YA, Malaweera Arachchi U, Gunatilake SR (2023) Clay-supported nanoscale zero-valent iron composites for water purification. In: Vithanage M, Lazzara G, Rajapaksha AU (eds) Clay composites, Advances in material research and technology. Springer, Singapore. https://doi.org/10.1007/978-981-99-2544-5_21

Aliahmad P, Miyake-Stoner SJ, Geall AJ et al (2023) Next generation self-replicating RNA vectors for vaccines and immunotherapies. Cancer Gene Ther 30:785–793. https://doi.org/10.1038/s41417-022-00435-8

Almansour I, Jermy BR (2024) Nucleic acid vaccine candidates encapsulated with mesoporous silica nanoparticles against MERS-CoV. Hum Vaccin Immunother 20(1). https://doi.org/10.1080/21645515.2024.2346390

Alshangiti DM, El-damhougy TK, Zaher A, Madani M, Ghobashy MM (2023) Revolutionizing biomedicine: advancements, applications, and prospects of nanocomposite macromolecular carbohydrate-based hydrogel biomaterials: a review. RSC Adv 13:35251–35291. https://doi.org/10.1039/D3RA07391B

Amekyeh H, Sabra R, Billa N (2024) A window for enhanced oral delivery of therapeutics via lipid nanoparticles. Drug Des Devel Ther 18:613–630. https://doi.org/10.2147/DDDT.S439975

An X, Yang J, Cui X, Zhao J, Jiang C, Tang M, Dong Y, Lin L, Li H, Wang F (2024) Advances in local drug delivery technologies for improved rheumatoid arthritis therapy. Adv Drug Deliv Rev 209:115325. Available from: https://doi.org/10.1016/j.addr.2024.115325. ISSN 0169-409X

Anju S, Prajitha N, Sukanya VS, Mohanan PV (2020) Complicity of degradable polymers in health-care applications. Mater Today Chem 16:100236. https://doi.org/10.1016/j.mtchem.2019.100236

Anjum A, Gupta D, Singhal R (2024) Introduction to nanocoatings and packaging. In: Sustainable approach to protective nanocoatings. IGI Global. Available at: https://www.igi-global.com/chapter/introduction-to-nanocoatings-and-packaging/340545

Anwer AH, Ahtesham A, Shoeb M, Mashkoor F, Ansari MZ, Zhu S, Jeong C (2023) State-of-the-art advances in nanocomposite and bio-nanocomposite polymeric materials: a comprehensive review. Adv Colloid Interf Sci 318:102955. https://doi.org/10.1016/j.cis.2023.102955

Arakelian T, Oosterhuis K, Tondini E, Los M, Vree J, van Geldorp M, Camps M, Teunisse B, Zoutendijk I, Arens R, Zondag G, Ossendorp F, van Bergen J (2022) Pyroptosis-inducing active caspase-1 as a genetic adjuvant in anti-cancer DNA vaccination. Vaccine 40(13):2087–2098. ISSN 0264-410X. https://doi.org/10.1016/j.vaccine.2022.02.028

Barua P, Hossain N, Siddiqui MTH, Nizamuddin S, Mazari SA, Mubarak NM (2022) Chapter 19 – Future development, prospective, and challenges in the application of green nanocomposites in environmental remediation. In: Koduru JR, Karri RR, Mubarak NM, Bandala ER (eds) Sustainable nanotechnology for environmental remediation. Elsevier, pp 483–511. https://doi.org/10.1016/B978-0-12-824547-7.00027-8. ISBN 9780128245477

Bhat DK, Bantawal H, Uma P, Kumar SP, Shenoy US (2024) Designing sustainable porous graphene-CaTiO3 nanocomposite for environmental remediation, vol 5. Sustain Chem Environ, p 100071

Biddeci G, Spinelli G, Colomba P, Di Blasi F (2023) Halloysite nanotubes and sepiolite for health applications. Int J Mol Sci 24(5):4801. https://doi.org/10.3390/ijms24054801

Bloom K, van den Berg F, Arbuthnot P (2021) Self-amplifying RNA vaccines for infectious diseases. Gene Ther 28:117–129. https://doi.org/10.1038/s41434-020-00204-y

Büyüktiryaki S, Ersöz A, Say R (2024) Chapter 9 – Green magnetic core–shell MIPs for environmental applications. In: Hussain CM, Keçili R (eds) Green imprinted materials. Elsevier, pp 269–329. https://doi.org/10.1016/B978-0-443-15431-7.00008-8. ISBN 9780443154317

Cai L, Mao J, Wang H, Chen G, Xu X, Yuan Q, Chen W (2023) Application of DNA-based hydrogels as a drug delivery system for immunomodulatory therapy. J Drug Deliv Sci Technol 86:104677. https://doi.org/10.1016/j.jddst.2023.104677

Cao P, Xu ZP, Li L (2022) Tailoring functional nanoparticles for oral vaccine delivery: recent advances and future perspectives. Compos Part B Eng 236:109826. https://doi.org/10.1016/j.compositesb.2022.109826

Cao L, Liu J, Meng Y, Hou M, Li J, Song Y, Wang Y, Song H, Zhang R, Liang R, Guo X (2024) A tear-free and edible dehydrated vegetables packaging film with enhanced mechanical and barrier properties from soluble soybean polysaccharide blending carboxylated nanocellulose. Int J Biol Macromol 264(Part 2):130707. https://doi.org/10.1016/j.ijbiomac.2024.130707

Chatterjee S, Mishra S, Chowdhury KD, Ghosh CK, Saha KD (2021) Various theranostics and immunization strategies based on nanotechnology against Covid-19 pandemic: an interdisciplinary view. Life Sci 278:119580. https://doi.org/10.1016/j.lfs.2021.119580

Chaudhari A, Vadodariya P, Vyas A, Patel D, Teli D (2024) Composition of nanoparticles. In: Chavda VP, Apostolopoulos V (eds) Nanocarrier vaccines: biopharmaceutics-based fast track development. https://doi.org/10.1002/9781394175482.ch2. Published online February 16, 2024

Chavda VP, Apostolopoulos V (2024) Nanocarrier vaccines: biopharmaceutics-based fast track development. Wiley. ISBN 1394174683, 9781394174683

Chen BZ, He YT, Zhao ZQ, Feng YH, Liang L, Peng J, Yang CY, Uyama H, Shahbazi MA, Guo XD (2023a) Strategies to develop polymeric microneedles for controlled drug release. Adv Drug Deliv Rev 203:115109. https://doi.org/10.1016/j.addr.2023.115109. ISSN 0169-409X

Chen BQ, Zhao Y, Zhang Y, Pan YJ, Xia HY, Kankala RK, Wang SB, Liu G, Chen AZ (2023b) Immune-regulating camouflaged nanoplatforms: a promising strategy to improve cancer nano-immunotherapy. Bioact Mater 21:1–19. https://doi.org/10.1016/j.bioactmat.2022.07.023

Cheng F, Wang Y, Bai Y, Liang Z, Mao Q, Liu D, Wu X, Xu M (2023) Research advances on the stability of mRNA vaccines. Viruses 15(3):668. https://doi.org/10.3390/v15030668

Chiang TL, Cheng AY, You JL, Chang CP, Ger MD (2024) Efficient copper plating on glass substrates through amine-functionalized nanocomposite-immobilized silver nanoparticles. Colloids Surf A Physicochem Eng Asp 688:133581. https://doi.org/10.1016/j.colsurfa.2024.133581

Chouhan AS, Rangi N (2023) A research on future scenario in the field of role of nanorobotics a device for diagnosis and treatment. Glob Acad J Med Scie [Internet]. Available from: www.gajrc.com. https://doi.org/10.36348/gajms.2023.v05i02.004. ISSN: 2706-9036 (Print) | ISSN: 2707-2533 (Online)

Colorado HA, Gutierrez-Velasquez EI, Gil LD, et al (2024) Exploring the advantages and applications of nanocomposites produced via vat photopolymerization in additive manufacturing: a review. Adv Compos Hybrid Mater 7(1):1. Available at: https://doi.org/10.1007/s42114-023-00808-z

Coltelli MB, Morganti P, Lazzeri A (2023) Sustainability assessment, environmental impact, and recycling strategies of biodegradable polymer nanocomposites. In: Deshmukh K, Pandey M (eds) Biodegradable and biocompatible polymer nanocomposites. Elsevier, pp 699–737. https://doi.org/10.1016/B978-0-323-91696-7.00016-7. ISBN 9780323916967

Constantino VRL, Figueiredo MP, Magri VR, Eulálio D, Cunha VRR, Alcântara ACS, Perotti GF (2023) Biomaterials based on organic polymers and layered double hydroxides nanocomposites: drug delivery and tissue engineering. Pharmaceutics 15(2):413. https://doi.org/10.3390/pharmaceutics15020413

Creighton R (2021) Integrated fiber microneedles for oral mucosal vaccine delivery. Dissertation. University of Washington. ProQuest Dissertations Publishing. ProQuest ID: 28643081

Das SK, Parandhaman T, Dey MD (2021) Biomolecule-assisted synthesis of biomimetic nanocomposite hydrogel for hemostatic and wound healing applications. Green Chem 23(2):706–719. Available from: https://doi.org/10.1039/D0GC03010D

de Moor WRJ (2020) The design, development and characterization of a self-replicating DNA expression technology. PhD thesis. University of Cape Town, Faculty of Health Sciences, Department of Pathology, Cape Town. http://hdl.handle.net/11427/36847. Supervisors: Rybicki E, Regnard G, Williamson AL

De Moraes IG, Fischbacher J, Hong Y, Naud C, Okuno H, Masseboeuf A, Devillers T, Schrefl T, Dempsey NM (2024) Nanofabrication, characterisation and modelling of soft-in-hard FeCo-FePt magnetic nanocomposites. Acta Mater 119970. Available at: https://doi.org/10.1016/j.actamat.2024.119970

Dehghani S, Esfahani SK, Ramezani M, Alibolandi M (2024) Chapter 5 – Hydrogel-based nanomedicines for cancer immunotherapy. In: Kesharwani P, editor. Nanomedicine in cancer immunotherapy. Academic Press. p. 139–174. Available from: https://doi.org/10.1016/B978-0-443-18770-4.00011-1. ISBN 9780443187704

Dejpasand MT, Rahemi Ardekani S, Sakhaei Ardekani A, Saievar-Iranizad E, Soleymani F (2024) Efficient green light-based photocatalytic removal of bilirubin by a biocompatible nanocomposite of nitrogen doped graphene quantum dots/Au. Mater Sci Semicond Process 171:108044. https://doi.org/10.1016/j.mssp.2023.108044. ISSN: 1369-8001

Ding M, Gao T, Song Y, Yi L, Li W, Deng C, Zhou W, Xie M, Zhang L (2024a) Nanoparticle-based T cell immunoimaging and immunomodulatory for diagnosing and treating transplant rejection. Heliyon 10. https://doi.org/10.1016/j.heliyon.2024.e24203

Ding X, Ma J, Fan T, Issa R, Li Y, Weng D, Zhang D, Chen Y (2024b) Inorganic nanoparticles-based strategies for the microbial detection in infectious diseases. Interdiscip Medi 2(2). https://doi.org/10.1002/INMD.20230045

Dowaidar M (2024) Uptake pathways of cell-penetrating peptides in the context of drug delivery, gene therapy, and vaccine development. Cell Signal 117:111116. https://doi.org/10.1016/j.cellsig.2024.111116

Dukarski W, Rykowska I, Krzyżanowski P, Gonsior M (2024) Flame retardant additives used for polyurea-based elastomers—a review. Fire 7(2):50. https://doi.org/10.3390/fire7020050

El-Sayed A, Kamel M (2020) Advances in nanomedical applications: diagnostic, therapeutic, immunization, and vaccine production. Environ Sci Pollut Res 27:19200–19213. https://doi.org/10.1007/s11356-019-06459-2

Eskandari A, Leow TC, Rahman MBA et al (2024) Advances in therapeutic cancer vaccines, their obstacles, and prospects toward tumor immunotherapy. Mol Biotechnol. https://doi.org/10.1007/s12033-024-01144-3

Eusébio D, Neves AR, Costa D, Biswas S, Alves G, Cui Z, Sousa Â (2021) Methods to improve the immunogenicity of plasmid DNA vaccines. Drug Discov Today 26(11):2575–2592. https://doi.org/10.1016/j.drudis.2021.06.008. ISSN 1359-6446

Eusébio D, Paul M, Biswas S, Cui Z, Costa D, Sousa Â (2024) Mannosylated polyethylenimine-cholesterol-based nanoparticles for targeted delivery of minicircle DNA vaccine against COVID-19 to antigen-presenting cells. Int J Pharm 654:123959. https://doi.org/10.1016/j.ijpharm.2024.123959. ISSN 0378-5173

Fan Y, Wen J, Shi L et al (2024) Engineering of smart nanostructured lipid carriers encapsulated in gold nanocomposites and camptothecin for synergistic photothermal and chemotherapy treatment of ovarian cancer. J Mater Sci 59:5518–5537. https://doi.org/10.1007/s10853-024-09481-z. Published March 17, 2024

Farkas C, Gál L, Csiszár A, et al. (2024) Sustainable printed circuit board substrates based on flame-retarded PLA/flax composites to reduce environmental load of electronics: quality, reliability, degradation and application tests. Sustainable Mater Technol 40:e00902. Available from: https://doi.org/10.1016/j.susmat.2024.e00902. ISSN 2214-9937

Feng ZX, Adolfsson KH, Xu YN, Fang HQ, Hakkarainen M, Wu MB (2021) Carbon dot/polymer nanocomposites: from green synthesis to energy, environmental and biomedical applications. Sustain Mater Technol 29. https://doi.org/10.1016/j.susmat.2021.e00304

Fonseca LMS, Machado BAS, Oliveira FO, Santos JRJ, Silva JW, Hodel KVS (2024) An overview on recent patents and technologies on nanoparticles for nucleic acid delivery. Nanotechnol Rev 3(2):171–186. https://doi.org/10.1080/13543776.2024.2338097

Gebre MS, Brito LA, Tostanoski LH, Edwards DK, Carfi A, Barouch DH (2021) Novel approaches for vaccine development. Cell 184(4):607–618. https://doi.org/10.1016/j.cell.2021.02.030

Gohar O, Khan MZ, Bibi I, Bashir N, Tariq U, Bakhtiar M, Abdul Karim MR, Ali F, Hanif MB, Motola M (2024) Nanomaterials for advanced energy applications: recent advancements and future trends. Mater Des 241:112930. https://doi.org/10.1016/j.matdes.2024.112930

Gomes Souza F Jr, Bhansali S, Pal K, Silveira Maranhão F, Santos Oliveira M, Valladão VS, Brandão e Silva DS, Silva GB (2024) A 30-year review on nanocomposites: comprehensive bibliometric insights into microstructural, electrical, and mechanical properties assisted by artificial intelligence. Materials 17(5):1088. https://doi.org/10.3390/ma17051088

Graf TP, Qiu SY, Varshney D, Laracuente ML, Euliano EM, Munnangi P, Pogostin BH, Baryakova T, Garyali A, McHugh KJ (2023) A scalable platform for fabricating biodegradable microparticles with pulsatile drug release. Adv Mater 35(22). https://doi.org/10.1002/adma.202300228

Gupta A, Choudhury AM, Meena J, Bauri S, Maiti P (2024) Ordered mesoporous silica delivering siRNA as cancer nanotherapeutics: a comprehensive review. ACS Biomater Sci Eng. https://doi.org/10.1021/acsbiomaterials.3c01749

Guyon R (2023) Development of a microfluidic-based platform for delayed vaccine delivery. PhD thesis. University of Oxford. Available at: https://ora.ox.ac.uk/objects/uuid:e74a5b19-9f68-433d-bae2-0c091c11fd06

Guzmán E, Ortega F, Rubio RG (2022) Layer-by-layer materials for the fabrication of devices with electrochemical applications. Energies 15(9):3399. https://doi.org/10.3390/en15093399

Hadi HS, Ali ZTA (2024) Green approach for the synthesis of nanocomposite used as reactive material to containment amoxicillin and lead in groundwater. Case Stud Chem Environ Eng 9:100707. https://doi.org/10.1016/j.cscee.2024.100707

Haider A, Khan S, Iqbal DN, Shrahili M, Haider S, Mohammad K, Mohammad A, Rizwan M, Kanwal Q, Mustafa G (2024) Advances in chitosan-based drug delivery systems: a comprehensive review for therapeutic applications. Eur Polym J 210:112983. https://doi.org/10.1016/j.eurpolymj.2024.112983

Hao M, Chen B, Zhao X, Zhao N, Xu F-J (2020) Organic/inorganic nanocomposites for cancer immunotherapy. Mater Chem Front 4:2571–2609. https://doi.org/10.1039/D0QM00323A

Harshitha M, Nayak A, Disha S, Akshath US, Dubey S, Munang'andu HM, Chakraborty A, Karunasagar I, Maiti B (2023) Nanovaccines to combat aeromonas hydrophila infections in warm-water aquaculture: opportunities and challenges. Vaccines 11(10):1555. https://doi.org/10.3390/vaccines11101555

Harun-Ur-Rashid M, Imran AB, Hasan SA (2023) Green polymer nanocomposites in automotive and packaging industries. Curr Pharm Biotechnol 24(1):145–163. https://doi.org/10.2174/1389201023666220506111027

Hasan M, Liu Q, Kanwal A et al (2024) A comparative study on green synthesis and characterization of Mn doped ZnO nanocomposite for antibacterial and photocatalytic applications. Sci Rep 14:7528. https://doi.org/10.1038/s41598-024-58393-0

Hassan MA (2024) Efficient synthesis of single-phase tin(II) oxide nanoparticles via microwave-assisted hydrothermal method. Int J Chem Chem Eng 2024:12–17

Hassan T, Salam A, Khan A et al (2021) Functional nanocomposites and their potential applications: a review. J Polym Res 28(1):36. https://doi.org/10.1007/s10965-021-02408-1. Published January 8, 2021

Hassan MH, Sajjad H, Zubair M (2024) Biomedical applications of graphene oxide: progress and challenges – a comprehensive review. Int Res J Educ Technol 6(3):2581–7795

Hassanzadeh P (2020) Nanotheranostics against COVID-19: from multivalent to immune-targeted materials. J Control Release 328:112–126. https://doi.org/10.1016/j.jconrel.2020.08.060

Hatae AC, Roque-Borda CA, Pavan FR (2023) Strategies for lipid-based nanocomposites with potential activity against mycobacterium tuberculosis: microbial resistance challenge and drug delivery trends. OpenNano 13:100171. https://doi.org/10.1016/j.onano.2023.100171

He H, Shi J, Yu S, Yang J, Xu K, He C, Li X (2024a) Exploring green and efficient zero-dimensional carbon-based inhibitors for carbon steel: from performance to mechanism. Constr Build Mater 411:134334. https://doi.org/10.1016/j.conbuildmat.2023.134334. ISSN: 0950-0618

He Y, Zhang S, She Y, Liu Z, Zhu Y, Cheng Q, Ji X (2024b) Innovative utilization of cell membrane-coated nanoparticles in precision cancer therapy. Exp Nanomed. Available from: https://doi.org/10.1002/EXP.20230164. Published online March 21, 2024

Hiremath A, Murthy AA, Thipperudrappa S, Bharath KN, Jones IP (2021) Nanoparticles filled polymer nanocomposites: a technological review. Cogent Eng 8(1). https://doi.org/10.1080/23311916.2021.1991229

Hsu CY, Kaur I, Ali E, Naeem YA, Alajeeli F, Younis NK, Faisal AN, Mohammed IH, Khalil NA, Jawad MA, Hussein MA, Ahmad I, Elawady A, Alsalamy A (2024) Emerging trends in environmentally sustainable synthesis of metal-organic frameworks incorporating copper nanoparticles for the treatment of gram-negative bacterial infections. J Organomet Chem 1010:123111. https://doi.org/10.1016/j.jorganchem.2024.123111. Available at: https://www.sciencedirect.com/science/article/pii/S0022328X24001062

Huang G, Huo S, Ren J, Chen W, Yang H, Xiao S, Wang T, Peng H, Song P (2024) Engineering Ce/P-functionalized g-C3N4 for advanced ABS nanocomposites exhibiting unparalleled fire retardancy, enhanced thermal and mechanical properties. Appl Mater Today 38:102191. https://doi.org/10.1016/j.apmt.2024.102191

Huda MN, Nafiujjaman M, Deaguero IG, Okonkwo J, Hill ML, Kim T, Nurunnabi M (2021) Potential use of exosomes as diagnostic biomarkers and in targeted drug delivery: progress in clinical and preclinical applications. ACS Biomater Sci Eng 7(6):2106–2149. https://doi.org/10.1021/acsbiomaterials.1c00217

Hughes KJ, Ganesan M, Tenchov R, Iyer KA, Ralhan K, Diaz LL et al (2024) Nanoscale materials at work: mapping emerging applications in energy, medicine, and beyond. ChemRxiv. https://doi.org/10.26434/chemrxiv-2024-s75wv. This content is a preprint and has not been peer-reviewed

Ingole SB, Sharma P, Verma R, Chowdhury S, Patil PP, Dwivedi SP, Khan AK (2024) Enhancing mechanical and thermal properties of polymer matrix nanocomposites through tailored nanomaterial architectures. E3S Web Conf 511, Article Number 01016, Pages 12. https://doi.org/10.1051/e3sconf/202451101016. Published online April 10, 2024

Jackson J, Camargos CHM, Liu C, Fonseca de Faria A et al (2024) Antimicrobial activity of thin-film composite membranes functionalized with cellulose nanocrystals and silver nanoparticles via one-pot deposition and layer-by-layer assembly. Environ Sci Water Res Technol 10(8):639–651. https://doi.org/10.1039/D3EW00826F

Jain N, Shahrukh S, Famta P, Shah S, Vambhurkar G, Khatri DK, Singh SB, Srivastava S (2023) Immune cell–camouflaged surface-engineered nanotherapeutics for cancer management. Acta Biomater 155:57–79. https://doi.org/10.1016/j.actbio.2022.11.001

Jamalipour Soufi G, Iravani S (2020) Eco-friendly and sustainable synthesis of biocompatible nanomaterials for diagnostic imaging: current challenges and future perspectives. Green Chem 22(9):2662–2687. https://doi.org/10.1039/D0GC00734J

Jin Z, Gao Q, Wu K, Ouyang J, Guo W, Liang X-J (2023) Harnessing inhaled nanoparticles to overcome the pulmonary barrier for respiratory disease therapy. Adv Drug Deliv Rev 202:115111. https://doi.org/10.1016/j.addr.2023.115111

Kadkhodaie Kashani G, Naghib SM, Soleymani S, Mozafari MR (2024) A review of DNA nanoparticles-encapsulated drug/gene/protein for advanced controlled drug release: current status and future perspective over emerging therapy approaches. Int J Biol Macromol 268(Part 1):131694. https://doi.org/10.1016/j.ijbiomac.2024.131694

Kalia S, Kumar R, Dhiman R, Singh RK (2024) Ion irradiation/implantation induced defect engineering and modification in graphene derivatives-based nanocomposites: energy storage/conversion and sensor. J Energy Storage 83:110650. https://doi.org/10.1016/j.est.2024.110650

Khalil A, Barras A, Boukherroub R, Tseng CL, Devos D, Burnouf T, Neuhaus W, Szunerits S (2024) Enhancing paracellular and transcellular permeability using nanotechnological approaches for the treatment of brain and retinal diseases. Nanoscale Horiz 9:14–43. https://doi.org/10.1039/D3NH00306J

Khan ZG, Agrawal TN, Bari SB, Nangare SN, Patil PO (2024) Application of surface nitrogen-doped graphene quantum dots in the sensing of ferric ions and glutathione: spectroscopic investigations and DFT calculations. Spectrochim Acta A Mol Biomol Spectrosc 306:123608. https://doi.org/10.1016/j.saa.2023.123608. ISSN: 1386-1425

Khosravi A, Zarepour A, Iravani S, Varma RS, Zarrabi A (2024) Sustainable synthesis: natural processes shaping the nanocircular economy. Environ Sci Nano 11(3):688–707. https://doi.org/10.1039/D3EN00973D

Kisakov DN, Belyakov IM, Kisakova LA, Yakovlev VA, Tigeeva EV, Karpenko LI (2023) The use of electroporation to deliver DNA-based vaccines. Vaccine:102–123. https://doi.org/10.1080/14760584.2023.2292772. Published online: 18 Dec 2023

Konwarh R, Singh AP, Varadarajan V, Cho WC (2024) Harnessing alginate-based nanocomposites as nucleic acid/gene delivery platforms to address diverse biomedical issues: a progressive review. Carbohydr Polym Technol Appl 7:100404. https://doi.org/10.1016/j.carpta.2023.100404

Koushki K, Keshavarz Shahbaz S, Keshavarz M, Bezsonov EE, Sathyapalan T, Sahebkar A (2021) Gold nanoparticles: multifaceted roles in the management of autoimmune disorders. Biomol Ther 11(9):1289. Available from: https://doi.org/10.3390/biom11091289

Kozak M, Hu J (2024) DNA vaccines: their formulations, engineering and delivery. Vaccines 12:71. Available from: https://doi.org/10.3390/vaccines12010071

Kulal D, Kodialbail VS (2021) Visible light mediated photocatalytic dye degradation using Ag2O/AgO-TiO2 nanocomposite synthesized by extracellular bacterial mediated synthesis – an eco-friendly approach for pollution abatement. J Environ Chem Eng 9(4):105389. https://doi.org/10.1016/j.jece.2021.105389

Kumar R, Kumar P (2024) Exploring the diverse biomedical applications of hybrid nanomaterials. In: Innovations and applications of hybrid nanomaterials, p 15. https://doi.org/10.4018/979-8-3693-3268-9.ch001

Kumar A, Jayeoye TJ, Mohite P, Singh S, Rajput T, Munde S, Eze FN, Chidrawar VR, Puri A, Prajapati BGP, Parihar A (2024) Sustainable and consumer-centric nanotechnology-based materials: an update on the multifaceted applications, risks and tremendous opportunities. Nano-Struct Nano-Objects 38:101148. https://doi.org/10.1016/j.nanoso.2024.101148

Kuna K, Baddam SR, Kalagara S, Akkiraju PC, Tade RS, Enaganti S (2024) Emerging natural polymer-based architectured nanotherapeutics for the treatment of cancer. Int J Biol Macromol 262(Part 1):129434. https://doi.org/10.1016/j.ijbiomac.2024.129434. ISSN 0141-8130

Kuperkar K, Atanase LI, Bahadur A, Crivei IC, Bahadur P (2024) Degradable polymeric bio(nano) materials and their biomedical applications: a comprehensive overview and recent updates. Polymers 16(2):206. https://doi.org/10.3390/polym16020206

Ledesma-Feliciano C, Chapman R, Hooper JW, Elma K, Zehrung D, Brennan MB, Spiegel EK (2023) Improved DNA vaccine delivery with needle-free injection systems. Vaccines 11(2):280. https://doi.org/10.3390/vaccines11020280

Li Z, Liu M, Young RJ (2024a) Dependence of the reinforcement of polymer-based nanocomposites upon the nanofiller geometry. Nano Mater Sci. https://doi.org/10.1016/j.nanoms.2024.04.014. ISSN 2589-9651

Li B, Ashrafizadeh M, Jiao T (2024b) Biomedical application of metal-organic frameworks (MOFs) in cancer therapy: stimuli-responsive and biomimetic nanocomposites in targeted delivery, phototherapy and diagnosis. Int J Biol Macromol 260(Part 2):129391. Available at: https://doi.org/10.1016/j.ijbiomac.2024.129391

Li M, Chen F, Yang Q, Tang Q, Xiao Z, Tong X, Zhang Y, Lei L, Li S (2024c) Biomaterial-based CRISPR/Cas9 delivery systems for tumor treatment. Biomater Res 28:0023. https://doi.org/10.34133/bmr.0023

Liu Y, Li M, Cui T, Chen Z, Xu L, Li W, Peng Q, Li X, Zhao D, Valencia CA, Dong B, Wang Z, Chow HY, Li Y (2024a) A superior heterologous prime-boost vaccination strategy against COVID-19: a bivalent vaccine based on yeast-derived RBD proteins followed by a heterologous vaccine. J Med Virol. https://doi.org/10.1002/jmv.29454. (Liu Y, Li M, Cui T contributed equally to this work)

Liu Y, Chen X, Lin X, Yan J, Yu DG, Liu P, Yang H (2024b) Electrospun multi-chamber core-shell nanofibers and their controlled release behaviors: a review. Wiley Interdiscip Rev Nanomed Nanobiotechnol 16(2). https://doi.org/10.1002/wnan.1954

Lu B, Lim JM, Yu B, Song S, Neeli P, Sobhani N, Pavithra K, Bonam SR, Kurapati R, Zheng J, Chai D (2024) The next-generation DNA vaccine platforms and delivery systems: advances, challenges and prospects. Front Immunol 15:1332939. https://doi.org/10.3389/fimmu.2024.1332939

Luisetto M, Almukthar N, Tarro G, Ahmadabadi NB, Edbey KEK, Khan FA, Hamid GA, Fiazza C, Cabianca L, Ilman I, Rafa YA, Rasool MG, Prince GG, Yurevich O (2022) Graphene and derivatives: physico-chemical and toxicology properties in the mRNA vaccine manufacturing strategy. Sci World J Pharm Sci 1(2). https://www.researchgate.net/publication/362806219

Madkour LH (2018) Biogenic–biosynthesis metallic nanoparticles (MNPs) for pharmacological, biomedical and environmental nanobiotechnological applications. Chron Pharm Sci 2(1):2572–7761

Maham M, Nasrollahzadeh M, Sajadi SM (2020) Facile synthesis of Ag/ZrO2 nanocomposite as a recyclable catalyst for the treatment of environmental pollutants. Compos Part B Eng 185:107783. https://doi.org/10.1016/j.compositesb.2020.107783

Malik S, Waheed Y (2024) Recent advances on vaccines against malaria: a review. Asian Pac J Trop Med 17(4):143–159. https://doi.org/10.4103/apjtm.apjtm_678_23

Malik S, Muhammad K, Waheed Y (2023) Emerging applications of nanotechnology in healthcare and medicine. Molecules 28(18):6624. https://doi.org/10.3390/molecules28186624

Malode SJ, Pandiaraj S, Alodhayb A, Shetti NP (2024) Carbon nanomaterials for biomedical applications: progress and outlook. ACS Appl Bio Mater 7(2):752–777. https://doi.org/10.1021/acsabm.3c00983

Manchanda N, Vishkarma H, Goyal M, Shah S, Famta P, Talegaonkar S, Srivastava S (2024) Surface functionalized lipid nanoparticles in promoting therapeutic outcomes: an insight view of the dynamic drug delivery system. Curr Drug Targets 25(4):278–300. https://doi.org/10.2174/0113894501285598240216065627

Meher MK, Naidu G, Mishra A, Poluri KM (2024) A review on multifaceted biomedical applications of heparin nanocomposites: progress and prospects. Int J Biol Macromol. Available from: https://doi.org/10.1016/j.ijbiomac.2024.129379. https://www.sciencedirect.com/science/article/pii/S014181302400182X

Mehta S, Suresh A, Nayak Y, Narayan R, Nayak UY (2022) Hybrid nanostructures: versatile systems for biomedical applications. Coord Chem Rev 460:214482. Available from: https://doi.org/10.1016/j.ccr.2022.214482

Miao Y, Wang S, Zhang B, Liu L (2023) Carbon dot-based nanomaterials: a promising future nano-platform for targeting tumor-associated macrophages. Front Immunol 14. https://doi.org/10.3389/fimmu.2023.1133238

Mirzaee S, Mirzaei-Kalar Z, Bayrami A et al (2024) Experimental and theoretical investigation on structure, toxicity, and DNA interaction of turmeric-coated cobalt ferrite nanocomposites. J Iran Chem Soc 21:387–398. https://doi.org/10.1007/s13738-023-02931-6

Mishra V, Sunil BDY, Judi HK, Jain AK, Anandhi RJ, Kaushik A (2024) A review on smart materials in biomedical applications: current trends and future challenges. E3S Web Conf 505:01023. Available from: https://doi.org/10.1051/e3sconf/202450501023

Mondal P, Sharma M, Chattopadhyay P, Datta S (2024) DNA vaccine: trials and possible therapeutic approaches. In: Ray S, Ghosh D (eds) Environment, development and sustainability. Vidyasagar College For Women, Kolkata. ISBN 978-93-95847-87-2. Available from: https://www.researchgate.net/publication/379565331

Mujtaba M, Negi A, King AWT, Zare M, Kuncova-Kallio J (2023) Surface modifications of nanocellulose for drug delivery applications; a critical review. Curr Opin Biomed Eng 28:100475. https://doi.org/10.1016/j.cobme.2023.100475. ISSN 2468-4511

Nguyen Dang BT, Kwon TK, Lee S, Jeong JH, Yook S (2024) Nanoparticle-based immunoengineering strategies for enhancing cancer immunotherapy. J Control Release 365:773–800. https://doi.org/10.1016/j.jconrel.2023.12.007. ISSN: 0168-3659

Nie R, Zhang QY, Feng ZY, Huang K, Zou CY, Fan MH, Zhang YQ, Zhang JY, Li-Ling JL, Tan B, Xie HQ (2024) Hydrogel-based immunoregulation of macrophages for tissue repair and regeneration. Int J Biol Macromol 268(Part 1):131643. https://doi.org/10.1016/j.ijbiomac.2024.131643

Nordin AH, Nur Husna SM, Ahmad Z et al (2023) Natural polymeric composites derived from animals, plants, and microbes for vaccine delivery and adjuvant applications: a review. Gels 9(3):227. https://doi.org/10.3390/gels9030227

Oladipo AO, Lebelo SL, Msagati TAM (2023) Nanocarrier design–function relationship: the prodigious role of properties in regulating biocompatibility for drug delivery applications. Chem Biol Interact 377:110466. Available from: https://doi.org/10.1016/j.cbi.2023.110466

Omidian H, Chowdhury SD (2023) Advancements and applications of injectable hydrogel composites in biomedical research and therapy. Gels 9(7):533. https://doi.org/10.3390/gels9070533

Omidian H, Akhzarmehr A, Dey Chowdhury S (2024) Hydrogel composites for multifunctional biomedical applications. J Compos Sci 8(4):154. https://doi.org/10.3390/jcs8040154

Osman AI, Chen Z, Elgarahy AM, Farghali M, Mohamed IMA, Priya AK, Hawash HB, Yap PS (2024) Membrane technology for energy saving: principles, techniques, applications, challenges, and prospects. 5(5). https://doi.org/10.1002/aesr.202400011. First published: 19 February 2024

Pagliero M, Comite A, Costa C (2024) Chapter 7 – Mixed matrix and nanocomposite membranes. In: Tavajohi N, Khayet M (eds) Polymeric membrane formation by phase inversion. Elsevier, pp 225–266. https://doi.org/10.1016/B978-0-323-95628-4.00013-6. ISBN 9780323956284

Pandiyarajan S, Manickaraj SSM, Liao AH, Baskaran G, Selvaraj M, Assiri MA, Zhou H, Chuang HC (2024) Supercritical CO2 mediated construction of aluminium waste recovered γ-Al2O3 impregnated Dracaena trifasciata biomass-derived carbon composite: a robust electrocatalyst for mutagenic pollutant detection. J Colloid Interface Sci 659:71–81. https://doi.org/10.1016/j.jcis.2023.12.117

Pandya A, Shah Y, Kothari N et al (2023) The future of cancer immunotherapy: DNA vaccines leading the way. Med Oncol 40:200. https://doi.org/10.1007/s12032-023-02060-3

Pengsomjit U, Alabdo F, Karuwan C, Kraiya C, Alahmad W, Ozkan SA (2024) Innovative graphene-based nanocomposites for improvement of electrochemical sensors: synthesis, characterization, and applications. Crit Rev Anal Chem:1–19. https://doi.org/10.1080/10408347.2024.2343854

Prasad A, Bakr MM, ElMeshad AN (2024) Surface-functionalized polymeric nanoparticles for breast cancer treatment: processes and advances. J Drug Target:1–28. https://doi.org/10.1080/1061186X.2024.2353359

Priyadarshi R, Ghosh T, Roy S, Rhim JW (2024) Biopolymer-based antimicrobial nanocomposite materials for food packaging and preservation. In: Jaiswal AK, Shankar S (eds) Food packaging and preservation. Academic Press, pp 33–52. https://doi.org/10.1016/B978-0-323-90044-7.00003-3. ISBN 9780323900447

Qamar SUR (2021) Nanocomposites: potential therapeutic agents for the diagnosis and treatment of infectious diseases and cancer. Coll Interface Sci Commun 43:100463. Available from: https://doi.org/10.1016/j.colcom.2021.100463

Qin T, Smink AM, de Vos P (2023) Enhancing longevity of immunoisolated pancreatic islet grafts by modifying both the intracapsular and extracapsular environment. Acta Biomater 167:38–53. Available from: https://doi.org/10.1016/j.actbio.2023.06.038

Rai M, Ingle AP, Yadav A, Golińska P, Trzcińska-Wencel J, Rathod S, Bonde S (2023) Nanotechnology as a promising approach for detection, diagnosis, and treatment of food allergens. Curr Nanosci 19:90–102. https://doi.org/10.2174/1573413718666220426101432

Raj A, Chowdhury A, Ali SW (2022) Green chemistry: its opportunities and challenges in colouration and chemical finishing of textiles. Sustain Chem Pharm 27:100689. https://doi.org/10.1016/j.scp.2022.100689

Rajput A, Shevalkar G, Pardeshi K, Pingale P (2023) Computational nanoscience and technology. OpenNano 12:100147. Available from: https://doi.org/10.1016/j.onano.2023.100147

Ravinayagam V, Jermy BR (2022) Nanodrug delivery systems for infectious diseases: from challenges to solutions. In: Hameed S, Rehman S (eds) Nanotechnology for infectious diseases. Springer. https://doi.org/10.1007/978-981-16-9190-4_13

Rehan F, Zhang M, Fang J, Greish K (2024) Therapeutic applications of nanomedicine: recent developments and future perspectives. Molecules 29(9):2073. https://doi.org/10.3390/molecules29092073

Ren S, Wang M, Wang C, Wang Y, Sun C, Zeng Z, Cui H, Zhao X (2021) Application of non-viral vectors in drug delivery and gene therapy. Polymers 13(19):3307. Available at: https://doi.org/10.3390/polym13193307

Rinoldi C, Zargarian SS, Nakielski P, Li X, Liguori A, Petronella F et al (2021) Nanotechnology-assisted RNA delivery: from nucleic acid therapeutics to COVID-19 vaccines. Methods Biosci 5(9):2100402. https://doi.org/10.1002/smtd.202100402

Ruman UE, Khan A, Fahad HM, Asif M, Shaheen F, Aziz MH, Ahmad R, Alam M, Sharif S, Afzal S (2024) Biogenic-ecofriendly synthesized SnO2/CuO/FeO/PVP/RGO nanocomposite for enhancing energy density performance of hybrid supercapacitors. J Energy Storage 89:111643. https://doi.org/10.1016/j.est.2024.111643. ISSN 2352-152X. Available at: https://www.sciencedirect.com/science/article/pii/S2352152X24012283

Saadh MJ, Al-dolaimy F, Alamir HTA, Kadhim O, Al-Abdeen SHZ, Sattar R, Jabbar AM, Abid MK, Jetti R, Alawadi A, Alsalamy A (2024) Emerging pathways in environmentally friendly synthesis of carbon-based quantum dots for exploring antibacterial resistance. Inorg Chem Commun 161:112012. https://doi.org/10.1016/j.inoche.2023.112012

Saddique A, Kim JC, Bae J, Cheong IW (2024) Low-temperature, ultra-fast, and recyclable self-healing nanocomposites reinforced with non-solvent silylated modified cellulose nanocrystals. Int J Biol Macromol 254(Part 3):127984. https://doi.org/10.1016/j.ijbiomac.2023.127984

Salabat A, Mirhoseini F (2022) Polymer-based nanocomposites fabricated by microemulsion method. J Appl Polym Sci 43(3):1282–1294. https://doi.org/10.1002/pc.26504

Salman A, Lupi S, Vaccari L, Piccirilli F, Eid MM (2023) FTIR microscopy evaluation of the immunogenicity of eco-friendly γFe2O3@Ag@Cs nanocomposite as a platform for the discovery and screening of vaccine adjuvants. Spectrochim Acta A Mol Biomol Spectrosc 287(Part 2):122087. https://doi.org/10.1016/j.saa.2022.122087

Sandbhor P, Palkar P, Bhat S, Johnb G, Godab JS (2024) Nanomedicine as a multimodal therapeutic paradigm against cancer: on the way forward in advancing precision therapy. Nanoscale 13. Available from: https://doi.org/10.1039/D3NR06131K

Sandua X, Rivero PJ, Esparza J, Fernández-Palacio J, Conde A, Rodríguez RJ (2022) Design of photocatalytic functional coatings based on the immobilization of metal oxide particles by the combination of electrospinning and layer-by-layer deposition techniques. Coatings 12:862. Available from: https://doi.org/10.3390/coatings12060862

Sapuan SM, Nurazzi NM, Shazleen SS, Norrrahim MNF, Ng LYF, Sabaruddin FA, Padzil FNM (2023) Production and performance properties of composites reinforced with environmentally benign nanomaterials. In: Nurazzi NM, Ilyas RA, Sapuan SM, Khalina A (eds) Synthetic and natural nanofillers in polymer composites. Woodhead Publishing, pp 143–171. https://doi.org/10.1016/B978-0-443-19053-7.00015-9. ISBN 9780443190537

Sarasati A, Wihadmadyatami H, Ana ID (2023) Carbonate apatite nanoparticles: a novel nano-adjuvant for oral mucosal vaccines and immunomodulator. OpenNano 12:100149. https://doi.org/10.1016/j.onano.2023.100149

Sarmadi M, Ta C, VanLonkhuyzen AM, De Fiesta DC, Kanelli M, Sadeghi I, Behrens AM, Ingalls B, Menon N, Daristotle JL, Yu J, Langer R, Jaklenec A (2022) Experimental and computational understanding of pulsatile release mechanism from biodegradable core-shell microparticles. Sci Adv 8(28). https://doi.org/10.1126/sciadv.abn5315

Schneider R, Facure MHM, Chagas PAM, Andre RS, dos Santos DM, Correa DS (2021) Tailoring the surface properties of micro/nanofibers using 0D, 1D, 2D, and 3D nanostructures: a review on post-modification methods. Adv Mater Interfaces 8(13):2100430. https://doi.org/10.1002/admi.202100430. Published June 1, 2021

Scotland BL, Shaw JR, Dharmaraj S, Caprio N, Cottingham AL, Lasola JJM, Sung JJ, Pearson RM (2023) Cell and biomaterial delivery strategies to induce immune tolerance. Adv Drug Deliv Rev 203:115141. Available from: https://doi.org/10.1016/j.addr.2023.115141

Selim MM, El-Safty S, Tounsi A, Shenashen M (2024) A review of magnetic nanoparticles used in nanomedicine. APL Mater. https://doi.org/10.1063/5.0191034. Published online January 31, 2024

Seth R, Meena A (2024) Enzymes-based nanomaterial synthesis: an eco-friendly and green synthesis approach. Clean Techn Environ Policy. https://doi.org/10.1007/s10098-024-02854-7

Shafaati M, Saidijam M, Soleimani M, Hazrati F, Mirzaei R, Amirheidari B, Tanzadehpanah H, Karampoor S, Kazemi S, Yavari B, Mahaki H, Safaei M, Rahbarizadeh F, Samadi P, Ahmadyousefi Y (2021) A brief review on DNA vaccines in the era of COVID-19. Future Virol 17(1). https://doi.org/10.2217/fvl-2021-0170

Shahid N, Erum A, Hanif S, Malik NS, Tulain UR, Syed MA (2024) Nanocomposite hydrogels – a promising approach towards enhanced bioavailability and controlled drug delivery. Curr Pharm Des 30(1):48–62. https://doi.org/10.2174/0113816128283466231219071151

Shakil F (2024) Reactive extrusion: filled polymer compounds and its applications. In: Sethi SK, Gupta HS, Verma A (eds) Polymer composites: from computational to experimental aspects, Materials horizons: from nature to nanomaterials. Springer, Singapore. https://doi.org/10.1007/978-981-97-0888-8_12

Shange MG, Khumalo NL, Mohomane SM, Motaung TE (2024) Factors affecting silica/cellulose nanocomposite prepared via the sol–gel technique: a review. Materials 17(9):1937. https://doi.org/10.3390/ma17091937

Shanmugam V, Mensah RA, Försth M, Sas G, Restás Á, Addy C, Xu Q, Jiang L, Neisiany RE, Singha S, George G, Tomlal JE, Berto F, Hedenqvist MS, Das O, Ramakrishna S (2021) Circular economy in biocomposite development: state-of-the-art, challenges and emerging trends. Compos C Open Access 5:100138. https://doi.org/10.1016/j.jcomc.2021.100138. Available at: https://www.sciencedirect.com/science/article/pii/S2666682021000335

Sharma P, Otto M (2024) Multifunctional nanocomposites modulating the tumor microenvironment for enhanced cancer immunotherapy. Bioact Mater 31:440–462. https://doi.org/10.1016/j.bioactmat.2023.08.022

Sharma R, Kumar H, Kumari R, Kumar G, Dhayal A, Yadav A, Yadav D, Yadav K, Saini C, Saloni, Kumar A, Pandit V (2024a) Innovative Al2O3-ZnO-TiO2@rGO nanocomposites: a versatile approach for advanced water purification, biomedical devices, and environmental remediation. Diam Relat Mater 145:111081. ISSN 0925-9635

Sharma M, Shah R, Saraf A et al (2024b) Current advances in the therapeutic potential of nanomedicines for pulmonary disorders. Emerg Mater. https://doi.org/10.1007/s42247-024-00708-y

Sheikhzadeh E, Beni V, Zourob M (2021) Nanomaterial application in biosensors for the detection of infectious diseases. Talanta 230:122026. https://doi.org/10.1016/j.talanta.2020.122026

Shi Y, Zhang Y, Zhu L, Miao Y, Zhu Y, Yue B (2024) Tailored drug delivery platforms: stimulus-responsive core–shell structured nanocarriers. Adv Healthc Mater 13(1). Available at: https://onlinelibrary.wiley.com/doi/abs/10.1002/adhm.202301726

Silva DF, Melo ALP, Uchôa AFC, Pereira GMA, Alves AEF, Vasconcellos MC, Xavier-Júnior FH, Passos MF (2023) Biomedical approach of nanotechnology and biological risks: a mini-review. Int J Mol Sci 24(23):16719. https://doi.org/10.3390/ijms242316719

Singh E, Kumar A, Lo SL (2024) Advancing nanobubble technology for carbon-neutral water treatment and enhanced environmental sustainability. Environ Res 252(Part 3):118980. https://doi.org/10.1016/j.envres.2024.118980

Sivanesan I, Gopal J, Muthu M, Shin J, Mari S, Oh J (2021) Green synthesized chitosan/chitosan nanoforms/nanocomposites for drug delivery applications. Polymers 13(14):2256. Available at: https://doi.org/10.3390/polym13142256

Song R, Murphy M, Li C, Ting K, Soo C, Zheng Z (2018) Current development of biodegradable polymeric materials for biomedical applications. Drug Des Devel Ther 12:3117–3145. Available from: https://doi.org/10.2147/DDDT.S165440

Sonowal L, Gautam S (2024) Self-assembly and nano-layering of apatite calcium phosphates in biomaterials. In: Advanced synthesis and medical applications of calcium phosphates, 1st edn. CRC Press, p 15. eBook ISBN 9781003360605

Tan J, Ding B, Teng B, Ma P, Lin J (2022) Understanding structure–function relationships of nanoadjuvants for enhanced cancer vaccine efficacy. Adv Funct Mater. https://doi.org/10.1002/adfm.202111670. Published January 5, 2022

Thakur A, Kumar A (2024) Strategies for enhancing biocompatibility of nanoparticles. In: Cruz JN (ed) Nanoparticle toxicity and compatibility. Materials research forum LLC. Available from: https://doi.org/10.21741/9781644902998-7. ISBN 1644902990, 9781644902998

Tibério MCSP, Reis CP (2023) A brief overview on nanomedicines' lifecycle from the product to the patient. Trabalho Final de Mestrado Integrado, Ciências Farmacêuticas, Universidade de Lisboa, Faculdade de Farmácia. Available from: http://hdl.handle.net/10451/64011

Tilwari A, Saxena R (2022) Biobased nanomaterials (plant-based for green materials) synthesis, properties and their application in biomedical science. In: Advanced nanocarbon materials, 1st edn. CRC Press, p 18. eBook ISBN 9781003110781

Tregoning JS, Stirling DC, Wang Z, Flight KE, Brown JC, Blakney AK, McKay PF, Cunliffe RF, Murugaiah V, Fox CB, Beattie M, Tam YK, Johansson C, Shattock RJ (2023) Formulation, inflammation, and RNA sensing impact the immunogenicity of self-amplifying RNA vaccines. Mol Ther Nucleic Acids 31:29–42. https://doi.org/10.1016/j.omtn.2022.11.024

Tripathi RM, Chung SJ (2021) Eco-friendly synthesis of SnO2-Cu nanocomposites and evaluation of their peroxidase mimetic activity. Nanomaterials 11(7):1798. https://doi.org/10.3390/nano11071798

Tripathy DB, Gupta A (2024) Nanocomposites as sustainable smart materials: a review. J Reinf Plast Compos. https://doi.org/10.1177/07316844241233162. First published online February 13, 2024

Uyanga KA, Li W, Daoud WA (2024) Exploiting cellulose-based hydrogels for sustainable, intelligent wearables in pandemic preparedness and control. Eur Polym J 212:113041. https://doi.org/10.1016/j.eurpolymj.2024.113041

Vafaeva KM, Chhetri A, Sudan P, Mishra M, Pakkiraiah B, Mohan C (2024) Polymer matrix nanocomposites for sustainable packaging: a green approach. E3S Web Conf 511:01008. Available from: https://doi.org/10.1051/e3sconf/202451101008. Published online April 10, 2024

Vagena IA, Gatou MA, Theocharous G, Pantelis P, Gazouli M, Pippa N, Gorgoulis VG, Pavlatou EA, Lagopati N (2024) Functionalized ZnO-based nanocomposites for diverse biological applications: current trends and future perspectives. Nanomaterials 14:397. https://doi.org/10.3390/nano14050397

Vasudevan M, Senthil Kumar B (2024) Recent advances in NPD: a comprehensive examination of process support for integrating nanotechnology. SSRN. Available from: https://doi.org/10.2139/ssrn.4814491. Published May 1. Accessed May 12, 2024

Verma C, Chauhan DS, Aslam R, Banerjee P, Aslam J, Quadri TW, Zehra S, Verma DK, Quraishi MA, Dubey S, AlFantazi A, Rasheed T (2024) Principles and theories of green chemistry for corrosion science and engineering: design and application. Green Chem 26(8):4270–4357. https://doi.org/10.1039/D3GC05207A

Verstappen K, Klymov A, Cicuéndez M, da Silva DM, Barroca N, Fernández-San-Argimiro FJ, Madarieta I, Casarrubios L, Feito MJ, Diez-Orejas R, Ferreira R, Leeuwenburgh SCG, Portolés MT, Marques PAA, Walboomers XF (2024) Biocompatible adipose extracellular matrix and reduced graphene oxide nanocomposite for tissue engineering applications. Mater Today Bio 26:101059. Available from: https://doi.org/10.1016/j.mtbio.2024.101059

Wallis J, Shenton DP, Carlisle RC (2019) Novel approaches for the design, delivery, and administration of vaccine technologies. Clin Exp Immunol 196(2):189–204. https://doi.org/10.1111/cei.13287

Wang H, Liang X, Wang J, Jiao S, Xue D (2020) Multifunctional inorganic nanomaterials for energy applications. Nanoscale 12:14–42. https://doi.org/10.1039/C9NR07008G

Wang X, Zhong X, Li J, Liu Z, Cheng L (2021) Inorganic nanomaterials with rapid clearance for biomedical applications. Chem Soc Rev. https://doi.org/10.1039/D0CS00461H

Wang EY, Sarmadi M, Ying B, Jaklenec A, Langer R (2023a) Recent advances in nano- and microscale carrier systems for controlled delivery of vaccines. Biomaterials 303:122345. https://doi.org/10.1016/j.biomaterials.2023.122345

Wang Y, Di S, Yu J, Wang L, Li Z (2023b) Recent advances of graphene–biomacromolecule nanocomposites in medical applications. J Mater Chem B. https://doi.org/10.1039/D2TB01962K

Wang Z, Wang X, Xu W, Li Y, Lai R, Qiu X, Chen X, Chen Z, Mi B, Wu M et al (2023c) Translational challenges and prospective solutions in the implementation of biomimetic delivery systems. Pharmaceutics 15(11):2623. https://doi.org/10.3390/pharmaceutics15112623

Wang J, Li L, Xu ZP (2024) Enhancing cancer chemo-immunotherapy: innovative approaches for overcoming immunosuppression by functional nanomaterials. Small Methods 8(1):2301005. Available from: https://doi.org/10.1002/smtd.202301005. First published 2023 Sept 24

Wei X, Song M, Jin G, Jia W, Wang J, Liang M, Zou L (2024) Multidimensional profiling of functionalized photothermal nanoplatforms for synergistic cancer immunotherapy: design, strategy, and challenge. Coord Chem Rev 499:215488. Available from: https://doi.org/10.1016/j.ccr.2023.215488. ISSN 0010-8545

Wu B, Huang J, Yu Y, Zhu H, Zhu K, Yang Y, Kuang T (2024) In-depth investigation of how carbon nanofiller dispersion affects microcellular foaming behavior in poly(butylene succinate) nanocomposites. J Supercrit Fluids 209:106252. https://doi.org/10.1016/j.supflu.2024.106252. ISSN 0896-8446

Xiao M, Tang Q, Zeng S, Yang Q, Yang X, Tong X, Zhu G, Lei L, Li S (2023) Emerging biomaterials for tumor immunotherapy. Biomater Res 27:47. https://doi.org/10.1186/s40824-023-00369-8
Xu Y, Zheng X, Li Y, Ye X, Cheng H, Wang H, Lyu J (2023a) Exploring patient medication adherence and data mining methods in clinical big data: a contemporary review. J Evid Based Med 16(3):342–375. https://doi.org/10.1111/jebm.12548
Xu W, Liu W, Yang J, Lu J, Zhang H, Ye D (2023b) Stimuli-responsive nanodelivery systems for amplifying immunogenic cell death in cancer immunotherapy. Immunol Rev. https://doi.org/10.1111/imr.13237. Published online July 5
Xu X, Wang X, Liao YP, Luo L, Xia T, Nel AE (2023c) Use of a liver-targeting immune-tolerogenic mRNA lipid nanoparticle platform to treat peanut-induced anaphylaxis by single- and multiple-epitope nucleotide sequence delivery. ACS Nano 17(5):4942–4957. https://doi.org/10.1021/acsnano.2c12420
Yashas SR, Shahmoradi B, Wantala K, Shivaraju HP (2021) Potentiality of polymer nanocomposites for sustainable environmental applications: a review of recent advances. Polymer 233:124184. https://doi.org/10.1016/j.polymer.2021.124184
Zdrehus R, Delcea C, Mocan L (2024) Role of biofunctionalized nanoparticles in digestive cancer vaccine development. Pharmaceutics 16(3):410. https://doi.org/10.3390/pharmaceutics16030410
Zhang K, Xu Z, Cheng S, Li Y, Zhang Y, Zhao X, Wang Y (2024a) Ambient synthesis of one-dimensional core-shell nanocomposites by encapsulating SiC nanowires with layered double hydroxide nanosheets for high-performance microwave absorption application. Mater Today Phys 40:101289. https://doi.org/10.1016/j.mtphys.2023.101289
Zhang C, Wang H, Shen J, Wang X (2024b) Dominant factors for moisture resistance of sol–gel silica coatings: surface chemical composition or inner microstructure? Appl Surf Sci 642:158616. https://doi.org/10.1016/j.apsusc.2023.158616. ISSN 0169-4332
Zhang JJ, Li Y, Guo S, Zhang W, Fang B, Wang S (2024c) Moving beyond traditional therapies: the role of nanomedicines in lung cancer. Front Pharmacol 15:1363346. Available from: https://doi.org/10.3389/fphar.2024.1363346

Asra Hamidi She known as **Jivam**, is a dedicated research specialist at Quantum Immunogenix under the mentorship of Professor Sunil Joshi. With a strong focus on medical biotechnology and quantum biology, her research explores innovative intersections between molecular genetics and quantum physics to address complex challenges in gene editing, epigenetics, and immunogenomics. Passionate about advancing the frontiers of science, Asra is particularly interested in leveraging quantum principles to enhance precision in gene targeting and engineering, designing innovative vaccines, and emphasizing ethical, sustainable approaches that minimize environmental and animal impact.

Masoumeh Poudineh She is a dedicated high school biology teacher with over 20 years of experience inspiring young minds. She holds a master's degree in biochemistry, reflecting her deep understanding of the molecular underpinnings of life. Her passion for integrating cutting-edge scientific research into her teaching has not only enriched her students' learning experiences but also positioned her as a valuable mentor in the field of biology education.

Biodegradable Materials for Customized Medical Devices Improve Personalized Medicine

12

Rishav Sharma, Rishabha Malviya, Mahendra Patel,
and Dhaval Raval

Abstract

Given the advancements in biodegradable materials, there is now a significant opportunity to transform healthcare technologies by utilizing sensors that decompose spontaneously after use. With the use of these materials, implantable electrical systems that don't need to be removed or repeated upon, lower chronic inflammatory reactions, and offer promising future directions for biomedical technology can be developed. Biodegradable materials are used to create ecologically friendly sensor systems that can assist in alleviating some of the key environmental concerns by reducing the quantity of electronic or medical waste generated and, subsequently, the carbon footprint. The last few decades have seen a notable surge in interest in biodegradable materials for use in biological applications. Biodegradable polymers (BDPs), both synthetic and natural, have been thoroughly explored for several biomedical applications. These consist of implants, scaffolds for tissue engineering, and screws for treating bone fractures, sutures, wound dressings, and other carriers for the targeted and prolonged release of medications. BDPs are commonly utilized as advantageous materials for the administration of medications to the eyes. Biodegradable materials for customized medical devices that improve personalized medicine are being made from a growing quantity of organically produced polymers, plant-based polysaccharides like alginate, cellulose, dextran, and starch, and protein-based polymers like collagen, chitosan, fibrin, silk, and gelatin. Most significantly, though, these methods avoided the requirement to remove the delivery systems following

R. Sharma · R. Malviya (✉) · M. Patel · D. Raval
Department of Pharmacy, School of Medical and Allied Sciences, Galgotias University, Greater Noida, Uttar Pradesh, India

Aculife HealthCare Pvt Ltd. (Nirma Group), Ahmedabad, Gujarat, India
e-mail: rishabha.malviya@galgotiasuniversity.edu.in

R. Malviya, S. Sundram (eds.), *Sustainable Nanocomposites with Green Biomaterials*, Biomaterials, Bioengineering and Sustainability 2,
https://doi.org/10.1007/978-3-031-79110-9_12

administration. This background allows the current chapter to provide a thorough summary of the structural and functional biodegradable materials that have been utilized to create a variety of electronic devices that are either biodegradable or bioresorbable. The latest developments in biodegradable materials for personalized medical devices that enhance personalized treatment were also covered in this chapter.

Keywords

Implants · Biomaterial · Biodegradable materials · Liposomes · Sensing device

Abbreviations

BDPs	Biodegradable polymers
AM	Additive manufacturing
RES	Reticuloendothelial system
PEG	poly(ethylene glycol)
FDA	Food Drug Administration
PLGA	poly(lactic-co-glycolic acid)
BBB	Blood-brain barrier
PBS	Phosphate Buffer Solution

12.1 Introduction

In the last 10 years, biodegradable sensors and electronic gadgets have become appealing substitutes for invasive and non-invasive health monitoring. These materials naturally break down or completely dissolve in their physiological settings (Kang et al. 2016; Tao et al. 2014). They offer a special chance for transient medical implants to be used for *in vivo* sensing and ongoing body state monitoring. Such sensor systems integrated into garments or wearables with flexible form factors could provide a sanitary way to monitor several physiological indicators (Dahiya et al. 2019; Manjakkal et al. 2019; Kim et al. 2011). Real-time monitoring of variables including strain, pressure, temperature, pH, oxygen, and other particular indicators will significantly enhance our understanding of tissue healing, early postsurgical infection identification, and customized treatment plans (Franco et al. 2022; Hosseini et al. 2020; Kafi et al. 2019; Navaraj et al. 2020; Dang et al. 2018). To improve patient comfort and do away with the need for disposable sensors, resorbable devices that provide short-term performance are excellent temporary implants for diagnostic and therapeutic applications. These devices must only function for a predetermined amount of time before failing naturally without the need for surgical removal (Zhai et al. 2022). Furthermore, due to their low cost, minimal environmental impact, and ability to replace disposable sensors with less waste, they are ideal for point-of-care diagnostic single-shot assessments (Hwang et al. 2012; Irimia-Vladu et al. 2012; Li et al. 2018; Bhattacharjee et al. 2020a, b).

Conventional medical sensors are mostly non-invasive or utilized externally; the drawback of implanted sensors is that they need to be removed surgically, which puts patients through extra challenges and discomfort during the retrieval procedure (Vajramani et al. 2005; Shakthivel et al. 2015). The primary advancements in the development of biodegradable sensors have occurred in the last ten years or so, beginning with partially degradable sensor materials. In the past few years, flexible form-factor wireless technologies, circuits, and power supplies have all been included in completely biodegradable systems made possible by functional materials and fabrication techniques (Kim et al. 2009). It is difficult to find appropriate materials for biodegradable sensors since they must be nontoxic, biocompatible, and biodegradable while still having high-performance electrical, optical, and mechanical qualities. Other crucial elements include flexibility and the right mechanical characteristics for conformal interface and low tissue stresses. The efficiency, precision, and accuracy of additive manufacturing (AM), also known as three-dimensional (3D) printing, has been demonstrated by scientific advancements in novel manufacturing approaches (Zhu et al. 2021). As a result, numerous engineering and biomedical applications have found their foundation, as demonstrated by Fig. 12.1. Implants tailored to each patient can be created using additive manufacturing (AM) technology without the use of molds or machining thanks to imaging techniques and computer-aided design (CAD) software (Arif et al. 2022; Touri et al. 2019; Singh and Ramakrishna 2017). Comparing this technology to traditional manufacturing procedures, it is very suitable for creating complex structures utilizing various materials (Liu et al. 2021; Javaid and Haleem 2018; Li et al. 2018). The automotive, aerospace, construction, jewelry, fast prototyping, and biomedical industries are just a few of the production sectors where this technology has shown promise throughout time (Hao et al. 2021; Nesic et al. 2020; Vanaei et al. 2021; Lamichhane et al. 2019; Rezvani Ghomi et al. 2019). Appropriate drug release is crucial for achieving the best possible therapeutic outcome in the human body. Various timed-release protocols are required for various clinical situations. Constant drug release is one significant form of release profile; this profile is significant since most drugs have a limited range of concentrations at which they work well in the human body. If the medication is taken below the range, it might not work as well as it could cause undesired side effects. The delivery of medication in pulses is another significant feature. It's necessary for some substances, including hormones, to be released in pulses at regular intervals. The release is timed to correspond with the human body's hormonal cycles (Kikuchi and Okano 2002; Stubbe et al. 2004). Other types include a rising release profile, which is crucial for patients who may grow tolerant to medicine over time. For example, a decreasing release profile is necessary to maintain health in cases of arthritis, when an excessive amount of medication is initially needed to alleviate pain in the morning, followed by lesser doses to prevent the pain from reoccurring. In these cases, the drug's initial large dose is needed to act against its targets quickly, and subsequent lower levels are then gradually reduced. For this reason, a broad range of release profiles is required for various situations. Furthermore, the notion of "individualized medicine" has garnered significant interest lately as a method that aims to tailor treatment to the needs

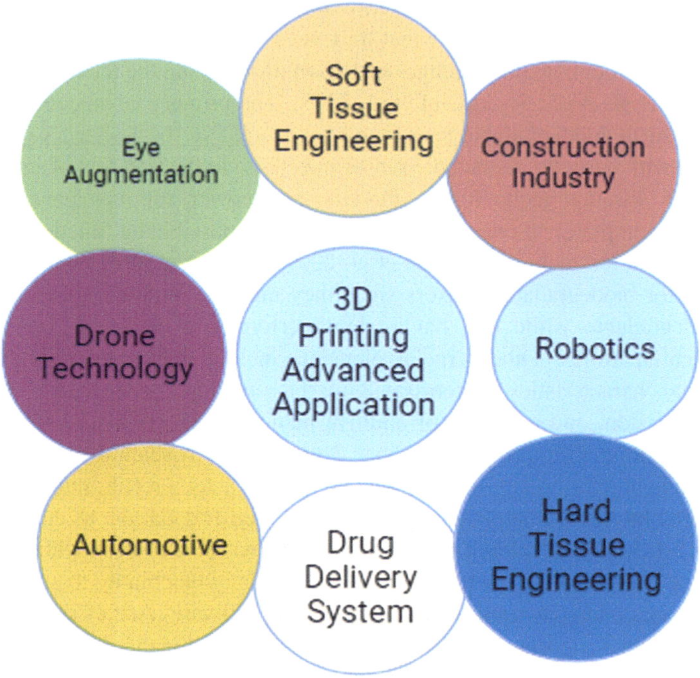

Fig. 12.1 Current scientific advancements in a range of engineering sectors

of certain patients (Stubbe et al. 2004; Hamburg and Collins 2010). Many factors must be taken into account to provide the most effective personalized treatment, including biological differences between individuals, variations in living and working environments, various illness types, and interactions between the drug being administered and other substances the body has consumed (such as food and herbs) (Wening and Breitkreutz 2011; Bates 2010). Therefore, it's critical to develop a universal, programmable method for creating drug carriers that may release medications according to a precise time frame and profile for a variety of applications.

12.2 Applications of Nano Biomaterials in Personalized Medicine

Drug delivery is currently the main focus of Bionanotechnology and nanomedicine research. Because just the precise amount of medication needed to treat an illness is delivered based on the needs of the patient, targeted drug delivery and controlled drug release are synonymous with personalized medicine. Considering nanoparticles can enter tissues more deeply through tiny capillaries and the epithelial lining, they can target particular tissues with minimal distribution to other tissues, increase diffusivity, and biodistribution, and lack immunogenicity (Arain et al. 2009). All of these properties allow for improved delivery efficiency. Systemic delivery of therapeutics should be accomplished by the use of 20–200 nm diameter nanoparticles. Larger particles quickly get eliminated from circulation by the reticuloendothelial system (RES) when being absorbed, but smaller particles typically pass through the hepatic sinusoidal endothelium's fenestration and cause hepatic accumulation rather than prolonged periods of circulation (Moghimi et al. 2001). In comparison to charged and hydrophobic particles, the surfaces of neutral and hydrophilic nanoparticles exhibit reduced opsonization (Zhang and Uludağ 2009; Roser et al. 1998). The failure of conventional therapies may be addressed with carefully crafted, site-specific targeted nanoparticles. Nanogels, liposomes, dendrimers, micelles, polymeric nanoparticles, and protein nanoparticles are examples of nanoparticle systems for drug delivery. The potential targets for active drug delivery encompass miserable cells with mutated cell surface receptors or elevated receptor levels. Nanoparticles are capable of higher concentration and/or focused cellular absorption when the right ligand is attached. Moreover, the circulatory behavior of nanoparticle drug carriers can be modulated by transforming their surface using polymer chains, such as poly(ethylene glycol) (PEG). PEG-ylation of nanotechnology leads to reduced elimination by phagocytes in the liver and spleen, despite PEG's low toxicity due to ethylene oxide, which considerably hinders the interaction of their outermost layer. Considering their surface interaction is significantly restricted, PEG-ylation of nanomaterials enhances inadequate clearance by phagocytes in the liver and spleen, because PEG has low toxicity due to ethylene oxide. Lower clearance prolongs the blood circulation of the carrier and the amount of time the drug is released from the body, both of which improve therapy. The delivery of contrasting chemicals for *in vivo* imaging and molecular therapy via target-specific injection may significantly influence the rate of disease identification, patient recuperation, and customized treatment (Ntziachristos et al. 2004). Bioimaging applications will be demonstrated, with an emphasis on the individual needs of the patient to improve diagnosis, treatment, prevention, and visualization. Biomedicine has been greatly impacted by nanotechnology. Many inorganic, polymeric, and natural nanomaterials and devices have been created with efficient treatment techniques in mind. The FDA has authorized more than 20 items based on nanotechnology for clinical use thus far. Commercially available controlled-release nanosystem, as previously discovered, offer several advantages over conventional drug delivery. These advantages include (i) prolonged drug half-life leading to enhanced therapeutic activity; (ii) improved solubility of hydrophobic drugs; and (iii) decreased immunogenicity. When it comes to bio-imaging, nanoparticles can make tissues (like tumors) more visible by improving their permeability and retention in that particular tissue (Davis et al. 2008).

12.3 Optimal Biomaterial

To preserve and improve human health, nanomedicine uses molecular instruments at the nanoscale for diagnosis, therapy, sickness prevention, traumatic injury rehabilitation, pain management, and other purposes. This technique entails precisely engineering molecules and/or functional assemblies at the nanoscale to create novel materials that will interact with tissues, organelles, and/or cells to produce therapeutic effects. In contrast with conventional materials with common composition, nanotechnology biomaterials exhibit unique chemical, physicochemical, immunoresponsive, and biological properties. However, the unique qualities that enable nanoparticles to perform unique tasks also give rise to worries about harmful consequences on biological systems and possible risks to human health. For example, endocytosis or diffusion over the cell membrane may be used by nanoparticles to induce biological reactions in cells (Ntziachristos et al. 2004). The metabolic processes and signaling pathways needed to regulate cell division, mortality, and proliferation might shift as a consequence of interaction. In the search for the perfect biomaterial, bioengineers, and polymer scientists in previous decades concentrated mostly on artificial polymers, liposomes, and biologically derived materials. However, because of its poor biocompatibility, immunogenicity, and cell toxicity, the initial generation of biomaterials was far from ideal for biomedical uses (Koutsopoulos 2012). The upcoming generation of nanosystem holds significance for personalized nanomedicine. Several methods that significantly outperform the conventional drug delivery pathways have been devised, and scaffolds made of nanomaterials have been created that replicate the majority of the extracellular matrix features found in natural organs. However, one of the biggest obstacles in medicine is still developing the perfect nano biomedical platform for tissue regeneration therapy or a customized drug delivery system (Koutsopoulos 2012).

12.4 Biocompatible Polymer Nanofabrications

Due to developments in polymer science, researchers created synthetic polymeric matrices in the 1960s to address many different kinds of medical issues (Koutsopoulos 2012). Throughout the years, numerous synthetic polymers have been developed and examined for potential use in biomedical settings. Triggers: (i) low biocompatibility; (ii) toxicity of synthesis chemicals; and/or (iii) toxicity of substances resulting from polymer degradation have been the main barriers to the practical use of polymers in clinical settings. Nevertheless, the advantages they offer and the absence of a competitive system usually outweigh these issues. Utilization in biomedical applications, and biocompatible, low-toxicity polymeric materials have come a long way in the past 20 years. The most suitable alternatives include biodegradable synthetic polymers for developing new kinds of prosthetic material devices, carriers for controlled-release medication delivery, and scaffolds for tissue engineering. Materials having well-identified physical, chemical, biological, biomechanical, and deteriorating qualities are required for each of these applications to give successful therapies. The biocompatible polymer matrices that are most common will be discussed in the following paragraphs.

12.4.1 Polymer Nanoparticles Interacting with Biomolecules

Nanoparticles of synthetic polymers having a diameter of about 50 nm were created by (Hoshino et al. 2008). To replicate the way that natural antibodies bind to particular proteins. When the patient is unable to develop an antibody against a particular antigen and the antigen is identified, plastic nanoparticle antibodies may be utilized for customized treatments. Plastic antibodies are functional biomaterials that are affordable, non-toxic, and stable. In biomedicine, compounds have potential applications in biomolecule separation, biosensors, antidotes for viruses and toxins, and diagnostics. This approach entails the binding sites of polymer nanoparticles through molecular imprinting when a target molecule is present. While polymer molecular imprinting was first reported in the 1950s (Mosbach and Ramström 1996), it was not until these conditions that its promise for the nanofabrication of synthetic antibodies was realized (Wulff 1995; Arshady and Mosbach 1981). To promote hydrogen bonding, hydrophobic interactions, and charge-charge couplings in the aqueous solution, more acrylamides were employed. N-isopropyl acrylamide was the polymer nanoparticles'backbone monomer and N, N′-methylene bisacrylamide performed the role of a cross-linker. Because heating and organic solvents are not used for polymer synthesis, there is less chance that the biomacromolecules utilized for imprinting will become denatured (Debord and Lyon 2003; Ogawa et al. 2003). After polymerization in the presence of melittin, prolonged dialysis is used to remove the melittin that is attached to the polymer nanoparticles. The polymer nanoparticles were designed to accurately attach to melittin and produce a population of melittin treatment recognition sites with an affinity constant comparable to that of a natural antibody. Plastic antibodies that have undergone PEG-ylation have blood circulation increases; these polymer nanoparticles are roughly the size of proteins.

12.4.2 Drug Delivery Using Polymer Nanoparticles

To enable active targeting of the particles, higher therapeutic efficacy, and fewer negative consequences compared to the drug combined with or non-targeted drug-carrying nanoparticles, drug-loaded polymeric nanoparticles can have their surfaces changed with ligands or antibodies. Personalized therapies may benefit from "smart" drug delivery systems, such as those that can selectively target particular body cells rather than any type of cell. The occurrence of targeting ligands on the surface of the nanoparticles confers advantages over conventional therapies, as it facilitates increased tumor permeability and accumulation in the cancerous tissue. Additionally, receptor-mediated endocytosis increases the uptake of drug-loaded nanoparticles by cancer cells. This mechanism leads to less harmful side effects, greater therapeutic activity, and higher intracellular drug concentration. For nano drugs to transcytose through strong endothelial and epithelial barriers, including the blood-brain barrier, ligand-mediated targeting is also essential (Gabathuler 2010). Most cell types do not allow bare siRNA to reach their cytoplasm at the same time, therapeutic siRNA treatments need to be effectively delivered into the target cells. By attaching cell binding or cell-penetrating peptides to the surface of polymer nanoparticles, nucleic acid transport can be improved (Nah et al. 2002; Parker and Seymour 2002). A range of hydrophobic and hydrophilic medicinal substances, including high molecular weight DNA, have been explored for encapsulation and delivery using poly (lactic-co-glycolic acid) (PLGA) nanoparticles. The preparation of PLGA nanoparticles is done by solvent displacement or emulsion solvent evaporation (Panyam and Labhasetwar 2003). Based on the relative uniformity in behavior, smaller size, and ease of dispersal of the particles formed in an aqueous environment, polyvinyl alcohol is a frequently utilized emulsifier for PLGA nanoparticle formulations. The use of organic solvents during manufacture has been found to negatively impact the bioactivity of medicinal substances, such as proteins (Crotts and Park 1998). Nevertheless, there are several concerns surrounding the use of PLGA in drug delivery.

12.4.3 Tissue Regeneration with Polymer Nanofiber Matrices

Biocompatible scaffolds that can be molded to closely resemble the shape and size of a specific patient's injured tissue offer an exciting possibility for customized nanomedicine applications, considering the scarcity of donor-supplied tissues and organs. Developing a scaffold requires awareness of the tissue's mechanical and physical attributes, as well as the dimension and type of pharmacological indicators attached to the material. The implanting of modified human ear tissue on a mouse's dorsal is arguably the most well-known example (Cao et al. 1997). Vacanti et al. established cartilage cells onto an artificial biodegradable polymer fashioned like an ear to demonstrate the feasibility of developing tissue-engineered cartilage to a specific form. To construct the polymer template, polyglycolic acid was soaked in a 1% solution of polylactic acid and then shaped into the shape of an ear. Chondrocytes from cow articular cartilage were injected into subcutaneous compartments on the mouse dorsal, and they were subsequently inserted into the polyglycolic acid–polylactic acid template. The separated chondrocytes replicated the body component after surviving insertion within the specially designed biocompatible, biodegradable polymer scaffold. Histology after 12 weeks revealed the development of cartilage, living and functioning chondrocytes, and a new extracellular matrix, which ultimately took the place of the destroyed synthetic scaffold. These results demonstrated that it is possible to make PLGA structures in the required shape and seed them with patient cells to create new tissues that may be used in reconstructive surgery. By utilizing an internal titanium wire skeleton encased in a porous collagen matrix, Vacanti's group has recently enhanced their technique. To address the PLGA scaffold's incapacity to maintain the construct's size and shape over extended periods, titanium wire was bent to resemble the ridges of a human auricle. Using patients' cells to create tissue-engineered urethras for urethral restoration is another example of personalized therapy (Raya-Rivera et al. 2011). After obtaining a tissue sample from each patient, polyglycolic acid: poly(lactide–co–glycolide acid) tabularized scaffolds were used to cultivate and seed muscle and epithelial cells. Endoscopic tests and further investigation flow assessments indicated that large urethral prostheses' absences of strictures were maintained for as long as 6 years following surgery. Furthermore, urethral biopsies tested 3 months after implantation confirmed that the targeted transplants had adopted a normal-looking structure.

12.5 Liposome

The concept of using liposomes as drug and protein delivery vehicles were initially put out in the 1960s, even though lipid-based systems were known to be harmful to cells in the 1950s (Rideal and Taylor 1957; Bangham and Horne 1962). Liposomes are made up of lipids that self-assemble into micelles in aqueous media. The hydrophilic micelle "head," which is in connection with the aqueous solvent, retains hydrophobic tail regions adjacent to the micelle center. Comparably, the bilayer of lipids forms around the aqueous center of small molecules. At the moment, several commercial pharmaceutical products are approved for use in clinical settings. They consist of the following: (i) Doxil, which is used to treat ovarian cancer; (ii) DaunoXome, which is used to treat Kaposi sarcoma; (iii) Myocet, which is used to treat women with metastatic breast cancer; and (iv) AmBisome, which is used to treat systemic fungal infections. These items include cholesterol liposomes encasing doxorubicin, hydrogenated soy phosphatidylcholine, and PEG-ylated stearoyl phosphatidylethanolamine. Some liposome-based formulations, however, exhibit increased toxicity against a variety of bodily cell types, an inflammatory response, and serum instability, which makes them less appealing for use in certain clinical settings. To avoid unwanted drug leakage, conventional pharmaceutical liposomes typically have moderately stable bilayers and an average diameter of more than or equal to 75 nm, which provides a sizable encapsulation volume.

12.5.1 Gene and Medication Delivery Using Liposomes

The versatility of liposomes, their openness to functionalization and molecular design, and their ability to encapsulate a broad variety of therapeutic agents. It was impossible to classify the first generation of liposomes as nanomedical platforms for customized treatment. Recent investigations suggest that liposomes with surface modifications and cell-targeting molecules, including monoclonal antibodies, would be a better option than standard liposomes for administering pharmaceutical material to target organs or particular cells to address a patient's health issues. It has long been known that monoclonal antibodies can be used in drug-targeting applications (Bernstein et al. 1980; Fatouros et al. 2006; Koutsopoulos et al. 2006). It was impossible to classify the first generation of liposomes as nanomedical platforms for customized treatment. When targeting cancerous cells with anti-cancer treatments, these systems can be utilized in a manner akin to that of regular liposomes. The lethal effects of arsenic, which can be imposed locally to destroy cancer cells, are combined with the beneficial properties of liposomes. Under certain circumstances, liposomes might be a better option than other capillary endothelia because blood-brain barrier (BBB) tight junctions are approximately 100 times less permeable than them. Precisely lipophilic compounds (b400 Da) can passively cross the blood-brain barrier, while other molecules are transported. Transport proteins on the surface of endothelial cells at the blood-brain barrier can convey the encapsulated payload once liposomes have been modified, for instance, with transferrin, antibodies, or triamcinol. Furthermore, polymeric and inorganic nanoparticles without PEG-ylation are less stable and stiffer than liposomes in circulation over long periods, which have mechanical and physicochemical characteristics that make them more suited to transporting medications across the blood-brain barrier.

12.5.2 Biodegradable vs Non-biodegradable Polymers

Polymeric materials that can break down into smaller pieces without harming the body are known as biodegradable polymers or BDPs for short. The polymeric chains are broken down by either enzymatic or nonenzymatic processes throughout the biodegradation process. Specialized enzymes are responsible for carrying out the enzymatic degradation process (Abbott et al. 2010; Tamariz and Rios-Ramírez 2013). Enzymatic breakdown typically occurs more easily with natural polymers. Nevertheless, enzymatic breakdown is a possibility for synthetic polymers. The degradation of poly(l-lactide) and poly(e-caprolactone) is exemplified by the actions of the enzyme's lipases and proteinase K. When a polymer undergoes nonenzymatic breakdown, its backbone splits apart in the presence of water, enabling water molecules to haphazardly break down the chemical bonds inside the polymer and lower its molecular weight (Lin and Anseth 2013; Niemelä and Kellomäki 2011). Non-BDPs, in contrast to BDPs, are made of lengthy hydrocarbon chains with chemical linkages that are incapable of being broken down by biological processes (Niemelä and Kellomäki 2011). Using BDPs has the apparent benefit of not requiring surgical removal following clinical application. BDPs are safe to metabolize and excrete from the human body through regular metabolic pathways; non-BDPs, on the other hand, may accumulate in various bodily tissues and cause toxicity. Both natural and synthetic types of BDPs are available (Imazato et al. 2017; Song et al. 2018). In tissue engineering, biological, and medication administration applications, both types are frequently utilized (Fig. 12.2) (George et al. 2019). Natural BDPs are frequently divided into two distinct groups depending upon their chemical composition and characteristics such as polysaccharides, which consist of chitosan, hyaluronic acid, alginate, and cellulose derivatives, and protein-based polymers, which include gelatine, albumin, and collagen (Esquivel et al. 2015). Furthermore, not only are natural polymers biodegradable and biocompatible but they can also be chemically

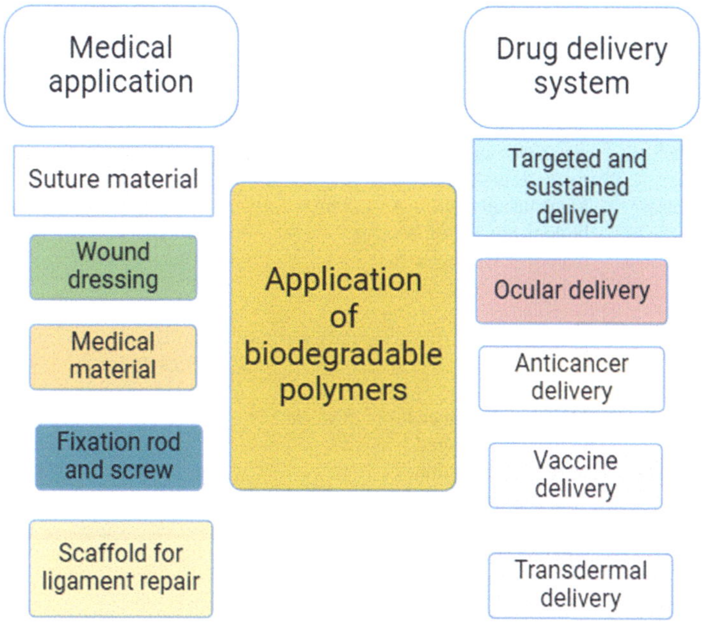

Fig. 12.2 Applications in medicine and different biodegradable polymer drug delivery systems

modified to improve their qualities. For example, chemical modification of chitosan is commonly utilized to improve bulk polymer characteristics and increase mucoadhesion capacity. However, the fields of tissue engineering, controlled medication delivery, and medical implants have found greater utility for synthetic BDPs (Gunatillake et al. 2003).

12.6 Biodegradable Sensors: Fabrication and Implementation

12.6.1 Physical Sensors

One of the most crucial types of physical sensors used to track bodily states is the pressure/strain sensor. For health monitoring, a variety of biodegradable pressure sensor types, including electronic skins and soft robotics, have been developed. These kinds of sensors consist of piezoelectric, piezocapacitive, or piezoresistive material in addition to capacitive structures, which are those that are filled with air or biodegradable dielectric materials. Antibiotics are injected into various bodily regions to avoid the accumulation of hazardous pressure inside of the skull in post-operative organs such as the brain, eyes, or muscles (Kang et al. 2016; Shin et al. 2019; Boutry et al. 2019. Resonant frequency mechanism-based, the hypothetical scenario consists of a biodegradable wireless capacitive pressure sensor encased in

Table 12.1 Comprehensive Overview of Various Biodegradable Pressure Sensor Types

Sensing device	Device design	Degradation time	*In vivo* and *In Vitro test*	References
Pressure sensor with a pseudocapacitive response	Ag nanowires deposited on the leaf skeleton as electrodes and rose petals as dielectric	75 days	PBS solution	Elsayes et al. (2020)
Pressure sensor with a pseudocapacitive response	Mechanically and biologically skin-like materials as a dielectric in between Au electrodes	8 h	NH 3 T3 fibroblasts seeded on PseD-U20 were assayed with MTT and live-dead assay. Lipase DPBS solution at 37 °C	Chen et al. (2020)
Pressure sensor with a pseudocapacitive response	MXene-impregnated tissue paper layered between a PLA thin sheet coated with interdigitated silver and a biodegradable polylactic acid thin sheet	8 weeks	PBS response at 37 °C, rhodamine-dyed NaOH solution	Sencadas et al. (2020)
Pressure sensor with a pseudocapacitive response	The approaches of electrospinning and magnetron sputtering were used to create silk fiber electrodes.	14 days	Papain bio enzymes aqueous solution at 37 °C	Hou et al. (2019)

polycaprolactone. It comprises a pair of parallel plates connected from each other by an electrodeposited conductor Zn/Fe bilayer layer of poly-L-lactic acid (Luo et al. 2013). A range of additional bioresorbable sensors are listed in Table 12.1, together with details on their stated composition, rate of degradation, and examination of samples *in vitro* or *in vivo*.

12.7 Biodegradable Materials for Improving Personalized Medicine

Preventative care techniques are becoming more popular as a result of personalized medicine, which adjusts therapies to each patient's unique physiology, genetic composition, and medical history. The advancement of low-modulus materials in medical equipment, among other personalized medicine techniques, is made possible by the development of surgical instruments, medicines, and medical devices. Thus, patient-specific 3D printed gadgets, point-of-care testing tools, and continuous and real-time health monitoring have become the forefront of biomedical research (Garcia et al. 2021). Protein-based polymers, including collagen, chitosan, fibrin, silk, and gelatin, as well as plant-based polysaccharides, like alginate, cellulose, dextran, and starch, are among the many naturally occurring polymers that are being used more and more in the development of biodegradable devices for improving personalized medicine. By using biodegradable materials, implantable biosensors can reduce chronic inflammatory reactions and do away with the necessity for extraction or reoperation (Xu et al. 2019). Biodegradable materials are typically either polymers (such as alginate, cellulose, and dextran), which are typically used as substrates, or metals (such as electrodes), which can dissolve or be broken down by bodily fluids, reducing the reaction to foreign bodies (Li et al. 2020b).

12.8 Future Perspectives

Recent developments in nanotechnology are making it possible to nanofabricate through minimally invasive treatments, nanosensors and biomaterials at the nanoscale are predicted to greatly facilitate diagnostics and personalized medical therapies. Using molecular tools at the nanoscale, nanomedicine focuses on maintaining and enhancing human health through diagnosis, therapy, disease prevention, trauma damage recovery, pain management, and more. To produce new materials with therapeutic benefits, this approach requires meticulously engineering molecules as well as functional assemblies at the nanotechnology that will have an impact on organelles, tissues, and/or cells. Nanoscale biomaterials exhibit unique chemical, physicochemical, immunoresponsive, and biological characteristics especially compared to conventional materials of identical composition. The primary objective of improving nanomaterial technologies is to develop an implantable medical device that depends on a set that the patient can use at home. The enclosed and regulated release of pharmaceuticals determined by the patient's genetic screening will be made possible by the use of the nanobiomedical device. The integration of cutting-edge drug delivery platforms with next-generation sequencing technology is necessary to save patient lives as this technology becomes a standard in clinical testing.

12.9 Conclusion

Biodegradable materials have two primary advantages over biostable polymers such as biodegradability and biocompatibility. The chemical and optical properties of these polymers may also be utilized to produce a variety of medical devices, comprising sutures, screws, wound healing, and bone fixation devices. Primarily in the field of ophthalmology, biodegradable polymeric substances are crucial to a customized and regulated drug delivery system. Different polymer combinations can be utilized to develop a range of ophthalmic drug delivery devices with desired properties, including stimuli-sensitive hydrogels, implants, scaffolds, nano/microparticles, inserts, and contact lenses. This chapter involves the advancement of biodegradable materials and monitoring technologies, which have become essential to sustainable healthcare systems and human health management. In the future, the way diseases are managed in healthcare systems may shift to preventative, predictive, and customized care due to the use of biodegradable wearable and implanted sensors to continuously monitor their physiological condition. Moreover, the development of sensors and electronic equipment that naturally disintegrates in the environment may contribute to the solution of the electronic waste problem. Materials can break down at different speeds and bioresorbable devices can dissolve in different ways in different biological systems. For practical applications, transient materials that are intended to regulate degradation at current rates are required. Finally, a deeper understanding of immune response and biological systems is needed to accurately design the device for the *in vivo* environment and to actualize these technologies for therapeutic usage. More communication and cooperation between biologists and engineers are therefore crucial.

References

Abbott NJ, Patabendige AA, Dolman DE, Yusof SR, Begley DJ (2010) Structure and function of the blood-brain barrier. Neurobiol Dis 37(1):13–25

Arain FA, Kuniyoshi FH, Abdalrhim AD, Miller VM (2009) Sex/gender medicine is the biological basis for personalized care in cardiovascular medicine. Circ J 73(10):1774–1782

Arif ZU, Khalid MY, ur Rehman E. (2022) Laser-aided additive manufacturing of high entropy alloys: processes, properties, and emerging applications. J Manuf Process 78:131–171

Arshady R, Mosbach K (1981) Synthesis of substrate-selective polymers by host-guest polymerization. Macromol Chem Phys 182(2):687–692

Bangham AD, Horne RW (1962) Action of saponin on biological cell membranes. Nature 196(4858):952–953

Bates S (2010) Progress towards personalized medicine. Drug Discov Today 15(3–4):115–120

Bernstein ID, Tam MR, Nowinski RC (1980) Mouse leukemia: therapy with monoclonal antibodies against a thymus differentiation antigen. Science 207(4426):68–71

Bhattacharjee M, Middya S, Escobedo P, Chaudhuri J, Bandyopadhyay D, Dahiya R (2020a) Microdroplet based disposable sensor patch for detection of α-amylase in human blood serum. Biosens Bioelectron 165:112333

Bhattacharjee M, Vilouras A, Dahiya RS (2020b) Microdroplet-based organic vapor sensor on a disposable GO-chitosan flexible substrate. IEEE Sensors J 20(14):7494–7502

Boutry CM, Beker L, Kaizawa Y, Vassos C, Tran H, Hinckley AC, Pfattner R, Niu S, Li J, Claverie J, Wang Z (2019) Biodegradable and flexible arterial-pulse sensor for the wireless monitoring of blood flow. Nat Biomed Eng 3(1):47–57

Cao Y, Vacanti JP, Paige KT, Upton J, Vacanti CA (1997) Transplantation of chondrocytes utilizing a polymer-cell construct to produce tissue-engineered cartilage in the shape of a human ear. Plast Reconstr Surg 100(2):297–302

Chen S, Sun L, Zhou X, Guo Y, Song J, Qian S, Liu Z, Guan Q, Meade Jeffries E, Liu W, Wang Y (2020) Mechanically and biologically skin-like elastomers for bio-integrated electronics. Nat Commun 11(1):1107

Crotts G, Park TG (1998) Protein delivery from poly (lactic-co-glycolic acid) biodegradable microspheres: release kinetics and stability issues. J Microencapsul 15(6):699–713

Dahiya R, Akinwande D, Chang JS (2019) Flexible electronic skin: from humanoids to humans [scanning the issue]. Proc IEEE 107(10):2011–2015

Dang W, Manjakkal L, Navaraj WT, Lorenzelli L, Vinciguerra V, Dahiya R (2018) Stretchable wireless system for sweat pH monitoring. Biosens Bioelectron 107:192–202
Davis ME, Chen Z, Shin DM (2008) Nanoparticle therapeutics: an emerging treatment modality for cancer. Nat Rev Drug Discov 7(9):771–782
Debord JD, Lyon LA (2003) Synthesis and characterization of pH-responsive copolymer microgels with tunable volume phase transition temperatures. Langmuir 19(18):7662–7664
Elsayes A, Sharma V, Yiannacou K, Koivikko A, Rasheed A, Sariola V (2020) Plant-based biodegradable capacitive tactile pressure sensor using flexible and transparent leaf skeletons as electrodes and flower petal as a dielectric layer. Adv Sustainable Syst 4(9):2000056
Esquivel R, Juárez J, Almada M, Ibarra J, Valdez MA (2015) Synthesis and characterization of new thiolated chitosan nanoparticles obtained by ionic gelation method. Int J Polym Sci 2015:502058
Fatouros DG, Ioannou PV, Antimisiaris SG (2006) Arsonoliposomes: novel nanosized arsenic-containing vesicles for drug delivery. J Nanosci Nanotechnol 6(9–10):2618–2637
Franco FF, Richard AH, Libu M (2022) Cu_2O-based electrochemical biosensor for non-invasive and portable glucose detection. Biosensors 12(3):174. https://doi.org/10.3390/bios12030174
Gabathuler R (2010) Approaches to transport therapeutic drugs across the blood-brain barrier to treat brain diseases. Neurobiol Dis 37(1):48–57
Garcia L, Kerns G, O'Reilley K, Okesanjo O, Lozano J, Narendran J, Broeking C, Ma X, Thompson H, Njapa Njeuha P, Sikligar D (2021) The role of soft robotic micromachines in the future of medical devices and personalized medicine. Micromachines 13(1):28
George A, Shah PA, Shrivastav PS (2019) Natural biodegradable polymers based nano-formulations for drug delivery: a review. Int J Pharm 561:244–264
Gunatillake PA, Adhikari R, Gadegaard N (2003) Biodegradable synthetic polymers for tissue engineering. Eur Cell Mater 5(1):1–6
Hamburg MA, Collins FS (2010) The path to personalized medicine. N Engl J Med 363(4):301–304
Hao W, Zheng Z, Zhu L, Pang L, Ma J, Zhu S, Du L, Jin Y (2021) 3D printing-based drug-loaded implanted prosthesis to prevent breast cancer recurrence post-conserving surgery. Asian J Pharm Sci 16(1):86–96
Hoshino Y, Kodama T, Okahata Y, Shea KJ (2008) Peptide imprinted polymer nanoparticles: a plastic antibody. J Am Chem Soc 130(46):15242–15243
Hosseini ES, Manjakkal L, Shakthivel D, Dahiya R (2020) Glycine–chitosan-based flexible biodegradable piezoelectric pressure sensor. ACS Appl Mater Interfaces 12(8):9008–9016
Hou C, Xu Z, Qiu W, Wu R, Wang Y, Xu Q, Liu XY, Guo W (2019) A biodegradable and stretchable protein-based sensor as artificial electronic skin for human motion detection. Small 15(11):1805084
Hwang SW, Tao H, Kim DH, Cheng H, Song JK, Rill E, Brenckle MA, Panilaitis B, Won SM, Kim YS, Song YM (2012) A physically transient form of silicon electronics. Science 337(6102):1640–1644
Imazato S, Kitagawa H, Tsuboi R, Kitagawa R, Thongthai P, Sasaki JI (2017) Non-biodegradable polymer particles for drug delivery: a new technology for "bio-active" restorative materials. Dent Mater J 36(5):524–532
Irimia-Vladu M, Głowacki ED, Voss G, Bauer S, Sariciftci NS (2012) Green and biodegradable electronics. Mater Today 15(7–8):340–346
Javaid M, Haleem A (2018) Additive manufacturing applications in medical cases: a literature-based review. Alexandria Journal of Medicine 54(4):411–422
Kafi MA, Paul A, Vilouras A, Hosseini ES, Dahiya RS (2019) Chitosan-graphene oxide-based ultra-thin and flexible sensor for diabetic wound monitoring. IEEE Sensors J 20(13):6794–6801
Kang SK, Murphy RK, Hwang SW, Lee SM, Harburg DV, Krueger NA, Shin J, Gamble P, Cheng H, Yu S, Liu Z (2016) Bioresorbable silicon electronic sensors for the brain. Nature 530(7588):71–76

Kikuchi A, Okano T (2002) Pulsatile drug release control using hydrogels. Adv Drug Deliv Rev 54(1):53–77

Kim DH, Kim YS, Amsden J, Panilaitis B, Kaplan DL, Omenetto FG, Zakin MR, Rogers JA (2009) Silicon electronics on silk as a path to bioresorbable, implantable devices. Appl Phys Lett 95(13):133701

Kim DH, Lu N, Ma R, Kim YS, Kim RH, Wang S, Wu J, Won SM, Tao H, Islam A, Yu KJ (2011) Epidermal electronics science 333(6044):838–843

Koutsopoulos S (2012) Molecular fabrications of smart nanobiomaterials and applications in personalized medicine. Adv Drug Deliv Rev 64(13):1459–1476

Koutsopoulos S, Fatouros DG, Ioannou PV, Antimisiaris SG (2006) Thermal behavior of novel non-sonicated arsonolipid-containing liposomes. Biophys Chem 121(2):150–154

Lamichhane S, Bashyal S, Keum T, Noh G, Seo JE, Bastola R, Choi J, Sohn DH, Lee S (2019) Complex formulations, simple techniques: can 3D printing technology be the Midas touch in the pharmaceutical industry? Asian J Pharm Sci 14(5):465–479

Li R, Wang L, Kong D, Yin L (2018) Recent progress on biodegradable materials and transient electronics. Biomed Mater 3(3):322–333

Li J, Wu C, Chu PK, Gelinsky M (2020b) 3D printing of hydrogels: rational design strategies and emerging biomedical applications. Mater Sci Eng R Rep 140:100543

Lin CC, Anseth KS (2013) The biodegradation of biodegradable polymeric biomaterials. In: Biomaterials science. Academic, pp 716–728

Liu G, Zhang X, Chen X, He Y, Cheng L, Huo M, Yin J, Hao F, Chen S, Wang P, Yi S (2021) Additive manufacturing of structural materials. Mater Sci Eng R Rep 145:100596

Luo M, Martinez AW, Song C, Herrault F, Allen MG (2013) A microfabricated wireless RF pressure sensor made completely of biodegradable materials. J Microelectromech Syst 23(1):4–13

Manjakkal L, Dang W, Yogeswaran N, Dahiya R (2019) Textile-based potentiometric electrochemical pH sensor for wearable applications. Biosensors 9(1):14

Moghimi SM, Hunter AC, Murray JC (2001) Long-circulating and target-specific nanoparticles: theory to practice. Pharmacol Rev 53(2):283–318

Mosbach K, Ramström O (1996) The emerging technique of molecular imprinting and its future impact on biotechnology. Bio/Technology 14(2):163–170

Nah JW, Yu L, Han SO, Ahn CH, Kim SW (2002) Artery wall binding peptide-poly (ethylene glycol)-grafted-poly (L-lysine)-based gene delivery to artery wall cells. J Control Release 78(1–3):273–284

Navaraj W, Smith C, Dahiya R (2020) E-skin and wearable systems for health care. In: Wearable bioelectronics. Elsevier, 178, p 133

Nesic D, Durual S, Marger L, Mekki M, Sailer I, Scherrer SS (2020) Could 3D printing be the future for oral soft tissue regeneration? Bioprinting 20:e00100

Niemelä T, Kellomäki M (2011) Bioactive glass and biodegradable polymer composites. In: Bioactive glasses. Woodhead Publishing, pp 227–245

Ntziachristos V, Schellenberger EA, Ripoll J, Yessayan D, Graves E, Bogdanov A Jr, Josephson L, Weissleder R (2004) Visualization of antitumor treatment using fluorescence molecular tomography with an annexin V–Cy5. 5 conjugate. Proc Natl Acad Sci 101(33):12294–12299

Ogawa K, Nakayama A, Kokufuta E (2003) Preparation and characterization of thermosensitive polyampholyte nanogels. Langmuir 19(8):3178–3184

Panyam J, Labhasetwar V (2003) Biodegradable nanoparticles for drug and gene delivery to cells and tissue. Adv Drug Deliv Rev 55(3):329–347. https://doi.org/10.1016/s0169-409x(02)00228-4

Parker AL, Seymour LW (2002) Targeting of polyelectrolyte RNA complexes to cell surface integrins as an efficient cytoplasmic transfection mechanism. J Bioact Compat Polym 17(4):229–238

Raya-Rivera A, Esquiliano DR, Yoo JJ, Lopez-Bayghen E, Soker S, Atala A (2011) Tissue-engineered autologous urethras for patients who need reconstruction: an observational study. Lancet 377(9772):1175–1182

Rezvani Ghomi E, Khalili S, Nouri Khorasani S, Esmaeely Neisiany R, Ramakrishna S (2019) Wound dressings: current advances and future directions. J Appl Polym Sci 136(27):47738

Rideal EK, Taylor FH (1957) On hemolysis by anionic detergents. Proc R Soc London Ser B 146(923):225–241

Roser M, Fischer D, Kissel T (1998) Surface-modified biodegradable albumin nano-and microspheres. II: effect of surface charges on in vitro phagocytosis and biodistribution in rats. Eur J Pharm Biopharm 46(3):255–263

Sencadas V, Tawk C, Alici G (2020) Environmentally friendly and biodegradable ultrasensitive piezoresistive sensors for wearable electronics applications. ACS Appl Mater Interfaces 12(7):8761–8772

Shakthivel D, Taube W, Raghavan S, Dahiya R (2015) VLS growth mechanism of Si-nanowires for flexible electronics. In: 2015 11th conference on Ph. D. Research in microelectronics and electronics (PRIME). IEEE, pp 349–352

Shin J, Yan Y, Bai W, Xue Y, Gamble P, Tian L, Kandela I, Haney CR, Spees W, Lee Y, Choi M (2019) Bioresorbable pressure sensors protected with thermally grown silicon dioxide for the monitoring of chronic diseases and healing processes. Nat Biomed Eng 3(1):37–46

Singh S, Ramakrishna S (2017) Biomedical applications of additive manufacturing: present and future. Curr Opin Biomed Eng 2:105–115

Song R, Murphy M, Li C, Ting K, Soo C, Zheng Z (2018) Current development of biodegradable polymeric materials for biomedical applications. Drug Des Devel Ther 24:3117–3145

Stubbe BG, De Smedt SC, Demeester J (2004) "Programmed polymeric devices" for pulsed drug delivery. Pharm Res 21:1732–1740

Tamariz E, Rios-Ramírez A (2013) Biodegradation of medical purpose polymeric materials and their impact on biocompatibility. In: Biodegradation-life of science. Intech, Croatia, pp 1–29

Tao H, Hwang SW, Marelli B, An B, Moreau JE, Yang M, Brenckle MA, Kim S, Kaplan DL, Rogers JA, Omenetto FG (2014) Silk-based resorbable electronic devices for remotely controlled therapy and in vivo infection abatement. Proc Natl Acad Sci 111(49):17385–17389

Touri M, Kabirian F, Saadati M, Ramakrishna S, Mozafari M (2019) Additive manufacturing of biomaterials−the evolution of rapid prototyping. Adv Eng Mater 21(2):1800511

Vajramani GV, Jones G, Bayston R, Gray WP (2005) Persistent and intractable ventriculitis due to retained ventricular catheters. Br J Neurosurg 19(6):496–501

Vanaei S, Parizi MS, Salemizadehparizi F, Vanaei HR (2021) An overview of materials and techniques in 3D bioprinting toward biomedical application. Eng Regener 2:1–8

Wening K, Breitkreutz J (2011) Oral drug delivery in personalized medicine: unmet needs and novel approaches. Int J Pharm 404(1–2):1–9

Wulff G (1995) Molecular imprinting in cross-linked materials with the aid of molecular templates—a way towards artificial antibodies. Angew Chem Int Ed Engl 34(17):1812–1832

Xu M, Obodo D, Yadavalli VK (2019) The design, fabrication, and applications of flexible biosensing devices. Biosens Bioelectron 124:96–114

Zhai Z, Du X, Long Y, Zheng H (2022) Biodegradable polymeric materials for flexible and degradable electronics. Front Electron 3:985681. https://doi.org/10.3389/felec.2022.985681

Zhang S, Uludağ H (2009) Nanoparticulate systems for growth factor delivery. Pharm Res 26:1561–1580

Zhu Y, Joralmon D, Shan W, Chen Y, Rong J, Zhao H, Xiao S, Li X (2021) 3D printing biomimetic materials and structures for biomedical applications. Bio-Des Manuf 4:405–428

Rishav Sharma is working as a scientist in F&D Department of Aculife HealthCare Pvt Ltd. (Nirma Group), Ahmedabad, Gujarat, India. He has authored two books and eight book chapters with reputed international publishers. He has published 15 SCI/Scopus manuscripts with reputed international publishers and attended more than 10 national and international conferences. He completed his BPharm at the Kanpur Institute of Technology and Pharmacy, Kanpur, Uttar Pradesh, India, and MPharm at Galgotias University, India.

Rishabha Malviya completed a B. Pharmacy from Uttar Pradesh Technical University and an M. Pharmacy (Pharmaceutics) from Gautam Buddha Technical University, Lucknow Uttar Pradesh. His PhD (Pharmacy) work was in Novel formulation development techniques. He has 14 years of research experience and has worked as an Associate Professor in the Department of Pharmacy, School of Medical and Allied Sciences, Galgotias University for the past 11 years. His area of interest includes formulation optimization, nanoformulation, targeted drug delivery, localized drug delivery, and characterization of natural polymers as pharmaceutical excipients. He has authored more than 150 research/review papers for national/international journals of repute. He has 51 patents (12 grants, 38 published, 1 filed) and publications in reputed National and International journals with a total of 140 cumulative impact factors. He has also received an Outstanding Reviewer award from Elsevier. He has authored/Edited/edited 91 books (Wily, CRC Press/Taylor and Francis, Springer Nature, Apple Academic Press, IOP Publishing, River Publisher, and OMICS publication) and authored 170 book chapters. His name has been included in Word's top 2% scientist list for the year 2020, 2021, 2022, 2023 and 2024 by Elsevier BV and Stanford University. He is a Reviewer/Editor/Editorial board member of more than 50 national and international journals of repute. He has been invited as an author for "Atlas of Science" and pharma magazine dealing with industry (B2B) "Ingredient South Asia Magazines".

Mahendra Patel Currently working as a Senior General Manager and R&D Head in Aculife HealthCare Pvt Ltd. (Nirma Group), Ahmedabad, Gujarat, India. He is Pharmaceutical development and manufacturing expert, with 20 plus years of experience in the pharmaceutical industry.

Dhaval Raval Currently working as a Senior Manager in the F&D Department of Aculife HealthCare Pvt Ltd. (Nirma Group), Ahmedabad, Gujarat, India. He has experience in pharmaceutical development and manufacturing, having worked in the pharmaceutical sector for more than 15 years

Antimicrobial Biomaterial for Dental Healthcare Application

13

Bhupinder Singh, Rishabha Malviya, and Christian Kaunert

Abstract

Dental diseases continue to be common even with improvements in oral care and cleanliness. Microbial infection is a significant contributing factor that can result in implant failure, periodontitis, and tooth decay. A potential remedy is offered by antimicrobial biomaterials, which offer infection control in addition to structural support. The growing clinical demands of dental patients have led to the development of a variety of dental biomaterials. For application in the oral environment, newly created dental biomaterials need to be both biocompatible and physically durable. At the moment, zirconia, titanium alloys, resin composites, and other materials are the most often utilized dental materials. Also, materials for tissue engineering that aim to regenerate oral tissues both soft and hard are gaining popularity. The design, modes of action, efficacy and clinical uses of antimicrobial biomaterials for dental applications are reviewed in this study along with the most recent advancements in this field. It cover a variety of antimicrobial agents, including as natural extracts mixed with dental materials, bioactive glass, metal nanoparticles and polymers. This chapter focuses on the diverse aspects on the Antimicrobial Biomaterial for Dental Healthcare Application.

Keywords

Biomaterial · Dental healthcare · Antimicrobial agents · Silver nanoparticles · Chitosan

---

B. Singh (✉)
Sharda University, Greater Noida, Uttar Pradesh, India

R. Malviya
Galgotias University, Greater Noida, Uttar Pradesh, India

C. Kaunert
Dublin City University, Dublin, Ireland

University of South Wales, Newport, UK

R. Malviya, S. Sundram (eds.), *Sustainable Nanocomposites with Green Biomaterials*, Biomaterials, Bioengineering and Sustainability 2,
https://doi.org/10.1007/978-3-031-79110-9_13

13.1 Introduction

Dental diseases such as decay and periodontitis are primarily caused by microbial infections. The traditional dental materials while providing structural integrity often fail to prevent microbial colonization. This has led to the exploration of antimicrobial biomaterials, aimed at supporting dental structures while preventing or killing pathogenic microorganisms (Montoya et al. 2023). Smart biomaterials have the ability to recognize and react to endogenous or extrinsic stimuli, including chemical, mechanical, electrical, or magnetic signals (Sam et al. 2023). The application and development of intelligent dental biomaterials for antimicrobial applications in dentistry has advanced significantly in the last several decades. The goal of these biomaterial systems is to increase the longevity of dental devices and avoid infections by offering improved efficacy and customizable bio-functionalities (Hakim et al. 2024).

Dental materials are essential for treating and enhancing oral health since teeth have a limited capacity for self-regeneration. Enamel is acellular and cannot self-regenerate, in contrast to bone and the dental pulp stem cell pool is necessary for the restricted capacity of dentin to self-regenerate. For example, in reaction to cavities, odontoblast cells deposit minerals and produce reactive dentin (Singh and Kaunert 2024a). With an atubular shape and cells trapped inside the matrix, this tissue is ill-organized. Dental materials have a crucial role in the diagnosis, treatment, prevention, and management of pathological disorders related to the mouth and teeth. For instance, following a bacterial infection that damaged the tissue in the teeth, resin composites are frequently utilized to restore function. Improved dentin regeneration and magnetic resonance imaging have been demonstrated by using colloidal gold nanoparticles (NPs) and superparamagnetic iron oxide as a theranostic agent for dental pulp capping (Hakim et al. 2024).

Even with the availability of dental materials, significant oral illnesses continue to be common, despite the fact that many of them may be avoided with basic self-care practices (Singh et al. 2024e). Dental materials are still an essential tool in dentistry because bacteria may create acids that demineralize hard tissues and contribute to the failure of both direct and indirect restorations, the oral cavity presents problems for dental materials. They accomplish this by interfering with salivary esterases which leads to the hydrolytic breakdown of dental resin adhesives (Mehrjou et al. 2023) (Fig. 13.1).

13.1.1 Background of the Study

The fields of materials science and pharmaceutical sciences are always working to counteract the spread of microbes while maintaining the healthy oral biota and preventing tissue damage. Because of this, there is now a lot of interest in restorative and preventative strategies to reduce the risk of secondary caries, oral microbial

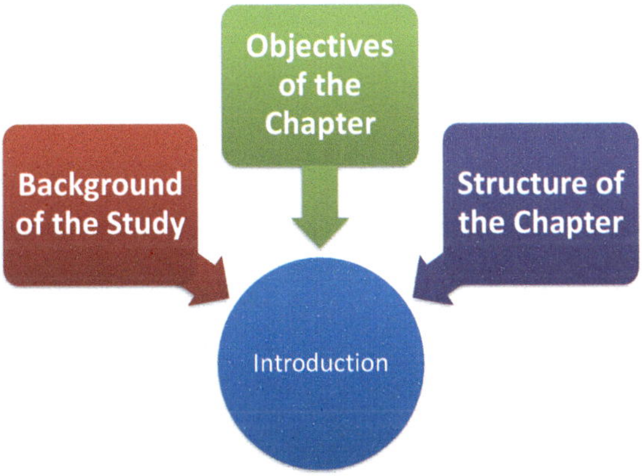

Fig. 13.1 Shows the landscapes of introduction split sections. (Source: Original)

biofilms, and tooth mineral retention (Chojnacka et al. 2023). As a result, novel tactics are being investigated; therefore, the focus of this study is on methods to prevent or restrict the development of bacterial biofilms. Pathogenic bacteria are the source of dental caries, or tooth decay, which affects over one-third of the world's population from early infancy to old age. In order to cure dental decay, cavities are commonly filled with filling materials; however, if tooth decay is not addressed, it can cause discomfort, infection, and ultimately tooth loss (Butler et al. 2023).

Many microorganisms can build bacterial biofilms on these filling materials, which can lead to secondary caries and the eventual failure of dental restorations. The oral cavity is home to many microbes. Antimicrobial dental materials are being researched because controlling caries requires limiting the growth of bacterial biofilms (Abraham and Venkatesan 2023). Thorough recommendations are vital to reduce the problems brought on by the prescription of needless antibiotics, especially the emergence of bacterial resistance. Studies reveal that just 12% of dentists accurately and appropriately prescribe antibiotics, highlighting the need of well-defined protocols. There are adverse effects from antibiotics might also include hypersensitivity responses, allergic reactions, and dermatological conditions (Singh and Kaunert 2024b).

13.1.2 Objectives of the Chapter

The purpose of this chapter is to analysis the types of antimicrobial agents used in dental biomaterials, their mechanisms of action, and their clinical effectiveness. It also examines the challenges faced in developing and implementing these materials in dental practice. This chapter has the following objectives to:

- create and enhance dental biomaterials that are inherently antimicrobial in order to stop bacteria from colonizing and forming biofilms on dental surfaces.
- adding antimicrobial agents that can prevent secondary caries and reduce restoration failures, we can extend the life and efficacy of dental restorations.

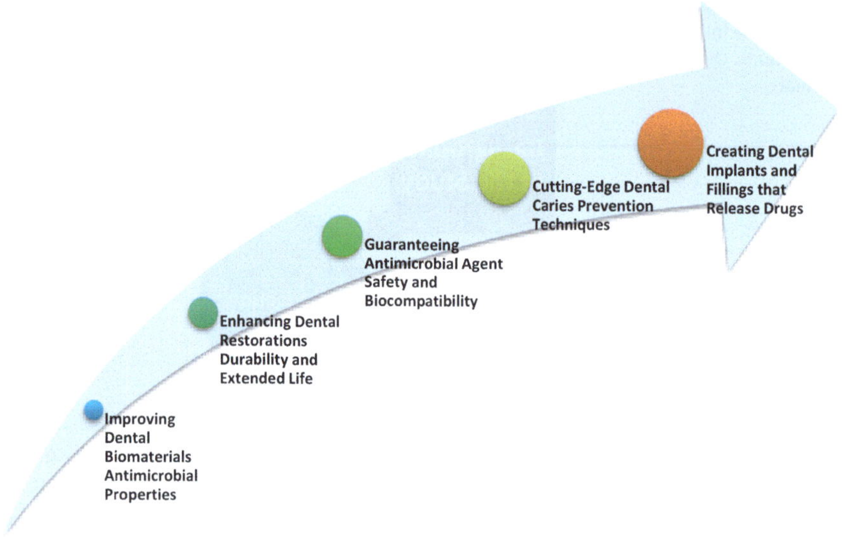

Fig. 13.2 Expresses the objectives of the chapter. (Source: Original)

- guarantee that the antimicrobial agents utilized in dental biomaterials are safe and biocompatible with oral tissues, preserving the oral microbiome's equilibrium and averting negative consequences.
- develop dental implants and fillings that, over time, release antimicrobial medications, preventing bacterial infections and promoting general oral health.
- investigate and provide novel preventative strategies that utilize antimicrobial biomaterials in order to lower the rate of dental caries and encourage the preservation of natural tooth structure (Fig. 13.2).

13.1.3 Structure of the Chapter

This chapter comprehensively explores the various provisions on Antimicrobial Biomaterial for Dental Healthcare Application. Section 13.2 elaborates the Dental biomaterials: Types of Antimicrobial Agent Information. Section 13.3 explores the Action Mechanisms: Microbial Cell Membrane Disruption. Section 13.4 shares the Challenges and Concerns. Finally, Sect. 13.5 Conclude the Chapter with Future Scope (Fig. 13.3).

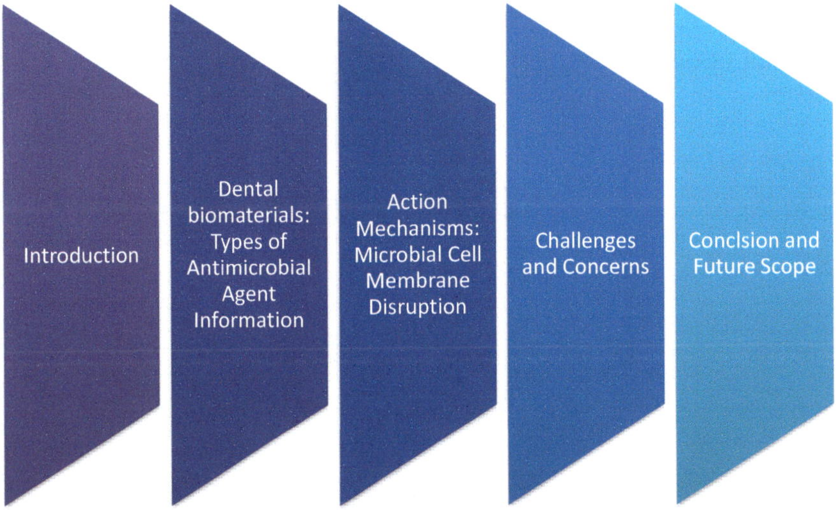

Fig. 13.3 Displays the flow of this chapter. (Source: Original)

13.2 Dental Biomaterials: Types of Antimicrobial Agent Information

Biomaterials and technology are promoting tissue regeneration along with restoring lost or injured tissues. Tooth stem cells and biomaterials have been the subject of much investigation because they may be able to regenerate tissue in the dentin, periodontal ligament, tooth pulp, and even enamel (Mace et al. 2024). Also, dental stem cells have been investigated as potential cell sources to support the restoration of non-dental tissues such bone and nerves. Materials that can promote the production of tertiary dentin, such as MTA or Biodentine have been the subject of several investigations. Still, few materials have been developed expressly to stimulate the regeneration of root-like structures, perhaps due to the lack of knowledge on the cellular processes underpinning effective tertiary dentin development (Singh 2024a).

Any biocompatible material that is utilized to replace, repair, or enhance damaged bodily tissues is referred to as a biomaterial. Biomaterials are used in many different medical specialties, including wound healing, orthopedics, dentistry, plastic surgery, and cardiology (Malheiros et al. 2023). With hundreds of microbes, enzymes, pH shifts, temperature swings, and intense mechanical loads, the mouth cavity poses a difficult environment. These circumstances, in addition to other mechanical, thermal, and biological changes, must be tolerated by biomaterials used in dentistry. The oral cavity uses more biomaterials than any other area of the body, including metals, polymers, ceramics, amalgams, and composites. This variety of materials is constantly being expanded by continuing research in this field (Singh et al. 2024e) (Fig. 13.4).

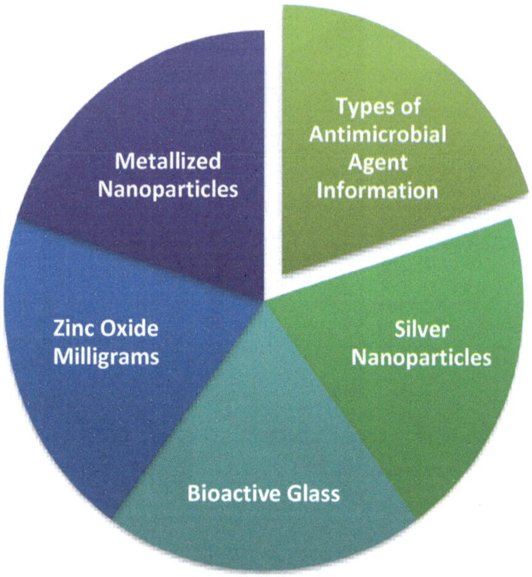

Fig. 13.4 Showcase the diverse aspects of dental biomaterials. (Source: Original)

13.2.1 Metallized Nanoparticles

There is different bio- and nanomaterials are currently used in dental implantology and dentistry. As, various biomaterials, such as polymers, dental cements, metal alloys, and dental ceramics, are used to restore and shield broken or missing teeth. Hydrogen peroxide and carbamide peroxide are frequently used for tooth whitening (Mace et al. 2024; Del Olmo et al. 2023). Every dental nano- and biomaterial has a unique mix of benefits and drawbacks. The development of nanotechnology has increased the quality of materials and broadened their uses. The addition of nanoparticles has lessened the negative aspects of the materials in many cases. With adding distinct functional groups, these nanoparticles graphene, carbon nanotubes, hydroxyapatite, silver, silica, titania and zirconia improve the quality of a range of bioproducts. With a low failure rate, titanium dental implants have been utilized for almost 50 years globally (de Campos et al. 2023). Because of its chemical makeup and capacity for crystal formation, hydroxyapatite, which covers the implant surface, may create chemical connections with inorganic materials, allowing titanium implants to be immediately integrated with the surrounding tissues. Several biocompatible nanocoatings are applied to dental implants to improve osseointegration and provide durable outcomes. With improving the osseointegration process, novel nano- and biocoatings applied to the implant surface might hasten the development of ankylosis. Drug-releasing implants may also be made using titania nanotubes, porous silicon, and nanoporous anodic alumina (Singh and Kaunert 2024c).

13.2.2 Silver Nanoparticles

Silver nanoparticles have a extensive variety of antibacterial activities and it can be added to composite resins and adhesives, among other dental materials to stop the growth of biofilms and the colonization of bacteria. Due to its antibacterial qualities, silver has been used in medicine for ages (Yin et al. 2023). Silver nanoparticles have been created recently and added to a variety of biomaterials. Even at low concentrations, they have a strong antibacterial impact due to their tiny size. As a result, dental material surfaces have been treated with these nanoparticles to inhibit or minimize the formation of biofilm (Atila and Kumaravel 2023).

13.2.3 Zinc Oxide Milligrams

Metal oxide nanoparticles called zinc oxide nanoparticles, or ZnO NPs, present fresh opportunities for biological uses ranging from diagnostics to therapy. These nanoparticles' antibacterial, regenerative, and mechanical qualities provide them a broad variety of uses (Bapat et al. 2023). Their applications include coating dental implants, improving the antifungal effects of denture bases in rehabilitative dentistry, remineralizing cervical dentinal lesions, boosting the stability of local drug delivery systems, improving the antibacterial properties of restorative materials, acting as an anti-sensitivity agent in toothpastes, and acting as an antimicrobial and antifungal agent against pathogenic oral microflora (Kunrath et al. 2023).

13.2.4 Bioactive Glass

Dentine, pulp tissue scaffolds, periodontal membranes, and bone cements are only a few examples of the oral tissue regeneration applications where the idea of tissue engineering in dentistry has previously proven successful. Bioactive glass (BAG) is one material that is frequently utilized. After discovering that inert metal and plastic materials were rejected in amputation situations, This substance in order to construct a transplant that would be compatible with the human body (Li et al. 2023). Glass known as BAG precipitates hydroxyapatite in aqueous solutions and forms a non-rejecting connection with both soft and hard tissues. BAG's bioactive qualities have transformed healthcare through its use in dentistry and medicine to regenerate hard tissues. Because nanotechnology makes it possible to synthesize BAG at the nanoscale which is helpful for covering dental (Singh 2024b) (Fig. 13.5).

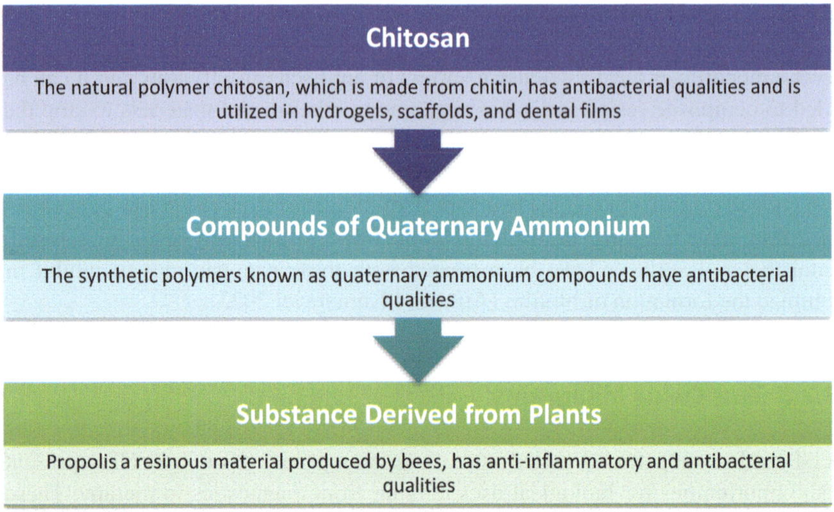

Fig. 13.5 Illuminates the types of antimicrobial agent information. (Source: Original)

13.3 Action Mechanisms: Microbial Cell Membrane Disruption

Biomaterials constitute a natural and artificial structural elemental analysis based on the interaction between structure, physicochemical qualities, and mechanical features, even if there isn't a single, widely accepted idea. The sequencing of the human genome at the end of the twentieth century served as a significant catalyst for advancement, even if the area of biomaterials is still being shaped by constant discussions and changes (Shineh et al. 2024). There has been a discernible movement in the last 20–25 years away from traditional synthetic biomaterials and toward the inclusion of really biological materials in everyday decisions. Dental biomaterials and natural goods are two rapidly expanding fields of study; items produced from plants have the potential to improve the physicochemical characteristics of biomaterials used in dentistry (Singh and Kaunert 2024d).

A number of antimicrobial substances including- chitosan and silver, cause leakage and microbial cell death by damaging the integrity of microbial cell membranes. For directed bone regeneration or tissue regeneration, biocompatible barriers and biomimetic compounds have been suggested, either separately or in combination (Sun et al. 2023). These materials are used because they have the ability to stimulate the development and differentiation of cells inside the periodontal apparatus, hence simulating the creation of teeth naturally and facilitating the establishment of periodontal attachment. The ability of several plant-derived compounds obtained from food and medicinal plants to treat periodontal disorders and promote periodontal healing has been emphasized (Paul et al. 2023).

13.3.1 Suppression of Enzymatic Function

A consortium of several species of microorganisms is immersed in a hydrated matrix of polysaccharides and polypeptides that the microbes have generated and gathered from their surroundings. The microbial resistance to the host's defense mechanism and antimicrobial or antibiotic treatments is strengthened by this biofilm matrix (Rodríguez-Hernández et al. 2023). Implant surfaces with microrough surfaces encourage bacterial adhesion and provide microbes with niches to occupy, which raises the possibility of microbial colonization and survival on the surface. Particularly during peri-implantitis treatment, this adhesion makes surface cleaning more difficult (Singh 2023).

13.3.2 Medical Uses: Restoration of Teeth

Dental restorations are used to restore natural dental function, replace lost teeth, fix damaged tooth structures, and improve the look of natural teeth. Different kinds of dental restorations are made from different materials (Hossain et al. 2023). The goals of these treatments are to restore lost teeth, preserve the integrity of the remaining tooth structures, stop neighboring teeth from moving, reduce the loss of supporting bone, and enhance the appearance of teeth that don't look well (Ortega et al. 2023). While cosmetic dentistry concentrates on improving the appearance of your smile, restorative dentistry seeks to improve dental health and function. Aesthetic restorative dentistry refers to dental operations that have both practical and aesthetic benefits (Khan and Shakoor 2023). For example, a dental crown enhances the regularity of your smile while restoring the ability to chew. The restorative dentistry is significant because it helps to maintain long-term dental health. Your dentist can fully restore your smile's functioning and enhance your ability to chew, speak, and consume by fixing broken or rotting teeth (Singh 2024c).

13.4 Challenges and Concerns

All nano- and biomaterials utilized in dentistry have their own advantages and drawbacks. Nanotechnology has significantly improved material quality and expanded their application possibilities (Saha et al. 2024). Many of the drawbacks associated with materials have been addressed by the introduction and utilization of nanoparticles (Singh et al. 2024f). Nanoparticles such as graphene, carbon nanotubes, hydroxyapatite, silver, silica, titania, zirconia, and others enhance the properties of various bioproducts by incorporating different functional groups into them. With a low failure rate, titanium dental implants have been used extensively around the world for almost 50 years (Thirumalai et al. 2024). Because of its chemical makeup and capacity to crystallize inorganic materials, hydroxyapatite coats the implant surface, promoting chemical bonding and prompt integration of titanium implants with surrounding tissues (Singh and Kaunert 2024b). A variety of biocompatible nanocoatings are applied to dental implants to improve osseointegration and provide durable outcomes (Thangaraju et al. 2023). With adding novel nano- and biocoatings to implant surfaces, one can accelerate the osseointegration process and encourage the formation of ankyloses more quickly. Also, titania nanotubes, porous silicon, and nanoporous anodic alumina are used in the creation of drug-releasing implants (Williams et al. 2023).

13.4.1 Quality of Biocompatibility

The word most commonly used to characterize a biomaterial's or biomaterials' appropriateness for a medical device with regard to biological needs is biocompatibility (Matusiak et al. 2023; Shineh et al. 2024). It can also mean that a substance has the capacity to evoke a suitable host response in a certain application (Correia et al. 2023). The assessment of safety or biocompatibility centers on figuring out the appropriate host response. The objective of this chapter is to present a workable method for determining safety and/or biocompatibility (Ersanli et al. 2023).

13.4.2 Development of Resistance

Dentists play a significant role, as they prescribe approximately 10% of all antibiotics for humans. This is of particular concern because antibiotics are the most commonly prescribed drugs by dental teams (Qu et al. 2023). It indicates that a significant number of these prescriptions are unnecessary. There is an urgent need for clear and pertinent guidelines to encourage dental teams to optimize their antibiotic prescribing practices, ensuring antibiotics are used only when absolutely necessary (Singh 2024d). The World Health Organization has classified antimicrobial resistance as a serious danger to world health and projects that by 2050, it will be responsible for ten million deaths yearly (Chen et al. 2023). This condition happens when antimicrobial medications are overused and misused, leading to a loss of efficacy in treating illnesses caused by bacteria, viruses, fungi, and parasites (Montoya et al. 2024). In the absence of prompt and concerted international action, we risk ushering in a post-antibiotic age in which ordinary diseases might potentially turn deadly (Xue et al. 2024). The extended-Duration Efficiency The durability of dental materials' antibacterial activity is crucial to their effectiveness. It is essential to conduct studies assessing these materials' long-term performance (Rajkumar et al. 2024).

13.4.3 Procedures for Regulation and Approval

Dental infections that arise from disorders such as periodontal diseases, dental caries, trauma, and pulpal necrosis can have detrimental effects on the oral cavity's hard and soft tissues (Taye et al. 2024). Fever, edema, and discomfort are typical symptoms (Budiul et al. 2024). Antibiotic medication is usually used after surgical and endodontic procedures for infected teeth. Infections can also be treated using other techniques including photodynamic treatment and low-level laser therapy (Chen et al. 2024). During dental operations, antibiotics are frequently administered to treat a variety of infections, including prophylactic, local, and focal infections as well as odontogenic and nonodontogenic infections (Liang et al. 2024). Patients with immunosuppressive illnesses, infective endocarditis, metabolic problems, and those with prosthetic joints may consider receiving antibiotic prophylaxis (Omid et al. 2024).

13.5 Conclusion and Future Scope

Antimicrobial biomaterials which offer the combined advantages of structural support and infection control which constitute a major breakthrough in dental treatment. Recent years have witnessed a substantial advancement in our understanding of dental biomaterials, with particular promise for biological engineering of organ systems and tissues. Historical viewpoints are being adjusted to provide innovative and intriguing new ideas as times change. Modern understanding of material structure has led to the development of the material science discipline known as biomaterials. The area of biomaterials is defined as the application of concepts from material science to biology. The surface characteristics of biomaterials and the biological integration of prostheses and tissue-engineering constructions affect how host cells and tissues react to implants. Despite their potential, there are obstacles in the way of their development and clinical use which calls for constant innovation and study. Antimicrobial biomaterials have the potential to greatly improve dental treatment results and patient quality of life by addressing these issues.

References

Abraham AM, Venkatesan S (2023) A review on application of biomaterials for medical and dental implants. Proc Inst Mech Eng Part L J Mater Des Appl 237(2):249–273

Atila D, Kumaravel V (2023) Advances in antimicrobial hydrogels for dental tissue engineering: regenerative strategies for endodontics and periodontics. Biomater Sci 11(20):6711–6747

Bapat RA, Bedia SV, Bedia AS, Yang HJ, Dharmadhikari S, Abdulla AM et al (2023) Current appraises of therapeutic applications of nanocurcumin: a novel drug delivery approach for biomaterials in dentistry. Environ Res 238:116971

Budiul MM, Vlase G, Negru D, Okolišan D, Bradu IA, Tăşală A et al (2024) Dental biopolymer composites with antibiotics, bisphosphonate, and hydroxyapatite for possible use in bone tissue regeneration. J Chem 2024:1–19

Butler J, Handy RD, Upton M, Besinis A (2023) Review of antimicrobial nanocoatings in medicine and dentistry: mechanisms of action, biocompatibility performance, safety, and benefits compared to antibiotics. ACS Nano 17(8):7064–7092

Chen Z, Chu Z, Jiang Y, Xu L, Qian H, Wang Y, Wang W (2023) Recent advances on nanomaterials for antibacterial treatment of oral diseases. Mater Today Bio 20:100635

Chen W, Zhang C, Peng S, Lin Y, Ye Z (2024) Hydrogels in dental medicine. Adv Ther 7(1):2300128

Chojnacka K, Moustakas K, Mikulewicz M (2023) Multifunctional cellulose-based biomaterials for dental applications: a sustainable approach to oral health and regeneration. Ind Crop Prod 203:117142

Correia BL, Malekshiri M, Bartolomeu M, Silva V, Oliveira A, Noites R et al (2023) P9-Antimicrobial properties and bioactivity potential of smart nanoparticles for dental applications. In: National meeting of the Center for Interdisciplinary Research in Health (CIIS), p 12

de Campos MR, Botelho AL, dos Reis AC (2023) Antimicrobial incorporation on 3D-printed polymers used as potential dental materials and biomaterials: a systematic review of the state of the art. Polym Bull 80(7):7313–7340

Del Olmo JA, Pérez-Álvarez L, Martínez VS, Cid SB, Ruiz-Rubio L, González RP et al (2023) Multifunctional antibacterial chitosan-based hydrogel coatings on Ti6Al4V biomaterial for biomedical implant applications. Int J Biol Macromol 231:123328

Ersanli C, Tzora A, Skoufos I, Voidarou C, Zeugolis DI (2023) Recent advances in collagen antimicrobial biomaterials for tissue engineering applications: a review. Int J Mol Sci 24(9):7808

Hakim LK, Yari A, Nikparto N, Mehraban SH, Cheperli S, Asadi A et al (2024) The current applications of nano and biomaterials in drug delivery of dental implant. BMC Oral Health 24(1):126

Hossain MI, Shams AB, Das Gupta S, Blanchard GJ, Mobasheri A, Hoque Apu E (2023) The potential role of ionic liquid as a multifunctional dental biomaterial. Biomedicines 11(11):3093

Khan SA, Shakoor A (2023) Recent strategies and future recommendations for the fabrication of antimicrobial, antibiofilm, and antibiofouling biomaterials. Int J Nanomedicine 18:3377–3405

Kunrath MF, Shah FA, Dahlin C (2023) Bench-to-bedside: feasibility of nano-engineered and drug-delivery biomaterials for bone-anchored implants and periodontal applications. Mater Today Bio 18:100540

Li J, Feng Y, Chen W, Zhang S, Ma J, Chen S et al (2023) Electroactive materials: innovative antibacterial platforms for biomedical applications. Prog Mater Sci 132:101045

Liang X, Yu B, Ye L, Lin D, Zhang W, Zhong HJ, He J (2024) Recent advances in quaternary ammonium monomers for dental applications. Materials 17(2):345

Mace MAM, Reginatto CL, Soares RMD, Fuentefria AM (2024) Three-dimensional printing of medical devices and biomaterials with antimicrobial activity: a systematic review. Bioprinting 38:e00334

Malheiros SS, Nagay BE, Bertolini MM, de Avila ED, Shibli JA, Souza JGS, Barão VA (2023) Biomaterial engineering surface to control polymicrobial dental implant-related infections: focusing on disease modulating factors and coatings development. Expert Rev Med Devices 20(7):557–573

Matusiak J, Przekora A, Franus W (2023) Zeolites and zeolite imidazolate frameworks on a quest to obtain the ideal biomaterial for biomedical applications: a review, vol 67. Mater Today, pp 495–517

Mehrjou B, Wu Y, Liu P, Wang G, Chu PK (2023) Design and properties of antimicrobial biomaterials surfaces. Adv Healthc Mater 12(16):2202073

Montoya C, Roldan L, Yu M, Valliani S, Ta C, Yang M, Orrego S (2023) Smart dental materials for antimicrobial applications. Bioact Mater 24:1–19

Montoya C, Baraniya D, Chen T, Al-Hebshi NN, Orrego S (2024) The effect of dental material type and masticatory forces on periodontitis-derived subgingival microbiomes. Biofilms 7:100199

Omid R, Banakar M, Mousavi SM, Lai CW (2024) Bio-based nanomaterials in dentistry. In: Biobased nanomaterials: applications in biomedicine, food industry, agriculture, and environmental sustainability. Singapore, Springer, pp 249–313

Ortega MA, De Leon-Oliva D, Boaru DL, Fraile-Martinez O, García-Montero C, Diaz R et al (2023) Unraveling the new perspectives on antimicrobial hydrogels: state-of-the-art and translational applications. Gels 9(8):617

Paul M, Pramanik SD, Sahoo RN, Dey YN, Nayak AK (2023) Dental delivery systems of antimicrobial drugs using chitosan, alginate, dextran, cellulose and other polysaccharides: a review. Int J Biol Macromol 247:125808

Qu S, Ma X, Yu S, Wang R (2023) Chitosan as a biomaterial for the prevention and treatment of dental caries: antibacterial effect, biomimetic mineralization, and drug delivery. Front Bioeng Biotechnol 11:1234758

Rajkumar DSR, Keerthika K, Vijayaragavan V (2024) Chitosan-based biomaterial in wound healing: a review. Cureus 16(2):e55193

Rodríguez-Hernández AP, Vega-Jiménez AL, Vázquez-Olmos AR, Ortega-Maldonado M, Ximenez-Fyvie LA (2023) Antibacterial properties in vitro of magnesium oxide nanoparticles for dental applications. Nano 13(3):502

Saha R, Roy P, Chakraborty J (2024) Synthesis and in-vitro evaluation of Zr-doped radiopaque bioactive glass: a possible biomaterial for endodontic application. Mater Chem Phys 320:129416

Sam S, Joseph B, Thomas S (2023) Exploring the antimicrobial features of biomaterials for biomedical applications. Results Eng 17:100979

Shineh G, Mobaraki M, Afzali E, Alakija F, Velisdeh ZJ, Mills DK (2024) Antimicrobial metal and metal oxide nanoparticles in bone tissue repair. Biomed Mater Devices 2:918

Singh B (2023) Unleashing alternative dispute resolution (ADR) in resolving complex legal-technical issues arising in cyberspace lensing E-commerce and intellectual property: proliferation of E-commerce digital economy. Revista Brasileira de Alternative Dispute Resolution-Brazilian Journal of Alternative Dispute Resolution-RBADR 5(10):81–105

Singh B (2024a) Evolutionary global neuroscience for cognition and brain health: strengthening innovation in brain science. In: Biomedical research developments for improved healthcare. IGI Global, pp 246–272

Singh B (2024b) Social cognition of incarcerated women and children: addressing exposure to infectious diseases and legal outcomes. In: Reddy K (ed) Principles and clinical interventions in social cognition. IGI Global, pp 236–251. https://doi.org/10.4018/979-8-3693-1265-0.ch014

Singh B (2024c) Lensing legal dynamics for examining responsibility and deliberation of generative AI-tethered technological privacy concerns: infringements and use of personal data by nefarious actors. In: Ara A, Ara A (eds) Exploring the ethical implications of generative AI, pp 146–167). IGI Global. https://doi.org/10.4018/979-8-3693-1565-1.ch009

Singh B, Kaunert C (2024a) Future of digital marketing: hyper-personalized customer dynamic experience with AI-based predictive models. In: Revolutionizing the AI-digital landscape: a guide to sustainable emerging technologies for marketing professionals, p 189

Singh B, Kaunert C (2024b) Salvaging responsible consumption and production of food in the hospitality industry: harnessing machine learning and deep learning for zero food waste. In: Singh A, Tyagi P, Garg A (eds) Sustainable disposal methods of food wastes in hospitality operations. IGI Global, pp 176–192. https://doi.org/10.4018/979-8-3693-2181-2.ch012

Singh B, Kaunert C (2024c) Revealing green finance mobilization: harnessing FinTech and blockchain innovations to surmount barriers and foster new investment avenues. In: Harnessing Blockchain-digital twin fusion for sustainable investments. IGI Global, pp 265–286

Singh B, Kaunert C (2024d) Harnessing sustainable agriculture through climate-smart technologies: artificial intelligence for climate preservation and futuristic trends. In: Exploring ethical dimensions of environmental sustainability and use of AI. IGI Global, pp 214–239

Singh B, Kaunert C, Vig K (2024e) Reinventing influence of artificial intelligence (AI) on digital consumer lensing transforming consumer recommendation model: exploring stimulus artificial intelligence on consumer shopping decisions. In: Musiolik T, Rodriguez R, Kannan H (eds) AI impacts in digital consumer behavior. IGI Global, pp 141–169. https://doi.org/10.4018/979-8-3693-1918-5.ch006

Singh B, Vig K, Kaunert C (2024f) Modernizing healthcare: application of augmented reality and virtual reality in clinical practice and medical education. In: Modern technology in healthcare and medical education: blockchain, IoT, AR, and VR. IGI Global, pp 1–21

Sun Z, Ma L, Sun X, Sloan AJ, O'Brien-Simpson NM, Li W (2023) The overview of antimicrobial peptide-coated implants against oral bacterial infections. Aggregate 4(3):e309

Taye MB, Ningsih HS, Shih SJ (2024) Exploring the advancements in surface-modified bioactive glass: enhancing antibacterial activity, promoting angiogenesis, and modulating bioactivity. J Nanopart Res 26(2):28

Thangaraju E, Varathan R, Duraisamy PS (2023) Antimicrobial nanosystems for dental applications. In: Antimicrobial nanosystems. Elsevier, pp 437–445

Thirumalai A, Girigoswami K, Harini K, Pallavi P, Gowtham P, Girigoswami A (2024) A review of the current state of probiotic nanoencapsulation and its future prospects in biomedical applications. Biocatal Agric Biotechnol 57:103101

Williams AG, Moore E, Thomas A, Johnson JA (2023) Graphene-based materials in dental applications: antibacterial, biocompatible, and bone regenerative properties. Int J Biomater 2023:8803283

Xue M, Chakraborty S, Gao R, Wang S, Gu M, Shen N et al (2024) Antimicrobial guanidinylate polycarbonates show oral in vivo efficacy against clostridioides difficile. Adv Healthc Mater 13:2303295

Yin Z, Liu Y, Anniwaer A, You Y, Guo J, Tang Y, ..., Huang C (2023) Rational designs of biomaterials for combating oral biofilm infections. Adv Mater 1:e2305633

Bhupinder Singh working as Professor at Sharda University, India. Also, Honorary Professor in University of South Wales UK and Santo Tomas University Tunja, Colombia. His areas of publications as Smart Healthcare, Medicines, fuzzy logics, artificial intelligence, robotics, machine learning, deep learning, federated learning, IoT, PV Glasses, metaverse and many more. He has 3 books, 139 paper publications, 163 paper presentations in international/national conferences and seminars.

Rishabha Malviya completed a B. Pharmacy from Uttar Pradesh Technical University and an M. Pharmacy (Pharmaceutics) from Gautam Buddha Technical University, Lucknow Uttar Pradesh. His PhD (Pharmacy) work was in Novel formulation development techniques. He has 14 years of research experience and has worked as an Associate Professor in the Department of Pharmacy, School of Medical and Allied Sciences, Galgotias University for the past 11 years. His area of interest includes formulation optimization, nanoformulation, targeted drug delivery, localized drug delivery, and characterization of natural polymers as pharmaceutical excipients. He has authored more than 150 research/review papers for national/international journals of repute. He has 51 patents (12 grants, 38 published, 1 filed) and publications in reputed National and International journals with a total of 140 cumulative impact factors. He has also received an Outstanding Reviewer award from Elsevier. He has authored/Edited/edited 91 books (Wily, CRC Press/Taylor and Francis, Springer Nature, Apple Academic Press, IOP Publishing, River Publisher, and OMICS publication) and authored 170 book chapters. His name has been included in Word's top 2% scientist list for the year 2020, 2021, 2022, 2023 and 2024 by Elsevier BV and Stanford University. He is a Reviewer/Editor/Editorial board member of more than 50 national and international journals of repute. He has been invited as an author for "Atlas of Science" and pharma magazine dealing with industry (B2B) "Ingredient South Asia Magazines".

Christian Kaunert is Professor of International Security at Dublin City University, Ireland. He is also Professor of Policing and Security, as well as Director of the International Centre for Policing and Security at the University of South Wales. In addition, he is Jean Monnet Chair, Director of the Jean Monnet Centre of Excellence and Director of the Jean Monnet Network on EU Counter-Terrorism Additionally, he was awarded a large Horizon 2020 research grant on Terrorist Radicalisation processes Mindb4Act. He is currently Editor of the Journal of Contemporary European Studies, was also previously an elected member of the national (UK) Executive Committee of the University Association for Contemporary European Studies (UACES).

Recent Trends in Polyhydroxyalkanoates: Bridging the Gap Between Bench and Bedside

14

Ashok Raja Chandrasekar

Abstract

Polyhydroxyalkanoates (PHA) have garnered considerable interest as a highly promising class of biomedical materials for diverse applications by virtue of their superior biocompatibility, biodegradability, and tuneable properties. The latest development of PHA biomaterials and its application in medicine will be discussed in this chapter. In addition, the development in few techniques for the production of PHA such as microbiological fermentation and chemosynthesis will be overviewed. These techniques enable researchers to accurately define the characteristics of polymers in terms of molecular weight, monomer composition, and mechanical properties. Furthermore, it addresses novel strategies for improving the biofunctionality of PHA materials through the methods of surface functionalization, blending with different materials, and bioactive agent loading technique. The biomedical applications of PHA as materials for tissue engineering scaffolds, drug delivery systems and wound dressing are discussed. It also provides the outcomes of research studies and preclinical animal trials that have been shown to be successful using PHA-based systems and lastly their clinical relevance at various levels. Overall, this chapter gives an insight of the issues and future developments regarding the biological performances of PHA biomaterials for medical uses, and it highlights the unmet needs for improved properties toward the regulatory approval for human use.

A. R. Chandrasekar (✉)
Department of Coating Processes, FunGlass, Alexander Dubček University of Trenčín, Trenčín, Slovakia
e-mail: ashokraja.chandrasekar@tnuni.sk

R. Malviya, S. Sundram (eds.), *Sustainable Nanocomposites with Green Biomaterials*, Biomaterials, Bioengineering and Sustainability 2,
https://doi.org/10.1007/978-3-031-79110-9_14

Keywords

Biomaterials · Biodegradable implants · Bioactivity · Tissue engineering · Sustainable materials · Biocompatibility · Bio scaffolds · Mechanical properties · Biomedical applications

Abbreviations

PHA	Polyhydroxyalkanoates
PHB	Polyhydroxybutyrate
PHV	Polyhydroxyvalerate
PHBV	Poly(3-hydroxybutyrate-co-3-hydroxyvalerate)
PHO	Polyhydroxyoctanoate
PHH	Polyhydroxyhexanoate
ROP	Ring Opening Polymerization
CAD	Computer-aided design
MTT	3-(4,5-dimethylthiazol-2-yl)-2,5-diphenyltetrazolium bromide
HPLC	High-pressure liquid chromatography
MS	Mass spectroscopy
AFM	Atomic force microscopy
SEM	Scanning electron microscopy
ELISA	Enzyme-linked immunosorbent assays
H&E	Hematoxylin and eosin
DSC	Differential scanning calorimetry
FDA	Food and Drug Administration
EMA	European Medicines Agency

14.1 Introduction to Polyhydroxyalkanoates (PHA) Biomaterials in Biomedicine

Polyhydroxyalkanoates (PHA) have been a critically important class of biomaterials owing to their distinguishing features such as biocompatibility, biodegradability, and tunability (Kalia et al. 2023). This chapter will be focusing on the synthesis, properties, and new applications of PHA for the biomedical field bridging the bench to bed. The ability of PHA many of which are produced by diverse microorganisms during nutrient limitation and are ecofriendly, inert, and biocompatible—signifies these materials as recent potential candidates (Butt et al. 2018). PHA provides various materials that meet specific biological applications with modified properties due to their structural diversity such as short-chain, medium-chain and long-chain PHAs (Pramanik 2023). Most PHAs are produced by microbial fermentation. This process calls upon the metabolic capabilities of microorganisms such as bacteria and archaea that can be grown on renewable carbon sources, converting them to granules of PHA inside their cell while growing (Singh et al. 2019; Carvalho et al. 2021). This is one of the best-studied and optimized processes by researchers for the attainable purity and consistency of PHA granules. Furthermore, chemical synthesis techniques have been studied to improve microbial fermentation, in order to obtain defined PHA with respect to molecular weight, monomer composition and mechanical properties (Zytner et al. 2023). The ability to tune PHA characteristics has become a major goal in the field of medical research, since it enables engineering properties of PHA suitable for different biomedical applications. As a result, PHA-based materials have been tailored for specific applications in which the surface morphology, degradation rate, and mechanical properties can be finely controlled by the researchers. As such, they are also now relevant in a wider range of biomedical applications.

14.1.1 Structure and Properties of Polyhydroxyalkanoates

PHAs possess wide range of molecular structures which leads to variety of physical and mechanical attributes. In combination these features are especially useful in a range of applications, particularly for such applications within the biomedical industry. To date, over 150 different monomer compositions have been identified for PHAs among which many of the corresponding polymers exhibit very diverse characteristics (Atarés et al. 2024). This diversity is particularly beneficial in the broad range of applications, especially within the extensive areas of biomedical applications where the material attributes are indispensable for the up-to-date performance and compatibility. They are composed of repetitive monomer units consisting usually of hydroxyalkanoic acid moieties. The basic configuration of a PHA polymer is shown below:

$$-\left[\mathrm{O-CH(R)-CH2-CO}\right]\mathrm{n}-$$

R, indicates a side chain, which can range from a hydrogen atom to bigger alkyl group All of these variations in the side chain lead to the production of different types of PHAs and hence having its own specific properties. List of the PHAs in different types is shown as:

14.1.2 Polyhydroxybutyrate (PHB)

- **Chemical Structure**: PHB is composed of 3-hydroxybutyrate monomers.

$$-\left[\mathrm{O-CH(CH3)-CH2-CO}\right]-\left[\mathrm{O-CH(CH3)-CH2-CO}\right]-$$

- **Properties:** The PHB is highly crystalline in nature which makes it stiff and brittle. The material has melting point approximately 177.1 °C and glass transition temperature is about 175.7 °C (Mandragutti et al. 2024).

14.1.3 Polyhydroxyvalerate (PHV)

- **Chemical Structure**: PHV is composed of 3-hydroxyvalerate monomers.

$$-\left[O-CH\left(C2H5\right)-CH2-CO\right]-$$

- **Properties:** PHV is more amorphous than PHB hence making it more ductile than PHB that causes higher biodegradability (Kim et al. 2023).

14.1.4 Poly(3-Hydroxybutyrate-Co-3-Hydroxyvalerate) (PHBV)

- **Chemical Structure**: PHBV is a copolymer of 3-hydroxybutyrate and 3-hydroxyvalerate.

$$-\left[O-CH\left(CH3\right)-CH2-CO\right]x-\left[O-CH\left(C2H5\right)-CH2-CO\right]y-$$

- **Properties:** PHBV is a polymer that can be designed for an array of strength, flexibility, and degradation rates with variations in the 3-hydroxyvalerate ratio. It shows better flexibility than PHB and lower melting point and glass transition temperature (Meereboer et al. 2020).

Longer side chains in the case of, e.g., polyhydroxyoctanoate (PHO) and polyhydroxyhexanoate (PHH), result in materials which exhibit completely different mechanical properties and degradation characteristics than the polymers described here.

Surface modifications are an important strategy to further enhance the biofunctionality of PHA biomaterials. They are aimed at improving cell adhesion (through micro- or nanoscale surface modifications, cell seeding methods, and cell encapsulation in the scaffold), proliferation, and tissue infiltration (Gregory et al. 2022). The scientists have tried to optimize the modification of PHA materials through a number of methods such plasma treatment (Esmail et al. 2021), chemical functionalization (Ke et al. 2017), and biomolecule immobilization (Smith et al. 2020). The modifications change the surfaces properties to have features, which address specific requirements for tissue engineering scaffolds and implant coatings. Research has also been conducted combining PHA with other polymers to enhance its overall biological performance (Kumar et al. 2021) including bioactive compounds (Misra et al. 2006). PHA can be blended with other polymers possessing complementary properties to develop composite materials with improved strength, flexibility, and bioactivity. Moreover, incorporating bioactive compounds like growth factors

(Masaeli et al. 2014), antimicrobial agents (Ladhari et al. 2023), drug-loaded nanoparticles (Elmowafy et al. 2019) in a PHA matrix will allow to intensify the therapeutic properties of PHA biomaterials. This enables their utilization in drug delivery and wound healing.

Rising global environmental apprehensions and government regulations to minimize plastic waste have been boosting the biodegradable polymers like PHAs market growth globally. PHAs due to various beneficial properties have gained great interest in the field of the biomedical industry, which seems largely related to perspective applications in medical devices, drug delivery systems, tissue-engineering scaffolds or other biomedical areas.

Market research reports presently the global PHA market to be worth USD 62 million in 2020, with growth rates expected of between 11.2 and 14.2% from 2020 to 2024 and 2024–2025 (Jaffur et al. 2021). Market growth is accelerated by advances in biotechnological manufacturing processes, greater investment in R&D, and the growing number of biocompatible material applications in medicine. The biomedical sector is a key driver in the growth of the market, as PHAs offers a differentiating set of properties that are fully in line with the demanding requirements of medical applications. They are also attractive as processes used in the manufacture of new medical devices and implants, as their nature to break down into non-toxic and similar metabolites as with living tissue. This book chapter provides a broad outline of synthesis and utilization of PHA, with an emphasis on challenges encountered in the biological applications of PHA.

14.2 Synthesis and Purification Methods for PHA

PHA is a diverse class of biodegradable biopolyesters that can be potentially used in various fields, including biomedicine, packaging, and environmental sustainability. Being complicated to synthesized, its polymers characterized with unique features and functional properties are obtained commercially through either chemical or microbiological means. This chapter provides insights into the different strategies for the synthesis of PHA, highlighting the progress, challenges, the different routes, and opportunities.

14.2.1 Microbial Fermentation

Microbial fermentation is the fastest growing method among the key methods for producing PHA, in which microbes use renewable carbon sources as a substrate for the synthesis of PHA polymers. The procedure aims at growing a bacterium that produces PHA in an optimally controlled chamber with minimal nutrients. Consequently, this leads to the in-situ accumulation of the PHA granules within the cells. The potential of many bacteria like Cupriavidus necator (Nygaard et al. 2021), Pseudomonas aeruginosa (Singh and Mallick 2008), and Bacillus species (Ali and Jamil 2018) have been investigated to synthesize PHAs from different carbon

Table. 14.1 Bacterial species involved in industrial scale PHA production (Sehgal and Gupta 2020)

S. No	Bacterial species	Carbon source	PHA yield (%)	PHA production (g/l)
1.	*Pseudomonas mendocina*	Coconut oil	58	–
2.	*Cupriavidus necator*	Palm oil	9.7	97
3.	Recombinant *E. coli*	Glycolate	1	–
4.	*Halomonas boliviensis*	Wheat bran	50	4
5.	*Haloferax mediterranei*	Rice bran	55.6	77.8
6.	*Ralstonia eutropha*	Potato starch	52.51	94
7.	*Pseudomonas pseudoflava*	Synthetic wastewater	57	20
8.	*Acinetobacter junii*	Rice mill waste	94.28	2.64
9.	*Burkholderia cepacia*	Coffee ground waste	54.78	2.69
10.	*Bacillus aerophilus*	*Aegle marmelos*	11.62	2.47

sources such as sugar (Dietrich et al. 2019b) to fatty acid (Razaif-Mazinah et al. 2017) and other resources such as agricultural (Razzaq et al. 2024) and industrial wastes (Kanzariya et al. 2023) have been explored. The choice of microorganism and its carbon sources will affect the productivity, composition, and material properties of PHA. Hence, it is important to choose them very carefully according to the requirement of the application. Some bacterial species and carbon sources involved in the industrial production of PHA is listed in the Table 14.1.

14.2.2 Genetic Engineering Strategies

An efficient approach along with microbial synthesis for increasing the PHA production yield. This productivity is achieved by manipulation of the microbial host genetic material through genetic engineering. This process leads to increase metabolic pathways and enhance the utilisation of nutrients. In PHA biosynthesis the overexpression of certain enzymes increases the conversion rate of the carbon sources to PHA precursors and at the same time prevents metabolic pathway competition (Solaiman et al. 2005). Pathway optimization by modulating the activity of the pathway in a microbial host via gene knockdown, overexpression, or engineering, has been employed as a strategy to improve the production of PHA monomers with desired monomer composition (Wang et al. 2023). With synthetic biology methods, novel PHA biosynthetic pathways and chemically engineered microorganism strains with better PHA production capabilities can be obtained.

14.2.3 Mixed Microbial Cultures

Mixed microbial cultures are an alternative to pure culture systems for the production of PHA. These utilize the synergistic environment of numerous microorganism types to maximize substrate interconversions and PHA production (Albuquerque et al. 2010). This approach uses the metabolic diversity present in natural microbial communities to make them capable of degrading complex carbon sources such as lignocellulosic biomass and wastewater streams (Bravo-Porras et al. 2024). To avoid or reduce the competition between microorganisms for the substrate and thereby metabolic waste, researchers have evolved mixed culture processes that similar to pure culture systems that can be modified according to the type of strains used in the consortium. Mixed-culture systems are more flexible in adapting to changes in the environment as well as substrate availability, allowing them to be suitable for global-scale PHA production within various bioprocessing scenarios.

14.2.4 Chemical Synthesis

Chemical synthesis is an alternative pathway to microbial fermentation in the production of PHA. As a consequence, the stereochemistry molecular weight, and monomer composition of PHA polymers can be manipulated to a much greater extent than in chemical synthesis (Li et al. 2019). The method of PHA monomer synthesis could regard many categories based on the chemical mechanisms such as Ring Opening Polymerization (ROP) and Condensation Polymerization. Among polyesters, PHAs are produced by ROP of β-lactone monomers, which is a versatile technique to synthesize PHA with specific structures and properties. This strategy facilitates functionalization, and stereochemistry (Barouti and Guillaume 2016) further, this could also enable the possibility of step-growth polymerization of hydroxyalkanoic acids via condensation polymerization to produce a high molecular weight PHA polymer (Westlie et al. 2023). Moreover, PHA polymers with specific characteristics are produced by a combination of microbial fermentation and chemical synthesis, which are ideal for controlled drug delivery and tissue engineering.

14.2.5 Advances in Downstream Processing

Recovery of PHA polymers from cells or chemical reaction mixtures is critical step during downstream processing of PHA polymers. This guarantees the yield PHA materials of superior quality for further applications (Saavedra del Oso et al. 2021). The main conventional downstream processing techniques involve, cell harvest, biomass extraction and polymer purification steps such as solvent extraction, precipitation and chromatography (Kosseva and Rusbandi 2018). But using solvent for separation of proteins and phenolic compounds is generally less efficient and more costly, in addition to environmental issues related to the use of solvent and the generation of waste (Mondal et al. 2023). New technologies in downstream processing like in situ extraction, membrane filtration and biorefinery methods are addressing these challenges (Wampfler et al. 2010). These technologies enhance the efficiency of the process, reduce environmental footprints and make biorefinery concepts in long-term PHA production applicable. While scaling of PHA production is still in development and other facilities required for scaling PHA are being discussed.

These advancements represent hope on PHA production could be addressed at a larger scale while also taking into consideration economic and environmental sustainability (Koller et al. 2013).

14.2.6 Towards Sustainable and Scalable PHA Production

The synthesis of PHAs has been continually improved and advanced through the deployment of microbial engineering, bioprocess optimisation, and sustainable biorefinery technologies. In the future, it is expected that more research efforts to be focused on coping with the main technical hurdles faced in the manufacturing of these solutions, in order to accelerate a higher adoption of PHA-based materials in the different industrial sectors (Dietrich et al. 2019a). Such challenges include securing adequate raw materials, developing a manufacturable process, and increasing downstream process efficiency. By integrating PHA production with waste recycling, the use of renewable feedstock sources, and circular economy concepts, sustainability and profitability of the PHA production process can be increased (Koller et al. 2017). Against this background, affordable laboratory scale synthesis of PHA is expected to innovate biodegradable polymers, and thus enhancing the sustainable and circular bioeconomy through deployment of cross-disciplinary techniques and technology.

14.3 Tuning Properties of PHA for Biomedical Applications

It is of critical importance that PHA meets specific performance criteria as dictated by the intended function for which it is being used for in biomedical applications. Since hundreds of different kinds of PHAs are possible to synthesis, they can be used to provide materials with differing properties suitable to these different biomedical needs. The ability to tailor the properties of PHA is essential for its beneficial applications in biomedicine, due to the necessity to modify mechanical properties if used in implantable devices, to alter degradation rates if used in tissue scaffolds, or to adjust surface properties if designed as drug delivery systems. The different properties of PHA are elucidated in the schematics depicted in Fig. 14.1.

14.3.1 Biodegradability

The main advantage of PHA for biomedical applications is its biodegradability. Biodegradability is the ability of a material to break down into no toxic compounds and pose no threat to the living environment. PHA has a molecular structure that can be metabolized by microbes and enzymes available in physiological condition. A property that is critical in case of biomedical implants and devices which are required to be body compatible over long durations. When PHA-based materials are implanted into the body for tissue engineering or drug delivery, they slowly degrade

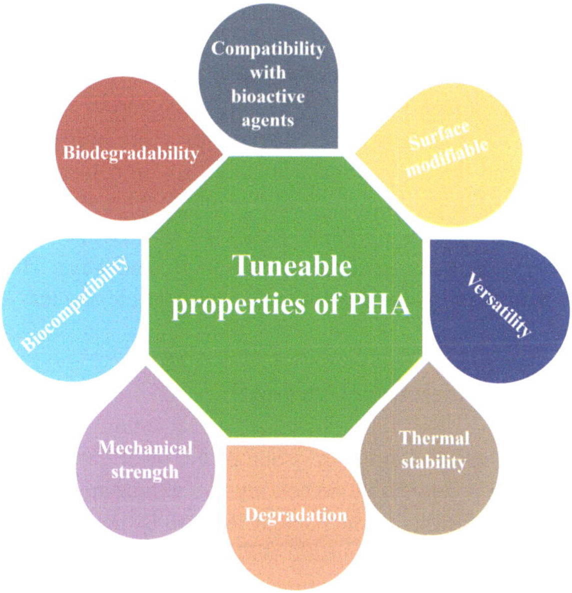

Fig. 14.1 Schematics on tuneable properties of PHA

over a period of time. After the purpose of the implanted substance is accomplished, its subtle breakdown makes sure that it does not stay forever, therefore preventing the difficulty of revision surgeries to remove them. Also, the compounds released by the degradation of PHA are non-toxic and can therefore be easily metabolised by the body, thereby reducing the risks of adverse effects or inflammatory reactions.

PHA is compatible to the essence of sustainable healthcare and serves the environment by their biodegradability. Therefore, healthcare professionals have the responsibility of addressing one of the most pressing issues, which is to reduce the adverse effects of medical waste and contribute to making the healthcare system more sustainable environmentally by using biodegradable materials such as PHA. One of the key reasons of PHA's candidature as a potential biomaterial is its, low cytotoxicity and biodegradability which are advantageous in healthcare applications.

14.3.2 Biocompatibility

PHA is an incredibly biocompatible polymer, making it a great option for biomedical use. More simply put, biocompatibility refers to a particular material biologically accepts with living tissue without causing any kind of adverse effect. PHA is biocompatible; the material exhibits good tolerance in regard to the body and the polymer does not cause immunological reactions, inflammation, and rejection of tissues.

PHAs are an appealing family of biomaterials because it is biocompatible with low immune responses. The above properties make them suitable for processing into several types of biomedical devices, implants, or scaffolds (Gregory et al. 2023; Kalia et al. 2023). This property is important in ensuring the biocompatibility and long-term performance of medical devices as it decreases the risks of issues, for instance infection, fibrosis, or implant rejection.

It is due to the similar chemical structure of PHA to naturally occurring biomaterials in the body which enhances its biocompatibility. PHA polymers must be degraded in the body to generate metabolites along with metabolic pathways. The biocompatibility of PHA comes from this phenomenon as PHA's metabolism largely coincide with normal metabolic pathways. This feature credits PHA as an excellent biomaterial for its safer, compatible and less adverse effect with host.

14.3.3 Mechanical Strength

The mechanical strength is one of the most important features of PHA to demonstrate its potential in load bearing biomedical application (i.e., implants, tissue engineering scaffolds). Mechanical strength is the ability of a material to withstand the force applied from outside without deforming or fracturing. PHA is a versatile material with respect to mechanical properties such as tensile strength, elasticity, toughness, etc., which can be achieved simply by making changes in chemical composition, molecular weight, and processing methods (Guessasma et al. 2020; Najah et al. 2022).

For this reason, PHA based materials destined to load-bearing applications (like orthopaedic implants or vascular stents) will have to cope also with the physiological stresses exerted by surrounding tissues and human day today activities. By perfecting the mechanical properties of PHA, researchers can ensure that implanted devices maintain their structural integrity and function for long periods in the body. PHA exhibits a versatile mechanical behaviour, which makes it possible to design and manufacture in numerous implants and scaffolds with customized designs and functionality to fit specific therapeutic demands. PHA-based bone scaffolds can be designed by researchers to be similar in mechanical properties to real bone tissue. This allows for the adhesion, reproduction and renovation of cells to form along the crushed cells.

Moreover, mechanical properties of PHA can be improved using nanoparticles or reinforcing fibres. Such approaches significantly improve the overall mechanical behaviour of PHA-based materials, expanding possibilities to use them for load-bearing biomedical devices and implants. Mechanical properties of PHA are an immanent characteristic, that can afford it an exceptional position in a variety of biotechnological applications, even simple as it is strength and the ability to serve as structural support in load-bearing situations.

14.3.4 Tuneable Degradation Rate

One of the more interesting properties of PHA—its ability to easily be 'tuned' to a slower degradation rate—is especially attractive for biomedical applications, as it allows researchers to control the rapid degradability of PHA-based materials under in biological conditions. This quality is important because different biological applications require different degradation kinetics of the macrostructures to match the need of tissue regeneration or drugs delivery.

The degradation of the PHA materials and its rate can be influenced through factors such as the composition of the PHA, the molecular weight or the crystallinity. For example, increasing the proportion of monomers with shorter chain lengths, or decreasing the molecular weight of the PHA, greatly enhances the rate of degradation (Wang et al. 2016). This superior property allows for the use of PHA as temporary implants as well as for drug delivery purposes owing to its fast degradation rate. Inversely, the inclusion of long-chain monomers or a high molecular weight of the PHA could make the degradation rate slower, which thus leads to a longer polymer stability and life in the body (Reddy et al. 2022). This is appreciable for applications that need long lasting structural support or sustained drug delivery, like tissue engineering scaffolds or implantable drug delivery devices.

The gradual degradation of PHA materials by tuning degradation rate allows the researchers to program the properties of PHA materials to meet specific needs in biomedical applications. Consequently, this property will depend on the biochemical healing kinetics to a therapeutic level of demand for material degradation in the designed implant. This trait of PHA-based materials demonstrates the material's versatility, which may suit a broad range of biomedical applications, including potential short-term implants and long-term tissue regeneration strategies.

14.3.5 Thermal Stability

Thermal stability allows the material to maintain its geometry and function appropriately in a variety of processing and physiological environments. The melting temperature of PHA is 40 — 180 °C which depends on the composition and the molar mass. The materials based on PHA present excellent thermal stability, that makes them capable of withstanding various processing methodologies like melt extrusion, solvent casting, and electrospinning, without reaching a significant decomposition or loss of properties (Kaniuk and Stachewicz 2021). Being processed, PHA retains its mechanical behaviour, molecular composition, and interactions with host — which lead to its interesting properties that make it suitable to develop medical devices, implants, and scaffolds for a multitude of applications.

Furthermore, thermal stability of PHA allows biomedical material sterilization by popular methods such as autoclaving, gamma irradiation or ethylene oxide gas treatment (Kalia et al. 2023). These sterilization techniques are indispensable to ensure the safety and sterility of PHA-based materials before implantation or application in clinical situations. Biocompatibility of PHA-based materials means that they are able to withstand the various conditions they would be subject to in the body, such as temperature, pH fluctuations, immunogenicity, mechanical stress, and so on. These traits are well-suited for a wide range of therapeutic applications in healthcare and improve the long-term resilience and efficiency of PHA implants, devices enzyme carriers, microspheres, functional and tissue engineering scaffolds regarding biofunction.

14.3.6 Versatility in Processing

PHA offers excellent versatility in processing which in turn an array of options for manufacture of tailored biomedical devices, implants, and scaffolds. PHA has been treated by different process including injection moulding, extrusion, compression moulding and 3D printing (Mehrpouya et al. 2021). Each of these technologies offer different strengths such as mouldings with complex geometry and close tolerances, are produced by injection moulding of PHA implants and devices. This method allows for the rapid production and custom modification of implant designs that perfectly match unique anatomy in each patient while maintaining high production yield.

Solvent casting is employed in the production of thin PHA films; electrospinning for PHA fibers and extrusion of PHA filaments for different applications such as wound dressings, drug delivery system, tissue engineering scaffolds (Mehrpouya et al. 2021). The uniform thickness and mechanical characteristics of extruded PHA materials make it suitable for a wide range of biomedical applications. Porous scaffolds for tissue engineering or slow-release drug systems can also be manufactured through compression moulding. This method is inherently capable of providing controlled porosity in the material along with mechanical strength. 3D-printing is also known as additive manufacturing, that enables the manufacture of complex 3D structures from computer-aided design (CAD) models. Specific 3D-printed structures allow a lot of control over the design of the scaffold, the size of pores and the potential to integrate bioactive agents. This feature makes them ideal for tissue engineering as well as for patient-specific implants implanted as unique body features.

In addition, that PHA-based materials lend it to be processed in various methods, this should allow scientist to fabricate PHA-based biomaterials with the required mechanical strength, degradation kinetics and surface properties for variety use in the biomedical fields. The processability of PHA makes it suitable for regenerative medicine, pharmaceuticals and individual healthcare solutions.

14.3.7 Surface Modifiability

Surface modifications are crucial for interaction of PHAs with the human system that include cell adhesion, proliferation, growth and differentiation. The physical, biological, and chemical characteristics of PHA surfaces can also be changed using surface modification to induce a better interaction with biological systems and making themselves more useful for medical applications. The desired properties of PHA surfaces can be achieved by surface modification such as plasma treatment, chemical grafting and incorporation of bioactive compounds. Plasma treatment modifies the surface chemical properties of PHA by creating reactive species, increasing hydrophilicity and surface energy, thereby increasing adhesion (Esmail et al. 2021). This also promotes cellular adhesion and proliferation on PHA surfaces, in turn, implants PHA surfaces can help integrate and regenerated. The covalent binding of functional groups or polymers to PHA surfaces is known as chemical grafting. This allows the incorporation of select chemical moieties or biomolecules. This method will modify the surface properties of PHA, thus regulating the cellular responses and tissue interactions (such as hydrophilicity, charge, and bioactivity) by the researchers.

The bioactivity of the PHA coatings has shown improvement especially through the application of bioactive components like growth factors, peptides, or extracellular matrix proteins on their surfaces, which drive some cellular responses (Chai et al. 2021). Surface constitute of PHA on bioactive coatings are needed to enable these surfaces to improve cellular adhesion, migration, and differentiation. Their properties allow them to be utilized in tissue engineering scaffolds, drug delivery systems, and implanted medical devices. Such a surface versatility of PHA facilitates their bio-interactions to be well-adjusted to biological systems, improving biocompatibility, tissue integration and effectiveness in therapeutic activities in biomedical applications. The ability to program the customisation of surface features of PHAs enables researchers to engineer materials that are tailored exactly to suit the unique requirements of multiple biomedical applications. Applications of these can be from tissue-repair strategies to drug delivery and diagnostics.

14.3.8 Compatibility with Bioactive Agents

PHA are able to interact with a vast variety of bioactive chemicals, drugs, growth factors, antimicrobials and imaging contrast agents. This makes them extremely useful in many biomedical applications. Bioactive molecules are interesting targets for loading and promotion of candidate properties in the case of various PHA materials, particularly for controlled release, when localized delivery and enhanced therapeutic efficacy are required. This compatibility of PHA with bioactive substances also allows for the development of multifunctional biomedical materials. They would be suitable materials for making drug delivery systems, better wound dressings, tissue engineering scaffolds and diagnostic devices (Thorat Gadgil et al. 2017). These bioactive molecules may either be physically entrapped, chemically conjugated, or surface adsorbed to PHA matrices depending on the requirements of each application.

Controlling the release of bioactive compounds from PHA scaffolds is a potential strategy of selectively delivering bioactive compounds to targeted tissues or cells, which minimizes their systemic side effects and increases therapeutic outcomes. By tuning the release kinetics of bioactive molecules, namely drugs or peptides, researchers can fine-tune the pharmacokinetics and pharmacodynamics of therapeutic therapies increasing patient compliance and efficacy. PHA matrices used in diagnostic applications can be combined with imaging contrast agents to non-invasively image tissues and organs or administer therapeutic agents. Providing biocompatibility and stability, as well as a range of release properties, PHA-derived imaging agents are suitable for various imaging approaches (e.g., magnetic resonance, MRI; computed tomography, CT; and ultrasound).

In addition, the compatibility of PHA with bioactive substances provides the biopolymer with wide applications in the field of biomedicine. This facilitates the synergy of therapeutic, diagnostic and regenerative modalities, thus enhancing therapeutic efficacy. The ability of PHA to be combined with bioactive materials could therefore be exploited by researchers to create new materials to help with these unmet clinical needs & could help with improving the patient experience in healthcare.

14.4 Biomedical Applications of PHA

PHA is biodegradable and biocompatible polymers that has been commonly used for biological applications. Owing to these properties, PHA based systems are appropriate for drug delivery systems which can control the release of therapeutic substances. PHAs can be employed as scaffolds in tissue engineering to promote cell proliferation and tissue regeneration. They are interesting because of their attractive mechanical properties and ability to be synthesised using microorganisms to produce medical implants and devices. PHA-based materials can be engineered by researchers to degrade at certain rates. It reduces the risk of chronic inflammation, and there is little less to no need for a revision surgery. Figure 14.2 shows schematics of myriad biomedical applications of PHAs.

14.4.1 Tissue Engineering Scaffolds

14.4.1.1 Bone Tissue Engineering

PHA scaffolds are widely used in bone tissue engineering because of its biocompatibility, biodegradability, and mechanical properties. The three-dimensional shape of these scaffolds mimics the bone extracellular matrix, to which osteoblasts strongly attach, proliferate, and differentiate. It is important to minimize local inflammation and any subsequent detrimental actions of PHA based scaffolds, which means the scaffold must either disappear after the tissue has regenerated or disappear without evoking any inflammatory response. Furthermore, it has been shown that PHA combined with bioactive molecules such as hydroxyapatite, may increase its osteoconductive properties, hence easing the bone regeneration process.

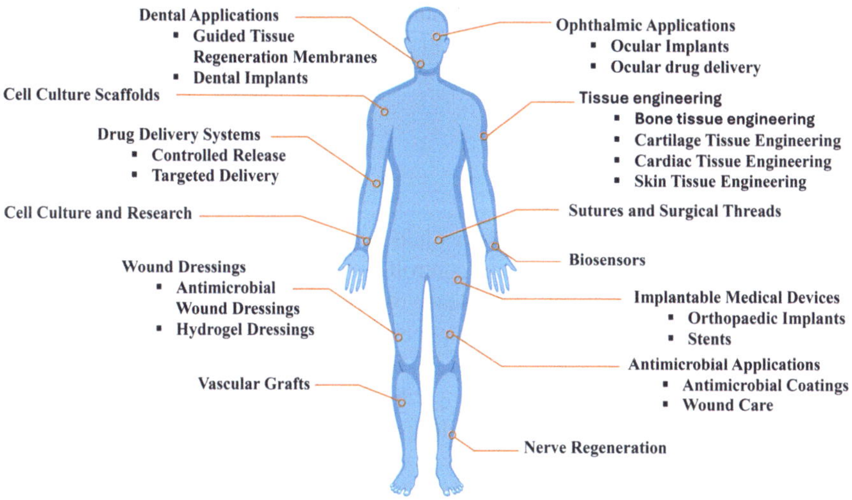

Fig. 14.2 Schematic representation of biomedical application of PHAs

14.4.1.2 Cartilage Tissue Engineering

PHA scaffolds played a vital role in the restoration and regeneration of diseased cartilage orthopaedic via cartilage tissue engineering. They have been shown to enhance cartilage formation by promoting adhesion and proliferation of chondrocyte compatible extracellular matrix. Adjusting the stiffness and flexibility of PHA to mimic the normal elastic modulus of cartilage. Furthermore, PHA scaffolds decorated with growth factors induce chondrogenesis and can therefore improve the repair value of cartilage (Zhao et al. 2003).

14.4.1.3 Cardiac Tissue Engineering

Certain findings demonstrate that PHA scaffolds can be used in cardiac tissue engineering, as they promote the proliferation and function of cardiac cells, which is important in heart tissue regeneration. These scaffolds can attract and help to proliferate heart muscle cells, or cardiomyocytes, grown on the scaffold to develop functional heart tissue. The biodegradability of the PHA ensures the scaffold degrades when new tissue forms, reducing the risk of long-term complications (Dhania et al. 2022).

14.4.1.4 Skin Tissue Engineering

Likewise, PHAs have been used as scaffolds in skin tissue engineering for wound healing and skin regeneration. Scaffolds provide a 3-D structure to attach keratinocytes and fibroblasts for proliferation and consequently new tissue forming for the skin. PHA scaffolds can be designed to degrade at a rate that is consistent with the healing rate, enabling researchers to control the release of bioactive compounds during the healing process. That means it can help regenerate tissues without any adverse effects. In addition, PHA is combined with antibiotics for the faster healing and infection free healing (Shishatskaya et al. 2016).

14.4.2 Drug Delivery Systems

14.4.2.1 Controlled Release

In drug delivery systems PHA matrices had been employed as carriers to entrap therapeutic molecules and provide a controlled release of the drugs. This feature ensures a sustained and controlled release of drugs for a specified duration, thus increasing the efficacy of treatment and patient compliance. By changing the degradation rate of PHA, the release kinetics can be controlled and thus the drug dosage and drug delivery time. PHA is biocompatible ensuring that the material does not elicit adverse reactions and hence suitable for drug delivery applications where sustained delivery is a necessity. PHA-based microspheres can deliver cancer drugs specifically to cancerous cells, they have been used as a targeted therapy, that also reduces the risk of side effects in other part of the body (Prakash et al. 2022).

14.4.2.2 Targeted Delivery

Researchers have explored and optimised to adjust PHA nanoparticles and microspheres to act as carriers for therapeutic drugs, an approach that enhances the efficacy of medication directed at specific tissues or cells. By altering the surface properties of PHA particles, researchers have been able to achieve specific targeting, such as transporting to specific kinds of cells (like cancer cells) or to inflammatory sites. Using a targeted intervention, the required dose is increased, and side effects are reduced, all leading to better immune therapy. Furthermore, it is possible to target drug release even better by specifically coupling the task of drug-releasing to selected triggers (environments): for instance, PHA-particles can be engineered into 'smart' devices that release the drug just when it is needed to achieve the desired therapeutic effect by releasing the drug in response to certain triggers (specific environmental conditions), such as changes in pH or temperature (Shrivastav et al. 2013).

14.4.3 Implantable Medical Devices

14.4.3.1 Orthopedic Implants

PHA is used to produce orthopaedic implants like screws, pins, plates. These biodegradable implants dissolve gradually so that patients do not require a second surgery to remove the implant once the fractured bone heals. PHA can be tailored to have the mechanical properties of bone such that there is adequate mechanical support in order to facilitate the healing process. Moreover, PHA bioactive molecule composites are able to enhance bone regeneration and accelerate bone fracture healing as well (Zhang et al. 2018).

14.4.3.2 Stents

PHAs are fabricated into biodegradable stents for use within cardiovascular applications to provide temporary structural scaffolding to keep blood vessels open and restore blood flow. These stents act as a temporary scaffold to hold open the blood vessel and are designed to slowly dissolve and disappear as the blood vessel remoulds and heals, which helps reduce the possibility of long-term complications that may be related to permanent stents. PHA- based stents will degrade in the body over time so it reduces the need for surgical extraction. Moreover, as with PHA stents, it is possible to couple pharmaceutical agents to suppress restenosis, hence further enhancing the overall outcome for patients (Zhang et al. 2018).

14.4.4 Wound Dressings

14.4.4.1 Antimicrobial Wound Dressings

The antimicrobial activity of the PHA wound dressings have also been described [54] and the versatility of the kind of antimicrobial agents that can be loaded in PHA have been highlighted before [20] pointing the great potential for including effective agents in the dressing to prevent infection and enhance wound healing [54]. Biocompatibility of PHA ensures that this kind of dressing does not induce any adverse reactions, this makes an appropriate candidate for treating chronic wounds. The dressing is biodegradable as PHA starts to break down, which means less need for dressing changes and less potential for disruption to healing. Furthermore, the designers could develop moisture retentive PHA dressings which heal faster and with significantly less scarring (Marcano et al. 2017).

14.4.4.2 Hydrogel Dressings

PHA hydrogels create a moist environment that is likely to help in wound healing [2]. Hydrogels rapidly absorb wound exudate, deliver bioactive agents, promote wound closure, and mitigate infection. PHA degradation can be carefully tailored to the rate at which the wound heals, ensuring that the dressing is enough to promote tissue regeneration, but avoids stimulating adverse reactions. PHA hydrogels are suitable for almost all types of wound repair such as burns, surgical wounds, and chronic wounds (Brelle et al. 2023).

14.4.5 Sutures and Surgical Threads

PHA is used in some biodegradable sutures that, along with the biodegradation over time, avoid the requirement for suture removal and minimise the risk of infection. When inserted, they provide a temporary closure of a wound and a support structure allowing tissue regeneration, eventually breaking down into products of complete resorption. PHA is mechanically tuneable to the requirements of different types of wounds and ensures adequate support during the healing interval. In addition, PHA sutures can be incorporated with active pharmaceutical ingredients effective in killing pathogens and thus aid in tissue repair and reduce infection risks (Koller 2018).

14.4.6 Dental Applications

14.4.6.1 Guided Tissue Regeneration Membranes

PHA membranes are used in dental procedures for tissue regeneration of the periodontal tissues. These membranes act as a barrier to both the penetration of epithelial cells into the periodontal defect and therefore may help promote bone and connective tissue repair. The biodegradability of PHA means that the membrane degrades naturally as new tissue forms, therefore reducing the need for surgically removing it. Finally, PHA membranes have the potential to incorporate growth factors for optimal perio therapy and tissue healing (Sun et al. 2018).

14.4.6.2 Dental Implants

These PHA based materials are also suitable for the design and construction of biodegradable dental implants for bone regeneration and integration. The implants act as "scaffolds" that provide a temporary structure to support the bone as it heals, and disintegrate after a few days to weeks, to allow the healing process to start, reducing the chances of potential long-term complications associated with permanent metal implants. Being biocompatible in nature, PHA does not harm the surrounding tissues at the site of implant or show any adverse side effects, which makes them a good candidate for dental therapy. Moreover, the use of PHA together with bioactive molecules might help not only bone regeneration, but also anchoring the implant to the bone tissue (Ansari et al. 2021).

14.4.7 Nerve Regeneration

The excellent biocompatibility of PHA could support nerve regeneration, and restore local function after peripheral nerve injury, providing an optimal microenvironment for nerve repair. The said conduits help the building of new nerve fibres at the site of injury and make possible the regeneration of nerve function. When the nerve tissue regenerates, the conduit biodegrades, thus mitigating against long-term complications. Moreover, the adjoined growth factors in PHA conduits could lead to a marked promotion of nerve regeneration and functional recovery (Lezcano et al. 2022).

14.4.8 Antimicrobial Applications

14.4.8.1 Antimicrobial Coatings

The biocompatibility of PHA makes use of this polymer attractive for manufacture of medical devices and for obtaining antimicrobial coatings preventing bacterial colonisation and infection. The release of antibacterial agents from these coatings occurs over time, leading to a lower risk of infections associated with implanted devices. PHA biocompatibility ensures that the coating does not have any side-effects on the coating and enables the coating to be applied to various medical devices from catheters, implants and surgical instruments (Rodríguez-Contreras et al. 2019).

14.4.8.2 Wound Care

To treat bacterial spread through infected wounds, these advanced PHA dressings can act as a successful dressing with antibacterial properties. Such dressings are intended to release antimicrobial agents that prevent infection and support healing, as a result reducing the risk of infection. The promise of biodegradability of PHA ensures that the dressing degrades in a natural manner as the wound heals with reduction in interference in wound healing that would have been posed by the dressing. Moreover, the PHB and PHA dressings provide a hydrophilic environment, which could result in the improved water retention of the wound bed and hence play a role in the accelerated wound healing process and possibly wound scar reduction (Li et al. 2020).

14.4.9 Ophthalmic Applications

14.4.9.1 Ocular Implants

PHA-based ocular implants used in ophthalmic surgeries with the option of biodegradability over traditional materials. These implants can help in regenerating eye tissues and disintegrate in the body over time preventing any long-term consequences to the eye. PHA also possesses excellent biocompatibility, thus the implants are non-toxic and non-immunogenic, which makes them suitable for a variety of ophthalmic applications such as corneal regeneration and glaucoma treatment (Mohtashami et al. 2020).

14.4.9.2 Ocular Drug Delivery

One advantage of these systems is that they are capable of delivering drugs to specific ocular tissues, expanding their use towards the treatment of eye-related diseases. Drug delivery systems have been widely used clinically in delivery therapeutic drugs for continuous release, by controlling the rate at which a drug is released and the best clinical effect to maintain a constant drug level in the body, such application has improved the effective treatment of the disease. The biocompatibility of PHA makes PHA safe for long term usage in ophthalmological applications. In addition, the degradation rate of PHA can be designed to match the healing rate by adjusting the hydrolysis time of the carrier, thus enabling an expected sustained effective drug concentration and therapeutic response (Mohtashami et al. 2020).

14.4.10 Cell Culture and Research

14.4.10.1 Cell Culture Scaffolds

PHA scaffolds are widely employed as an artificial environment for researchers to investigate cellular behaviour, differentiation and tissue organization. They build a three-dimensional platform akin to our own extracellular matrix, that allows cells to attach and grow. PHA is biocompatible, thus eliminating matrix from provoking any immune response, making these scaffolds ideal for in vitro cell culture applications. Furthermore, bioactive compound infused PHA scaffolds may be used to induce the desired cellular responses by tissue engineers and regenerative medicine experts (Dhania et al. 2022).

14.4.11 Biosensors

Biological indicators can be detected by the biosensors developed using PHA in medical diagnostics, which are compatible with living organisms and can be biodegraded naturally over time. The biosensors can measure physiological parameters in the regions of interest and provide real-time data for multiple health indicators. The biodegradable PHA used for the biosensors means that after they have accomplished their task, they will break down naturally avoiding any long-term problems. In addition to this the PHA biosensors can also be manipulated to sense specific stimuli thereby eliciting accurate and targeted clinical feedback (Kelwick et al. 2021).

14.4.12 Vascular Grafts

Naturally derived vascular grafts have long-term complications and PHA-based vascular grafts which replaces or repairs damaged blood vessels promote tissue regeneration and reduces long-term complications. This graft provides temporary support for the blood vessel and then dissolves over time so that natural tissue regrowth can take place. PHA is biocompatible, therefore, no adverse reactions should be expected from the grafts, which is an important characteristic of the grafts to be used for vascular applications. In addition to reinforcement, bioactive compounds can also be incorporated into PHA grafts to enhance tissue regeneration and facilitate the integration between the grafted material and the surrounding tissue (Hou et al. 2024).

14.5 Performance Assessment of PHA in Research and Animal Testing

14.5.1 Biocompatibility

14.5.1.1 In Vitro Cytotoxicity

Cell culture testing is the primary for evaluating the biocompatibility of PHAs. This includes techniques like MTT assay (measures cell metabolic activity) and live/dead staining (differentiate between living and dead cells). These evaluations examine the potential toxicity of PHA materials caused to cells as they check viability and proliferation. In order to continue further in vivo studies, it is required to demonstrate desirable good results without relevant cytotoxicity (Siew et al. 2007).

14.5.1.2 In Vivo Biocompatibility

PHA materials are implanted in animal models to evaluate in vivo biocompatibility, which is mainly focused on rodents. Tissue responses are monitored by the researchers for inflammatory, fibrotic, and other immune responses at certain time intervals for specific durations. Histologic and immunohistochemical analysis are performed for in-depth information on the tissue integration, inflammatory response at a cellular and tissue level. Good biocompatibility allows negligible adverse effects of the material, and it becomes more integrated with the host tissue (Valappil et al. 2006).

14.5.2 Biodegradability

14.5.2.1 Degradation Rate

A key parameter for biodegradable implant applications wherever temporary support is desired is the biodegradation rate. In case of PHA this rate can be changed by modulating the copolymer-structure of PHA. Typical studies include incubation of PHA samples in physiological conditions (e.g., phosphate-buffered saline at 37 °C) and measurement of weight loss, molecular weight changes, and alteration of mechanical properties with time. It is built to degrade in a controlled manner, provide support for tissue regeneration and harmlessly degrade (Meereboer et al. 2020).

14.5.2.2 Degradation Products

Degradation products of PHAs are compounds released during the breakdown of PHA which must be identified as its identification open many possibilities for applications. Methods such as high-pressure liquid chromatography (HPLC) and mass spectroscopy (MS) are used to assess these degradation products. It is essential that the byproducts are non-toxic and can be excreted or transformed into waste without causing the damage. For example, PHA degradation products are mainly short chain fatty acids which can be easily metabolized in the human body as endogenous monomers available for biomedical applications (Meereboer et al. 2020).

14.5.3 Mechanical Properties

14.5.3.1 Tensile Strength and Elasticity

PHA is precisely engineered to exactly mimic the tensile strength, elasticity, and other properties of the tissue it is intended to repair. Tensile testing measures how a material responds to uniaxial stress and provides information on both its strength and elasticity. Especially in load-bearing applications, as bone scaffolds, where the material should be able to bear the targeted physiological loads without deforming or breaking (Žiganova et al. 2023).

14.5.3.2 Flexibility and Toughness

Flexibility and strength are tested by bending and impact testing. PHAs bending and deformation properties are important for its application in human soft tissues as well as dynamic environments. By adjusting the copolymer ratio, the combination of different properties can be optimized to maintain good mechanical properties of the material for various applications (Žiganova et al. 2023).

14.5.4 Surface Properties

14.5.4.1 Surface Roughness and Topography

The surface properties of PHA are important for cell attachment and growth. Surface roughness and topography can be studied with techniques such as atomic force microscopy (AFM) and scanning electron microscopy (SEM). A rough surface is a more suitable cell adhesive and proliferation matrix, which increase the surface area, as well as corresponds to the native extracellular matrix (Tang et al. 2023).

14.5.4.2 Surface Chemistry

Functionalisation strategies targeting the surface chemistry of PHA may not only mediate a more direct interaction with biological tissues, but can also shape the molecular conformation, orientation, and dynamics that is prerequisite for other functions of deposited protein layers. Some of the techniques for improving surface chemistry are plasma treatment, bioactive molecules grafting and peptides and proteins coating. These modifications typically improve cell adhesion, proliferation, and differentiation processes, required for instance in tissue engineering applications where cell-material interactions are key in the process of tissue regeneration (Tang et al. 2023).

14.5.5 Hemocompatibility

14.5.5.1 Blood Compatibility

Hemocompatibility of PHA is evaluated by estimating its interactions with blood components. Hemolysis assays, coagulation time measurements and platelet adhesion experiments are inclusive in these studies. The hemolysis study is conducted with red blood cells in the blood and the coagulation tests are to monitor for an effect of PHA on blood clotting. Platelet adhesion test evaluates the thrombogenesis of the material. High degree of hemocompatibility makes these materials a suitable candidate for applications where direct contact with blood is inevitable such as vascular grafts and stents (Qu et al. 2006).

14.5.6 Immunogenicity

14.5.6.1 Immune Response

The immunogenicity is assessed by in vitro and in vivo studies of PHA. In in vitro tests, PHAs are incubated with immune cells and tested for the generation of cytokines and cell activation. For in vivo study, the responses of PHA were constructed by implanted PHA into animal models to detect the histological appearance accompanied by cytokine profiling. The minimal immunogenicity is important to prevent PHAs from being responsible for adverse immune reactions resulting in chronic inflammation or rejection (Prakash et al. 2022).

14.5.7 Drug Release Profiles

14.5.7.1 Controlled Release

The controlled release ability of PHA to encapsulate and release drugs makes it attractive for drug delivery applications. Release profiles are investigated by loading drugs into PHA matrices and examining the decrease in the release rate of the drugs over time utilising High-Performance Liquid Chromatography (HPLC) techniques. Sustained therapeutic levels are then maintained for long period of times with the controlled release leading to reduced dosing frequency and treatment efficacy. Overall, the rate at which PHA degrades can be modified to achieve desired release kinetics and thus make it relevant for many different therapeutic applications (Prakash et al. 2022).

14.5.7.2 Bioactivity Retention

The maintenance of the biological activity of the agents, both in PHA and after their release from the polymer system is a fundamental issue. In stability studies, PHA-drug formulations are allowed to incubate at physiological conditions and the drug activity is monitored as a function of time after release. Bioactivity is most commonly measured via techniques such Enzyme-linked immunosorbent assays (ELISA) or Cell-based assays etc. Maintenance of bioactivity is critical for the drugs' therapeutic potential in PHA systems.

14.5.8 Cellular Response

14.5.8.1 Cell Adhesion and Proliferation

The ability of PHA to promote adhesion of cells and its effect on cell growth was evaluated in vitro by cell culture studies. Once these PHA scaffolds are seeded with cells, cell attachment and subsequent growth on these scaffolds can be studied using light microscopy or assays such as MTT or Alamar Blue for metabolic activity of cells. The high level of adhesion and proliferation of cells shows that PHA is an environment suitable for the growth of cells and hence is quite important in tissue engineering applications (Masaeli et al. 2014).

14.5.8.2 Cell Differentiation

The ability of PHA scaffolds to support cell differentiation can be analyzed by seeding stem cells on a PHA scaffold and differentiation of them to obtain cell line of interest. Alternatively, markers for differentiation can be measured by PCR or immunostaining; for example, alkaline phosphatase for osteoblasts and collagen type II for chondrocytes. The successes in tissue differentiation on PHA scaffolds illustrate the potential for targeted tissue regeneration (Esmail et al. 2021).

14.5.9 Tissue Integration

14.5.9.1 Histological Analysis

Histological assessment assesses sections of the tissue surrounding the PHA implants to assess integration and new tissue formation. Staining techniques such hematoxylin and eosin (H&E) and immunohistochemistry are particularly relevant to the study of cell infiltration, extent of vascularization and tissue organization. For any implant to function for a long period of time, successful integration with the host tissue must occur, which means that the PHA material supports the formation of new tissue in vivo and is not rejected.

14.5.9.2 Mechanical Integration

Mechanical integration was assessed by measuring the strength of adhesion between PHA implants and the surrounding tissues. For example, the force necessary to separate the implant (pull-off tests) is a technique for assessing implant-to-host tissue bonding strength, which is a direct measurement of the osseointegration. These are critical to the mechanical integration that will maintain the stability and function of the implant in load bearing applications.

14.5.10 Sterilization and Stability

14.5.10.1 Sterilization Methods

Proper sterilization of PHA biomaterials is paramount but without loss of PHA properties. These methods include gamma irradiation, ethylene oxide, and autoclaving. Innovations is determining the impact each method has on the mechanical and chemical properties of PHA to verify that the material remains robust and fully functional after sterilization. Sterilization is necessary to ensure freedom from microbial contamination of PHA-based medical devices with respect to patient safety.

14.5.10.2 Long-Term Stability

For this, Long-term stability studies should be performed by storing PHA materials under physiological conditions evaluating periodically their mechanical, chemical, and biological properties. Methods as differential scanning calorimetry (DSC) or mechanical testing are employed to follow these modifications. Stability ensures the performance of PHA remains consistent throughout the planned use life of the medical device or implant—key to enabling long-term reliability.

14.5.11 Toxicology

14.5.11.1 Acute and Chronic Toxicity

Acute toxicity studies refer to the use of PHA or its breakdown products into the animal models to observe any immediate side effects. Chronic toxicity extends a similar study wherein the animals are exposed for a longer duration to evaluate for possible prolonged effects. Such studies can range from histopathological analysis of main organs, blood biochemistry, as well as behavioural assessments. In medical applications, such non-toxicity evaluations are absolutely necessary for safe use of PHA.

14.5.11.2 Carcinogenicity and Genotoxicity

Long-term studies are made to determine possible carcinogenicity, and in vitro assays such as the Ames test provide information on mutagenicity. Short-term or long-term in vivo studies may involve, e.g., monitoring animal models for tumour development. Evidence of no carcinogenic and genotoxic effects is a mandatory requirement for the regulatory approval and safe applications of PHA-based products in clinical application.

14.5.12 Animal Models

14.5.12.1 Small Animal Studies

Small animal models such as mice, or rats are often used for initial assessments of biocompatibility, degradation, and basic functional performance of PHA. These studies serve to provide initial data on the material behaviour in a living system, and to identify any potential problems before pursuing in larger and more complex higher animal models.

14.5.12.2 Large Animal Studies

On the other hand, large animals, including pigs or dogs, enable researchers to gain a better understanding of new implants such as PHA based material as these animals are more closely related to human physiology. Such studies confirm the function of PHA-based implants or devices in more realistic clinical conditions. By testing the PHA-based material in large animals the researchers can determine if the PHA achieves the required degree of integration, mechanical stability and functional outcomes necessary to advance a material toward clinical trials and human use.

14.6 Challenges of PHA as a Biomaterial Towards Clinical Transition

14.6.1 Cost of Production

PHA synthesis is carried out through intricate fermentation processes using microbe strains that respond to certain environmental requirements and specific nutrients. The expense of natural resources as well as the strenuous characteristics of such processes cause high costs of production. Moreover, the cost of downstream processing—extraction, purification, and drying—is tacked on to the total as well. These high costs are a major constraint to the broader use of PHA in biomedical application, in which economic efficiency are supposed to be necessary (Kaur and Roy 2015).

14.6.1.1 Economic Viability

For PHA to be useful and economically attractive, production must be scaled up. These shifts included, among other things, efforts to increase microbial efficiency, using high-yielding genetically engineered strains, and linking with renewable resources. Strategies for cost reduction are optimisation of fermentation process and minimisation the need for expensive nutrients. Moreover, academic-industry collaborations can spur innovation to reduce PHA production cost and improve scalability, rendering PHA more suited for biomedical applications (Liu et al. 2016).

14.6.2 Batch-to-Batch Variability

14.6.2.1 Consistency in Quality

It is difficult, however, to produce quality PHA in the same way from the same strain of microorganisms from batch to batch, or from one strain to the other, because each has different life cycles, behaves differently under different fermentation conditions, and creates PHA that must be purified differently from each other. These discrepancies can cause variations in the physical, chemical, and mechanical properties of the PHA with corresponding unpredictability in its performance in biomedical applications. One of the most important things is standardizing production processes, in order to reduce variability in PHA materials and make outputs that will use to stringent quality standards.

14.6.2.2 Standardization

Optimizing the standardization protocols involves defining precise guidelines in every step of production, each one, from the microbial culture conditions to purification and final product formulation. It is important for regulatory agencies to set bounds that dictate the production of PHAs because that would ensure that companies would adhere to a constant level. Application of quality control such as routine testing and validation are required to make certain that each batch of PHA complies with clinical requirements.

14.6.3 Degradation Control

14.6.3.1 Predictable Degradation Rates

Although PHA biodegradability is a key advantage, the challenge is to tune the degradation rate matching a particular medical application. Varying degrees of material degradation are required depending on the intended medical application, e.g. temporary scaffolds should degrade more quickly than long-term implants (Nygaard et al. 2021). This control is achieved by tuning the degradation behaviour, which can be done by varying the polymer composition and processing conditions.

14.6.3.2 Uniform Degradation

Safe clinical implementation of PHA requires that it degrades homogeneously while exhibiting minimal inflammatory responses and determining minimal or no toxic byproducts. Researcher will have to design PHA copolymers which are more constant in their degradation and does not produce toxic byproducts for both temporary and long-term implants. While additional in vivo studies need to be conducted which characterize the degradation pathways that may occur, and measure the long-term cellular responses, these studies will aid in the performance of PHA for a variety of medical applications, while maintaining patient safety and material dependability.

14.6.4 Mechanical Properties

14.6.4.1 Tailoring Mechanical Properties

Mechanically, PHA need to be comparable to the mechanical properties of the tissues they are to be replaced or to support. This means a balance of tensile, flexible and elastic force; bone scaffolds, for example, have to be more tensile whereas soft tissue applications need more flexibility. This could be achieved by alteration of polymer composition and structure as well as the addition of fillers or additives to the biocomposites.

14.6.4.2 Long-Term Stability

PHA must satisfy the necessary structural properties over the period of time for which the material will be used, without degradation or loss of functionality prematurely. The properties of material under physiological conditions, and long-term connective tissues are an important issue for the study (Zhao et al. 2003). This is a challenge that is being addressed by researchers on developing PHA formulations that are resistant to environmental degradation factors such as moisture and temperature variations enabling the material to remain stable and functional for the whole working life.

14.6.5 Surface Modifications

14.6.5.1 Functionalization Techniques

Improving PHA-cell (host) interactions can be achieved by reasonable surface modification strategies. Surface modification of PHA the surfaces prior cell seeding has also been found to be useful to promote cell adhesion, proliferation, and differentiation. Most prefered techniques for surface modification are plasma treatment, chemical grafting and coating materials with bioactive molecules. These are designed to generate a bioactive surface in order to mimic the extracellular matrix of tissue, which may improve integration with host tissues.

14.6.5.2 Long-Term Bioactivity

Among the biggest challenges is ensuring that any bioactive surface modifications are effective over the long term. Over time the bioactive molecule may degrade or become inactive, so potentially the material loses efficacy. Therefore, researchers must address techniques through which functionalization can be stabilized to retain bioactivity over the material lifespan. This includes testing how well surface modifications hold up within your body and working details on extending their lifespan.

14.6.6 Regulatory Approval

14.6.6.1 Regulatory Hurdles

Acquiring regulatory clearance for new biomaterials requires a lot of testing to show stability, efficiency, and quality. This is a lengthy and costly process that requires lots of preclinical and clinical data. Typical, it should be in line with global regulatory standards from the FDA or EMA evaluating the biocompatibility, toxicity, performance of PHA.

14.6.6.2 Clinical Trials

Clinical evaluation of performance of PHA-based products in human through clinical studies is necessary but difficult. These trials need to prove PHA to be both safe and effective in its intended application, which demands high investments and laborious planning. Engaging regulatory bodies early in the preclinical development journey is a key approach to accelerating approval to mitigate risks by troubleshooting potential issues in advance on the journey from benchtop to bedside.

14.7 Summary and Future Directions of PHA as a Biomaterial

14.7.1 Advancements in Production Techniques

14.7.1.1 Biotechnological Innovation

Implementation of genetic engineering for developing more efficient microbial strains would be a better option for PHA production in high yields with reduced costs. Improvements on fermentation can also be achieved by optimizing fermentation processes, such as changing carbon sources or fermentation conditions. These improvements could help drive down the cost and scale-up of PHA production and broaden its biomedical application base.

14.7.1.2 Renewable Resource

The sustainable development and economic feasibility of PHA production may be influenced by the use of renewable, low-cost feedstocks, such as agricultural residues or non-food crops. The strategy minimises the need for traditional, costlier resources and corresponds with the trend towards sustainable biomaterials. The potential for lower costs and higher process economics for PHA will enable research into new feedstocks and production methods.

14.7.2 Enhanced Functionalization

14.7.2.1 Advanced Surface Modifications

Advance techniques like plasma treatment, electrospinning and nanopatterning are employed in strikingly improving the surface properties of PHA. These techniques enable the features in the nano- to micro-scale in order to improve cell behaviours, such as attaching behaviours, proliferating behaviours, and differentiating behaviours. Modifications of such nature are required for any applications in tissue engineering and regenerative medicine.

14.7.2.2 Multi-functional PHA

Immobilization of growth factors, antimicrobial agents, or other bioactive molecules can furnish PHA composites with multiple functionalities. The controlled release of these biologically active composites can enable PHA-based medical devices with new properties to be designed for, among other things to induce tissue regeneration or avoid infections. Key to unlocking these multi-functional capabilities will be research into new composite formulations and delivery mechanisms.

14.7.3 Controlled Degradation

14.7.3.1 Customizable Degradation Profiles

Development of new copolymers and blends based on PHA that can control this degradation time in a applications-specific way, is mandatory. This includes the effects of monomer composition, polymer chain length, and processing conditions on degradation behaviour. Degradation profiles may be customized to support the temporary cavities and degrade harmlessly after the tissue fills the cavity to ensure that PHA materials do not cause any secondary reactions.

14.7.3.2 Biodegradation Modelling

Establishing predictive models to better predict the in vivo degradation behavior of PHA materials is needed to enable to better control their degradation profile in clinical applications. The models are used to mimic a range of biological scenarios that might promote the degradation of PHA over time in different environments, which could help researchers systematically tailor PHA properties for specific tasks. Such models can also aid in identifying potential problems, fine-tuning the PHA formulations to achieve good processability and safety.

14.7.4 Improved Mechanical Properties

14.7.4.1 Composite Materials

The addition of other polymers, ceramics or nanoparticles to PHA would not only provide good mechanical properties but also still retain PHA good biodegradable and biocompatibility properties. Such composites may offer strengths, flexibilities, and toughnesses are eminently suitable for a variety of biomedical applications. In different applications they can be used even if they have good mechanical characteristics and research into new composite formulations and manufacturing techniques will be able to overcome the current mechanical properties limitations.

14.7.4.2 Material Engineering

New polymerization and crosslinking approaches could allow to develop PHA materials with improved mechanical properties. All of this involves developing techniques that will enhance such long-term material properties as its tensile strength, flexibility, and resilience to environmental elements. Materials science can revolutionize and enhance the versatility, reliability of PHA as a biomaterial for biomedical applications.

14.7.5 Clinical Translation

14.7.5.1 Preclinical Studies

The gathering of verified preclinical data around toxicity, efficacy, and performance in animal models through large scale preclinical studies is necessary. Novel studies as mentioned above can potentially point to problems and improve the PHA formulations before actual clinical trials. In vivo studies can supply useful information on the material functioning, allowing to move to a transition stage of their application in humans.

14.7.5.2 Collaborative Research

The promotion of academic-industry-regulatory partnerships can simplify the process of PHA-based biomaterial development and validation for regulatory approval. Structural partnerships such as combining various partners, or stakeholder types, to bring complementary expertise, resources, and funding to bear on the problem, moving more quickly as efforts are prioritized and solutions are scaled. Meanwhile collaboration can also spur knowledge and innovation to underpin a successful clinical translation of PHA.

14.7.6 Personalized Medicine

14.7.6.1 Customizable Implants

Utilizing 3D printing and other high-tech manufacturing methods such as implantable and scaffolds of PHA custom made for each patient from their own MRI data could be a future goal. Allowing for better fit, form and patient outcomes-tailoring implants to meet each patient's unique anatomical and medical needs. Further advancements in 3D printing equipment and materials will make not only more cost-efficient and personalised medicine but also longer lasting and more efficient equipment.

14.7.6.2 Regenerative Medicine

Increased tissue remodeling and healing may come from combining PHA with stem cell therapy and other regenerative medicine approaches. Hence, in theory, the PHA scaffold can serve as a niche for stem cells that will enable them to grow, proliferate and differentiate in order to create new tissues. As PHA has been demonstrated to interact with stem cells, combination therapies will become the future of regenerative medicines with an increase in patient outcomes.

14.7.7 Regulatory Pathways

The pursuit of harmonized international standards for PHA biomaterials development will help securing regulatory approval and market access. Harmonised standards should facilitate approval, so that the same core safety and performance specifications are met no matter where the PHA products are presented for authorisation. PPP (public-private partnerships) that involve regulatory authorities, research organizations, and industrial stakeholders may lead to the definition of such criteria and to a position of prominence of PHA-based biomaterials on the global scale.

14.7.8 Accelerated Pathways

This highlights the need for research into fast-track regulatory pathways for high-impact biomaterials. The access to these pathways can decrease times and costs for the introduction of new PHA-based products and thereby enable, for instance, earlier availability of advanced medical treatments. Engagement with regulatory authorities at an early stage and comprehensive preclinical data may serve to make the case for an accelerated approval, and thus the benefit of such could be passed through to the patients and healthcare providers.

Addressing these problems and perspectives will empower PHA to more effectively shift focus from research and development to clinical usage, thereby providing a renewable, versatile, and sustainable choice for a variety of biomedical applications.

References

Albuquerque MGE, Torres CAV, Reis MAM (2010) Polyhydroxyalkanoate (PHA) production by a mixed microbial culture using sugar molasses: effect of the influent substrate concentration on culture selection. Water Res 44:3419–3433. https://doi.org/10.1016/j.watres.2010.03.021

Ali I, Jamil N (2018) Biosynthesis and characterization of Poly3-hydroxyalkanote (PHA) from newly isolated bacterium bacillus sp. AZR-1. Iran J Sci Technol Trans A: Sci 42:371–378. https://doi.org/10.1007/s40995-016-0132-6

Ansari S, Sami N, Yasin D et al (2021) Biomedical applications of environmental friendly poly-hydroxyalkanoates. Int J Biol Macromol 183:549–563. https://doi.org/10.1016/j.ijbiomac.2021.04.171

Atarés L, Chiralt A, González-Martínez C, Vargas M (2024) Production of polyhydroxyalkanoates for biodegradable food packaging applications using Haloferax mediterranei and agrifood wastes. Food Secur 13:950

Barouti G, Guillaume SM (2016) Polyhydroxybutyrate (PHB)-based triblock copolymers: synthesis of hydrophobic PHB/poly(benzyl β-malolactonate) and amphiphilic PHB/poly(malic acid) analogues by ring-opening polymerization. Polym Chem 7:4603–4608. https://doi.org/10.1039/C6PY00910G

Bravo-Porras G, Fernández-Güelfo LA, Álvarez-Gallego CJ et al (2024) Influence of the total concentration and the profile of volatile fatty acids on polyhydroxyalkanoates (PHA) production by mixed microbial cultures. Biomass Convers Biorefinery 14:239–253. https://doi.org/10.1007/s13399-021-02208-z

Brelle L, Faÿ F, Ozturk T et al (2023) Hydrogel based on polyhydroxyalkanoate sulfonate: control of the swelling rate by the ionic group content. Biomacromolecules 24:1871–1880. https://doi.org/10.1021/acs.biomac.3c00059

Butt FI, Muhammad N, Hamid A et al (2018) Recent progress in the utilization of biosynthesized polyhydroxyalkanoates for biomedical applications—review. Int J Biol Macromol 120:1294–1305. https://doi.org/10.1016/j.ijbiomac.2018.09.002

Carvalho LT, Vieira TA, Zhao Y et al (2021) Recent advances in the production of biomedical systems based on polyhydroxyalkanoates and exopolysaccharides. Int J Biol Macromol 183:1514–1539. https://doi.org/10.1016/j.ijbiomac.2021.05.025

Chai JM, Amelia TS, Mouriya GK et al (2021) Surface-modified highly biocompatible bacterial-poly(3-hydroxybutyrate-co-4-hydroxybutyrate): a review on the promising next-generation biomaterial. Polymers (Basel) 13:950

Dhania S, Bernela M, Rani R et al (2022) Scaffolds the backbone of tissue engineering: advancements in use of polyhydroxyalkanoates (PHA). Int J Biol Macromol 208:243–259. https://doi.org/10.1016/j.ijbiomac.2022.03.030

Dietrich K, Dumont M-J, Del Rio LF, Orsat V (2019a) Sustainable PHA production in integrated lignocellulose biorefineries. New Biotechnol 49:161–168. https://doi.org/10.1016/j.nbt.2018.11.004

Dietrich K, Dumont M-J, Orsat V, Del Rio LF (2019b) Consumption of sugars and inhibitors of softwood hemicellulose hydrolysates as carbon sources for polyhydroxybutyrate (PHB) production with Paraburkholderia sacchari IPT 101. Cellulose 26:7939–7952. https://doi.org/10.1007/s10570-019-02664-x

Elmowafy E, Abdal-Hay A, Skouras A et al (2019) Polyhydroxyalkanoate (PHA): applications in drug delivery and tissue engineering. Expert Rev Med Devices 16:467–482. https://doi.org/10.1080/17434440.2019.1615439

Esmail A, Pereira JR, Zoio P et al (2021) Oxygen plasma treated-electrospun polyhydroxyalkanoate scaffolds for hydrophilicity improvement and cell adhesion. Polymers (Basel) 13:1056

Gregory DA, Taylor CS, Fricker ATR et al (2022) Polyhydroxyalkanoates and their advances for biomedical applications. Trends Mol Med 28:331–342. https://doi.org/10.1016/j.molmed.2022.01.007

Gregory DA, Fricker ATR, Mitrev P et al (2023) Additive manufacturing of polyhydroxyalkanoate-based blends using fused deposition modelling for the development of biomedical devices. J Funct Biomater 14:40

Guessasma S, Belhabib S, Nouri H (2020) Thermal cycling, microstructure and tensile performance of PLA-PHA polymer printed using fused deposition modelling technique. Rapid Prototyp J 26:122–133. https://doi.org/10.1108/RPJ-06-2019-0151

Hou J, Zhang X, Wu Y et al (2024) Amphiphilic and fatigue-resistant organohydrogels for small-diameter vascular grafts. Sci Adv 8:eabn5360. https://doi.org/10.1126/sciadv.abn5360

Jaffur N, Jeetah P, Kumar G (2021) A review on enzymes and pathways for manufacturing polyhydroxybutyrate from lignocellulosic materials. 3 Biotech 11:483. https://doi.org/10.1007/s13205-021-03009-x

Kalia VC, Patel SKS, Lee J-K (2023) Exploiting polyhydroxyalkanoates for biomedical applications. Polymers (Basel) 15:1937

Kaniuk Ł, Stachewicz U (2021) Development and advantages of biodegradable PHA polymers based on electrospun PHBV fibers for tissue engineering and other biomedical applications. ACS Biomater Sci Eng 7:5339–5362. https://doi.org/10.1021/acsbiomaterials.1c00757

Kanzariya R, Gautam A, Parikh S et al (2023) Formation of polyhydroxyalkanoates using agro and industrial waste as a substrate—a review. Biotechnol Genet Eng Rev 39:897–936. https://doi.org/10.1080/02648725.2023.2165222

Kaur G, Roy I (2015) Strategies for large-scale production of polyhydroxyalkanoates. Chem Biochem Eng Q 29:157–172. https://doi.org/10.15255/CABEQ.2014.2255

Ke Y, Liu C, Zhang X et al (2017) Surface modification of polyhydroxyalkanoates toward enhancing cell compatibility and antibacterial activity. Macromol Mater Eng 302:1700258. https://doi.org/10.1002/mame.201700258

Kelwick RJR, Webb AJ, Wang Y et al (2021) AL-PHA beads: bioplastic-based protease biosensors for global health applications. Mater Today 47:25–37. https://doi.org/10.1016/j.mattod.2021.02.018

Kim J, Gupta NS, Bezek LB et al (2023) Biodegradation studies of polyhydroxybutyrate and polyhydroxybutyrate-co-polyhydroxyvalerate films in soil. Int J Mol Sci 24:7638

Koller M (2018) Biodegradable and biocompatible polyhydroxy-alkanoates (PHA): auspicious microbial macromolecules for pharmaceutical and therapeutic applications. Molecules 23:362

Koller M, Niebelschütz H, Braunegg G (2013) Strategies for recovery and purification of poly[(R)-3-hydroxyalkanoates] (PHA) biopolyesters from surrounding biomass. Eng Life Sci 13:549–562. https://doi.org/10.1002/elsc.201300021

Koller M, Maršálek L, de Sousa Dias MM, Braunegg G (2017) Producing microbial polyhydroxyalkanoate (PHA) biopolyesters in a sustainable manner. New Biotechnol 37:24–38. https://doi.org/10.1016/j.nbt.2016.05.001

Kosseva MR, Rusbandi E (2018) Trends in the biomanufacture of polyhydroxyalkanoates with focus on downstream processing. Int J Biol Macromol 107:762–778. https://doi.org/10.1016/j.ijbiomac.2017.09.054

Kumar V, Sehgal R, Gupta R (2021) Blends and composites of polyhydroxyalkanoates (PHAs) and their applications. Eur Polym J 161:110824. https://doi.org/10.1016/j.eurpolymj.2021.110824

Ladhari S, Vu N-N, Boisvert C et al (2023) Recent development of polyhydroxyalkanoates (PHA)-based materials for antibacterial applications: a review. ACS Appl Bio Mater 6:1398–1430. https://doi.org/10.1021/acsabm.3c00078

Lezcano MF, Álvarez G, Chuhuaicura P et al (2022) Polyhydroxybutyrate (PHB) scaffolds for peripheral nerve regeneration: a systematic review of animal models. Biology (Basel) 11:706

Li K, Zhang W-W, Shi X-J et al (2019) Preparation of novel mcl-PHA by green synthesis: influence of active small molecule via esterification effect on side chain. Mater Res Express 6:75328. https://doi.org/10.1088/2053-1591/ab166c

Li W, Cicek N, Levin DB et al (2020) Bacteria-triggered release of a potent biocide from core-shell polyhydroxyalkanoate (PHA)-based nanofibers for wound dressing applications. J Biomater Sci Polym Ed 31:394–406. https://doi.org/10.1080/09205063.2019.1693882

Liu C-C, Zhang L-L, An J et al (2016) Recent strategies for efficient production of polyhydroxyalkanoates by micro-organisms. Lett Appl Microbiol 62:9–15. https://doi.org/10.1111/lam.12511

Mandragutti T, Jarso TS, Godi S et al (2024) Physicochemical characterization of polyhydroxybutyrate (PHB) produced by the rare halophile brachybacterium paraconglomeratum MTCC 13074. Microb Cell Factories 23:59. https://doi.org/10.1186/s12934-024-02324-1

Marcano A, Bou Haidar N, Marais S et al (2017) Designing biodegradable PHA-based 3D scaffolds with antibiofilm properties for wound dressings: optimization of the microstructure/nanostructure. ACS Biomater Sci Eng 3:3654–3661. https://doi.org/10.1021/acsbiomaterials.7b00552

Masaeli E, Wieringa PA, Morshed M et al (2014) Peptide functionalized polyhydroxyalkanoate nanofibrous scaffolds enhance Schwann cells activity. Nanomedicine 10:1559–1569. https://doi.org/10.1016/j.nano.2014.04.008

Meereboer KW, Misra M, Mohanty AK (2020) Review of recent advances in the biodegradability of polyhydroxyalkanoate (PHA) bioplastics and their composites. Green Chem 22:5519–5558. https://doi.org/10.1039/D0GC01647K

Mehrpouya M, Vahabi H, Barletta M et al (2021) Additive manufacturing of polyhydroxyalkanoates (PHAs) biopolymers: materials, printing techniques, and applications. Mater Sci Eng C 127:112216. https://doi.org/10.1016/j.msec.2021.112216

Misra SK, Valappil SP, Roy I, Boccaccini AR (2006) Polyhydroxyalkanoate (PHA)/inorganic phase composites for tissue engineering applications. Biomacromolecules 7:2249–2258. https://doi.org/10.1021/bm060317c

Mohtashami Z, Javar HA, Tehrani MR et al (2020) Fabrication, optimization, and in vitro and in vivo characterization of intra-vitreal implant of budesonide generally made of PHBV. AAPS PharmSciTech 21:314. https://doi.org/10.1208/s12249-020-01828-4

Mondal S, Syed UT, Gil C et al (2023) A novel sustainable PHA downstream method. Green Chem 25:1137–1149. https://doi.org/10.1039/D2GC04261D

Najah MI, Aisyah R, Adzila S, Haq RHA (2022) Mechanical properties of calcium phosphate reinforced polyhydroxyalkanoate (PHA) biocomposite. J Thermoplast Compos Mater 36:3294–3311. https://doi.org/10.1177/08927057221128188

Nygaard D, Yashchuk O, Hermida ÉB (2021) PHA granule formation and degradation by Cupriavidus necator under different nutritional conditions. J Basic Microbiol 61:825–834. https://doi.org/10.1002/jobm.202100184

Prakash P, Lee W-H, Loo C-Y et al (2022) Advances in polyhydroxyalkanoate nanocarriers for effective drug delivery: an overview and challenges. Nano 12:175

Pramanik N (2023) A tool for biomedical application: synthesis and modification of polyhydroxyalkanoates. Sustain Chem Pharm 32:101041. https://doi.org/10.1016/j.scp.2023.101041

Qu X-H, Wu Q, Chen G-Q (2006) In vitro study on hemocompatibility and cytocompatibility of poly(3-hydroxybutyrate-co-3-hydroxyhexanoate). J Biomater Sci Polym Ed 17:1107–1121. https://doi.org/10.1163/156856206778530704

Razaif-Mazinah MRM, Anis SNS, Harun HI et al (2017) Unusual poly(3-hydroxyalkanoate) (PHA) biosynthesis behavior of pseudomonas putida Bet001 and Delftia tsuruhatensis Bet002 isolated from palm oil mill effluent. Biotechnol Appl Biochem 64:259–269. https://doi.org/10.1002/bab.1482

Razzaq S, Shahid S, Farooq R et al (2024) Sustainable bioconversion of agricultural waste substrates into poly (3-hydroxyhexanoate) (mcl-PHA) by Cupriavidus necator DSM 428. Biomass Convers Biorefinery 14:9429–9439. https://doi.org/10.1007/s13399-022-03194-6

Reddy VU, Ramanaiah SV, Reddy MV, Chang Y-C (2022) Review of the developments of bacterial medium-chain-length polyhydroxyalkanoates (mcl-PHAs). Bioengineering 9:225

Rodríguez-Contreras A, Guillem-Marti J, Lopez O et al (2019) Antimicrobial PHAs coatings for solid and porous tantalum implants. Colloids Surf B Biointerfaces 182:110317. https://doi.org/10.1016/j.colsurfb.2019.06.047

Saavedra del Oso M, Mauricio-Iglesias M, Hospido A (2021) Evaluation and optimization of the environmental performance of PHA downstream processing. Chem Eng J 412:127687. https://doi.org/10.1016/j.cej.2020.127687

Sehgal R, Gupta R (2020) Polyhydroxyalkanoate and its efficient production: an eco-friendly approach towards development. 3 Biotech 10:549. https://doi.org/10.1007/s13205-020-02550-5

Shishatskaya EI, Nikolaeva ED, Vinogradova ON, Volova TG (2016) Experimental wound dressings of degradable PHA for skin defect repair. J Mater Sci Mater Med 27:165. https://doi.org/10.1007/s10856-016-5776-4

Shrivastav A, Kim H-Y, Kim Y-R (2013) Advances in the applications of polyhydroxyalkanoate nanoparticles for novel drug delivery system. Biomed Res Int 2013:581684. https://doi.org/10.1155/2013/581684

Siew EL, Rajab NF, Bin OA et al (2007) In vitro biocompatibility evaluation of poly(3-hydroxybutyrate-co-4-hydroxybutyrate) copolymer in fibroblast cells. J Biomed Mater Res A 81A:317–325. https://doi.org/10.1002/jbm.a.31000

Singh AK, Mallick N (2008) Enhanced production of SCL-LCL-PHA co-polymer by sludge-isolated pseudomonas aeruginosa MTCC 7925. Lett Appl Microbiol 46:350–357. https://doi.org/10.1111/j.1472-765X.2008.02323.x

Singh AK, Srivastava JK, Chandel AK et al (2019) Biomedical applications of microbially engineered polyhydroxyalkanoates: an insight into recent advances, bottlenecks, and solutions. Appl Microbiol Biotechnol 103:2007–2032. https://doi.org/10.1007/s00253-018-09604-y

Smith S, Goodge K, Delaney M et al (2020) A comprehensive review of the covalent immobilization of biomolecules onto electrospun nanofibers. Nano 10:2142

Solaiman DKY, Catara V, Greco S (2005) Poly(hydroxyalkanoate) synthase genotype and PHA production of pseudomonas corrugata and P. mediterranea. J Ind Microbiol Biotechnol 32:75–82. https://doi.org/10.1007/s10295-005-0205-6

Sun J, Shen J, Chen S et al (2018) Nanofiller reinforced biodegradable PLA/PHA composites: current status and future trends. Polymers (Basel) 10:505

Tang L, Hou X, Wang J, Pan L (2023) Effect of different chemical surface treatments on interfacial compatibility and properties of polyhydroxyalkanoates/coffee grounds composites. Polym Compos 44:1175–1187. https://doi.org/10.1002/pc.27162

Thorat Gadgil BS, Killi N, Rathna GVN (2017) Polyhydroxyalkanoates as biomaterials. Medchemcomm 8:1774–1787. https://doi.org/10.1039/C7MD00252A

Valappil SP, Misra SK, Boccaccini AR, Roy I (2006) Biomedical applications of polyhydroxyalkanoates, an overview of animal testing and in vivo responses. Expert Rev Med Devices 3:853–868. https://doi.org/10.1586/17434440.3.6.853

Wampfler B, Ramsauer T, Kehl K et al (2010) Application of activated charcoal in the downstream processing of bacterial Olefinic poly(3-hydroxyalkanoates). Chimia (Aarau) 64:784. https://doi.org/10.2533/chimia.2010.784

Wang S, Chen W, Xiang H et al (2016) Modification and potential application of short-chain-length polyhydroxyalkanoate (SCL-PHA). Polymers (Basel) 8:1774

Wang J, Liu S, Huang J et al (2023) Genetic engineering strategies for sustainable polyhydroxyalkanoate (PHA) production from carbon-rich wastes. Environ Technol Innov 30:103069. https://doi.org/10.1016/j.eti.2023.103069

Westlie AH, Hesse SA, Tang X et al (2023) All-Polyhydroxyalkanoate triblock copolymers via a stereoselective-chemocatalytic route. ACS Macro Lett 12:619–625. https://doi.org/10.1021/acsmacrolett.3c00162

Zhang J, Shishatskaya EI, Volova TG et al (2018) Polyhydroxyalkanoates (PHA) for therapeutic applications. Mater Sci Eng C 86:144–150. https://doi.org/10.1016/j.msec.2017.12.035

Zhao K, Deng Y, Chun Chen J, Chen G-Q (2003) Polyhydroxyalkanoate (PHA) scaffolds with good mechanical properties and biocompatibility. Biomaterials 24:1041–1045. https://doi.org/10.1016/S0142-9612(02)00426-X

Žiganova M, Merijs-Meri R, Zicāns J et al (2023) Visco-elastic and thermal properties of microbiologically synthesized polyhydroxyalkanoate plasticized with triethyl citrate. Polymers (Basel) 15:2896

Zytner P, Kumar D, Elsayed A et al (2023) A review on polyhydroxyalkanoate (PHA) production through the use of lignocellulosic biomass. RSC Sustainability 1:2120–2134. https://doi.org/10.1039/D3SU00126A

Dr. Ashok Raja Chandrasekar earned his Ph.D. in medical bionanotechnology from the University of Madras, Tamil Nadu, India, in 2020 for the thesis "Fabrication of bioactive glass-based nanocomposites with biomedical applications." At present, he has been working as a researcher at FunGlass, Alexander Dubcek University of Trencin, Slovakia, since 2021. His area of expertise involves the preparation, characterization, and evaluation of bioactive nanomaterials and coatings for biomedical implant applications. Currently, he is working on controlling the degradation of magnesium and its alloys via Plasma Electrolytic Oxidation and sol-gel coatings for biodegradable implant applications as part of his R2 scholarship for excellent researchers, Slovakia.

Attractive Potential of Alginate-Based Oil-in-Water Emulsion Liquid Marbles for the Promising Carrier of the Controlled Release of Limonene

15

Homare Tsugifuji and Masanao Imai

Abstract

Alginate-based oil-in-water (O/W) emulsion Liquid Marbles (LMs) were successfully created by hybridizing stearate calcium and calcium chloride microparticles. The controlled release of a bio-functional hydrophobic volatile component (limonene) by the emulsion LM shells was achieved. The outer diameter of the LMs was uniform from 3.0 to 3.3 mm. Four types of LMs were examined: O/W emulsion droplet (NED), O/W emulsion LM (ELM), sodium-alginate O/W emulsion LM (AlgLM), and sodium alginate-shelled O/W emulsion LM (GELM). The amount of released limonene initiated after 60 min from the ELM was ca. 0.70 folds lesser than that from the NED. The released amount from the AlgLM was 0.47 folds less than that from the ELM. The participation of the limonene between the dispersed limonene phase and the aqueous alginate phase in the AlgLM was influenced by solubilized sodium alginate. The released amount from the GELM was significantly low. It was 0.032 folds less than that from the ELM. On the outer surface of the GELM, calcium chloride powder coexisted with calcium stearate. The calcium chloride powder on the outer

H. Tsugifuji
Department of Food Bioscience and Biotechnology, Faculty of Bioresource Sciences, Nihon University, Fujisawa, Kanagawa, Japan

Goka-plant, Kewpie Cooperation, Goka, Ibakaki, Japan
e-mail: homare04@outlook.jp

M. Imai (✉)
Department of Food Bioscience and Biotechnology, Faculty of Bioresource Sciences, Nihon University, Fujisawa, Kanagawa, Japan

Lab. of Chemical Engineering, Institute of Science and Technology, Kanto Gakuin University, Yokohama, Kanagawa, Japan
e-mail: XLT05104@nifty.com

R. Malviya, S. Sundram (eds.), *Sustainable Nanocomposites with Green Biomaterials*, Biomaterials, Bioengineering and Sustainability 2,
https://doi.org/10.1007/978-3-031-79110-9_15

surface gradually solubilized into the aqueous sodium alginate phase and quickly formed a cross-linked alginate shell. The mass transfer of the released limonene was inhibited both by the high viscosity of the alginate phase and the cross-linked alginate shell. LMs prepared from biocompatible materials showed promising abilities for the controlled release of limonene. The alginate-based O/W emulsion LMs can facilitate the controlled release of comprehensive volatile bioactive components in the forthcoming sustainable-material era.

Keywords

Liquid marble · LM · Carrier · Controlled release · O/W emulsion · Volatile component · Limonene · Sodium alginate · Sustainable green materials · Calcium stearate · Calcium chloride

Nomenclature

c	concentration of limonene in the LM [mol m $^{-3}$]
c_0	initial concentration of limonene in the LM [mol m $^{-3}$]
ΔH	evaporation enthalpy of limonene [J K $^{-1}$]
k	overall rate coefficient of the released limonene [s $^{-1}$]
p	vaper pressure of limonene [Pa]
T_b	boiling point of limonene [K]
t	time [s]

Abbreviations

AlgLM	The LM body contained the oil in water emulsion. Sodium alginate was solubilized in the water phase, as illustrated in Fig. 15.2
ELM	The LM body contained the oil in water emulsion, as illustrated in Fig. 15.2
GELM	Calcium alginate gel film covered on the outer edge of the AlgLM, as illustrated in Fig. 15.2
LM	liquid marble
NED	Emulsion droplet, not LM, as illustrated in Fig. 15.2
O/W	oil in water

15.1 Introduction

Liquid Marble (LM) is a spherical droplet that is stabilized by the adsorption of insoluble micro- or nano-sized solid particles onto an air-liquid interface. The pioneer work has been presented by Aussillous and Quere (2001). This original droplet is introduced as an LM. As its most unique characteristic, the LM is a nonstick liquid droplet that can freely roll on a solid flat plane without deforming the spherical shape. LM is an easily prepared and low-cost carrier for chemical reactions (Wang et al. 2016), aerobic bacteria cultivation (Tanaka et al. 2020; Yamakawa et al. 2022), and physicochemical stress-responsible micro devices (Fujii et al. 2012; Zang et al. 2015). Fabrication of LM and various applications have been reported (Bormashenko 2011; Roy et al. 2020; Murakami et al. 2022; Tenjimbayashi et al. 2023).

Limonene is a typical monoterpene found in citrus fruits. At room temperature and normal atmospheric pressure, it exists as a colorless and transparent liquid. Limonene has two structural isomers: D-isomer is abundant in the peel of fruits, and L-isomer is found in peppermint oil. They are preferably applied for daily used fragrances (Vieira et al. 2018). Limonene is reportedly highly toxic to aquatic organisms and is not approved for use in humans (Ravichandran et al. 2018). However, the D-isomer is attracting attention as a safe solvent for expanded polystyrene because its structure is similar to that of a styrene monomer (Monte and Landau-West 1982). Furthermore, it is used in detergents and adhesives for plastic models.

Alginate acid is a well-known typical oceanic biopolymer found in brown algae, such as kelp and other seaweeds. The polymeric structure of the alginate acid is composed of a D-mannuronic acid (M) domain and an L-guluronic acid (G) domain. The aqueous sodium alginate phase is viscous, biocompatible, and safe; it has been commonly applied for food, cosmeceuticals, and other industrial supplements. An ion exchange reaction is quickly caused mainly in a G-G domain by the addition of polyvalent cations (such as calcium ions) (Grant et al. 1973). A water-insoluble calcium alginate gel is sufficiently stable in the aqueous phase, with a wide range of temperatures, pH, and coexistent materials. Hence, it has been applied as a carrier of controlled release (Almeida and Almeida 2004). A calcium alginate gel membrane has excellent sensitivity for molecular size screening function (Julian et al. 1988; Kashima and Imai 2011). Calcium stearate has been employed as a food and pharmaceutical additive in functional food supplements.

The controlled release of functional components is important in medical, cosmetic, and agricultural fields. Aromatherapy involved with the controlled release of functional volatile components is commonly applied in daily psychological health care (Anandakumar et al. 2020). A biocompatible carrier for the controlled release is favorable because no waste product is obtained after carrier utilization.

This paper depicts the advantageous potential of alginate-based emulsion LM as a carrier for the controlled release of volatile functional chemicals. The viscous alginate aqueous phase in LM contains micro-sized dispersed hydrophobic volatile components. The ability to release is readily regulated by the alginate concentration and the hybridization of the specific coating powder to form a gel film on the outer edge of the LM.

15.2 Materials and Methods

15.2.1 Materials

Calcium stearate microparticles were employed to produce the LMs. They were kindly donated by Taihei Chemical Industrial Co., Ltd. (Osaka, Japan). The median particle diameter was 9.86 μm for Lot No.1902801. Limonene was employed as the hydrophobic functional component. It was purchased from FUJIFILM Wako Pure Chemical Corporation (Osaka, Japan). The purity and density of the limonene were 95.0 wt % and 842 kg m^{-3}, respectively. The molecular weight of limonene (aglycone) was 136.24. It is a volatile hydrophobic component (Log P 4.52), and its vaper pressure is reportedly 0.19 kPa (293 K). Sodium alginate and calcium chloride were purchased from FUJIFILM Wako Pure Chemical Corporation, and employed to prepare the alginate-based emulsion LMs.

15.2.2 Preparation of a Simple Water Liquid Marble

Calcium stearate particles were uniformly dispersed onto a flat rubber board. A pure water droplet was carefully dropped on the board. A syringe needle was made from stainless steel. Its outer diameter and inner diameter were 0.288 and 0.153 mm, respectively. Calcium stearate powder was easily adsorbed on the surface of the water droplet not solubilized in the aqueous phase. The aqueous droplet was gently rolled on the powdered board using a stainless steel micro-spatula until the calcium stearate particles were completely adsorbed on the water droplet surface, affording the simple water LM. The LM was gently rolled again and again on the clean rubber board to remove the excess adsorbed amount of calcium stearate powder from the LM surface.

15.2.3 Preparation of an Oil-in-Water Emulsion Droplet (NED) and an Emulsion Liquid Marble (ELM)

A pure water (10 mL) and limonene (0.5 mL) mixture was introduced into a test tube, which was completely sealed. It was immersed in a water bath and irradiated ultrasonically for 5 min at room temperature (298 ± 2 K). Limonene was dispersed in the pure water phase without a surfactant. The diameter of the dispersed phase was 10–20 μm in water, as shown in Fig. 15.1a.

The constant volumetric fraction of limonene in the O/W emulsion was 0.0476. The O/W emulsion droplet is referred to as NED herein, and it was not an LM. Its outer surface was entirely exposed to atmospheric air.

Stearate calcium powder was uniformly dispersed on a rubber board. The O/W emulsion was carefully dropped onto the rubber board. Thereafter, an O/W emulsion LM (ELM) was prepared. The preparation method was the same as that for the simple water LM, as mentioned in Sect. 15.2.2. The schematic illustration of the NED, ELM and other LMs are presented in Fig. 15.2, and their characteristics are summarized in Table 15.1.

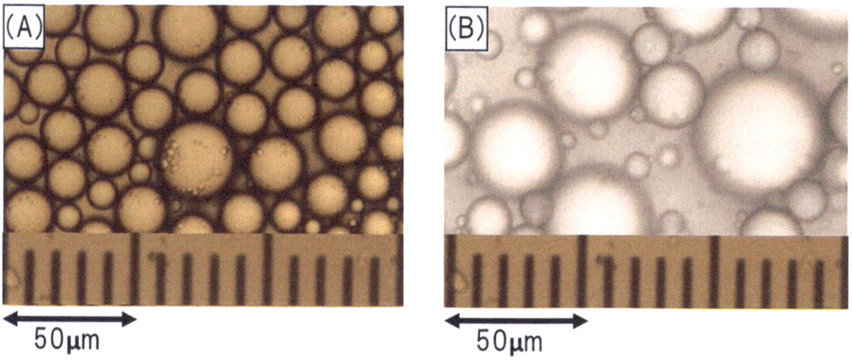

Fig. 15.1 Microphotograph of O/W emulsion contained in the LM body. (**a**) Limonene in the pure water, (**b**) limonene in the aqueous alginate phase

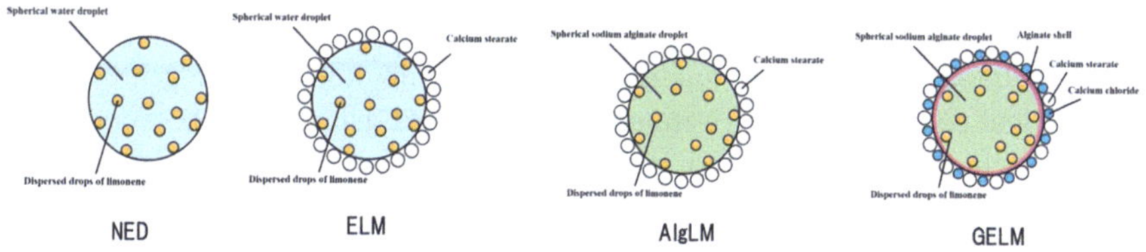

Fig. 15.2 Category of the emulsion droplet (NED) and the three Liquid Marbles, (ELM), (AlgLM) and GELM. NED Oil-in-water emulsion simple droplet not Liquid Marble, ELM Oil-in-water (O/W) emulsion Liquid Marble. Calcium stearate particles are adsorbed, AlgLM (O/W) emulsion Liquid Marble. Aqueous phase contained sodium alginate. Calcium stearate particles are adsorbed on the LM's surface, GELM (O/W) emulsion Liquid Marble. Aqueous phase contained sodium alginate. Calcium stearate and calcium chloride particles are adsorbed on the LM's surface

Table 15.1 Characteristics of the experimental emulsion droplet and the various liquid marbles

Inner phase			Microparticles adsorbed on the LM's surface	
Bulk		Dispersed	Non	Non
NED	Water	Limonene	StCa	Non
ELM	Water		StCa	Non
ALgLM	Water+ Na Alginate		StCa	$CaCl_2$
GELM	Water+ Na Alginate		StCa	Non

15.2.4 Preparation of Sodium Alginate Oil-in-Water Emulsion LM (AlgLM)

Sodium alginate (0.025 g) was completely dissolved in 10 mL of pure water. The concentration of sodium alginate was 0.25 wt %. Limonene (0.5 mL) was dispersed into the aqueous sodium alginate phase by ultrasonic irradiation for 5 min. The diameter of the dispersed phase was 20–50 μm, as shown in Fig. 15.1b.

An alginate aqueous emulsion droplet was carefully deposited onto the stearate-calcium-dispersed rubber board. Thereafter, a sodium alginate O/W emulsion LM (AlgLM) was prepared. The preparation method was the same as that for the simple water LM, as mentioned in Sect. 15.2.2.

15.2.5 Preparation of Sodium Alginate Shelled Oil-in-Water Emulsion LM (GELM)

To prepare the sodium alginate shelled O/W emulsion LM (GELM), calcium chloride was employed as a cross-linking agent for forming the alginate gel. Calcium stearate and calcium chloride were completely mixed using an agate mortar. The mixed calcium chemical powder was applied to make the GELM. The constant mass fraction of calcium chloride in the hybridized particles was 0.5. The outer surface of the GELM became a cross-linking structure of the alginate gel because of the calcium chloride.

15.2.6 Characteristics of the Liquid Marbles Based on the Release of Limonene

The release of limonene from various LMs was characterized based on the time course of the moles of limonene released in a thermo-controlled glass chamber (32 mL). Three NEDs or LMs were softly placed on a thermo-controlled plate (PC-400D; Taitec Corporation, Koshigaya Japan). The schematic illustration of the release examination is presented in Fig. 15.3.

The initial limonene content of the NED or LMs was 1.09×10^{-5} mol. This was the total content of the three examined droplets. The constant temperature in the chamber was set to 313 K (40 °C) . The base plate of the chamber attached to the LMs was continuously thermo-controlled. The sample gas (10 mL) was taken from the chamber at the desired time using a gas-use micro-syringe. The amount of limonene was analyzed by gas chromatography (GC-148; Shimadzu Corporation, Kyoto Japan). An analytical column (DB-WAX 122–7032) was applied (Agit Technology Corporation, Santa Clara US). The column temperature was set to 323 K. The temperatures of the injection inlet and the detector were 483 K and 513 K, respectively. The carrier gas was analytical-grade nitrogen. The flow rate of the nitrogen was 2.4 mL min^{-1}.

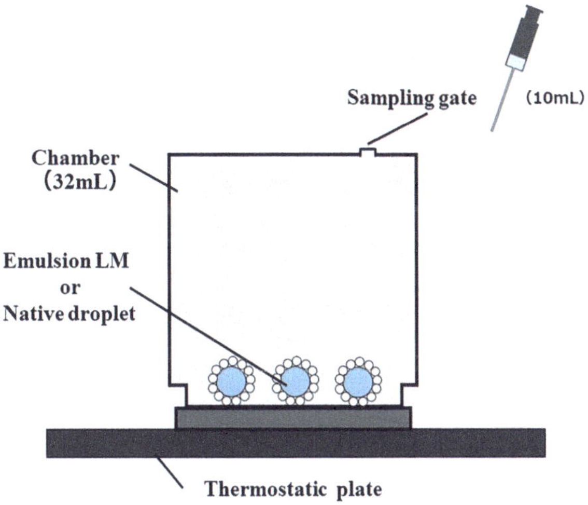

Fig. 15.3 Experimental apparatus for the release of limonene from the liquid marbles

15.3 Results and Discussion

15.3.1 Preparation of the Simple Water Liquid Marble

Figure 15.4a presents a typical simple water LM. The simple water LM was well spherical, with a constant diameter of approximately 3 mm. No adhesive characteristic was observed on any flat and smooth solid board. The LMs were easily rolled on the slightly sloped flat board without any shape deformation

15.3.2 Released Amount of Limonene from the O/W Emulsion Droplet and the Three Types of Liquid Marbles

Figure 15.5 presents the time course of the accumulated amount of limonene released into the chamber. The moles of limonene released gradually increased with time. There was no lag time regardless of the type of LMs. The accumulated amount of limonene from the NED at 60 min was 2.77 × 10 $^{-6}$ mol. It was 25.4% of the initial content in the NEDs.

Based on Trouton's law presented by Eq. (15.1), the evaporation enthalpy of limonene (ΔH) was estimated to be 39,700 J K^{-1}, where the boiling point of limonene (T_b) was 451 K.

$$\frac{\Delta H}{T_b} \fallingdotseq 88\ \mathrm{J\ K^{-1}} \tag{15.1}$$

The saturated vaper pressure (p) of limonene at 313 K (40 °C) was estimated by following the Clausius-Clapeyron equation.

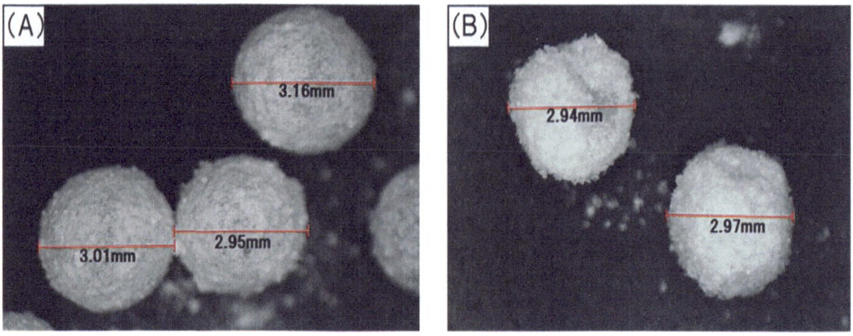

Fig. 15.4 Microphotographs of the simple water LM (**a**) and the alginate shelled LM (GELM) (**b**)

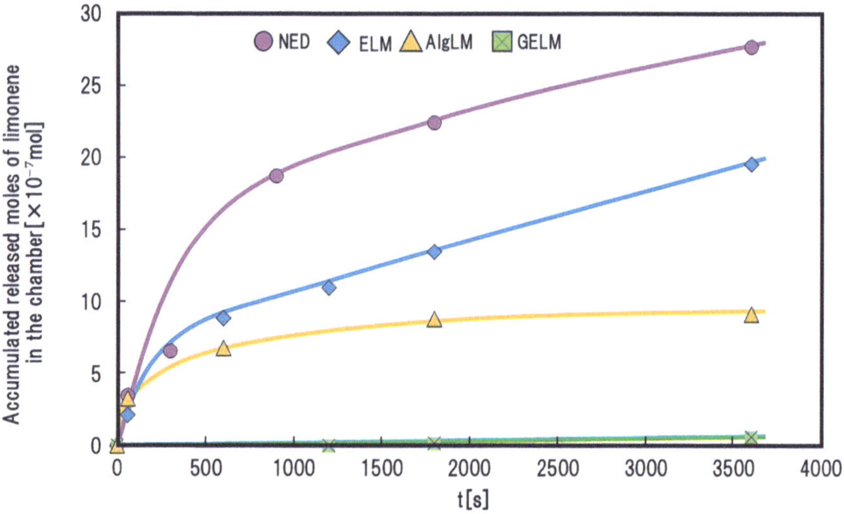

Fig. 15.5 Time course of the accumulated amount of the released limonene from LMs in the chamber

$$\ln\frac{p}{190} = -\frac{\Delta H}{R}\left(\frac{1}{273+40} - \frac{1}{273+20}\right) \quad (15.2)$$

where the saturated vapor pressure of limonene at 293 K (20 °C) was 190 Pa and at 313 K (40 °C) was estimated to be 541 Pa.

The partial pressure of limonene in the chamber at 60 min for the NED was calculated to be 224 Pa based on the equation of the state for ideal gas. This was 41.4% of the saturated vapor pressure of limonene at 313 K. The surface of the NED was fully exposed; it was not covered by calcium stearate microparticles. Limonene was easily vaporized in the chamber. The apparent shape of the NED at 60 min was fully degraded, and the NED was adhesively attached to the base of the chamber.

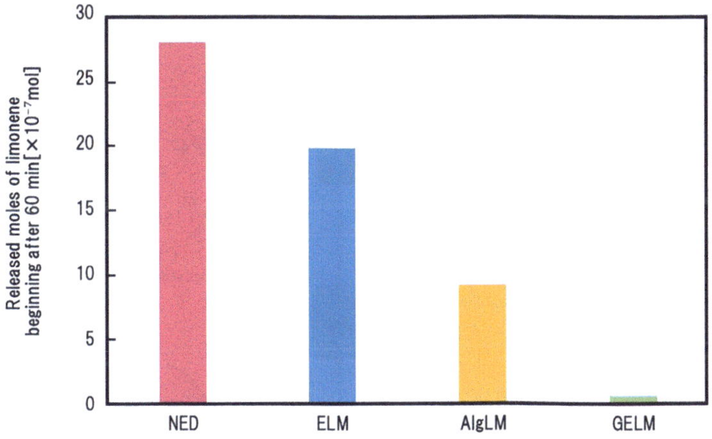

Fig. 15.6 Released amount of the limonene from the LMs in the chamber initiated after 60 min

The amount of limonene released at 60 min is presented in Fig. 15.6.

The amount of limonene released from the ELM at 60 min was 1.96×10^{-6} mol, which was 70.8% of that released from the NED at 60 min. The released amount of limonene was effectively reduced by ca. 29%. This decrease was caused by a reduction in the vaper pressure of limonene contained in the ELM. A small amount of calcium stearate solubilized in the O/W emulsion phase of ELM influenced the vaporization of limonene. Detailed measurement and discussion on the change in the vaper pressure of limonene affected by the adsorbed component of the LM surface were necessary for the controlled release of functional volatile compounds.

The outer surface area of the ELM was wholly covered by calcium stearate powder. The effective surface area available for the vaporization of limonene from the ELM was reduced by the adsorbed calcium stearate powder. It was also a considerable factor in the decrease in the released amount of limonene.

The amount of limonene released from the sodium-alginate O/W emulsion LM (AlgLM) at 60 min was 9.15×10^{-7} mol. It was 46.7% of that of the ELM at 60 min. The reduction in the vaper pressure of limonene was caused by the solubilization of sodium alginate in the aqueous phase of AlgLM.

The viscosity of the aqueous sodium alginate of aqueous phase was 27.8 mPa s at room temperature (298 ± 2 K). The released limonene was mainly supplied and vaporized from the dispersed microdroplets of limonene located at the outer edge of the AlgLM body. The dispersed limonene located inner region of the AlgLM hardly transported to the outer edge of the AlgLM because of the viscous resistance of the aqueous alginate phase. The limonene located in the inner core of the AlgLM hardly contributed to the vaporization and release.

The amount of limonene released from the GELM at 60 min was 6.21×10^{-8} mol. It was 3.17% of that of the ELM at 60 min. The limonene release was strongly suppressed by the formation of a calcium alginate gel film on the outer edge of the

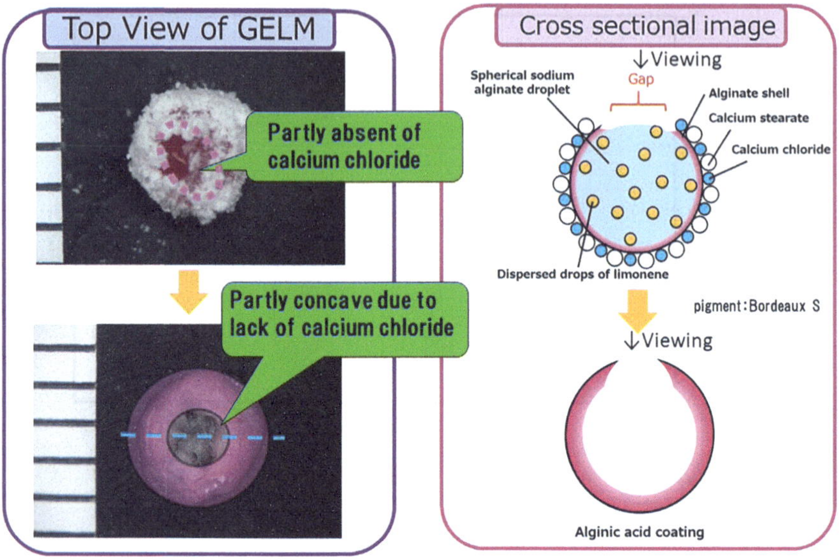

Fig. 15.7 Alginate gel film formation by calcium chloride in the outer edge of GELM

GELM. The calcium chloride powder adsorbed on the surface of the GELM quickly cross-linked the alginate polymer chain, and the alginate gel film was strongly formed. The cross-linked gel film of calcium alginate formed on the outer edge of the GELM also reduced the vaper pressure of the limonene.

Figure 15.7 presents the microphotographs of the GELM surface. To examine the role of calcium chloride, the powder was intentionally absent from the GELM partial surface. Calcium alginate gel coating was completely formed on the LM surface, except on the local part without calcium chloride powder. A small concave was observed on a part of the GELM surface, owing to the lack of calcium chloride.

The difference in the surface tension between the alginate gel film and the original aqueous alginate phase caused the partial deformation of the spherical shape of the GELM. The photographic image of the partial deformation is presented in Fig. 15.4b.

15.3.3 Release Rate of Limonene from the Simple Emulsion Droplet and the Three Types of Liquid Marbles

The release rate of limonene was assumed to be the first-order kinetics as follows;

$$-\frac{dc}{dt} = k\,c \tag{15.3}$$

where, c indicates the residual concentration of limonene in the NED or various LM bodies.

The k represents the overall rate coefficient of the released limonene. For the integration of Eq. (15.3), the initial concentration of the limonene contained in the LM was termed c_0.

$$\ln \frac{c}{c_0} = -k\,t \tag{15.4}$$

According to Eq. (15.4), the overall rate coefficient (k) was evaluated by graphical analysis; $\ln \frac{c}{c_0}$ vs. t. The residual concentration of limonene in the LM (c) was calculated from the mass balance of the limonene released. As presented in Fig. 15.5, in the initial period of less than 600 s, the limonene was rapidly released from the LMs. In the late period of over 600 s, the limonene was gradually released from the LMs. The rate processes of the release were discussed according to two regions in the time course of the released limonene.

The overall rate coefficient is presented in Fig. 15.8. The hatching bars indicate the overall coefficient of the initial period of less than 600 s. The solid bars indicate the overall coefficient of the late period of more than 600 s.

15.3.3.1 The Overall Rate Coefficient of the Limonene Release in the Initial Period (t < 600 s)

The overall initial rate coefficient strongly depended on the type of LMs. The initial rate coefficient of the NED was $2.42 \times 10^{-4}\ s^{-1}$, and that of the ELM was $1.21 \times 10^{-4}\ s^{-1}$, which was 50% of that of the NED.

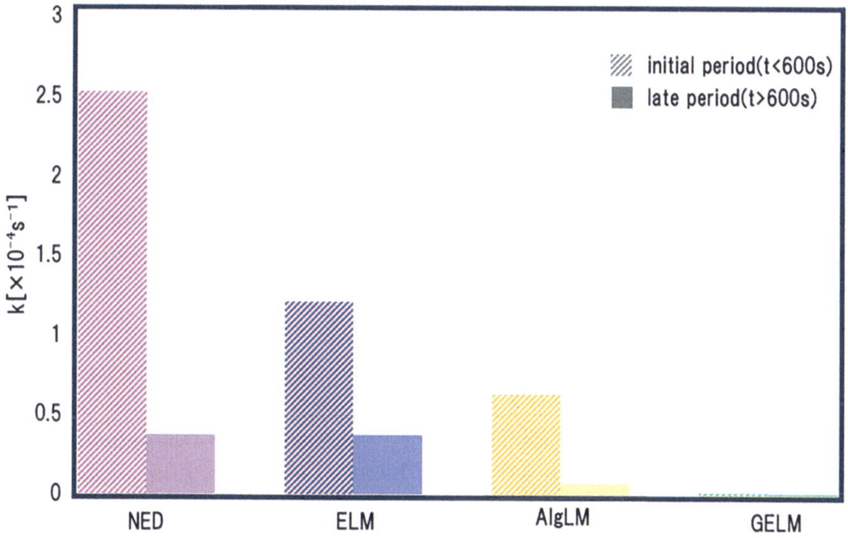

Fig. 15.8 Overall release rate coefficient of the LMs. The coefficient was presented by twin bars according to two regions in the time course

The ratio of LM's outer surface covered by calcium stearate particles was estimated by the mass balance between a simple liquid droplet and a prepared LM. In our experimental results, the mean number of the calcium stearate particle covering the outer surface of one LM was 1.82 × 10^5. The sum of the cross-sectional area caused by the particles was 1.39 × 10^{-5} m^2, according to the median particle diameter 9.86 μm. This indicated that the covered surface area of the LM inhibited the vaporization and release of limonene from the LM surface. The mean surface area of one simple liquid droplet was 2.83 × 10^{-5} m^2, under the mean diameter was assumed to be 3.0 × 10^{-3} m based on our microscopic photograph. The ratio of the covered LM surface was calculated to be 49.1%. This considerably corresponded with the decrease in the initial rate coefficient between the NED and ELM.

The overall initial rate coefficient for the release of the AlgLM was 6.34 × 10^{-5} s^{-1}. A viscous sodium alginate (27.8 mPa s) aqueous phase formed the AlgLM body. The limonene dispersed inside of the AlgLM participated in the continuous aqueous sodium alginate phase and was released into the atmospheric air. The participation of limonene was suppressed by the hydrophilicity of the sodium alginate molecular chain. Furthermore, the diffusion coefficient of limonene was reduced by the relatively high viscosity of the aqueous sodium alginate phase. Therefore, the overall initial release rate coefficient of the AlgLM was reduced compared with that of the ELM.

The initial rate coefficient for the release of the GELM was 1.78 × 10^{-6} s^{-1}. It was only 1.47% of that of the ELM, and is hardly obvious as shown in Fig. 15.8. A remarkable reduction in the initial release rate was observed. The calcium alginate gel film formed at the outer surface of the GELM strongly inhibited the vaporization and release of limonene. The diffusion coefficient of limonene in the calcium alginate film was speculated to be considerably low. A detailed measurement of the diffusion coefficient of limonene in the alginate gel film is necessary for further study. The desired initial release rate was achieved by the adequate regulation of the sodium alginate concentration and the amount of additive calcium chloride powder.

15.3.3.2 The Overall Rate Coefficient of the Release in the Late Period ($t \geqq 600$ s)

The overall release rate coefficient of limonene in the late period for the ELM was 3.78 × 10^{-5} s^{-1}, which was almost the same as that of the NED (3.88 × 10^{-5} s^{-1}). The NED and ELM had deformed spherical shapes in the late period for the limonene release because of the gradual drying process in the experimental chamber. Their shapes looked like those of an adhesive flat pie. No significant difference in their final shape was observed. This was consistent with the approximation of the overall release rate coefficient of limonene in the late period.

The overall release rate coefficient of the limonene in the late period for the AlgLM was 7.73 × 10^{-6} s^{-1}. It was merely 20.4% of that of the ELM. The sodium alginate concentration of the AlgLM body was continuously increased by the dehydration, and the sodium alginate because increasingly viscous during the releasing period. The diffusion coefficient gradually decreased. The spherical shape of the AlgLM was gradually shrunk with the increase in the dehydration level in the chamber.

The overall release rate coefficient of the limonene in the late period for the GELM was 9.83×10^{-7} s^{-1}. It was only 2.60% of that of ELM and is hardly evident, as shown in Fig. 15.8. Limonene was slightly released throughout the alginate gel film. Water vaporization with the limonene release from the GELM occurred simultaneously. Hence, the outer calcium alginate film became a rigid shell. The release of limonene was significantly reduced because of the stiffness of the outer alginate film.

The controlled release of the volatile component from bioresources (limonene) was objectively realized by the LM combined with the controlled viscosity of the LM body and the formation of a gel film on the outer surface of the LM. The LM is a promising biocompatible carrier with no apprehension after practical utilizations.

15.4 Conclusion

LMs containing an oil-in-water microemulsion were successfully prepared. The controlled release of the volatile component (limonene) was successfully achieved. The concentration of the sodium alginate in the LM body influenced the viscous parameters and the diffusion coefficient. The hybridization of stearate calcium and calcium chloride microparticles effectively afforded stable LMs and the gel film coating of the outer surface of the LMs. Sodium alginate-based emulsion LM is a prospective sustainable green device for the controlled release of volatile components.

Acknowledgement The authors are sincerely grateful to Professor Dr. Kazumitsu Naoe, National Institute of Technology, Nara College, for his generous assistance and fruitful scientific information.

References

Almeida PF, Almeida AJ (2004) Cross-linked alginate–gelatine beads: a new matrix for controlled release of pindolol. J Control Release 97:431–439. https://doi.org/10.1016/j.jconrel.2004.03.015

Anandakumar P, Kamaraj S, Vanitha MK (2020) D-limonene: a multifunctional compound with potent therapeutic effects. J Food Biochem 45:e13566. https://doi.org/10.1111/jfbc.13566

Aussillous P, Quere D (2001) Liquid marbles. Nature 411:924–927. https://doi.org/10.1038/35082026

Bormashenko E (2011) Liquid marbles: Properties and applications. Curr Opin Colloid Interface Sci 16:266–271. https://doi.org/10.1016/j.cocis.2010.12.002

Fujii S, Aono K, Suzaki M, Hamasaki S, Yusa S, Nakamura Y (2012) pH-responsive hairy particles synthesized by dispersion polymerization with a macroinitiator as an inistab and their use as a gas-sensitive liquid marble mtabilizer. Macromolecules 45:2865–2873. https://doi.org/10.1021/ma300048m

Grant GT, Morris ER, Rees DA, Smith PJC, Thom D (1973) Biological interactions between polysaccharides and divalent cations: the egg-box model. FEBS Lett 32:195–198. https://doi.org/10.1016/0014-5793(73)80770-7

Julian TN, Radebaugh GW, Wisniewski SJ (1988) Permeability characteristics of calcium alginate films. J Control Release 7:165–169. https://doi.org/10.1016/0168-3659(88)90008-9

Kashima K, Imai M (2011) Dominant impact of the α-L-guluronic acid chain on regulation of the mass transfer character of calcium alginate membranes. Desalin Water Treat 34:257–265. https://doi.org/10.5004/dwt.2011.2892

Monte WC, Landau-West D (1982) Solubility of polystyrene in certain vegetable oils, essential oils and their constituents. J Food Sci 47:1832–1835. https://doi.org/10.1111/j.1365-2621.1982.tb12893.x

Murakami R, Kanda N, Yamamoto M (2022) Mechanism of multilayer formation on liquid marbles. Colloid Interface Sci Commun 47:100607–100612. https://doi.org/10.1016/j.colcom.2022.100607

Ravichandran C, Badgujar PC, Gundev P, Upadhyay A (2018) Review of toxicological assessment of d-limonene, a food and cosmetics. Food Chem Toxicol 120:668–680. https://doi.org/10.1016/j.fct.2018.07.052

Roy PK, Binks BP, Bormashenko E, Legchenkova I, Fujii S, Shoval S (2020) Manufacture and properties of composite liquid marbles. Curr Opin Colloid Interface Sci 575:35–41. https://doi.org/10.1016/j.cis.2020.04066

Tanaka S, Okao H, Matsuda N, Sawai J, Naoe K, Imai M (2020) Preparation of biocompatible liquid marbles stabilized by food-grade stearate microparticle for aerobic bacteria cultivation. App Biochem Biotechnol 191:1684–1694. https://doi.org/10.1007/s12010-020-03299-6

Tenjimbayashi M, Mouterde T, Roy PK, Uto K (2023) Liquid marbles: review of recent progress in physical propertis, formation techniques, and lab-in-a-marble applications in microreactors and biosensors. Nanoscale 15:18980–18998. https://doi.org/10.1039/d3nr04966c

Vieira AJ, Beserra FP, Souza MC, Totti BM, Rozza AL (2018) Limonene: aroma of innovation in health and disease. Chem Biol Interact 283:97–106. https://doi.org/10.1016/j.cbi.2018.02.007

Wang D, Zhu L, Chen J-F, Dai L (2016) Liquid marbles based on magnetic upconversion nanoparticles as magnetically and optically responsive miniature reactors for photocatalysis and photodynamic therapy. Angew Chem 55:10795–10799. https://doi.org/10.1002/anie.201604781

Yamakawa K, Hirobe A, Honjo S, Higashio K, Sawai J, Naoe K, Imai M (2022) Stearate liquid marbles for bacterial cellulose production: influence of the liquid marble interface on bacterial cellulose properties. J Phys Chem C 126:1611–1622. https://doi.org/10.1021/acs.jpcc.1c08597

Zang D, Li J, Chen Z, Zhai Z, Geng X, Binks BP (2015) Switchable opening and closing of a liquid marble via ultrasonic levitation. Langmuir 31:11502–11507. https://doi.org/10.1021/acs.langmuir.5b02917

Homare Tsugifuji, BSc. In March 2017, he graduated from Tokyo Metroporitan Harumi Sogo High School. In March 2021, he graduated from College of Bioresource Sciences, Nihon University, under the supervision of Professor Masanao Imai. In April 2021, he has entered the Kewpie Cooperation. His typical academic activities are as followed; "Preparation of Liquid Marble and volatile properties from the interface involved with correlation between surface characteristic of supporting flat plate and droplet stability (Oral)". The annual meeting of student, organized by the Society of Chemical Engineers Japan (on the web) on 6 Mar. 2021. Especially on 28 Feb. 2021, he has received honorable the award "The Commendation of Selected Presentation" on behalf of The Colloquium of Advanced Chemical Engineering and Material Science. His research interest focuses on biochemical engineering involved with physicochemical aspects of O/W emulsion system. He is currently concerned with Botany (Functional ingredients derived from plants), Extraction of essential oil and aroma from flowers, and Landscape studies as well.

Masanao Imai, PhD. He is a Research Fellow at the Institute of Science and Technology, Kanto Gakuin University. In 1976, he Graduated from Dept. of Industrial Chemistry, National Institute of Technology, Toyama College. In1980, he graduated from Dept. of Chemical Engineering, The University of Tokyo. In 1987, he has received PhD from The University of Tokyo under the supervision of Professor Shintaro Furusaki. In April 1985, he entered the Fuji Oil Co. Ltd. In 1987, he moved to Tokyo University of Agriculture and Technology. In 1998, he moved to Nihon University and work as Professor of Chemical Engineering (2012–2022) and Director of Food Manufacturing and Educational Center (2017–2021). In 2021, he moved to Institute of Science and Technology, Kanto Gakuin University and worked as Research Fellow of Chemical Engineering. He has received honorable the award "Academic Cultural Award of Takaoka City" (Mayor Prize) in Feb, 2001. His research activities are presented as followed; 22 scholarly books, 109 refereed papers, 249 international presentations and 73 other academic publications. He specializes in enzyme reaction in microemulsion phase and supercritical carbon dioxide concerned about separation of functional component from natural resources. Biopolymer sciences involved with characteristics of gels and membranes are currently investigated.

Suitability of 3D-Printed Cellulose-Based Polymer Materials for Electrical Insulation Applications

16

Morgan Lecoublet, Mohamed Ragoubi, Nathalie Leblanc, and Ahmed Koubaa

Abstract

This chapter presents a suitability analysis of 3D-printed cellulose-based polymeric materials for electrical insulation applications. Polylactic acid (PLA) and nanocrystalline cellulose (NCC) bionanocomposites were produced, and their multiphysical, dielectric, and mechanical properties were analyzed. The addition of nanofillers increased the mechanical properties of the bionanocomposites through reinforcement effects and the increased crystallinity due to nanofiller nucleation. The nanofiller incorporation, however, slightly reduced the electrical insulation properties by increasing the dielectric constant, dielectric loss, and electrical conductivity of 3D-printed bionanocomposites. This effect was associated with the polarity of the nanofillers, as demonstrated by chemical surface analysis. By adjusting the sample architectures, we have rendered the dielectric and mechanical properties of NCC-based bionanocomposites competitive with synthetic polymers used in electrical insulation, such as LDPE, HDPE, PP, and PVC. A final 5-month aging stage in water and at room temperature showed that, despite the increased polarity of the bionanocomposites due to the addition of NCCs, the nanofiller had very little effect on water absorption and consequently little effect on long-term electrical insulation performance. This investigation

M. Lecoublet (✉) · A. Koubaa
Forest Research Institute, Université du Québec en Abitibi-Témiscamingue (UQAT), Rouyn-Noranda, QC, Canada
e-mail: lecm57@uqat.ca; ahmed.koubaa@uqat.ca

M. Ragoubi · N. Leblanc
Unité de Recherche Transformations et Agro-Ressources, UniLaSalle, (ULR 7519 UniLaSalle Université d'Artois), Mont-Saint-Aignan, France
e-mail: mohamed.ragoubi@unilasalle.fr; nathalie.leblanc@unilasalle.fr

R. Malviya, S. Sundram (eds.), *Sustainable Nanocomposites with Green Biomaterials*, Biomaterials, Bioengineering and Sustainability 2,
https://doi.org/10.1007/978-3-031-79110-9_16

will provide insights into the design of bio-based materials for long-term electrical insulation applications.

Keywords

3D printing · Dielectric properties · Dynamic mechanical properties · Electrical insulation application · Water aging · Bionanocomposites · Sample architecture

Abbreviations

APTES	(3-aminopropyl) triethoxysilane
BDS	Broadband dielectric spectroscopy
DMA	Dynamic mechanical analysis
DSC	Differential scanning calorimetry
FFF	Fused filament fabrication
FTIR	Fourier transform infrared spectroscopy
MCC	Microcrystalline cellulose
NCC	Nanocrystalline cellulose
PLA	Polylactic acid
RAF	Rigid amorphous phase
SEM	Scanning electron microscopy
TGA	Thermal gravimetry analysis
C	Effectiveness coefficient (without units)
E'	Storage modulus (in Pa)
E''	Loss modulus (in Pa)
ε'	Dielectric constant (without units)
ε''	Dielectric loss (without units)
σ_{AC}	Electrical conductivity in AC mode (in $S{\cdot}cm^{-1}$)
N	Entanglement density (in $mol{\cdot}m^{-3}$)
r	Reinforcing factor (without units)
R	Boltzmann constant (8.314462 $J{\cdot}mol^{-1}{\cdot}K^{-1}$)
Tan δ	Loss tangent (without units)
T_{cc}	Cold crystallization temperature (in °C)
T_g	Glass transition temperature (in °C)
T_m	Melting temperature (in °C)
T_{max}	Temperature corresponding to the derived weight peak (in °C)
W_f	Weight content of fillers (in %)
$W_{upt\,\infty}$	Infinite uptake value (in %)
X_{cr}	Degree of crystallinity (in %)

16.1 Introduction

The world is consuming more and more dielectric materials. The global electronic materials market has been estimated at USD 65.7 billion in 2022 (TechSci Research 2023). However, to ensure sustainable development, it is important to find biobased and renewable alternatives to current synthetic materials such as polyethylene and polypropylene. Polylactic acid (PLA) is among the most widely produced biopolymers worldwide, reaching 676 kilotons in 2023 (European Bioplastics 2023). Its current industrial production is mainly based on biosynthesis and lactide ring-opening synthesis (Yu et al. 2023). Its rheological, mechanical, thermal properties and its processability make it a viable alternative to polystyrene (PS), polyethylene terephthalate (PET), and polypropylene (PP) (Hamad et al. 2016) for biomedical, packaging, and green chemistry applications (Scaffaro et al. 2020). It is also a very promising candidate for application in low-k materials, as it has a low dielectric constant and electrical conductivity, suitable for application in electrical insulation (Lecoublet et al. 2023, 2024). However, PLA's poor thermomechanical stability in its rubbery state might limit its potential for such applications.

Adding cellulosic fillers such as microcrystalline cellulose (MCC) or nanocrystalline cellulose (NCC) is among the solutions to improve the biopolymers' mechanical properties, and to guarantee a 100% compostable material. A research team demonstrated a positive effect of low MCC content on PLA static mechanical properties (Xian et al. 2018). Adding 4% of MCC increased the tensile modulus of PLA by 26% and the tensile strength by 8%. Shojaeiarani et al. (2019) fabricated nanocomposites based on NCC and PLA and characterized their dynamic mechanical properties. At 1 Hz and 30 °C, neat PLA exhibited a storage modulus (E′) of 2.41 GPa, while the nanocomposites with 3% NCC ($W_{f\ NCC}$) exhibited a storage modulus (E′) of 2.71 GPa. According to the authors, the stiffening effect of the nanofiller was stronger in the rubbery state (T > 70 °C), where the reinforcement effect reached 236%. The authors also highlighted the importance of a nanofiller compatibilization step to improve mechanical properties. The esterification of NCCs gave an E′ at 30 °C of 3.8 GPa with a $W_{f\ NCC} = 3\%$ through improved filler/matrix interaction, representing a 40% improvement over unesterified bionanocomposite and a 60% improvement over neat PLA. Ten's research team noted that adding NCC into a polyhydroxybutyrate-co-valerate (PHBV) at 2.3% by weight improved the storage modulus of neat PHBV up to 80% in the glassy state (Ten et al. 2012). However, these cellulosic fillers were more polar than the matrix and increased the dielectric constant (ε′) and electrical conductivity (σ_{AC}) of PHBV.

The opportunities offered by Fused Filament Fabrication (FFF) 3D printing for rethinking component design are considerable. They should be exploited, considering the desire to pursue the integration of biopolymers in the dielectric field. Barbosa et al. (2018) produced a 3D-printed PLA electrical insulator and an optical voltage sensor to develop a new generation of smart grid electrical insulators. Dielectric strength measurement showed that 3D-printed electrical insulators could undergo voltages of 35 kV without degradation, validating these new biobased smart isolators for applications targeting a voltage of 13.8 kV. Furthermore, Zhang et al. (2016) exploited the linear relationship between the dielectric constant and the infill ratio of 3D-printed PLA to produce Fresnel lenses capable of manipulating electromagnetic waves in the GHz range.

One of the key factors to be considered when planning the long-term life of composite materials is their water uptake capacity. A material may be subjected to damp outdoor conditions or water immersion during its lifetime. This water uptake will negatively affect the multi-physical performances of the proposed material. Moreover, biomaterials and biobased materials are more prone to alteration by water immersion than their synthetic counterparts. PLA, for example, can undergo degradation in the presence of water (Badia et al. 2017). Moreover, the two proposed material improvements (adding cellulosic fillers and using 3D printing process) will affect water absorption and negatively influence their dielectric properties. Song et al. demonstrated that water absorption in polyimide increased its dielectric constant, as water possess a much higher dielectric constant (~80) than PLA (Song et al. 2019). Lay et al. noted that 3D-printed PLA had higher water absorption than injection-molded PLA (Lay et al. 2019), because the 3D printing process can generate defects such as porosities and decohesion between printed filaments and microcracks, areas where water absorption can be facilitated.

To pursue the search for biobased alternatives in electrical insulation applications, this study investigated the physical, dielectric, and mechanical properties of PLA/NCC bionanocomposites obtained by 3D printing. The effect of NCC content and 3D printing on dielectric and mechanical properties were studied using a broadband dielectric spectrometer (BDS) and dynamic mechanical analysis (DMA), respectively. NCC will be compared to microcrystalline cellulose (MCC) to study the effect of filler size. Their mechanical and dielectric performances will be compared to conventional polymers used in electrical insulation. A water aging study will investigate how the material's electrical insulation properties vary over time. This investigation will provide insights into the design of biobased materials for long-term electrical insulation applications.

16.2 Preparation of 3D-Printed Bionanocomposites

16.2.1 Raw Materials

PLI-005 supplied by Natureplast (Montdeville, FRA) was used. It has an MFI given by the technical data sheet of 25-35 g·10 min^{-1} (ISO 1133 @ 190 °C). NCC (Dextra Cel Nano HP) was supplied by Anomera (Montreal, QC, CAN). The bulk density given by suppliers was 0.3–0.7 g·cm^{-3}. MCC was supplied by J. Rettenmaier (Fosston, USA) and obtained from bleached kraft pulp. The density given by the suppliers was 0.27-0.33 g·cm^{-3}.

16.2.2 Filler Silanization and Masterbatch Processing

Silanization was carried out on NCC and MCC fillers, following a proposed protocol (Dammak 2022), by magnetic stirring at room temperature and in 1 L Erlenmeyer flasks. (3-aminopropyl) triethoxysilane (APTES) was used, supplied by Sigma Aldrich. The protocol consisted of a preliminary hydrolysis of 2% by mass of APTES in a 1:1 solution of ethanol (95% by volume) and distilled water for 1 h at room temperature. In the second step, acetic acid (99.7%) was added to obtain a pH between 4 and 5. For 1 L of solution, 20-22 ml of acetic acid achieved a pH of 4.5. After 15 min, the fillers were added and left in solution for 3 h. Figure 16.1 summarizes the reaction. Once the reaction was completed, the fillers were filtered and washed, then dried in an oven at 60 °C for 24 h. Once silanized, microscopic analysis of the MCC gave a median and a mean length of 34.1 and 51.9 μm, respectively, and a median and a mean width of 16.6 and 22.9 μm, respectively.

The masterbatch processing was carried out for NCC bionanocomposites. The melt route was chosen for industrial purposes and achieved on a Haake Minilab II (Thermo Scientific, Massachusetts, USA). This microcompounder was operated by batch extrusion of 5 g of PLA/NCC blend. The raw material was conditioned at 60 °C until the moisture content of the components fell below 1%. Preliminary tests showed that a temperature of 180 °C was sufficient to fully melt the PLA while limiting the thermal degradation of the extrudate. The protocol used consisted of a screw speed of 70 RPM at 180 °C for 1 min. Once obtained, the extrudate was finally granulated and stored in a desiccator before the extrusion processing.

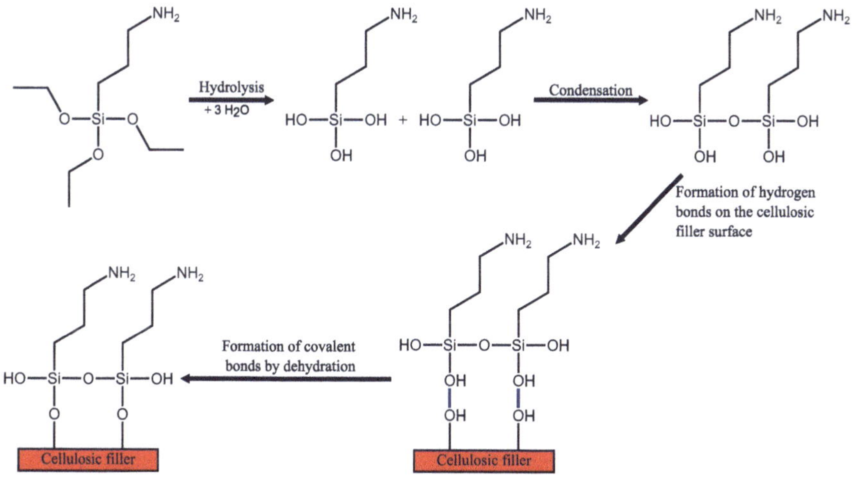

Fig. 16.1 Silanization reaction of cellulose filler through the use of APTES

16.2.3 Extrusion Processing

A twin-screw extruder was used to prepare the pellets at 190 °C and a rotational speed between 70 and 100 RPM. Die pressure was measured between 22 and 24 bars. Table 16.1 shows the different conditions produced. The maximal NCC and MCC weight contents ($W_{f\ NCC}$ and $W_{f\ NCC}$) were fixed at 5% to avoid excessive aggregates, especially for the NCC-based composites.

16.2.4 3D Printing Processing

An intermediate filament-making step was carried out to print the resulting pellets on a 3DEVO filament maker (3DEVO, Utrecht, NLD). A temperature profile of 170-185-185-180 °C was applied, with a screw speed of 3.5 RPM and a cooling power of 40%. 3D printing was performed on a CR-10 Max (Creality, Shenzhen, CHN) (Fig. 16.2a). Table 16.2 shows the parameters applied for the 3D printing. These parameters were chosen as they provided the best compromise for all conditions. Neat PLA posed no printing problems, but the bed had to be sprayed with a lacquer to improve adhesion and prevent delamination while printing the high-filled bionanocomposite. Furthermore, the filaments were dried at 60 °C just before use to remove any traces of moisture that might alter the print quality. Two shapes were produced: 10 mm diameter, 4 mm thick disks for dielectric tests, and 60 × 10 × 4 mm bricks for DMA tests (Fig. 16.2b).

16.2.5 Composites Characterization

16.2.6 Scanning Electron Microscopy (SEM) Analysis

The microstructure of the NCC and MCC biocomposites was investigated by SEM conducted using a Jeol JSM IT200 (Jeol, Akisima, JAP). The tests were conducted on hot-pressed samples, which were cryofractured and carbon-coated. Analyses were performed at 5 kV with a magnification of ×200.

Table 16.1 Experimental conditions for the production of MCC and NCC biocomposites and bionanocomposites

	PLA (g)	MCC (g)	NCC (g)	$W_{f\ MCC}$ (%)	$W_{f\ NCC}$ (%)
PLA	1000	0	0	0	0
N1	950	0	10	0	1
N3	850	0	30	0	3
N5	750	0	50	0	5
M5	950	50	0	5	0

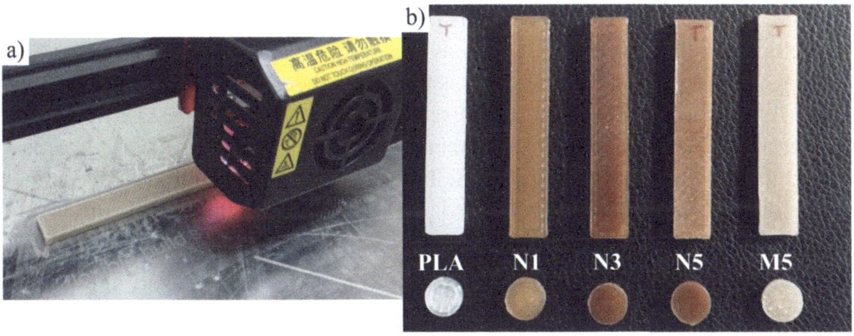

Fig. 16.2 (**a**) Creality CR-10 Max 3D printer, (**b**) Produced samples

Table 16.2 3D Printing parameters of NCC and MCC filament composites

Nozzle temperature (°C)	200
Nozzle diameter (mm)	0.8
Nozzle speed (mm·s^{-1})	40
Bed temperature (°C)	60
Sample thickness	4
Layer thickness (mm)	0.4
Infill ratio (%)	50; 100
Infill pattern	± 45°

16.2.7 Chemical Surface Analysis

Fourier transform infrared spectroscopy (FTIR) analysis was carried out on both fibers to qualify the grafting and monitor chemical property variations of the samples during water aging with a ITracer 100 (Shimadzu, Kyoto, JAP). Samples were analyzed with an attenuated total reflectance (ATR) accessory in the wavenumber range 500 cm^{-1} to 4000 cm^{-1}. The fillers were oven-dried overnight at 60 °C to remove residual water.

16.2.8 Microstructural Analysis

Differential scanning calorimeter (DSC) analysis assessed the materials' thermal properties and crystallinity. The measurements were performed on a DSC Q20 (TA Instruments, New Castle, USA) on 15 mg of material inserted in Al pans. The used protocol consisted of one heating and one cooling ramp between 20 and 200 °C with a heating rate of 10 °C·min^{-1} and an argon flow of 20 ml·min^{-1}. The degree of crystallinity (X_{cr}) was calculated with Eq. 16.1 (Zhu et al. 2024):

$$X_{cr} = \frac{\Delta H_m - \Delta H_{cc}}{\Delta H_m^0 * \left(1 - W_{filler}\right)} * 100 \quad (16.1)$$

where $\Delta H_m - \Delta H_{cc}$ is the difference between the cold crystallization enthalpy and the melting enthalpy, and $\Delta H^0{}_m$ is the polymer's theoretical fully crystalline melting enthalpy, both expressed in $J{\cdot}g^{-1}$. W_{filler} is the weight content of the added cellulosic fillers. A $\Delta H^0{}_m$ of 93 $J{\cdot}g^{-1}$ has been fixed for PLA (Zhu et al. 2024). Regarding the thermal stability characterization, thermal gravimetry analysis (TGA) was performed using a Tg 209 F1 (Netzsch, Selb, DEU). A scanning rate of 10 $°C{\cdot}min^{-1}$ under nitrogen up to 600 °C was applied. For each sample, 20 mg of materials was characterized.

16.2.9 Dynamic Mechanical Analysis (DMA)

DMA tests were realized in dual cantilever mode on a DMA 242 E Artemis (Netschz, Selb, DEU). The dual cantilever characterization mode was chosen to avoid the collapsing phenomena due to the PLA rubbery state. Tests were carried out in a ramp from 30 °C to 140 °C at a heating rate of 5 $°C{\cdot}min^{-1}$ and a controlled force of 1 N. All measurements were performed at 1 Hz, the most common frequency used for PLA measurements in the literature (Cristea et al. 2020).

16.2.10 Broadband Dielectric Spectroscopy (BDS) Analysis

BDS analyses were carried out with a dielectric spectrometer Keysight E4991A (Agilent Technologies, Santa Clara, USA). Two different analyses were made. A room-temperature analysis was first realized on five samples per condition with a frequency range of 1 MHz-1 GHz and a measurement precision of 10 measurements per decade. An isothermal analysis between −40 and 160 °C was also carried out on one sample per condition to determine the dielectric properties of our materials as a function of temperature at 1 MHz.

16.2.11 Water Aging Process

Regarding the water aging experiment, five samples per condition were placed in a distilled water bath at room temperature. Once removed from the bath, test specimens were carefully dried with paper towels and weighed using a Mettler Toledo XS205 analytical balance (Mettler-Toledo, Greifensee, CHE) with an accuracy of 0.1 mg to determine the water uptake. Eq. 16.2 gives the water uptake of the sample (Wan Ishak et al. 2020):

$$Water\,uptake(\%) = \frac{m_t - m_0}{m_0} * 100 \tag{16.2}$$

Where m_0 corresponds to the initial mass of the sample and m_t corresponds to the mass of the sample at time t. The specimens were then inserted into the dielectric spectrometer to determine the dielectric properties variations with the water aging. The tests were performed in the same condition as explained in part 16.2.10 for room-temperature analysis. Finally, FTIR tests were carried out at the end of the analysis (at $\sqrt{t} = 60\ h^{0.5}$) to monitor the chemical evolution of the samples after aging. Once the test battery had been completed, the sample was re-immersed. The water was renewed for each measurement.

16.3 Efficiency of the APTES Grafting

Figure 16.3 shows the FTIR spectra of MCC and NCC before and after grafting to verify the APTES grafting efficiency. The expected peaks were present for all the tested conditions (Supeno et al. 2014; Li et al. 2015; Hossain et al. 2020; Rathnayake et al. 2020; Buzón-Durán et al. 2021). The broad peak at approx. 3333 cm^{-1} corresponded to the cellulose surface's O–H bond stretching. Furthermore, the 1365, 1316, and 1201 cm^{-1} peaks came from OH in-plane bending. The literature also suggested that the peak at 1365 cm^{-1} may be caused by the deformation of the C–H bonds (Hossain et al. 2020). The peaks at 2895 and 1427 cm^{-1} corresponded to C–H bond stretching and symmetrical deformations of the CH_2 chemical group in the cellulose crystal phase, respectively. The series of peaks between 1160 and

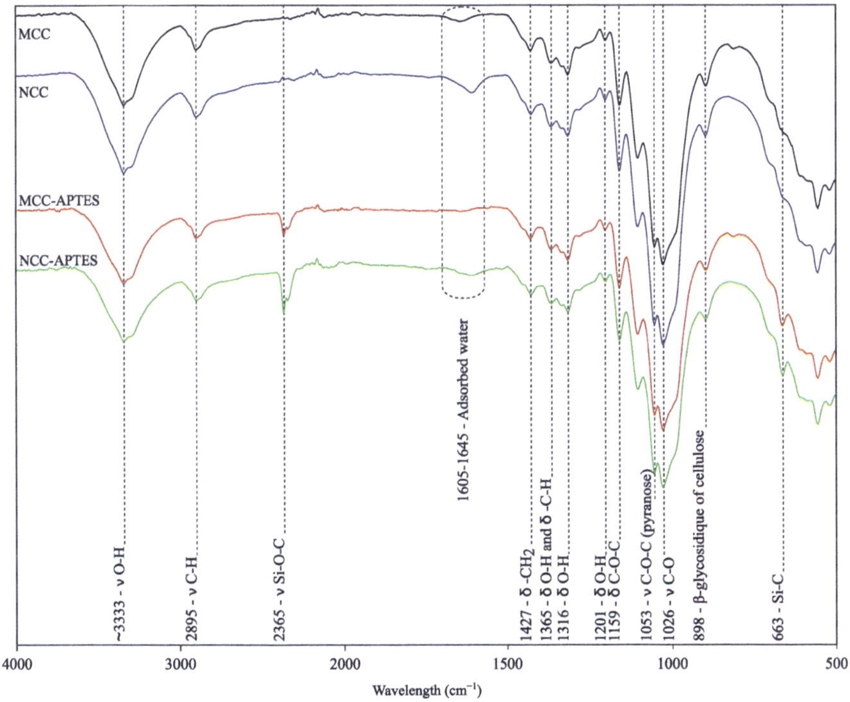

Fig. 16.3 FTIR spectra of grafted and ungrafted NCC and MCC

1025 cm^{-1} corresponded to the groups present on the cellulose ring. Finally, the peak at 898 cm^{-1} corresponded to the β-glycosidic bonds of cellulose. It may be noted that grafted fillers presented two new peaks at 2365 and 663 cm^{-1}, respectively, attributable to the presence of Si–O–C and Si–C bonds. Moreover, the weak peak at 1641 cm^{-1} was more pronounced for the ungrafted fillers. As this peak can be associated with the vibration of the water molecule adsorbed by the fillers, it can be deduced that the grafted MCC was more hydrophobic than the ungrafted MCC and, therefore, the APTES has been correctly grafted onto the MCC. The Si-based peak of APTES was also present for grafted NCC; it has a higher intensity than that of MCC-APTES. The peak associated with water adsorption was still present but at a lower intensity than for ungrafted NCC. These findings suggest that APTES has been at least partially grafted onto NCC.

TGA analyses revealed a change in the NCC and MCC degradation kinetics due to the grafting process (Fig. 16.4). A slight improvement in the NCC and MCC

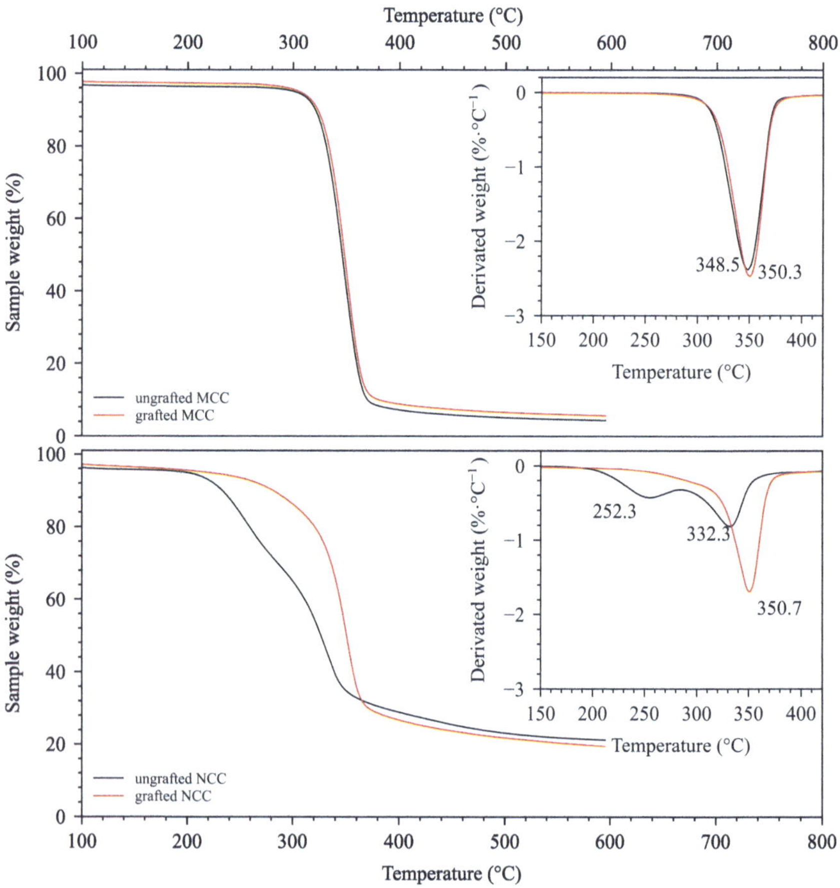

Fig. 16.4 TGA analysis of grafted and ungrafted NCC and MCC fillers

thermal stability was observed with grafting, with a 2 °C shift of the degradation peak. However, the grafting on NCC was much more noticeable, as it suppressed the first degradation phase of neat NCC at 252 °C and delayed the second degradation phase of NCC by 20 °C. The improvement in thermal stability could be associated with a reduction in NCC oxidation through fewer available hydroxyl sites.

16.4 Physicochemical Analysis of NCC-Based 3D Printed Bionanocomposites

Figure 16.5 shows the DSC results for the neat PLA, the bionanocomposites, and the reference biocomposite. Neat PLA presented a glass transition temperature (T_g) at 67.3 °C. A cold crystallization phenomenon was measured at 112.9 °C, and an endothermic peak was present at 177.6 °C, associated with the melting phenomenon of the crystalline phase. These values were consistent with the literature (Wang et al. 2020; Thumsorn et al. 2022). The neat PLA presented a low-crystalline behavior, with an initial X_{cr} of 10.1%, similar to the results of Thumsorn in 2022 for a commercial 3D-printed PLA (Thumsorn et al. 2022). Although the degree of crystallinity was relatively low, the heat capacity change of the glass transition phenomenon was much lower than expected for a low-X_{cr} polymer (Limsukon et al. 2023). This observation could indicate that the tested PLA contains a significant part of the rigid amorphous phase (RAF), hindering the mobility of the mobile amorphous phase (Limsukon et al. 2023). Regardless of the type of cellulose filler used, its addition altered the morphology and thermal behavior of the samples. Cellulose fillers slightly decreased the T_g, a phenomenon also observed in the literature with

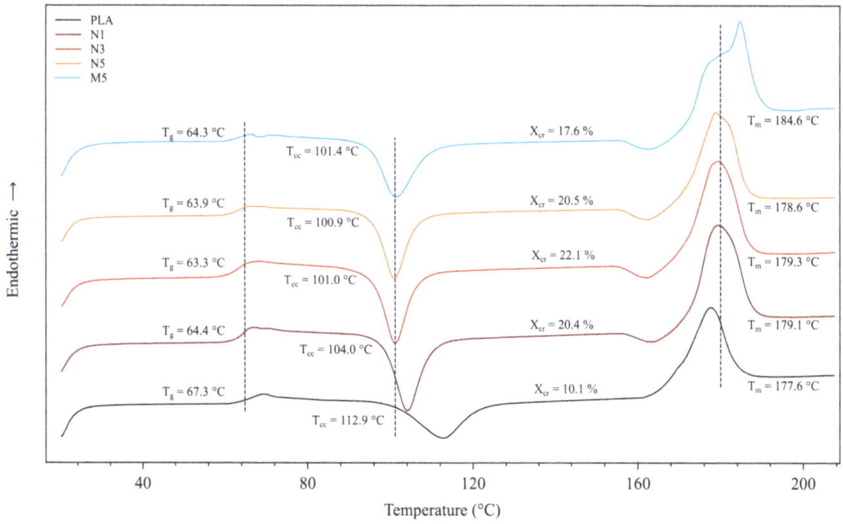

Fig. 16.5 DSC analysis of 3D-printed bionanocomposites and reference biocomposite

cellulose nanofibrils (Wang et al. 2020). Adding fillers also impacted the cold crystallization temperature (T_{cc}). Adding 1% by weight of NCC decreased the T_{cc} PLA by 9 °C. The lowest was measured for N5 with a T_{cc} PLA at 101.4 °C. This effect was attributed to the nucleation effect of the fillers. Cellulosic fillers altered the mobility of PLA chains and facilitated their crystallization (Wang et al. 2020). By analogy, this effect was also positive for the X_{cr}, with the highest value measured for N3 at 22.1%. The nucleating effect was greater for NCC than MCC, as M5 presented an X_{cr} of 17.6%. The melting temperature (T_m) slightly increased with the addition of fillers. The NCC-based bionanocomposites had a T_m 1 °C higher than neat PLA. Finally, it is worth noting that the M5 sample presented the highest melting peak at 184.5 °C and seemed to have two distinct melting peaks, indicating the coexistence of two different crystalline phases.

Figure 16.6 shows the TGA and derived TGA curves of the 3D printed materials. Neat NCC was added as a reference. The neat PLA presented a typical behavior of a monocomponent material with a T_{max} (temperature corresponding to the derived weight peak) measured at 376.2 °C. All the bionanocomposites and the reference biocomposite also exhibited a single degradation phenomenon. NCC and PLA had relatively similar temperature degradation, resulting in an overlap of component degradation between PLA and cellulose (Lecoublet et al. 2023). However, the incorporation of NCC has led to complex behavior. Compared to PLA, N1 presented a lower T_{max} by 8 °C, reaching 368 °C. This observation was also made by Wang et al. for PLA/nanofibrillated cellulose (NFC) at W_{NCF} = 1% (Wang et al. 2020). The thermal stability decreasing with the addition of fillers could come from a good interface between the filler and the matrix, increasing the possible heat transfer between the components (Qian et al. 2018). M5 and N1 presented similar thermal stability, demonstrating that both samples displayed a good fiber/matrix quality.

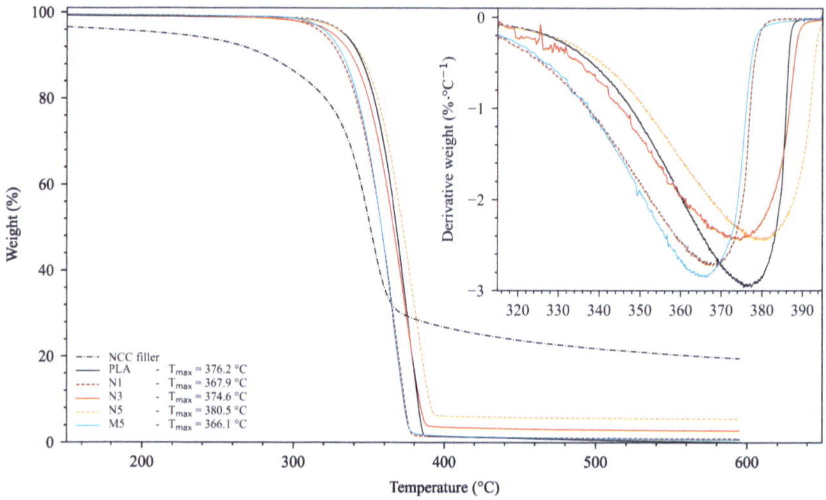

Fig. 16.6 TGA analysis of 3D-printed bionanocomposites and reference biocomposite

However, at higher NCC levels, the thermal stability was improved. N5 presented a thermal stability similar to neat PLA. This increased thermal stability can be attributable to a decrease in filler/matrix quality, possibly due to the presence of aggregates. Regarding the char yield at 600 °C, neat PLA, N1, N3, and N5 showed a char yield of 0.1, 0.8, 2.8, and 5.6% respectively. This strong increase may be due to the silanization of the fillers, releasing silica-based residues (Jin et al. 2020), or possibly remnants from the reaction itself. This assumption was confirmed by the M5 char yield of 0.7%, as MCC has a much lower specific surface area than NCC.

The chemical analysis of the samples is presented in Fig. 16.7. All the expected peaks of PLA were observed (Lecoublet et al. 2023). The neat PLA presented the typical C=O carbonyl vibration peak at 1749 cm^{-1}. The double peak between 2950 and 3000 cm^{-1} corresponded to the C–H vibration. In addition, the double peaks between 1350 and 1380 cm^{-1} were linked to the C–H bending. The peak at 1452 cm^{-1} may correspond to the deformation of the methyl group of PLA, and the two peaks at 1180 cm^{-1} and 1080 cm^{-1} may correspond to the deformation of the C–O stretching. No peaks around 3300 cm^{-1} and 1650 cm^{-1} were observed, demonstrating that neat PLA was free of hydroxyl functions and adsorbed water. The addition of

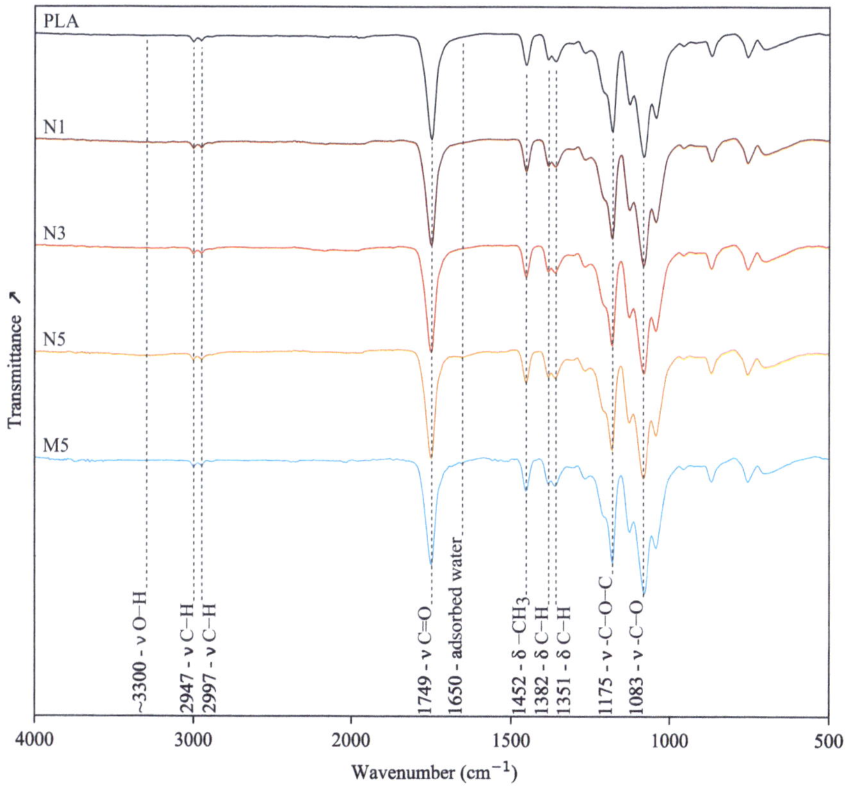

Fig. 16.7 FTIR analysis of 3D-printed bionanocomposites and reference biocomposite

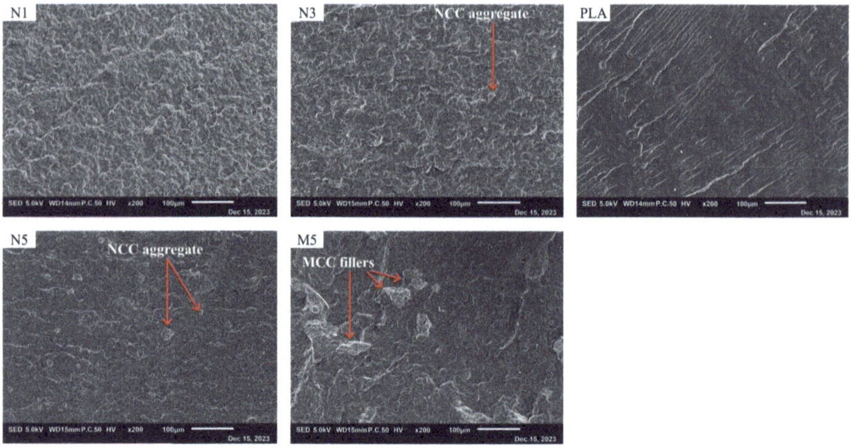

Fig. 16.8 SEM analysis of 3D-printed bionanocomposites and reference biocomposites

cellulosic fillers showed almost no chemical variation compared with the neat matrix. Samples N1 and N3 did not exhibit peaks around 3300 cm^{-1} and 1650 cm^{-1}, respectively, associable with hydroxyl groups and adsorbed water. Only N5 showed weak hydroxyl-associated peaks. It is in good agreement with the previous TGA observations, where we suspected the formation of aggregates at high NCC content. These aggregates can reduce the filler/matrix quality and allowed polar functions to be presented more easily on the surface, thus facilitating water absorption.

The SEM results are presented in Fig. 16.8. Neat PLA presented a brittle and mono-component morphology. The MCC particles were visible on the M5 sample. However, the MCC/PLA interface appeared to be well executed, as no void between the filler and the matrix was visible, confirming the TGA results regarding the quality of the interface. The N1 bionanocomposites were free of aggregates, assuming a good dispersion of the nanofillers. N3 presented few scattered aggregates, demonstrating a relatively good dispersion. For N5, however, several aggregates were noted, indicating that the NCC content may be too high to ensure a proper dispersion. One remarkable point was the increase in roughness of all bionanocomposites compared with neat PLA, suspected to be caused by the rearrangement of PLA matrix into a more crystalline structure.

16.5 Dielectric Analysis of NCC-Based 3D Printed Bionanocomposites

Figure 16.9a presents the dielectric constant measured at room temperature. The neat 3D-printed PLA presented a typical behavior broadly detailed in the literature (Dichtl et al. 2017; Kuzmanić et al. 2023; Lecoublet et al. 2024). At 1 MHz, the neat PLA presented an ε' of 2.56 and decreased to 2.12 when the frequency reached 1 GHz. According to Khouaja et al., this decrease is due to the reduced ability of the

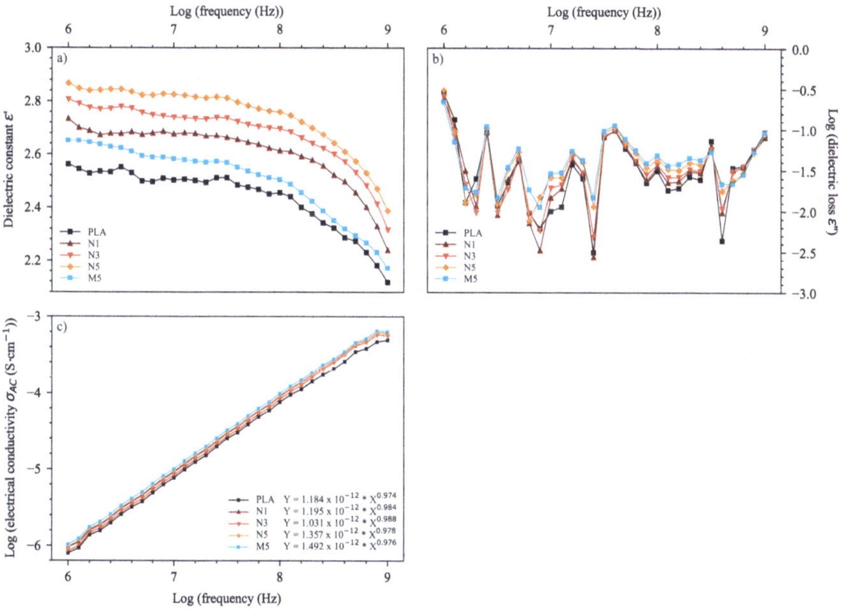

Fig. 16.9 Dielectric properties 3D-printed bionanocomposites and reference biocomposite, (**a**) dielectric constant, (**b**) dielectric loss, (**c**) electrical conductivity σ_{AC}

material's polar functions to interact with the electric field, lowering its polar potential (Khouaja et al. 2021). These dielectric constant values place the material in the apolar polymer range (Dubois 2001). Adding cellulosic fillers increased the ε′ independently of the used W_f and filler size. N5 showed the highest enhancement of ε′, with a value of 2.87 at 1 MHz. Adding polar fillers increased the material's polarizability, increasing the dielectric constant (Khouaja et al. 2021). In addition, the size of the cellulosic fillers had an impact on ε′. At $W_f = 5\%$ and 1 MHz, the bionanocomposites showed a higher dielectric constant increase compared to the biocomposite, respectively 12 and 4% higher than the neat PLA. The observation could be attributed to the specific surface area of the fillers used. Yakubu et al. (2020) have demonstrated that the filler size was inversely correlated to the dielectric constant and dielectric loss of composites. Due to their smaller size, NCCs had a greater specific surface area and relatively more silane and hydroxyl functions on the surface to interact with the applied field. Despite the increase in NCC, all the produced samples also remained in the apolar polymer range (Dubois 2001).

Figure 16.9b shows the dielectric loss ε″ of the samples measured at 20 °C. Dielectric loss refers to the part of the applied electromagnetic energy absorbed by the material and converted into heat (Khouaja et al. 2021). It is, therefore, an important parameter to be kept as low as possible to avoid accelerated degradation of biobased materials. The results showed important data noises, but some tendencies can still be observed. The ε″ of PLA was relatively low, below 0.1. Dichtl et al. (2017) also reported a dielectric loss in the same range for a 3D-printed

PLA. However, the addition of cellulosic fillers increased the ε″ as expected. As the presence of polar fillers increased the polarization potential, the dielectric loss logically increased. Adding cellulosic fillers also increased the material heterogeneity and thus favored the presence of interfacial polarization phenomenon (Khouaja et al. 2021). The influence of NCC content was statistically significant on dielectric loss (P-Value = 4.8×10^{-8}). The M5 showed the most significant increase, with a measured value at 100 MHz 53% higher than PLA (P-Value = 5.1×10^{-5}). But this observation contrasts with the findings of Yakubu et al., who observed that dielectric loss was inversely proportional to the size of micrometric rice husk particles in the PCL matrix. In our case, this difference in observation could be linked to the much larger difference in scale of size.

Electrical conductivity in AC mode (σ_{AC}) is displayed in Fig. 16.9c. Measured at 1 MHz and 1 GHz, the PLA presented a σ_{AC} value of respectively 7.95×10^{-7} and 4.92×10^{-4} S·cm^{-1}. The PLA values were close to those reported in the literature (Dichtl et al. 2017; Dalai and Parida 2022). The increase in electrical conductivity with frequency was present because higher electrical frequencies would favor the hopping carrier phenomenon, carrying the electric current into the insulating materials (Hussien 2011). The σ_{AC} increase was purely frequency-dependent, a typical behavior of an insulating material (Wu et al. 2019). The addition of cellulose fillers slightly increased σ_{AC}. This behavior could be related to the more polar nature of the fibers used, similar to the results discussed above. The addition of fillers did not result in any sudden increase in electrical conductivity, indicating that no electrical percolation was present (Spinelli et al. 2020). Like dielectric loss, the highest electrical conductivity was found for the M5 biocomposite, with a σ_{AC} of 1.04×10^{-6} S·cm^{-1} measured at 1 MHz, 30% higher than the neat PLA. It is worth noting, however, that despite a noticeable increase in the dielectric properties measured, this polarity and electrical conductivity increase remained very limited compared to other types of nanofillers used in the literature, such as multiwall carbon nanotube (MWCNT) (Spinelli et al. 2020) or nanoceramics (Wu et al. 2019).

The dynamic mechanical section shows that PLA's physical properties vary according to temperature. As the dielectric properties of polymeric materials are directly dependent on the mobility of macromolecular chains, it is also important to perform temperature tests to qualify the proposed materials from a dielectric point of view. Figure 16.10a) shows the dielectric constant of PLA, M5, and N5 versus temperature at 1 MHz. As expected, temperature had a significant influence on all the samples tested. Moreover, the different materials all displayed similar behavior. At sub-zero temperatures ($-40 < T < 0$ °C), the ε′ slightly decreased. For example, between −40 and 0 °C, PLA decreased from 2.41 to 2.36. Between 0 and 50 °C, ε′ slightly increased, from 2.36 to 2.57. This temperature variation is typical of glassy-state PLA (Badia et al. 2017; Dichtl et al. 2017; Lecoublet et al. 2023). The increase sharply accelerated between 50 and 90 °C, peaking at 90 °C with a value of 3.21. This sudden increase can be attributed to the α-relaxation of PLA and the transition to the rubbery state, where macromolecular chains gain a higher mobility (Lecoublet et al. 2023). An increase in the dielectric constant of PLA at 60 °C at low frequencies, linked with an increase in ionic mobility after the α-relaxation of PLA, was

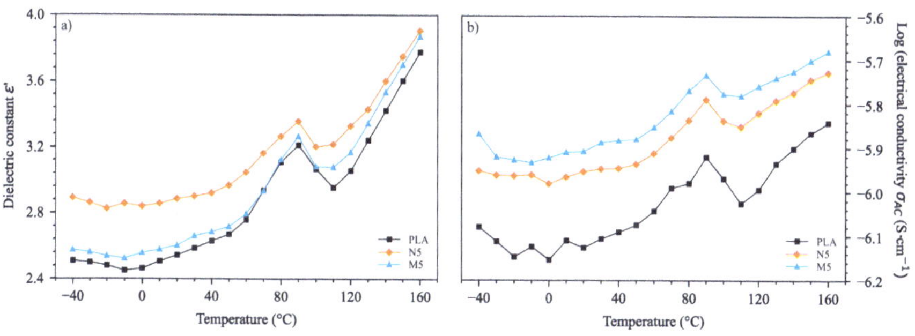

Fig. 16.10 Dielectric properties of 3D-printed bionanocomposites and reference biocomposite versus temperature at 1 MHz, (**a**) dielectric constant, (**b**) electrical conductivity σ_{AC}

also reported in the literature (Badia et al. 2017). The cold crystallization of the matrix was also clearly visible between 90 and 110 °C, where ε′ decreased to 2.85. The dielectric constant then increased again, reaching a maximum value of 3.78 at 1680 °C due to the temperature-induced softening of the rubbery polymer just before the melting of PLA. Over the whole temperature range tested, the addition of cellulosic fillers increased the ε′ as expected, with N5 showing the highest value, in agreement with the results shown in Fig. 16.9a. The electrical conductivity versus temperature behaved very similarly to the dielectric constant, with the same phenomena visible at the same temperatures (Fig. 16.10b). The highest σ_{AC} was measured for M5, in agreement with previous results illustrated in Fig. 16.9b. Furthermore, the variation in electrical conductivity between −40 and 160 °C, like the dielectric constant, was very limited and had very little impact on the electrical insulation properties, indicating the suitability of such materials for electrical insulation.

16.6 Mechanical Analysis of NCC-Based 3D Printed Bionanocomposites

The development of biobased materials for dielectric applications also requires the determination of their mechanical properties to identify their suitable range of applications. Among the available tools for determining mechanical properties, dynamic mechanical analysis (DMA) is particularly relevant to our case, as it allows the determination of solid-state viscoelastic properties across a wide temperature range. As the Joule effect can easily heat materials close to an electric field, it is capital to know their properties as a function of temperature. Figure 16.11 plots the DMA results over temperature for all samples. E′ is the storage modulus, representing the elastic phase of the material, and E″ is the loss modulus, representing the viscous phase (Cristea et al. 2020; Bagotia and Sharma 2019). The tan δ, also called loss tangent, represents the damping factor and is expressed as the ratio between the loss modulus and the storage modulus. First, we can note that PLA exhibited a

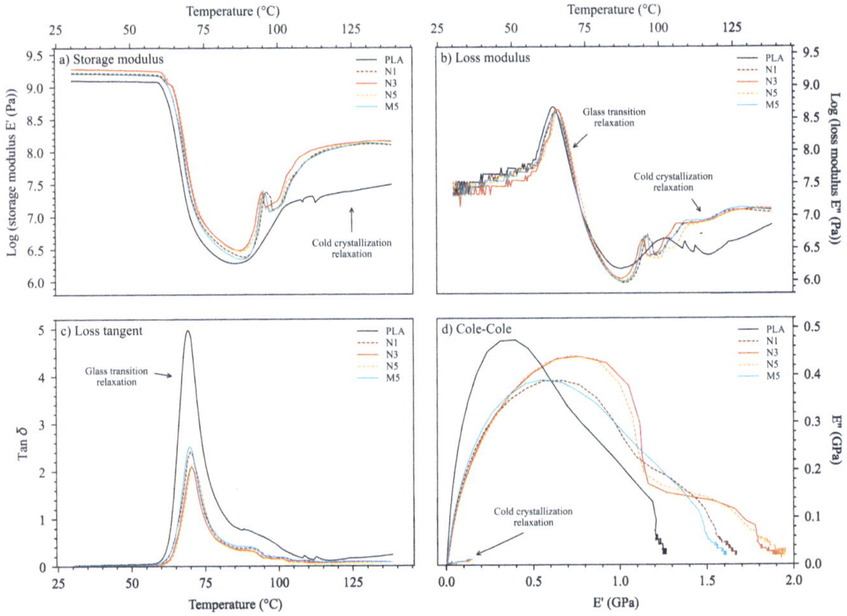

Fig. 16.11 DMA analysis of 3D-printed bionanocomposites and reference biocomposite

typical behavior widely observed in the literature (Bouakaz et al. 2015; Cristea et al. 2020). At low temperatures (T < 60 °C), E′ was virtually independent of temperature, with a measured value of 1.27 GPa at 30 °C, in agreement with the literature (Bouakaz et al. 2015). E″ was more than 10 times lower than E′, which is the expected behavior of a polymer in a glassy state. Between 55 and 60 °C, E′ drastically decreased, and E″ increased as PLA shifted from a glassy to a rubbery state. The maximum E″ peak was observed at 65 °C. Moreover, the tan δ peak appeared 5 °C later, with a maximum tan δ value of 5. Once the α-relaxation was reached, E′ continued to decrease slower than during the previous phase. The storage modulus reached the lowest value at 87 °C at 1.6×10^{-3} GPa. Beyond 90 °C, E′ increased intensely up to 100 °C, then gradually up to the end-of-test temperature at 140 °C. This increase is due to the formation of a complementary crystal network to the existing one, in line with previous DSC results. E″ also increased, and tan δ remained at values below 0.25.

All the 3D-printed bionanocomposites and the reference biocomposite showed similar behavior to neat PLA. The addition of cellulosic filler increased the storage modulus and reduced the damping factor of PLA across the entire temperature range. M5 values were comparable to the N1 condition, with a perfect overlap of curves. At 30 °C, N5 showed the highest storage modulus at 1.95 GPa, 50% higher than neat PLA. The bionanocomposites and the reference biocomposite showed a slight shift in T_g (determined by the in tan δ and E″ peaks methods, presented in Table 16.3 towards values 1 to 2 °C higher than for neat PLA. In the rubbery state,

Table 16.3 Glass transition temperature (T_g) determined by DMA

	$T_{peak\ E''}$ (°C)	$T_{peak\ \tan\delta}$ (°C)
PLA	64.5	69.2
N1	65.2	70.1
N3	65.4	70.3
N5	65.9	71.0
M5	65.5	69.9

adding cellulosic filler did, as expected, minimize the loss of mechanical properties. N3 exhibited the same reinforcing effect, with E′ almost doubled in the rubbery state compared to the neat PLA. In addition, N3 and N5 exhibited a tan δ peak 60% lower than neat PLA. One PLA drawback is its dramatic mechanical loss when subjected to temperatures above its glass transition, which was partially prevented by cellulosic fillers. The addition of cellulose fillers can therefore improve the thermomechanical stability of PLA.

The Cole-Cole curves, depicted in Fig. 16.11d), can provide relevant information about the homogeneity of the material (Bagotia and Sharma 2019): the more circular the Cole-Cole curve, the better the homogeneity of the tested specimen. In our case, PLA has an irregular arc-shaped curve. According to Thumsorn, 3D printing is an anisotropic printing process, generating many interface zones between the layers of printed material and, consequently area of heterogeneity (Thumsorn et al. 2022). M5 and N1 curves also showed an imperfect semi-circular character, indicating a rather good dispersion of fibers into the matrix and a suitable compatibility between fiber and matrix. On the other hand, the N3 and N5 bionanocomposites showed non-circular curves, suggesting the materials were heterogeneous, probably caused by aggregates.

The DMA is also a powerful tool for advanced composites analysis and assessing the impact and effectiveness of the used reinforcement. It can be used to determine numerous quality parameters such as the effectiveness coefficient C, the adhesion factor A, the reinforcing factor r, and the entanglement density N. The effectiveness coefficient *C* corresponds to the effectiveness of the reinforcement in the storage modulus. The effectiveness of the reinforcement is inversely proportional to the calculated C coefficient (Akindoyo et al. 2021). The coefficient can be obtained by Eq. 16.3: (Akindoyo et al. 2021; Madathinal Kunjappan et al. 2021; Thumsorn et al. 2022):

$$C = \frac{\dfrac{E'_{glass}}{E'_{rubber}}(bionanocomposite)}{\dfrac{E'_{glass}}{E'_{rubber}}(neatPLA)} \tag{16.3}$$

where E'_{glass} and E'_{rubber} are, respectively, the storage modulus in glassy and rubbery states, expressed in Pa; the glassy state was set at 30 °C and the rubbery at 75 °C. The adhesion factor *A* also assesses the quality of the fiber-matrix interface by focusing on the intensity of the tan δ peak. Similar to the effectiveness coefficient, the lower the factor, the greater the reduction in matrix molecular mobility due to the addition of reinforcements (Akindoyo et al. 2021). The coefficient can be calculated from Eq. 16.4 (Akindoyo et al. 2021; Madathinal Kunjappan et al. 2021; Thumsorn et al. 2022):

$$A = \left(\frac{1}{1-V_f}\right) * \left(\frac{\tan\delta_{bionanocomposite}}{\tan\delta_{neat\ PLA}}\right) - 1 \tag{16.4}$$

where V_f corresponds to the volumic content of fillers, the PLA density was measured at 1.25, and the cellulosic filler density was set at 1.53 to calculate the $V_{f\ MCC}$ and $V_{f\ NCC}$ (Ganesan et al. 2019). *Tan* $\delta_{biocomposite}$ and *tan* $\delta_{neat\ PLA}$ were the maximum tan δ peaks associated with the α relaxation highlighted in Fig. 16.11c. The reinforcement efficiency factor *r* can be obtained by Eq. 16.5, directly derived from the Einstein equation (Thumsorn et al. 2022):

$$\mathrm{r} = \frac{\left(\frac{E'_{biocomposite}}{E'_{neat\ PLA}} - 1\right)}{V_{f.}} \tag{16.5}$$

where $E'_{biocomposite}$ et E'_{rubber} were measured in the glassy state at 30 °C. The entanglement degree *N* corresponds to the entanglement degree of the PLA macromolecular chains and is expressed according to Eq. 16.6 (Thumsorn et al. 2022):

$$N = \frac{E'}{6RT} \tag{16.6}$$

Where *E′* was measured at 75 °C. *R* is the Boltzmann constant (8.314462 J·mol^{-1}·K^{-1}), and *T* is the absolute temperature of the measured storage modulus (in K). All the calculated and needed parameters are given in Table 16.4. In every case, the entanglement degree N increased with the addition of cellulosic fillers. It suggests an overall good dispersion and compatibility of the fillers with the matrix, as the micro-sized and nano-sized fillers hinder the mobility of the macromolecular chains of PLA. The N factor was optimal for condition N5, with a calculated value of 6.25×10^2 mol·m^{-3}. In addition, the best effectiveness coefficient C was also calculated for N5, with a value of 0.673. These values demonstrate that even though N5 had aggregates, this does not yet degrade its mechanical properties. However, despite the mechanical properties' improvement by using 5% NCC, the literature warns that excessive nanofillers can reduce the mechanical properties of the associated nanocomposite and render them not economically viable. For example, for a

Table 16.4 DMA properties of 3D printed samples

	W_f (%)	V_f (%)	E′ (GPa @ 30 °C)	E′ (GPa @ 75 °C)	Tan δ	C	A	r	N (mol·m^{-3})
PLA	–	–	1.27	3.73×10^{-3}	4.99	–	–	–	2.72×10^2
N1	1	0.83	1.67	6.65×10^{-3}	2.41	0.739	−0.513	38.5	4.86×10^2
N3	3	2.48	1.93	8.02×10^{-3}	2.03	0.708	−0.583	21.0	5.86×10^2
N5	5	4.16	1.95	8.55×10^{-3}	2.18	0.673	−0.544	13.0	6.25×10^2
M5	5	4.16	1.61	6.15×10^{-3}	2.53	0.772	−0.471	6.55	4.49×10^2

PLA/multiwalled carbon nanotube (MWCNT) nanocomposite, increasing the W_{MWCNT} from 3% to 5% reduced the nanocomposite's room-temperature storage modulus from 4.5 to 2.8 GPa (Younus et al. 2023). It was also noted that there was a significant increase in factor C when a nanofiller was excessively incorporated into a polymer matrix, raising suspicions of an excessive nanofiller/nanofiller interaction (Bagotia and Sharma 2019). Regarding the adhesive factor A, the best filler/PLA interface quality was observed for N3, probably due to the lower presence of aggregates than N5, which can reduce the interface quality. Reinforcement efficiency r was the highest for condition N1, with a value reaching 38.5, probably due to a very good NCC diffusion into the matrix and the slight increase in PLA crystallinity by adding a low level of NCC. Finally, comparing the M5 condition with the different nanocomposites produced showed that the tested parameters were below the NCC-based composites. These results demonstrate the relevance of using a low percentage of nanocellulose for production but also underline the importance of continuing research in search of the optimum $W_{f\,NCC}$.

16.7 Influence of Sample Architecture on Dielectric and Mechanical Properties

Adding cellulose fillers enhanced the mechanical properties and reduced the dielectric characteristics of 3D-printed PLA-based composites. A key benefit of the 3D-printing process lies in its ability to control sample architecture. By controlling the infill ratio, for example, it is possible to produce lighter parts with less material, which favors the material's sustainability. According to the literature, the infill ratio respectively positively and negatively impacted the electrical insulation and mechanical properties (Zhang et al. 2016; Abeykoon et al. 2020; Öteyaka et al. 2021; Kuzmanić et al. 2023). Producing 3D-printed composites reinforced with cellulose fillers offers the opportunity to manufacture lightweight parts with a lower infill ratio, yet relevant mechanical and dielectric properties.

To verify this statement, Table 16.5 presents the mechanical and dielectric properties of 3D-printed specimens with an infill ratio of 50% and 100%. This table also presents the density of the DMA specimens produced and the property loss between the 100 and 50% infill (named Δ_{infill}). As expected, lowering the infill ratio reduced the ε', σ_{AC}, and E′ of all samples. PLA, for example, reduced its E′ by 30% and ε' by 40%. Halving the infill ratio reduced the dielectric constant and storage modulus by 30-40%, which is in good agreement with density variations (between 30 and 40%). The electrical conductivity losses were higher, between 40 and 50%. According to the literature, electrical conductivity is strongly affected by the quality of the layer interfaces (Roumy et al. 2023). Reducing the infill ratio decreased the number of interfaces and, therefore, the electron transfer capacity between them. It is worth noting that the N3 and N5 conditions with a 50% infill ratio exhibited storage moduli similar to 3D printed PLA with a 100% infill ratio while offering a dielectric constant 15% lower through the combined effect of the infill ratio and the cellulose reinforcements.

Table 16.5 DMA properties of 3D printed samples. Δ_{infill} was based on the minimum theoretical value of 1 achievable for the dielectric constant

		BDS analysis		DMA analysis	
Condition	Infill ratio (%)	ε′	σ_{AC} (S·cm^{-1})	E′ (MPa)	Density (g·cm^{-3})
PLA	100	2.56	7.95×10^{-7}	1265	1.250
	50	1.94	4.66×10^{-7}	880	0.808
	Δ_{infill}	–39.9%	–41.4%	–30.5%	–35.4%
N1	100	2.73	9.72×10^{-7}	1668	1.253
	50	2.10	5.12×10^{-7}	1043	0.797
	Δ_{infill}	–36.6%	–47.3%	–37.5%	–36.4%
N3	100	2.81	8.61×10^{-7}	1927	1.259
	50	2.16	5.48×10^{-7}	1200	0.790
	Δ_{infill}	–35.6%	–36.4%	–37.7%	–37.3%
N5	100	2.87	9.76×10^{-7}	1951	1.265
	50	2.21	5.69×10^{-7}	1238	0.801
	Δ_{infill}	–35.2%	–41.7%	–36.6%	–36.7%
M5	100	2.65	1.04×10^{-6}	1610	1.265
	50	2.00	6.19×10^{-7}	1096	0.813
	Δ_{infill}	–39.3%	–40.5%	–31.9%	–35.7%

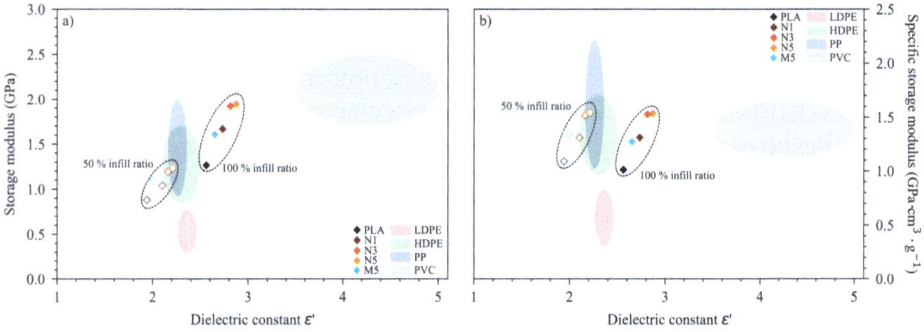

Fig. 16.12 Mechanical and dielectric performances of the conceived 3D-printed materials and their comparison with conventional polymers in the electrical insulation field. Ellipsoids are drawn according to the literature

To extend the performance analysis, Fig. 16.12a) and 16.12b) compare the results of the storage modulus of the produced materials with conventional polymers broadly used in electrical insulation. The materials selected were low-density polyethylene (LDPE), high-density polyethylene (HDPE), polypropylene (PP), and polyvinyl chloride (PVC). The E′ and ε′ results for these polymers were based on published values (Foley et al. 2012; Mike Chung 2012; Mofokeng et al. 2012; Ndlovu et al. 2013; Rimdusit et al. 2014; Mohagheghian et al. 2015; Anh et al. 2016; Mansour et al. 2016; Wang et al. 2017; Izci and Bowler 2019; Turnhout 2019; Chen et al. 2020, 2021; He et al. 2020; Jiang et al. 2021; Salakhov et al. 2021; Xu et al. 2021; Janík et al. 2022; Elsad et al. 2023; Jones et al. 2023; Vidakis et al. 2024). Figure 16.12a shows that the storage modulus of N3 and N5 with infill ratio = 100% were in the same range as PVC but had a more attractive dielectric constant for electrical insulation, making them ideal PVC replacements. N3 and N5 with infill ratio = 50% could replace LDPE, HDPE, and PP as they possess similar E′ and more attractive ε′. In applications requiring lightweight performances, it may be interesting to consider the specific storage modulus obtained by dividing the storage modulus by the density of the material (given in Table 16.5). The higher the value, the better its applicability in lightweight applications such as aeronautics. Figure 16.12b shows the specific storage modulus vs dielectric constant. These results underline the strong relevance of using N3 and N5 conditions with an infill ratio of 50% even to replace PVC, thanks to their low mass.

16.8 Water Aging Analysis

16.8.1 Chemical Alteration of 3D-Printed Samples

To check the chemical alteration of samples caused by water aging, Fig. 16.13 presents the chemical properties of the neat PLA, M5, and N5 before and after 5 months of water aging. After aging, neat PLA showed a very weak O–H stretching and bending peak at respectively ~3400 and 1640 cm^{-1}, indicating a low uptake. Banjo et al. (2022) also reported increased O–H peaks after water aging, linked with water adsorption during the immersion. This observation is consistent with the FTIR

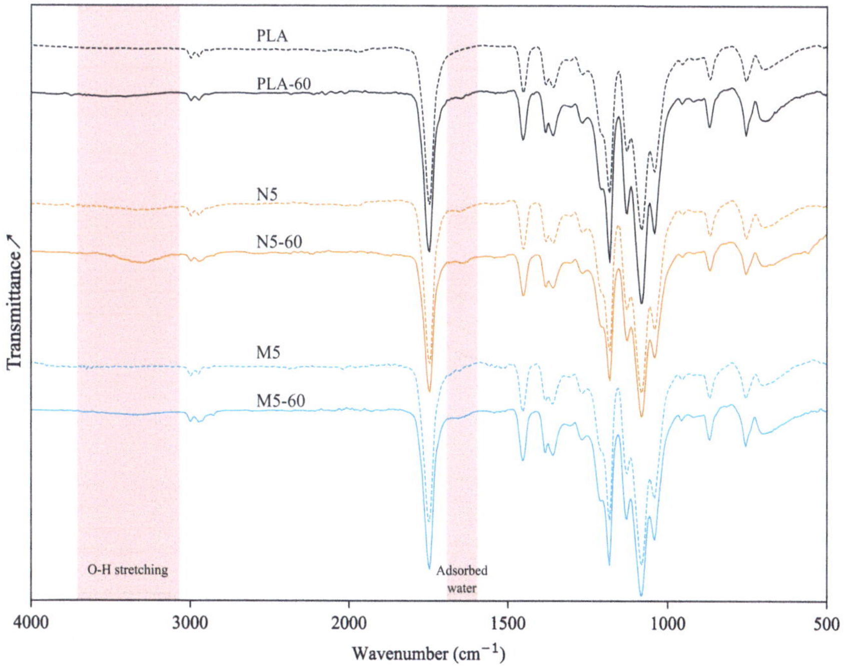

Fig. 16.13 FTIR spectra of 3D-printed PLA, N5, and M5 before and after water aging

results presented in Fig. 16.7, as PLA was free of polar functions apart from its carbonyl function. Water-aged M5 and N5 presented a small and broad peak around 3350 cm^{-1}, more pronounced than their unaged versions. A small and narrow peak was also present at 1640 cm^{-1}, corresponding to water adsorption due to silane and hydroxyl groups on the fiber surface, indicating that water was assimilated by the specimen more intensely than for neat PLA thanks to hydroxyl groups at the fiber's surfaces (Xian et al. 2018; Dalai and Parida 2022). Condition N5 showed the greatest intensity of peaks associated with hydroxyl functions, indicating that this condition had a great water uptake.

16.8.2 Water Uptake of 3D-Printed Samples

Figure 16.14 presents the water uptake of the bionanocomposites and the reference biocomposite. Moreover, Table 16.6 presents the uptake value when time tend to an infinite value ($W_{upt\ \infty}$), calculated using OriginLab's asymptotic1 function, and the slope of the curve during the linear uptake phase. The curves were very close to Fickian curves, i.e., a linear increase in water uptake during the first phase of water aging, followed by an inflection towards a terminal value during the second absorption phase. The value has not yet stabilized after 5 months of water aging. After $60^{0.5}$ h, the highest uptake was noted for M5 and N5, while the lowest was noted for PLA. PLA absorbed 0.8% by weight of water. This relatively low rate was due to the carbonyl functions in the PLA backbones, moderately attracting polar

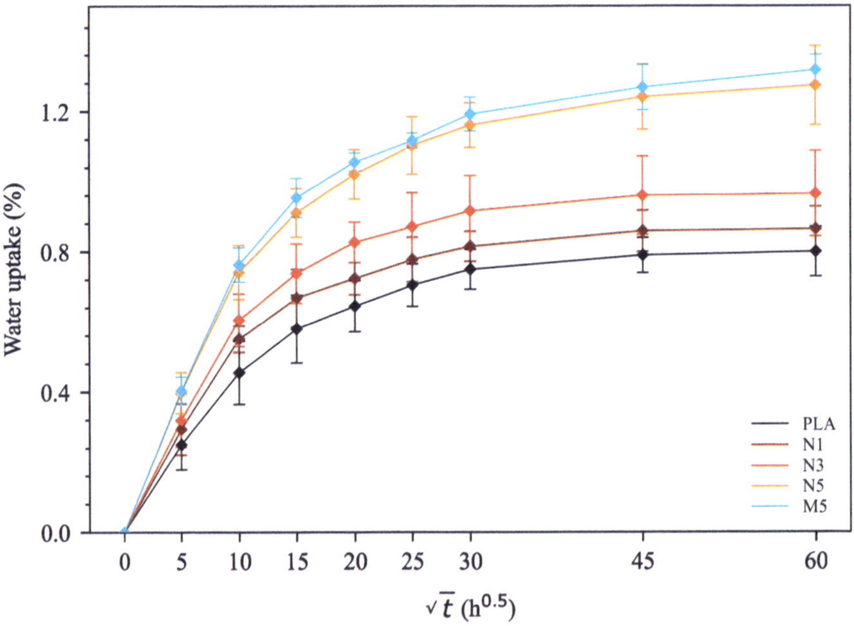

Fig. 16.14 Water uptake of bionanocomposites and reference biocomposite

Table 16.6 Water uptake parameters

Condition	$W_{upt\,\infty}$ (%)	Slope at low time ($\sqrt{t} \leq 10$)
PLA	0.811	0.0455
N1	0.867	0.0551
N3	0.976	0.0607
N5	1.276	0.0743
M5	1.308	0.0763

water molecules, as shown by the FTIR results. As the fillers contain silane and hydroxyl functions on their surfaces, they can more easily attract water molecules and increase the retention capacity of cellulose-based composites (Wan Ishak et al. 2020). The addition of cellulose fillers slightly increased the water absorption capacity. At $W_f = 5\%$, N5 and M5 showed a similar water uptake value around 1.3%. Since NCCs were much smaller than MCCs, they present a larger specific surface area and a greater polar function capacity and should, therefore, have produced higher water uptake results than MCC. To explain this phenomenon, we can argue that NCC was better dispersed on the microscopic scale in PLA than MCC, which could hinder water diffusion and slow the absorption kinetics (Amjad et al. 2022). In addition, NCC‘s good compatibility with PLA may limit the number of polar functions available and reduce the material's absorption capacity.

16.8.3 Dielectric Properties Alteration with Water Aging of 3D-Printed Samples

Figure 16.15 presents the dielectric constant of PLA, M5, and N5 before and after water aging. At 1 MHz, water-aged PLA, M5, and N5 exhibited an ε' of 3.16, 3.58, and 3.69, i.e., respectively 23, 35, and 25% higher than before aging. The notable increase was linked to the absorption of polar water (Torgovnikov 1993). The addition of cellulosic fillers positively influenced the dielectric constant of the water-aged samples. As Fig. 16.15 shows, adding fillers increased the water uptake capacity and, therefore, the polarization capacity of the biocomposite and bionanocomposites. After water aging and regarding the electrical conductivity, the lowest σ_{AC} was measured for PLA at 1.20×10^{-6} S·cm^{-1}, and the highest σ_{AC} was measured for M5 at 1.61×10^{-6} S·cm^{-1} (Table 16.7). The water-aged PLA, M5, and N5 presented a 39, 60, and 49% increase in σ_{AC}, respectively. As seen previously, the addition of cellulosic fillers increased water absorption and retention capacity, consequently enhancing ionic mobility and electrical conductivity. However, no abrupt increase in electrical conductivity was observed. The σ_{AC} versus frequency curves retained their linear shapes, with no characteristic inflection at low frequencies (Spinelli et al. 2020). These observations show that although water aging influences electrical conductivity, no percolation thresholds were reached and attest to the insulating nature of these materials.

To better understand the dielectric performance variation over time, Figs. 16.16a and 16.6b show the variation of dielectric constant and electrical conductivity measured at 1 MHz with water aging. As expected, all conditions showed an increase in

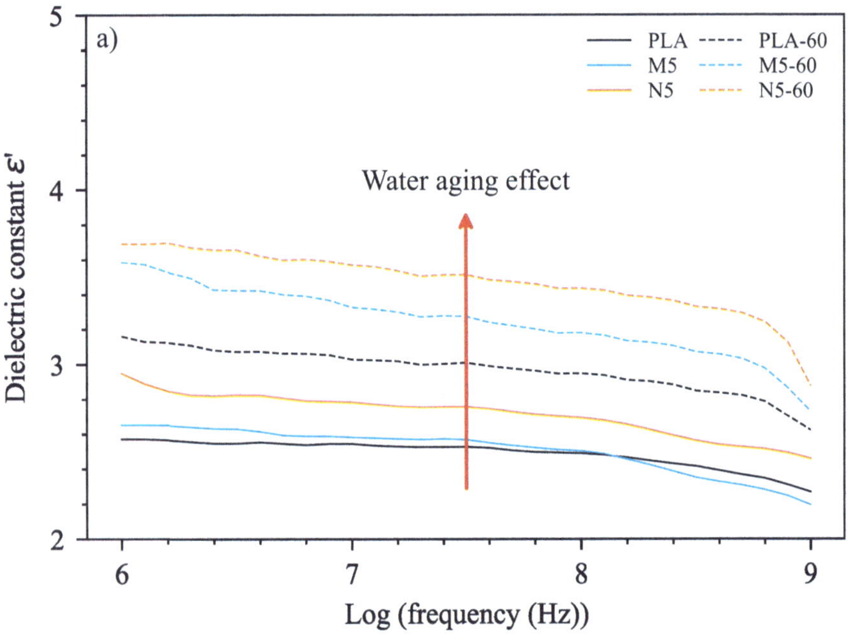

Fig. 16.15 Dielectric constant versus frequency of unaged and water-aged neat PLA, M5, and N5

Table 16.7 Electrical conductivity versus frequency of unaged and water-aged neat PLA, M5, and N5

	Log (electrical conductivity σ_{AC} ($S{\cdot}cm^{-1}$))			
Condition	1 MHz	10 MHz	100 MHz	1 GHz
PLA	−6.064 ± 0.010	−5.058 ± 0.009	−4.068 ± 0.009	−3.098 ± 0.015
M5	−5.996 ± 0.013	−5.007 ± 0.021	−4.016 ± 0.019	−3.208 ± 0.015
N5	−6.017 ± 0.020	−5.075 ± 0.034	−4.090 ± 0.035	−3.180 ± 0.032
PLA-60	−5.921 ± 0.017	−4.957 ± 0.039	−3.966 ± 0.035	−3.092 ± 0.027
M5-60	−5.793 ± 0.013	−4.832 ± 0.022	−3.853 ± 0.021	−2.975 ± 0.032
N5-60	−5.843 ± 0.009	−4.904 ± 0.015	−3.924 ± 0.016	−3.039 ± 0.014

dielectric constant and electrical conductivity over time due to adding water molecules. Moreover, ε' and σ_{AC} variations validate the water absorption curves, i.e., they both possess a first linear phase, which tends to stabilize towards a finite value. Furthermore, PLA, N1, N3, and N5 show similar variations in ε′ and σ_{AC}. However, M5 does not yet appear to be approaching a finite value.

To verify the relationship between mass and polarity variation, Fig. 16.17 shows water uptake versus dielectric constant. First, neat PLA had a more pronounced dielectric sensitivity than all the other samples tested. For the same mass of water absorbed, PLA had its ε′ altered faster than the bionanocomposites and the reference biocomposite. We can argue that, as PLA is a low-polarity polymer compared to cellulose, the water molecules infiltrating between PLA's macromolecular chains

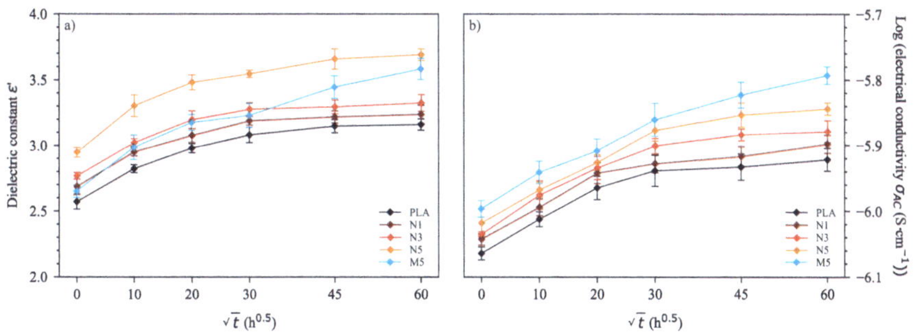

Fig. 16.16 Variation in dielectric constant and electrical conductivity over water aging time of 3D-printed bionanocomposites and reference biocomposite

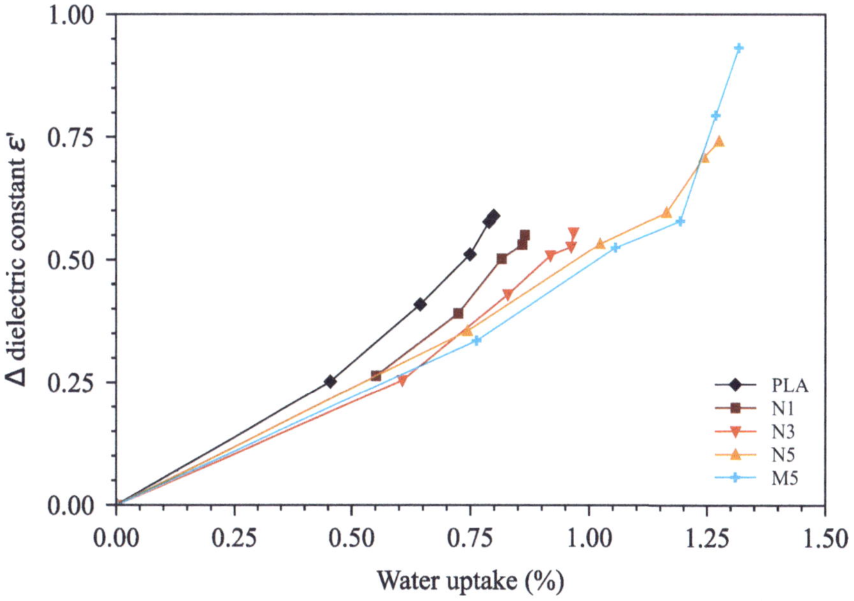

Fig. 16.17 Variation of dielectric constant as a function of water uptake of 3D-printed bionanocomposites and reference biocomposite

were not subject to any interactions and, therefore, have relatively high molecular mobility, increasing the polarization of water-aged PLA. In contrast, cellulose-based composites showed lower dielectric sensitivity. We could argue that when a water molecule penetrates the composite, it can more easily form a hydrogen bond with the fillers. These hydrogen bonds limit the mobility of water and, therefore, the polarizability of a material with internal moisture (Torgovnikov 1993). It's also worth pointing out that the curves obtained seemed to be exponential and not linear. Furthermore, condition N5 presented a sharp rise in water uptake above 1.25%. It might indicate that, as the MCC surface fillers become saturated and can no longer accept more water molecules, the additional water molecules were converted to free water (Torgovnikov 1993). Moreover, Lay et al. noted that the absorption of water molecules might increase the risk of creating fiber/matrix decohesion and micro-cracks due to relative heterogeneity at the microscale, increasing the capacity of water reception (Lay et al. 2019).

16.9 Conclusion

This book chapter focused on developing new 3D-printed biobased materials to replace conventional polymeric materials for electrical insulation applications. Bionanocomposites up to a W_f of 5% of NCC were produced and characterized mainly by broadband dielectric spectroscopy (BDS) and dynamic mechanical analysis (DMA). BDS and DMA measurements showed that cellulose nanofillers altered PLA properties. NCC mechanically stiffened the samples due to its very presence and increased crystallinity of PLA. However, the BDS results showed that nanofillers enhanced dielectric properties as nanocellulose had polar functions on its surface, increasing the polarization potential of the produced samples. A PLA/microcrystalline cellulose (MCC) reference was added to this research. It demonstrated that at equivalent levels (W_f = 5%), NCC's much higher specific surface area has significantly enhanced its mechanical and dielectric properties. Knowing the reducing effect of the electrical insulation caused by NCC, a sample architectural optimization step was also carried out by reducing the infill ratio. With an infill ratio of 50%, bionanocomposites showed improved electrical insulation properties compared to a neat and 100% infill ratio PLA, making them competitive alternatives for synthetic polymers used in electrical insulation, such as LDPE, HDPE, PP, and PVC. The study also highlighted that, although bionanocomposites were more polar, they lost very little electrical insulation properties after 5 months of water aging, indicating their revelence for long-term application in outdoor environments.

References

Abeykoon C, Sri-Amphorn P, Fernando A (2020) Optimization of fused deposition modeling parameters for improved PLA and ABS 3D printed structures. Int J Lightweight Mater Manuf 3:284–297. https://doi.org/10.1016/j.ijlmm.2020.03.003

Akindoyo JO, Beg MDH, Ghazali S et al (2021) Synergized high-load bearing bone replacement composite from poly(lactic acid) reinforced with hydroxyapatite/glass fiber hybrid filler—mechanical and dynamic mechanical properties. Polym Compos 42:57–69. https://doi.org/10.1002/pc.25807

Amjad A, Anjang Ab Rahman A, Awais H et al (2022) A review investigating the influence of nanofiller addition on the mechanical, thermal and water absorption properties of cellulosic fibre reinforced polymer composite. J Ind Text 51:65S–100S. https://doi.org/10.1177/15280837211057580

Anh TT, Vanga-Bouanga C, David E, Frechette M (2016) Dielectric response and breakdown strength improvement of LDPE/TiO2 nanocomposites by using POSS as compatibilizing agent. In: 2016 IEEE Electrical Insulation Conference (EIC). IEEE, Montreal, pp 65–68

Badia JD, Reig-Rodrigo P, Teruel-Juanes R et al (2017) Effect of sisal and hydrothermal ageing on the dielectric behaviour of polylactide/sisal biocomposites. Compos Sci Technol 149:1–10. https://doi.org/10.1016/j.compscitech.2017.05.026

Bagotia N, Sharma DK (2019) Systematic study of dynamic mechanical and thermal properties of multiwalled carbon nanotube reinforced polycarbonate/ethylene methyl acrylate nanocomposites. Polym Test 73:425–432. https://doi.org/10.1016/j.polymertesting.2018.12.006

Banjo AD, Agrawal V, Auad ML, Celestine A-DN (2022) Moisture-induced changes in the mechanical behavior of 3D printed polymers. Composite Part C 7:100243. https://doi.org/10.1016/j.jcomc.2022.100243

Barbosa KM, Martinez GA, Pereira F d C et al (2018) High voltage optical sensor insulation using PLA (Polylactic acid). In: 2018 13th IEEE International Conference on Industry Applications (INDUSCON). IEEE, São Paulo, pp 1362–1367

Bouakaz BS, Pillin I, Habi A, Grohens Y (2015) Synergy between fillers in organomontmorillonite/graphene–PLA nanocomposites. Appl Clay Sci 116–117:69–77. https://doi.org/10.1016/j.clay.2015.08.017

Buzón-Durán L, Martín-Gil J, Martín-Ramos P (2021) Chapter 4 – characterization of cellulose nanoparticles. pp 86–112

Chen J-Q, Wang X, Sun W-F, Zhao H (2020) Improved water-tree resistances of SEBS/PP semi-crystalline composites under crystallization modifications. Molecules 25:3669. https://doi.org/10.3390/molecules25163669

Chen J, Li B, Sun Y et al (2021) Greatly enhanced breakdown strength and energy density in ultraviolet-irradiated polypropylene. IET Nanodielectr 4:223–228. https://doi.org/10.1049/nde2.12022

Cristea M, Ionita D, Iftime MM (2020) Dynamic mechanical analysis investigations of PLA-based renewable materials: how are they useful? Materials 13:5302. https://doi.org/10.3390/ma13225302

Dalai S, Parida C (2022) Interpretation of Cole-Cole dielectric dispersion of green composites from medical LINAC modified luffa fiber/PLA. J Mater Sci Mater Electron 33:6911–6925. https://doi.org/10.1007/s10854-022-07871-0

Dammak Y (2022) Propriétés mécaniques et stabilité dimensionnelle des biocomposites fabriqués par impression 3d par extrusion de filament. UQAT

Dichtl C, Sippel P, Krohns S (2017) Dielectric properties of 3D printed polylactic acid. Adv Mater Sc Eng 2017:6913835. https://doi.org/10.1155/2017/6913835

Dubois J (2001) Propriétés diélectriques des plastiques. Techniques de l'Ingénieur – Plastiques et composites 1–20. 10.51257/a-v1-am3140

Elsad RA, Habashy MM, Izzularab MA, Abd-Elhady AM (2023) Evaluation of dielectric properties for PVC/SiO2 nanocomposites under the effect of water absorption. J Mater Sci Mater Electron 34:786. https://doi.org/10.1007/s10854-023-10199-y

European Bioplastics (2023) Bioplastics market data. European Bioplastics e.V. https://www.european-bioplastics.org/market/. Accessed 11 Feb 2024

Foley JR, Jordan JL, Siviour CR (2012) Probabilistic estimation of the constitutive parameters of polymers. Eur Phys J Conf 26:04041. https://doi.org/10.1051/epjconf/20122604041

Ganesan K, Barowski A, Ratke L (2019) Gas permeability of cellulose aerogels with a designed dual pore space system. Molecules 24:2688. https://doi.org/10.3390/molecules24152688

Hamad K, Kaseem M, Deri F, Ko YG (2016) Mechanical properties and compatibility of polylactic acid/polystyrene polymer blend. Mater Lett 164:409–412. https://doi.org/10.1016/j.matlet.2015.11.029

He L, Zeng J, Huang Y et al (2020) Enhanced thermal conductivity and dielectric properties of h-BN/LDPE composites. Materials 13:4738. https://doi.org/10.3390/ma13214738

Hossain MA, Roy CK, Sarkar SD et al (2020) Improvement of the strength of poly(acrylic acid) hydrogels by the incorporation of functionally modified nanocrystalline Cellulose. Mater Adv 1:2107–2116. https://doi.org/10.1039/D0MA00478B

Hussien B (2011) The DC and AC electrical properties of (PMMA -Al2O3) composites. Eur J Sci Res 52:236–242

Izci E, Bowler N (2019) Dielectric properties of isotactic polypropylene and montmorillonite nanocomposites. In: 2010 annual report conference on electrical insulation and dielectic phenomena. West Lafayette, pp 1–4

Janík R, Kohutiar M, Dubec A et al (2022) DMA analysis of plasma modified PVC films and the nature of initiated surface changes. Materials 15:4658. https://doi.org/10.3390/ma15134658

Jiang H, Zhang X, Gao J, Guo N (2021) Dielectric and AC breakdown properties of SiO2/MMT/LDPE micro–nano composites. Energies 14:1235. https://doi.org/10.3390/en14051235

Jin K, Tang Y, Zhu X, Zhou Y (2020) Polylactic acid-based biocomposite films reinforced with silanized nanocrystalline cellulose. Int J Biol Macromol 162:1109–1117. https://doi.org/10.1016/j.ijbiomac.2020.06.201

Jones H, McClements J, Ray D et al (2023) Thermomechanical properties of virgin and recycled polypropylene – high-density polyethylene blends. Polymers 15:4200. https://doi.org/10.3390/polym15214200

Khouaja A, Koubaa A, Daly H (2021) Dielectric properties and thermal stability of cellulose high-density polyethylene bio-based composites. Ind Crop Prod 171:113928. https://doi.org/10.1016/j.indcrop.2021.113928

Kuzmanić I, Vujović I, Petković M, Šoda J (2023) Influence of 3D printing properties on relative dielectric constant in PLA and ABS materials. Prog Addit Manuf 8:703–710. https://doi.org/10.1007/s40964-023-00411-0

Lay M, Thajudin NLN, Hamid ZAA et al (2019) Comparison of physical and mechanical properties of PLA, ABS and nylon 6 fabricated using fused deposition modeling and injection molding. Composite Part B 176:107341. https://doi.org/10.1016/j.compositesb.2019.107341

Lecoublet M, Ragoubi M, Leblanc N, Koubaa A (2023) Dielectric and rheological performances of cellulose acetate, polylactic acid and polyhydroxybutyrate-co-valerate biobased blends. Polymer 285:126358. https://doi.org/10.1016/j.polymer.2023.126358

Lecoublet M, Ragoubi M, Leblanc N, Koubaa A (2024) Dielectric and viscoelastic properties of 3D-printed biobased materials. Ind Crop Prod 212:118354. https://doi.org/10.1016/j.indcrop.2024.118354

Li H, Liu H, Li J et al (2015) Dielectric behaviors of carbon nanotube/silicone elastomer composites. Sci Eng Compos Mater 22:215–221. https://doi.org/10.1515/secm-2013-0280

Limsukon W, Rubino M, Rabnawaz M et al (2023) Hydrolytic degradation of poly(lactic acid): Unraveling correlations between temperature and the three phase structures. Polym Degrad Stab 217:110537. https://doi.org/10.1016/j.polymdegradstab.2023.110537

Madathinal Kunjappan A, Reghunadhan A, Ramachandran AA et al (2021) Discussion on degree of entanglement, chain confinement, and reinforcement efficiency factor of PTT/PE blend nanocomposite embedded with MWCNTs. Polym Adv Technol 32:2916–2928. https://doi.org/10.1002/pat.5303

Mansour SA, Elsad RA, Izzularab MA (2016) Dielectric properties enhancement of PVC nanodielectrics based on synthesized ZnO nanoparticles. J Polym Res 23:85. https://doi.org/10.1007/s10965-016-0978-5

Mike Chung T (2012) Functionalization of polypropylene with high dielectric properties: applications in electric energy storage. Green Sustainable Chem 2:29–37. https://doi.org/10.4236/gsc.2012.22006

Mofokeng JP, Luyt AS, Tábi T, Kovács J (2012) Comparison of injection moulded, natural fibre-reinforced composites with PP and PLA as matrices. J Thermoplast Compos Mater 25:927–948. https://doi.org/10.1177/0892705711423291

Mohagheghian I, McShane GJ, Stronge WJ (2015) Impact perforation of monolithic polyethylene plates: projectile nose shape dependence. Int J Impact Eng 80:162–176. https://doi.org/10.1016/j.ijimpeng.2015.02.002

Ndlovu SS, Van Reenen AJ, Luyt AS (2013) LDPE–wood composites utilizing degraded LDPE as compatibilizer. Composites Part A 51:80–88. https://doi.org/10.1016/j.compositesa.2013.04.005

Öteyaka MÖ, Aybar K, Öteyaka HC (2021) Effect of infill ratio on the tensile and flexural properties of unreinforced and carbon fiber-reinforced polylactic acid manufactured by fused deposition modeling. J Mater Eng Perform 30:5203–5215. https://doi.org/10.1007/s11665-021-05694-4
Qian S, Sheng K, Yu K et al (2018) Improved properties of PLA biocomposites toughened with bamboo cellulose nanowhiskers through silane modification. J Mater Sci 53:10920–10932. https://doi.org/10.1007/s10853-018-2377-2
Rathnayake WSM, Karunanayake L, Samarasekara AMPB, Amarasinghe DAS (2020) Sunflower oil-based MCC surface modification to achieve improved thermomechanical properties of a polypropylene composite. Cellul 27:4355–4371. https://doi.org/10.1007/s10570-020-03053-5
Rimdusit S, Wongmanit P, Damrongsakkul S et al (2014) Characterizations of poly(vinyl chloride)/acrylonitrile styrene acrylate blends for outdoor applications. Eng J 18:105–118. https://doi.org/10.4186/ej.2014.18.1.105
Roumy L, Touchard F, Marchand D et al (2023) Durability of Joule effect of 3D printed carbon black/polylactic acid: electrical cyclic tests and analytical modelling. Int J Fatigue 173:107677. https://doi.org/10.1016/j.ijfatigue.2023.107677
Salakhov II, Shaidullin NM, Chalykh AE et al (2021) Low-temperature mechanical properties of high-density and low-density polyethylene and their blends. Polymers 13:1821. https://doi.org/10.3390/polym13111821
Scaffaro R, Maio A, Nostro A (2020) Poly(lactic acid)/carvacrol-based materials: preparation, physicochemical properties, and antimicrobial activity. Appl Microbiol Biotechnol 104:1823–1835. https://doi.org/10.1007/s00253-019-10337-9
Shojaeiarani J, Bajwa DS, Hartman K (2019) Esterified cellulose nanocrystals as reinforcement in poly(lactic acid) nanocomposites. Cellul 26:2349–2362. https://doi.org/10.1007/s10570-018-02237-4
Song N, Yao H, Ma T et al (2019) Decreasing the dielectric constant and water uptake by introducing hydrophobic cross-linked networks into co-polyimide films. Appl Surf Sci 480:990–997. https://doi.org/10.1016/j.apsusc.2019.02.141
Spinelli G, Kotsilkova R, Ivanov E et al (2020) Dielectric spectroscopy and thermal properties of poly(lactic) acid reinforced with carbon-based particles: experimental study and design theory. Polymers 12:2414. https://doi.org/10.3390/polym12102414
Supeno S, Daik R, El-Sheikh SM (2014) The synthesis of a macro-initiator from cellulose in a zinc-based ionic liquid. Bioresources 9:1267–1275. https://doi.org/10.15376/biores.9.1.1267-1275
TechSci Research (2023) Electronic materials market size and trends 2028
Ten E, Bahr DF, Li B et al (2012) Effects of cellulose nanowhiskers on mechanical, dielectric, and rheological properties of poly(3-hydroxybutyrate-co-3-hydroxyvalerate)/cellulose nanowhisker composites. Ind Eng Chem Res 51:2941–2951. https://doi.org/10.1021/ie2023367
Thumsorn S, Prasong W, Kurose T et al (2022) Rheological behavior and dynamic mechanical properties for interpretation of layer adhesion in FDM 3D printing. Polymers 14:2721. https://doi.org/10.3390/polym14132721
Torgovnikov GI (1993) Dielectric properties of wood and wood-based materials. Springer, Berlin/Heidelberg
Turnhout J (2019) The polymeric materials encyclopedia poly(vinyl chloride) (dielectric properties). pp 1–24
Vidakis N, Michailidis N, Petousis M, et al (2024) Multifunctional HDPE/Cu biocidal nanocomposites for MEX additive manufactured parts: perspectives for the defense industry. Def Technol S2214914724000692. https://doi.org/10.1016/j.dt.2024.03.004
Wan Ishak WH, Rosli NA, Ahmad I (2020) Influence of amorphous cellulose on mechanical, thermal, and hydrolytic degradation of poly(lactic acid) biocomposites. Sci Rep 10:11342. https://doi.org/10.1038/s41598-020-68274-x
Wang Q, Wang Y, Meng Q et al (2017) Preparation of high antistatic HDPE/polyaniline encapsulated graphene nanoplatelet composites by solution blending. RSC Adv 7:2796–2803. https://doi.org/10.1039/C6RA26458A
Wang Q, Ji C, Sun J et al (2020) Structure and properties of polylactic acid biocomposite films reinforced with cellulose nanofibrils. Molecules 25:3306. https://doi.org/10.3390/molecules25143306

Wu W, Liu T, Zhang D et al (2019) Significantly improved dielectric properties of polylactide nanocomposites via TiO2 decorated carbon nanotubes. Composites Part A 127:105650. https://doi.org/10.1016/j.compositesa.2019.105650

Xian X, Wang X, Zhu Y et al (2018) Effects of MCC content on the structure and performance of PLA/MCC biocomposites. J Polym Environ 26:3484–3492. https://doi.org/10.1007/s10924-018-1226-3

Xu W, Tan L, Qin S et al (2021) Efficient fabrication of highly exfoliated and evenly dispersed high-density polyethylene/expanded graphite nanocomposite with enhanced dielectric constant and extremely low dielectric loss. Composite Part A 142:106242. https://doi.org/10.1016/j.compositesa.2020.106242

Yakubu A, Abbas Z, Abdullahi S (2020) Mechanical, dielectric and shielding performance of rice husk/polycaprolactone composites enhanced via rice husk particles inclusion. Open Access Libr J 7:1–12. https://doi.org/10.4236/oalib.1106514

Younus MM, Naguib HM, Fekry M, Elsawy MA (2023) Pushing the limits of PLA by exploring the power of MWCNTs in enhancing thermal, mechanical properties, and weathering resistance. Sci Rep 13:16588. https://doi.org/10.1038/s41598-023-43660-3

Yu J, Xu S, Liu B et al (2023) PLA bioplastic production: From monomer to the polymer. Eur Polym J 193:112076. https://doi.org/10.1016/j.eurpolymj.2023.112076

Zhang S, Arya RK, Pandey S et al (2016) 3D-printed planar graded index lenses. IET Microwaves Antennas Propag 10:1411–1419. https://doi.org/10.1049/iet-map.2016.0013

Zhu S, Sun M, Meng X, Cheng R (2024) Crystallization improvement of PLA by the talc with "grafting from" method of polymerization of lactide. J Polym Res 31:20. https://doi.org/10.1007/s10965-023-03857-6

Morgan Lecoublet was born in Saint-Malo, France in 1995. He was graduated with the M.Sc. Degree in Eco-conception of polymers and composites, Université de Bretagne Sud, France, in 2019. He completed a PhD on the dielectric properties of biobased materials under the joint supervision of the Laboratoire Biomatériaux of the Université du Québec en Abitibi Téminscamingue (Qc, CAN) and the Unité de Transformation et Agroressources of Unilasalle, France in June 2024. His research interests include topics in new biobased materials and their advanced characterization. He has published articles on the dielectric properties of biobased materials and flax-epoxy composites. He is currently working as a post-doctoral fellow for the COREPAN-Bois consortium on the challenges of recycling medium-density fiberboard.

Mohamed Ragoubi is an Associate Professor of Engineering and Applied Materials science with profound expertise in teaching (Basic Engineering Physics and Mechanics of Polymer Composites Materials), research, and Innovation. My area of interest covers teaching, innovative pedagogical tools, and research axis, focusing on advanced Materials sciences, natural fiber reinforcement, new materials functionality, multiphysics and multiscale behavior of composite materials, durability, and life cycle assessment. In 2015, he joined the Polytechnic Institute UniLaSalle, France, to work as an Assistant Professor in Physics and Materials Science. Before that, he worked as a post-doctoral fellow between 2010 and 2013 in the physics and mechanics textile materials laboratory LPMT in Mulhouse and at Lorraine University. He also integrated the CTT Group in Montreal,

Canada, and its Geosynthetics and Building Materials division, where he became a senior researcher. My expertise includes Engineering polymer composite materials, smart textiles, and recycled materials. He received his Master's degree in Mechanics and Materials Science from ENSISA in France in 2007 and his Ph. D. in Polymers and Composite Materials from the Lorraine University in France in 2010. He managed several industrial projects, published over 50 scientific and conference papers in highly renowned journals, and mentored many postdoc fellowships, Ph.D students' programs, and internships.

Nathalie Leblanc Native from Normandie, I obtained my Ph.D. in 2000 and my diploma of ability to conduct research from the University of Rouen in 2010 in the scientific discipline of the chemistry of materials with a specialty in physical chemistry of biopolymers. My scientific centers of interest are to develop knowledge about intrinsic characteristics (biochemical composition, morphology, porosity) of plant materials and their consequences on the performance of agro-based materials to value by-products of crops generated during agro-industrial transformations. This research aims to contribute to the current challenges of developing a bioeconomy linked to the use of bioresources. My areas of expertise are non-food valorization of agro-resources, functionalization of agro-based materials with specific performances by formulation and shaping, implemented by thermocompression or by extrusion, for example, according to the targeted application sector (building, transport, packaging), and study of relationships, between composition/structure/properties of biobased materials (thermal, acoustic, mechanical performance, hygroscopic behavior, fire resistance). Working at UniLaSalle since 2002, I am currently joint head of the research unit Transformations & Agro-Resources (ULR 7519 UniLaSalle—Artois University) and head of the research team "Valorization in Molecules and Materials for Innovation" (VAM²IN) » of this unit.

Ahmed Koubaa is a professor at the Université du Québec en Abitibi-Témiscamingue and has held the Canada Research Chair in Wood Valorization, Characterization, and Transformation since 2004. He holds bachelor's (1989) and master's (1991) degrees in wood science from Université Laval and a Ph.D. degree in paper engineering (1996) from Université du Québec à Trois-Rivières. He has over 30 years of experience in research on wood materials and wood-based composites. His research focuses on wood quality characterization and the valorization of wood industry residues to produce biomaterials and bioenergy to mitigate the effects of climate change and reduce greenhouse gas emissions. It has an extensive network of national and international collaborations. He is a member of several research groups and consortia in Quebec, Canada and internationally. For further information, visit https://orcid.org/my-orcid?orcid=0000-0002-7895-1901; https://scholar.google.com/citations?user=867vqm0AAAAJ&hl=en; Ahmed.koubaa@uqat.ca.

Advances in Green Biomaterials for Biomedical Implants and Tissue Engineering

17

Asib Ridwan, Md. Nizam Uddin, Asif Karim Neon, and Md. Aliahsan Bappy

Abstract

Green biomaterials have emerged as attractive and competitive alternative options in the fields of biomedical implants and tissue engineering. Biomaterials find extensive applications in supporting, enhancing, restoring, or replacing damaged tissues and organs within the human body. Furthermore, they maintain continuous interaction with the human body. Developing multifunctional green biomaterials that are compatible with human physiology is essential for achieving the necessary biological function while minimizing negative biological responses. The exploration of green biomaterials is gaining momentum in response to growing global concerns about environmental sustainability. This research area focuses on harnessing natural, renewable resources to create groundbreaking green biomaterials. The development of sustainable biomaterials encompasses the entire lifecycle of materials, considering factors such as desired properties, environmental impact during production, and strategies for

A. Ridwan
Department of Mechanical Engineering, Arkansas State University-Jonesboro, Jonesboro, AR, USA
e-mail: asib.ridwan1@smail.astate.edu

M. N. Uddin (✉)
Division of Engineering , Texas A & M University-Texarkana, Texarkana, TX, USA
e-mail: muddin@tamut.edu

A. K. Neon
Department of Mechanical Engineering, Khulna University of Engineering and Technology, Khulna, Bangladesh

M. A. Bappy
Department of Mechanical Engineering, Lamar University, Beaumont, TX, USA
e-mail: mbappy@lamar.edu

R. Malviya, S. Sundram (eds.), *Sustainable Nanocomposites with Green Biomaterials*, Biomaterials, Bioengineering and Sustainability 2,
https://doi.org/10.1007/978-3-031-79110-9_17

recycling or disposal. This study shows an overview of the latest developments in synthetic and natural biomaterials covering their preparation, uses, potential applications, and future challenges in biomedical implants and tissue engineering applications. This study aims to guide designing future research, emphasizing the critical importance of selecting green multifunctional biomaterials to achieve optimal performance in biomedical implants and tissue engineering.

Keywords

Biomedical implants · Biomedical applications · Tissue engineering · Biocompatibility · Biodegradability

Abbreviations

ACM	cellular Matrice
3D	Three Dimentional
ECM	Extracellular Matrix
GelMA	Gelatin Methacryloyl
HIV	Human Immunodeficiency Virus
HAp	Hydroxyapatite
MPS	Metal-Polymer composite Stents ()
MOF	Metal-organic framework
PEEK	polyetheretherketone
PEO	poly(ethylene oxide)
PLLA	Poly-L-lactic acid
PLA	polylactide
PNIPAM	poly(N-isopropyl acrylamide) ()
ROS	Reactive Oxygen Species

17.1 Introduction

Ensuring global well-being necessitates the maintenance of excellent health to preserve the quality of life. Presently, there is a sustained and increasing interest in the research and progress of biomaterials, with a particular emphasis on their application in the medical and healthcare fields. This is driven by the need for biomaterials in areas such as medication delivery, tissue engineering, the medical device sector, and the development of both permanent and temporary implants. With the increasing demand for precise and personalized treatments, researchers are focused on enhancing biomaterials for traditional uses and developing new bioactive materials for advanced applications such as regenerative medicine and theragnostic biomaterials, which combine diagnosis, monitoring, and therapy (Gu et al. 2023).

Biomaterials play a crucial role in modern medicine by promoting healing and restoring the original biological activity. Biomaterials can be natural or manufactured state. Due to their biocompatibility, they are utilized in various applications within the medical profession. Biocompatibility refers to the examination and understanding of the interactions between living organisms and inanimate substances (Assad and Jackson 2019). Biomaterials are defined as materials intended to interact with biological processes within the human body, to assess, assist, treat, improve, enhance, restore, or replace any impaired tissue, organ, or bodily function. Due to biomaterials'direct exposure to human bodily fluids and tissues, they must possess certain characteristics, including stability, biocompatibility, and safety. Corrosion in bio-metallic implants degrades surface integrity and biocompatibility by reacting with tissues that cause the release of corrosion byproducts and ultimately, premature failure. For a bioimplant, about 15 to 20 years for older patients and over 20 years for younger patients minimum lifespan is expected (Rodriguez-Gabella et al. 2017).

Historical data suggest that biomaterials were initially utilized in ancient times. Ancient Egyptians utilized various materials for medical procedures, such as animal tendons for sutures, coconut shells for skull repairs, and wood and ivory for artificial teeth. In Greece and India, surgeons have been performing plastic surgery on injured troops since the first century BC, utilizing biomaterials (Hu et al. 2022). The initial instances of utilizing natural biomaterials in a more detailed manner may be traced back to the modern era. In 1891, Germany saw its first documented surgical operation involving the substitution of a hip joint with ivory. From the nineteenth century onwards, bone plates were introduced into surgical procedures to fix bone fractures and accelerate the healing process. In 1951, the first successful transplantation of tissue from a deceased individual took place. Then, in 1954, a synthetic artery replacement was used for the first time to treat 10 patients. In 1958, the Dacron vascular prosthesis was adopted as a viable alternative to harvested human grafts, which had shown challenges during implantation (Todros et al. 2021).

The development of artificial heart valves began in the early 1950s. The final stage in the development of valves is characterized by advancements in innovative valve alternatives known as regenerative or tissue-engineered valves (Fig. 17.1).

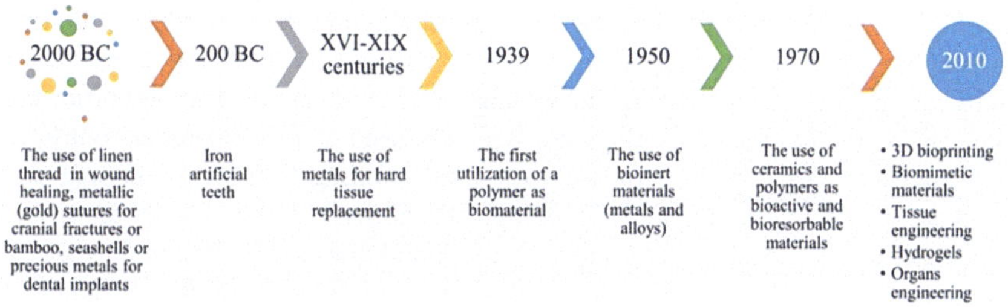

Fig. 17.1 Materials used as biomaterials at different times

17.2 Sources of Biomaterials for Biomedical Implants

Based on the properties of medicinal materials, biomaterials can be classified as metals, nonmetallic inorganics, hydrogels, other polymers, and bioderived materials.

Based on the properties of medicinal materials, we categorize regenerative biomaterials in the following order: metals, nonmetallic inorganics, hydrogels, other polymers, and bioderived materials. Bio-derived materials represent a distinctive composite medical material, each offering distinct advantages and limitations.

17.2.1 Metals

Metallic biomaterials find extensive applications in medical settings for repair and regeneration because they are very strong and durable. Titanium (Ti) is an excellent material for orthopedic implants due to its outstanding mechanical properties and high corrosion resistance. Current research is mostly centered on improving material performance through surface modification, alloying, and other methods (Wu et al. 2022). For example, the process of roughening surfaces is thought to improve osteointegrational bonding between bone tissues and implants by increasing contact and helping epithelial cells attach. Titanium can get micro/nanostructures through methods like electrochemical processes, laser techniques, acid-etching, and grit-blasting (Zhang et al. 2022). Modifying the surface of titanium implants can improve their interaction with cells, as well as alter their texture (Chen et al. 2021). For example, the Sun and Wang research team utilized lipid bilayer vesicle coating to confer biological functionalities to a titanium implant (Chen et al. 2020). Titanium demonstrates inadequate durability and is unsuitable for surfaces that require movement. However, the titanium oxide layer on the surface could come off (Fig. 17.2).

Titanium implants absorb elements from their environment and exhibit excellent resistance to corrosion. Incorporating alloying elements into titanium to create binary titanium alloys enhances the material's properties. For instance, adding niobium (Nb) to titanium can improve its tensile strength (Zhang et al. 2020). Ti6Al4V is the most often used medical alloy that contains titanium. It offers outstanding corrosion resistance and strong mechanical properties which is ideal for percutaneous or transcutaneous devices. Like titanium implants, Ti6Al4V requires surface modification to enhance the interface between cells and the material, promoting osseointegration and reducing inflammation (Liu et al. 2021). Specifically, surface modification can enhance biological sealing (Gao et al. 2022). Unlike permeable frameworks, titanium and its alloys are typically used as solid materials but have significant surfaces that are exposed to the surrounding tissue environment and encounter the difficulty of achieving biological sealing.

Wang et al. (2021) produced a Ti6Al4V implant that was treated with fibrinogen, leading to enhanced biological sealing and osseointegration. Due to the challenge of attaching proteins to metals, a polydopamine (PDA) coating was employed to facilitate the covalent interaction between the protein and the metal, illustrated in

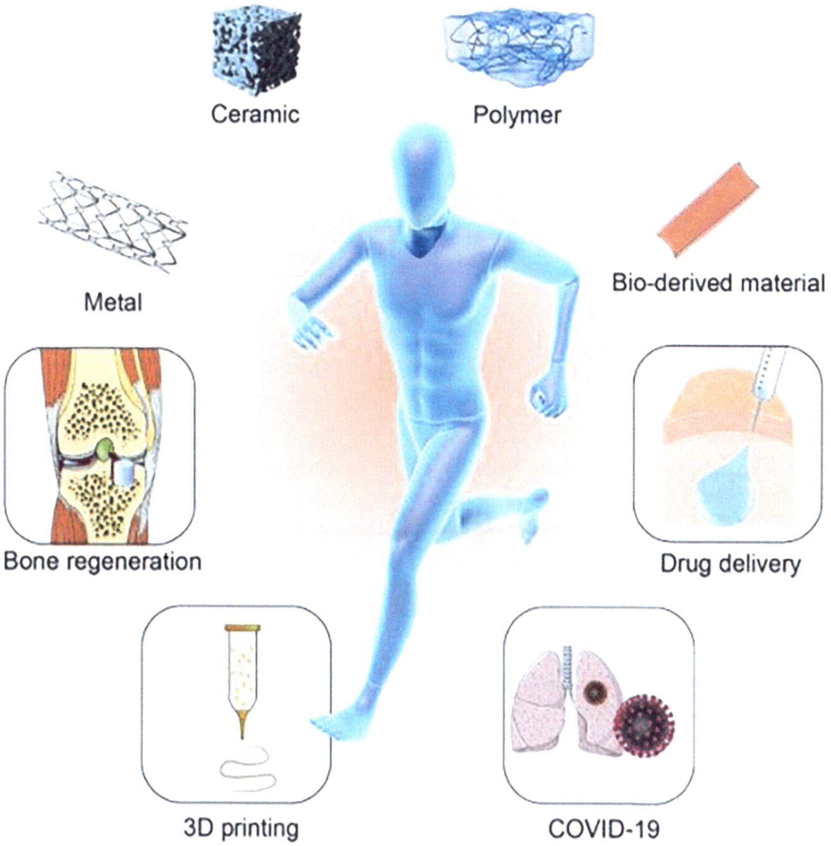

Fig. 17.2 Illustration of various sources and uses of biomaterials

Fig. 17.3a (Wang et al. 2021). Following the pioneering research by the Messersmith group, mussel-inspired surface chemistry has been developed further to create multifunctional coatings. This is due to the presence of amino and catechol groups, along with their ability to self-polymerize and spontaneously deposit (Yang et al. 2021). Altering the surface of metals can significantly affect various cellular activities on the materials.

In numerous instances, materials with high stability such as Titanium (Ti) are not compatible with the dynamic process of tissue regeneration. As a result, it might be necessary to remove these materials through a further surgical procedure, resulting in elevated levels of discomfort and increased costs for patients. Hence, biodegradable metals have the potential for tissue regeneration. Some metals possess corrosion and biosorption properties in living organisms, for use in degradable biomedical applications. However, Zhang et al. (2021) performed a thorough review of various publications and discovered the effectiveness of certain degradable metals tested for

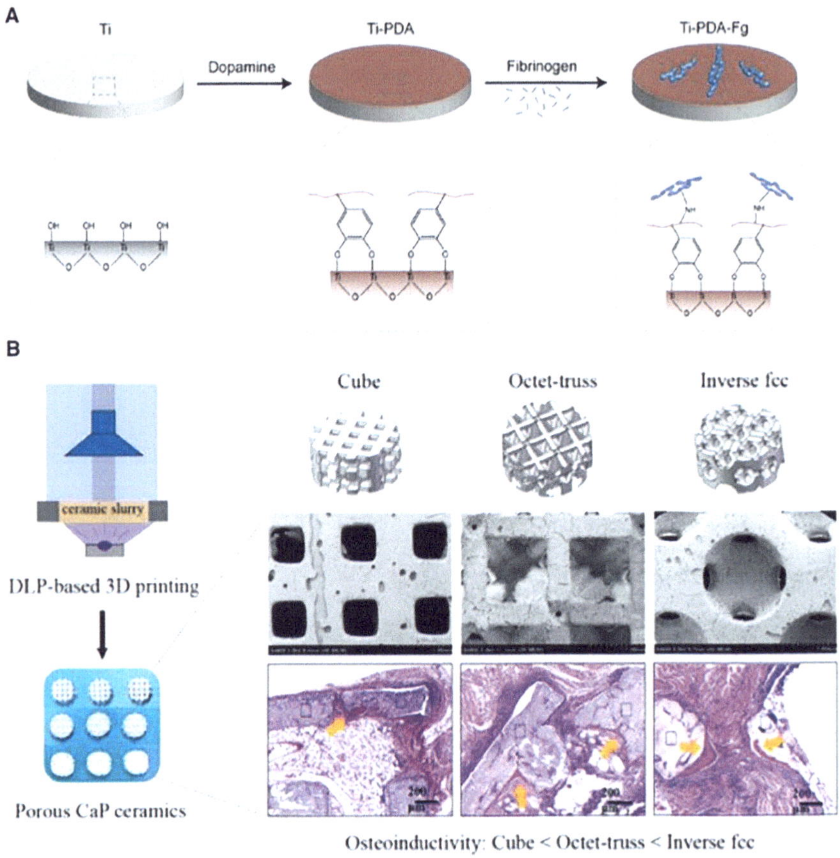

Fig. 17.3 Metal and ceramic implants. (**a**) A fibrinogen-modified titanium alloy at orthopedic percutaneous medical device. (**b**) Porous CaP ceramic scaffolds fabricated using DLP-based 3D printing

bone healing in animal studies may have been exaggerated. This is supported by the observation that degradable metals initially showed weaker fracture repair performance compared to non-degradable metals because of their low fixation ability. The deterioration time varies depending on the type of metal. For instance, magnesium (Mg) minerals undergo degradation in several months, but raw iron (Fe) materials degrade within a couple of years. Put simply, the rapid deterioration could hinder the healing of bone fractures when used in bone regeneration. Likewise, the ability of metal biomaterials to break down naturally could be impeded by their excessively sluggish corrosion when employed in cardiovascular therapy. Hence, it is crucial to adjust the rate at which metal degrades to a suitable timeframe for biomedical purposes. Wang et al. (2022) used a magnesium–strontium alloy to improve the material's resistance to corrosion in a canine mandibular defect model. According to

Zhang et al. (2021), it remains challenging for an alloy to undergo in vivo deterioration at a rate that precisely aligns with the regeneration process of bone tissue.

Developing further techniques to modify corrosion rates of metals in biomimetic conditions is a frontier. A significant advancement in recent times is the development of metal-polymer composite stents (MPS). Zhang et al. (2020) discovered that a polylactide (PLA) coating on iron did not prevent corrosion. It sped up the corrosion process. This was because of the interference of calcium-phosphate (Ca-P) precipitates on the surface of the iron, and the presence of a localized acidic microenvironment. Additionally, the hydrolysis of the aliphatic polyester also contributed to the accelerated corrosion. Additional approaches for controlling the corrosion of biometals have been proposed (Qi et al. 2019) and are currently being developed. Qin et al. (2022) used a Ca-P coating to tackle the problem of rapid degradation in Mg-based implants. Clinical research subsequently validated the efficacy and safety of this approach. Material degradation causes the release of metal ions from the metals. Several metals, such as copper (Cu), magnesium (Mg), calcium (Ca), and iron (Fe), contain essential components that are required by the human body. Thus, harnessing the liberated ions can enhance the capabilities of biomaterials. For instance, magnesium (Mg) possesses the ability to stimulate the growth of bone-forming cells (osteoblasts) and promote the development of blood vessels (angiogenesis). Magnesium can enhance the process of bone regeneration by increasing the number of osteoblasts, encouraging their transformation into bone-forming cells, and facilitating the deposition of minerals in the bone (Xie et al. 2021).

Although the ions produced from metallic biomaterials can be beneficial for the biological functioning of the human body, an excessive release of metallic ions can be detrimental. For instance, biomaterials containing magnesium might lead to adverse effects including cytotoxicity (Zhang et al. 2022). Furthermore, it was shown that magnesium alloy can significantly influence the adherence and movement of endothelial cells, hence impeding the re-endothelialization process of magnesium stents. Nevertheless, the US Food and Drug Administration (FDA) has issued a warning about the possible risk posed by the released nickel ions. Extensive collaboration between a company, university, and hospital led to the discovery of a nanocoating containing TiN, which has been found to efficiently decrease the release of Ni ions. As a result, a cutting-edge occlude using a Chinese core material technology has recently been successfully launched in multiple countries (Wu et al. 2021).

Suboptimal cytocompatibility does not necessarily have negative implications. The dissolution of bioactive ions, such as silver (Ag), copper (Cu), and magnesium (Mg), can exhibit strong antibacterial effects (Shen et al. 2021). This is advantageous for tissue regeneration due to its potential to prevent infections. Xu et al. (2021) incorporated copper into titanium alloys using powder metallurgy techniques to mitigate bacterial infection occurring with surgical implants. Alqattan et al. (2021) employed a nanocomposite coating made of Ag–selenium (Ag–Se) to prevent bacterial adherence on implants.

Metal ions can be used for anti-inflammatory purposes, in addition to their antibacterial properties. As an illustration, Liang et al. (2022) synthesized nanoscale zinc-based metal-organic frameworks (MOFs) that gradually release Zn ions, resulting in beneficial antibacterial and anti-inflammatory effects. Few metallic materials may be used for tumor ablation because of their excellent efficiency in converting light to heat. In addition, Researchers have shown interest in several rare-earth metals as biomaterials because of their unique electronic structures and capacity to exist in various valence states. For example, As an illustration, ceria has been used in industrial catalysis and recently progressed into nanoceria to eliminate nitrogen and reactive oxygen compounds (Chen et al. 2022). Cerium may switch between Ce3+ (reduced) and Ce4+ (oxidized) valence states, enabling ceria to generate oxygen gaps within its lattice structure. Besides, nanoceria has a higher number of oxygen vacancies because of its reduced ratio of surface area to volume. In addition, many materials containing rare-earth metals exhibit magnetic properties suitable for use in magnetic resonance imaging (MRI). Examples of such metals include dysprosium, holmium, and gadolinium.

Furthermore, certain metallic compounds can act as catalysts for proteins, enabling them to engage in vital biological processes. For instance, copper can act as a catalyst for the release of nitric oxide (NO) from NO donors. Yang et al. (2022) developed a PDA coating infused with copper to be used on blood contact materials that are implanted in the body. The purpose of this coating is to prevent blood clotting by continuously releasing nitric oxide. Therefore, the advancement of regenerative medicine has transformed metals from mere mechanical supports to also serve as functional sources.

Inorganic compounds that are not classified as metals Nonmetallic inorganics can be classified into two categories: amorphous glass and crystalline ceramics. They are predominantly advantageous for the regeneration of orthopedic tissue. Ceramics can be divided into two types: bioactive ceramics and bioinert ceramics. Bioactive ceramics may stimulate biological responses in the body after implantation.

Bioactive ceramics possess the ability to directly stimulate the difference between mesenchymal stem cells into bone cells. Additionally, they can sometimes modulate the immune response to create a suitable environment for the regeneration of bone by promoting beneficial inflammation. Following the occurrence of a fracture, the inflammatory macrophage of the M1 phenotype can enhance the inflammatory cascade, which plays a vital role in the primary stages of healing of bone. The inflammatory milieu undergoes alterations during the process of bone regeneration. Thus, it is necessary to effectively control and transform the M1 macrophage into the M_2 phenotype on time to enhance bone repair. Zhou et al. (2020) discovered that the the immunosuppressive properties of stem cells generated from bone marrow can be enhanced by the ionic products of Ca_2SiO_4 (calcium silicate). Li et al. (2021) discovered that dicalcium silicate (Ca_2SiO_4) can help bone formation in mesenchymal stem cells derived from the bone marrow of mice by instigating a specific type of inflammation commonly known as macrophagic inflammation (M1 phenotype). More precisely, the release of silica and calcium ions from Ca_2SiO_4 may affect the functionality of mitochondria and instigate autophagy in RAW264.7 cells, resulting in M1 polarization.

The in vivo effectiveness of ceramics has been verified based on fundamental investigations of mechanisms. Recently, calcium phosphate ceramics have successfully facilitated the regeneration of bone tissue to heal massive segmental bone lesions in a goat (Luo et al. 2022). A porous structure is essential because the pores inside it help neighboring cells to grow and allow for the dissolution of ionic substances. The size, permeability, and interconnectedness of a porous scaffold directly affect its effectiveness in bone regeneration. In summary, the scaffold's pore size must be adequate to facilitate the movement, growth, and specialization of cells. Greater porosity reduces the mechanical effectiveness of the scaffold, while insufficient porosity results in inadequate regeneration. The configuration of the pores may serve as a mechanical stimulus that affects cellular responses (Zhi et al. 2022). Wang et al. (2022) conducted a comparison of the osteoinductive effects of different calcium phosphate (CaP) scaffolds with unique structures, including octet-truss, cube, foam, and inverse face-centered cube (fcc). The researchers discovered that foam-like structured scaffolds had the highest level of osteoinduction. This was due to the presence of smaller pore width and non-through holes, which created a localized high ionic microenvironment. The fcc group with an inverse structure exhibited superior osteoinductive capability due to its spherical pore architecture, in contrast to the cube and octet-truss structures.

Bioactive glasses are a significant category of nonmetallic biomaterials that can attach to bone and produce biological responses by discharging physiologically active ions of certain elements such as calcium (Ca), silicon (Si), and phosphate (P). The initial iteration of bioactive glass is known as 45S5 (BioglassVR). The invention of this technique can be attributed to Larry Hench in 1969. Since then, it has been widely employed in clinical settings for bone healing purposes for an extended period. Recently, advancements have been made in biomaterials that are derived from bioactive glass to enhance their ability to perform different biological tasks. Additional biologically active components were stimulated in the bioactive glass. An example of this is the development of bioactive glass that contains strontium, which is intended to enhance the function of bone regeneration by gradually releasing strontium. Zhu et al. (2020) created Cu/Ca-impregnated bio-active glass nanoparticles by adding calcium and copper to the SiO_2 frameworks to treat osteosarcoma. The presence of released calcium ions and copper ions can synergistically increase the quantity of reactive oxygen species (ROS) in tumor cells through a Fenton-like process. Simultaneously, bioactive glass triggers the synthesis of hydroxyapatites and the release of calcium ions, which work together to promote calcification. Moreover, the increased ROS level and calcification resulted in the demise of tumor cells.

17.2.2 Hydrogels

A hydrogel is a type of polymer material that retains a significant water quantity within a 3-D network. Hydrogels possess distinct characteristics that have been utilized in the area of biomaterials (Han et al. 2022). One example of a soft material is hydrogel, which has moderate mechanical strength and is highly compatible with soft tissues (Shi et al. 2021). Hydrogels have been employed as substances to fill and stick tissues together. Hydrogels are biocompatible due to their high water content. Certain polymeric hydrogels exhibit exceptional degradability and biocompatibility. The 3D network of hydrogels permits the transportation, storage, and controlled release of pharmaceuticals. The researchers Yang et al. (2021) developed a thermogel that is injectable and can be used in conserving surgery of breast. This thermogel serves as a short-term filler material and also prevents local deterioration by continuously releasing Herceptin. A highly practical method for incorporating a medication into a hydrogel is through direct mixing, and how the drug is released is influenced by several factors (Chen et al. 2019). In general, macromolecular medications, like proteins, can be delivered in a controlled and continuous way because their sizes are similar to the mesh structure of a hydrogel. On the other hand, water-soluble small molecular pharmaceuticals typically result in a significant and rapid release. Occasionally, small molecule medicines can undergo a transition of their condensed physical forms, such as crystallization, in a hydrogel, resulting in a sustained release (Richbourg et al. 2021).

The release mechanism of a loaded medicine is determined by the hydrophilic-hydrophobic properties, except for cargo size. For instance, gemcitabine is a highly effective medicine against tumors. However, its hydrophilic nature and short half-life have significantly restricted its use in clinical settings. The researchers Cao et al. (2022) enhanced the hydrophobicity of gemcitabine by including fatty acids and then encapsulating it within a hydrogel. The modified gemcitabine demonstrated a protracted release from the hydrogel for 37 days. In contrast, the unmodified gemcitabine saw a rapid initial release, with nearly 70% of the drug being delivered on the very first day. Hydrophilic drugs can also attain the prolonged release of inappropriate hydrogels through their interaction with the networks of hydrogel. Lixisenatide, a type of hydrophilic peptide, has been approved by the FDA for application in the treatment of Type-2 diabetes mellitus. They utilized the lixisenatide and contrasting electrical charges of biodegradable polymer, which was created by them, to generate a continuous release by mixing the peptide with the solution of the polymer.

The hydrogel network can undergo crosslinking by either physical interactions or chemical bonds, leading to the formation of a physical hydrogel or a chemical hydrogel, respectively (Yang et al. 2020). Certain hydrogels can be created through physical stimulation without involving chemical reactions, making them very valuable in the field of biomedicine. An instance of chemical hydrogel is the photo-cross-linkable hyaluronic acid hydrogel, which was created through exposure to ultraviolet radiation. In a pig model, the combination of this hydrogel with platelet-rich plasma successfully facilitated the regeneration of porcine cartilage. If an aqueous solution that can flow easily solidifies quickly in response to a stimulus, it can be used in the process of 3D printing. Besides, Gelatin methacryloyl (gelMA) is another macromonomer generated from biomacromolecules. When it undergoes polymerization, it forms a light-curable hydrogel known as GelMA, as proposed by Zheng et al. (2022). The macromonomer with the resulting hydrogel has extensive utilization in the 3D printing field (Gao et al. 2021). By using a bilayered scaffold, Zheng et al. (2022) recently created a 3D-printed continuous structure. This was achieved by combining the transition of an aqueous solution of gelMA and the gelMA macromonomer photocrosslinking. This process utilizes the unique thermo-responsive rheological and non-Newtonian features of the aqueous solution of gelMA. The scaffold of hydrogel produced has great potential as a biomaterial for the regeneration of articular cartilage.

The traditional method of creating 3D-printed biomaterials involved putting them layer by layer in a laboratory setting, and then surgically implanting them in a living organism. In a recent study, Ding et al. (2022) utilized photoactive polymers to accomplish intravital 3D bioprinting. This involved creating 3D structures within live animals without the need for surgical operations. After conducting a thorough screening of various coumarin derivatives, 7-hydroxycoumarin-3-carboxylate–poly(– ethylene glycol) conjugates have been selected as the primary material for 3D bio-printing. Figure 17.3b demonstrates the utilization of multi-photon excitation to precisely create a photocrosslinking hydrogel within the skeletal muscle, dermis, and brain. In contrast to traditional methods of 3D printing, 3D bioprinting can accomplish the printing without the need for surgery and with real-time imaging. Urciuolo et al. (2020) described a method for non-invasive 3D bioprinting. The researchers administered bioink, which consisted of a solution including stem cells and monomers, through a subcutaneous injection. Subsequently, they provided the computer with photographs of the scaffold. The micromirror device chip generated digital near-infrared light and projected it on time to trigger, noninvasively, the three-dimensional polymerization of the bioink which is locally injected. Following several months, the previously administered stem cells successfully developed into a sophisticated tissue or organ. Significantly, they utilized this method to create a construct in the shape of an ear and successfully implanted it into BALB/c mice.

Thermosensitive hydrogels have been the subject of intense research since 1997 when Chen et al. (2020) introduced a block copolymer made of PLLA and poly(ethylene oxide) (PEO). This copolymer distinguishes itself from other thermo-sensitive polymers, such as poly(epoxy oxide)-b-poly(propylene oxide)-b-poly(epoxy oxide) (PEO-b-PPO-b-PEO) and poly(N-isopropyl acrylamide) (PNIPAM) due to its biodegradable characteristic. A variety of block copolymers derived from polyether polyester have been created and demonstrate the potential to undergo physical gelation when heated. These copolymers can exist as low-viscosity solutions at ambient temperature and convert into gels at body temperature (about 37 °C) (Jeong et al. 1997) (Fig. 17.4).

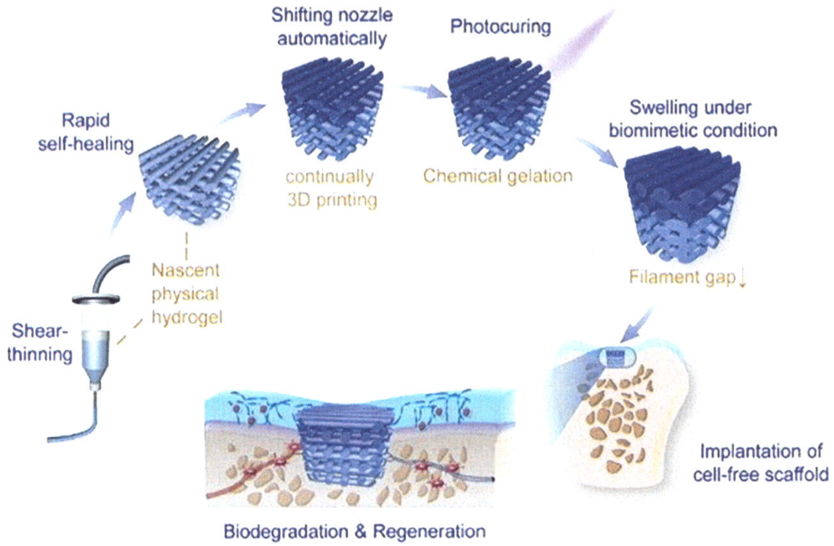

Fig. 17.4 Schematic diagrams of the major properties of material enabling the uninterrupted 3D-printing of the bilayered hydrogel scaffold

17.2.3 Other Polymers

Among all the categories of biomaterials, polymer is the largest (Li et al. 2021). Almost all the polymers demonstrate exceptional ability to be processed into different macroscopic physical shapes. One example is poly(methylmethacrylate), commonly known as bone cement, which is a filler material that aids in the restoration of bone. The utilization of synthetic polyetheretherketone (PEEK) in orthopedics is phenomenal. Besides, due to having exceptional mechanical and biocompatibility characteristics, PEEK is considered a suitable material for implantation. Biodegradable polymers, in parallel with non-biodegradable polymers, have also been utilized in clinical settings (Ding 2022). PLGA and PLLA, two highly successful biodegradable polymers, have found extensive use in medication delivery and cardiovascular stents.

Several natural polymers provide favorable characteristics such as excellent minimal immunogenic effects, biodegradability, and biocompatibility and cost-effectiveness. Common examples are chitosan, silk fibroin, alginate, and hyaluronic acid (Sharma et al. 2021). The distinctive mechanical qualities of silk threads led to their initial use as biomaterials for sutures. Understanding the structure of silk fibers reveals that silk fibroin, the main component of silk, is a significant natural polymer with promising applications. Aside from its use in commercial silk suture products, it can also be utilized in medication administration and repairing of tissue because of its excellent biocompatibility and controlled degradation properties. Zhai et al. (2020) constructed a microneedle patch with silk fibroins containing stabilized HIV immunogen and adjuvant. The release rate of these substances can be controlled by adjusting the b-sheet crystallinity level in the silk proteins. Alginate, an anionic polymer, has been scientifically demonstrated to be effective as a food additive and medicinal excipient. When divalent or multivalent cations are present like Ca2+, the water-based solution can change into a hydrogel. Alginate hydrogels have been thoroughly studied in the field of tissue repair and medication delivery. A group led by Moony has created a cryogel made of methacrylated alginate, which has interconnecting pores formed through cryogelation. The resulting alginate cryogel differs from normal alginate hydrogels as it may be easily injected using a needle for delivery because of its elastic characteristics. The cryogel was utilized to create a microporous vaccine by pre-loading cytokines or tumor cells which enhanced the immune response (Boopathy et al. 2019). In addition to that, some other biopolymers, such as nucleic acids and polypeptides, have shown substantial promise in the area of biomaterials (Najibi et al. 2022).

Polymers, in comparison to other material types, particularly metals, can encounter inadequate mechanical qualities (Mahmood et al. 2022). To improve the mechanics of polymers, there are three approaches: developing novel polymers, optimizing condensed-state properties, and introducing polymer-based composites. In a 3D-printed polycaprolactone scaffold, Zhao et al. (2021) employed laponite as the filler to control bone regeneration stiffness. Xu et al. (2021) created a scaffold by combining nanohydroxyapatite with collagen for enhancing mechanical properties. In the past, patients had only the option of choosing braces made of metal featuring wires made of alloy for orthodontic therapy, which caused damage to patients throughout the treatment process. The adaptation of polymers as the material for aligner films has been widely accepted in contemporary orthodontic treatments for invisible braces. In their study, Zheng et al. (2022) described the production of a transparent orthodontic aligner using polyurethane films that were shaped on a dental model created through 3D printing. The application of 3D printing technology in manufacturing transparent aligners ensures exceptional accuracy during the entire process of orthodontic treatment, leading to a favorable outcome.

17.2.4 Bio-derived Materials

Regenerative medicine has given rise to the concept of bio-derived materials, which are composite materials made from structural tissues or acellular matrices (ACMs) (Li et al. 2021). ACM can be categorized as cell ACM, tissue ACM, and organ ACM, where tissue ACM currently being the most prevalent. Moura et al. (2022) examined the disparity in bone regeneration between artificial, allograft, and autogenous bone substitutes. They concluded that allogenic bone grafts have a comparatively inferior ability to repair bone compared to artificial and autogenous bone substitutes, likely because of immunotoxicological reactions. This indicates that allogeneic cells can provoke an immune response, and hence it is necessary to eliminate such cells (Fig. 17.5).

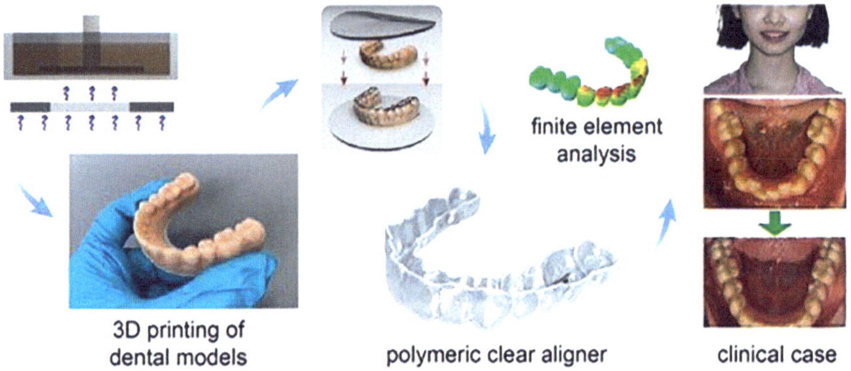

Fig. 17.5 Illustration of medical translation of polymer-based biomaterials. An event of polymeric clear aligners

Decellularization is a technique used to obtain the extracellular matrix (ECM) structure by removing all cellular components, resulting in a diminished immune response. The notion of decellularization was introduced in 1948 and has since become the most commonly employed method for eliminating immunogenic compounds. The decellularization methods encompass physical, chemical, and enzymatic techniques, either individually or in combination. It is important to note that decellularization results in varying degrees of disturbance to the original ECM structure, which can be reduced but not eliminated. Trypsin is a potent decellularization enzyme capable of effectively dissociating cells from proteins. Although the procedure is beneficial for preparing a vascular graft, it results in a reduction in artery strength and is detrimental to blood pressure resistance. Crosslinking improves the natural biomaterial's mechanical properties through physical and chemical means. According to reports, the process of photooxidation crosslinking has the potential to greatly enhance the remaining strength of decellularized vessels, occasionally surpassing the effectiveness of crosslinking of glutaraldehyde. In recent times, Ding's team suggested a two-step decellularization strategy that includes a biosurfactant to alter the network of bovine pericardium biomacromolecular crosslinked with glutaraldehyde. The subsequent bioprosthetic heart valve (BHV) showed solid anti-calcification properties based on their experiments (Yu et al. 2022).

17.3 Bioactivity of the Materials of Bio-medical Implant

Bioactive materials can be referred to as biomaterials that elicit a reaction from the human body, such as bonding of tissue and modification of the adjacent tissue. Following the introduction of implant materials into the human body, numerous reactions occur in the adjacent tissues, which are influenced by the specific characteristics of the material and its surface structure. Materials with bioinert properties, such as stainless steel, alumina, and titanium, exhibit limited adhesion to the adjacent tissues. Materials showing bioactivity have demonstrated the capacity to adhere to adjacent tissues and establish a bond. Bone is formed by collagen, an amalgamation of hydroxyapatite (HAp), and some other proteins, resulting in a composite (Gao et al. 2023).

There exist two distinct bioactive material categories: osteoproductive and osteoconductive. Osteoconductive, one adheres to the bone and stimulates bone regeneration on the bioactive material surface. Therefore, some other calcium phosphate composites combined with synthetic HAp are primarily utilized as substitutes for bone minerals. Tricalcium phosphate ceramics and synthetic HAp are materials that help in the new bone tissue growth. These materials have many medical applications, such as being utilized as a filler for defects of bone and as a coating on different metal alloys. Additionally, these materials can be absorbed by the body and are employed to fill bone deficiencies (Gautam et al. 2022).

Osteoproductive materials stimulate the formation of new bone in areas that are not directly connected to the bone/implant contact. Additionally, these materials can adhere to soft tissues in the body, such as the gingival and cartilage. Bioactive glasses are a type of substance that promotes bone growth. They are composed of 46.1% SiO_2, 26.9% CaO, 24.4% Na_2O and 2.6% P_2O_5. It is utilized in clinical settings to treat periodontal disorders and serves as a substance for filling bone. Additionally, these implants are utilized to substitute broken ear bones, hence facilitating the restoration of auditory function for a vast number of individuals. An optimal bioactive material must possess all the distinct attributes of osteogenicity, osteoinductivity osteoconductivity, and osteoproductivity which have been illustrated in Fig. 17.6. Developing a material that meets the entire requisite criteria is a difficult task, especially when it comes to bone (Ferraris et al. 2020). Surface

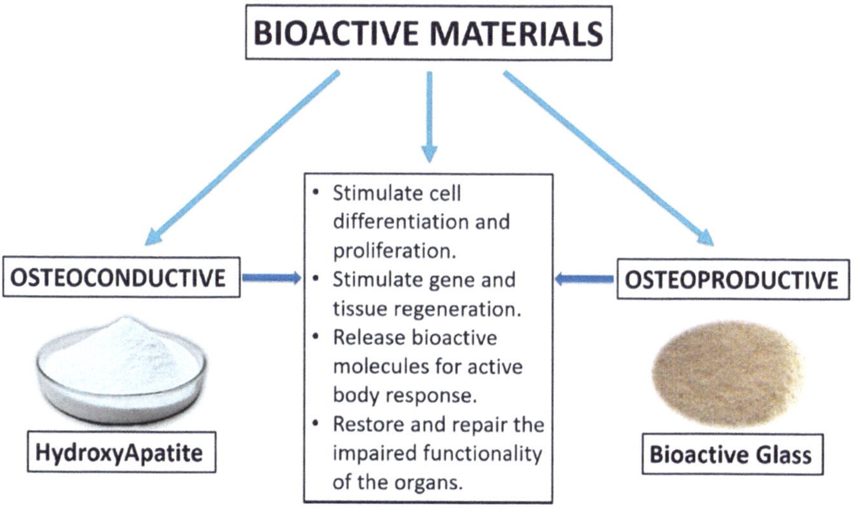

Fig. 17.6 Different bioactive materials and functions

alterations and composite preparation are two crucial methods for enhancing and augmenting the bioactivity of bone material. The primary purpose of doping HAp and surface coating is to enhance its bioactive nature. This is because HAp constitutes 70% of bone and is known as a super bioactive material (Najdanović et al. 2018).

17.4 Techniques of Metals and Metal Alloy Surface Modification

In the recent past, there has been a growing use of coatings made from diverse bioceramics and polymers to modify the surface of medical implants. The objective of this process is to create a surface with enhanced properties that possess unique features not found in uncoated implants. The technique of modifying the surface, including coatings, can effectively mitigate and surpass various limitations associated with biomaterials. Multiple production methods can enhance the performance and desirable features, such as roughness and porosity of materials. The application of these methods involves the deposition and synthesis of a polymeric layer on the surfaces of metal implants, resulting in improved performance of the implants shown by Joyce et al. (2021). The approaches are typically categorized into a couple of main segments, i,e; chemical modification and physical alteration, as seen in Fig. 17.7. In physical alteration processes, the entire surface is subjected to charges that are energized and flame or plasma. In contrast, chemical processes involve immersing the entire surface in chemical solutions, resulting in the formation of the coating on the surfaces, Liu et al. (2019). Figure 17.7 again illustrates the benefits and drawbacks of many crucial surface modification approaches, which are extensively covered in this section.

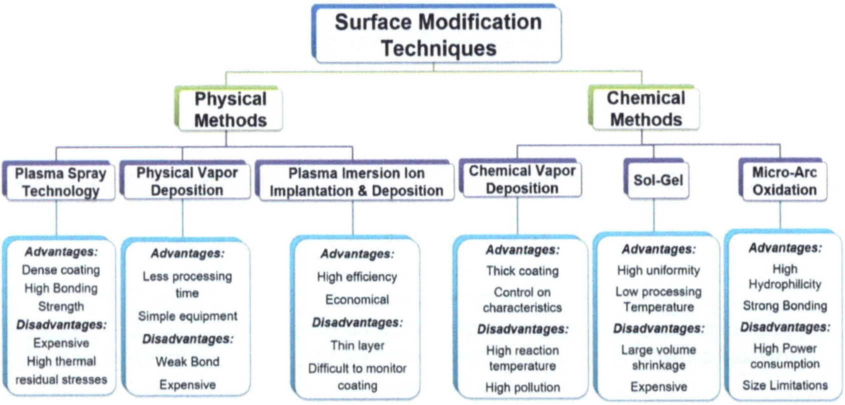

Fig. 17.7 Advantages and disadvantages of different surface modification techniques

17.5 Applications of Green Biomaterials in Biomedical Implants

Biomaterials are employed for the physical replacement and restoration of damaged tissues. While the tissues of the body typically operate efficiently throughout a person's lifespan, they can occasionally be affected by accidents, fractures, infections, and diseases that result in pain, deformity, and impaired functionality. Therefore, because of these circumstances, it is necessary to eliminate the affected tissues and substitute them with suitable biomaterials (Asmatulu et al. 2019; Dhanasekaran et al. 2020; García et al. 2018; Ijaola et al. 2023; Jin et al. 2021; Uddin et al. 2019). An implant is inserted either internally or externally to restore the natural functioning of the body. Figure 17.8 illustrates the diverse usefulness of medical implants in various domains, including Orthopedics, Dental, Cardiovascular, Ophthalmic, and others. Some important applications are discussed below:

17.5.1 Orthopedics

There are several uses of medical implants in orthopedics. Several disorders, such as rheumatoid arthritis and osteoarthritis, damage the integrity of mobile joints such as the shoulder, hip, ankle, knee, and so on. The excruciating discomfort

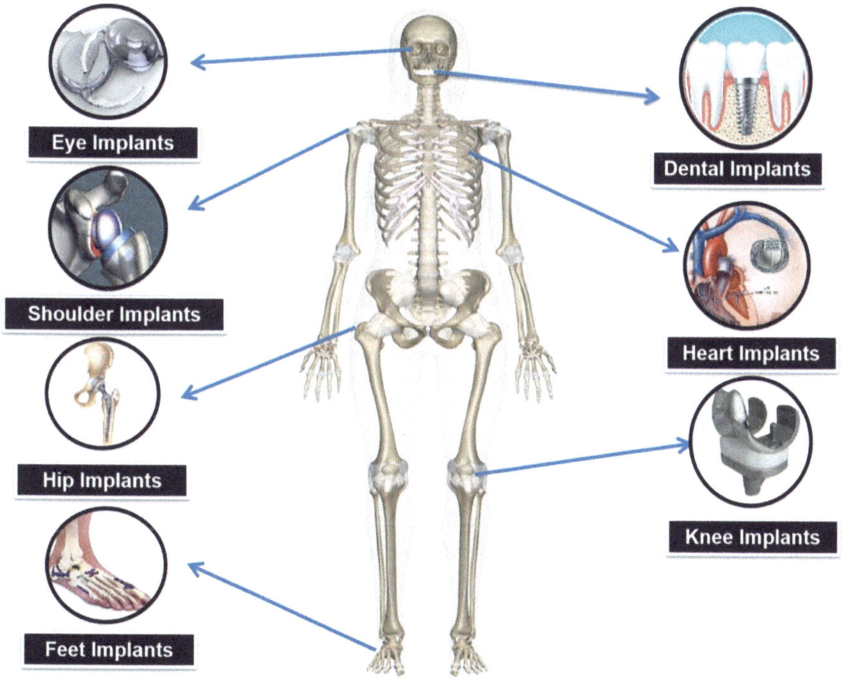

Fig. 17.8 Illustration of diverse medical implant applications in the body

experienced in weight-bearing joints necessitates the replacement of these joints with prostheses and artificial implants. Occasionally, unforeseen mishaps can result in fractures in tissues and bones. Subsequently, the biomaterial screws and plates are substituted with the impacted bone (Uddin et al. 2020). If a bone sustains a slight fracture, it will mend within a certain period, after which the implants are detached from the body. However, if the fracture is severe, the implants will remain inside the body for an extended period until the affected bone has fully healed. These medical implants are crucial for people's daily life.

17.5.2 Dental

Dental sciences extensively utilize medical implants. Our teeth can be damaged by bacterial contamination in our mouth because of our carelessness. The occurrence of cavities, demineralization, and tooth disintegration can lead to tooth loss (Kumar et al. 2018). The lost teeth can be substituted and restored with fabricated teeth from appropriate biomaterials. These prosthetic teeth closely resemble natural teeth and are secured in place without the need for adhesive (Javaid and Haleem 2019). Some individuals may have misaligned and malformed natural teeth. Shape-memory braces are utilized to modify the shape and positioning. Once it is restored, the braces are subsequently removed from the place it was placed.

17.5.3 Cardiovascular

In the cardiovascular system, which includes the heart and vessels, complications and obstructions may occur because of the viscosity of the blood, which can be attributed to a range of disorders. The arteries and vessels in the lower leg get obstructed because of the accumulation of fatty deposits. Consequently, these obstructed vessels must be substituted by artificial vessels (Rokaya et al. 2018). Occasionally, the heart valve has structural alterations because of aging, impeding its ability to completely open or close. The diseased valves are substituted with specifically fabricated valves, which reinstate the normal heart operation (Toong et al. 2020).

17.5.4 Ophthalmic

Occasionally, our ocular tissues are afflicted by various ailments and extended periods of screen usage, resulting in diminished eyesight. Contact lenses restore the visual acuity of the eye (Bussooa et al. 2018). Cataracts, an example of eye illness, result in opacity in the lens of the eye, leading to impaired visual acuity. During this procedure, the damaged eye lens is substituted by an intraocular lens through surgical intervention, leading to the restoration of the typical functionality of our eyes (Kim et al. 2020; Poomathi et al. 2019).

17.6 Conclusions

In conclusion, biomaterials are a promising and developing branch of material science that strives to meet technical requirements while considering environmental sustainability through implementing "green" constituents. Continued work in this area will allow the creation of sustainable materials that will revolutionize iconic sectors such as biomedicine and commuting, thereby facilitating a green future. A life-cycle approach complemented by green technology will be instrumental in the further development of materials and their fields of application. Such materials balance the expectation of high performance and the dire need to protect the environment. Besides, biomaterials such as proteins and polysaccharides have been successful in redefining sustainability and improving the quality of materials in different applications. In the end, the future of the industry is likely to involve tackling both technical and material challenges. For instance, achieving a degree of performance that can be controlled based on environmental conditions is an ongoing process. In addition, viable solutions to the recycling or biodegradability of the materials must be developed. Given that the trend of sustainability is expected to become more popular globally, the need for materials that allow making ethical decisions is bound to increase, pushing the development process. Advanced biomaterials promise a future of a technology-driven world that is simultaneously sustainable. Before they can realize their full value in supporting environmental sustainability worldwide, continuous and compulsory interdisciplinary research in material science, chemistry, and environmental engineering is urgently needed.

References

Alqattan M, Peters L, Alshammari Y, Yang F, Bolzoni L (2021) Antibacterial Ti-Mn-Cu alloys for biomedical applications. Regener Biomater 8(1):rbaa050

Asmatulu R, Veisi Z, Uddin MN, Mahapatro A (2019) Highly sensitive and reliable electrospun polyaniline nanofiber based biosensor as a robust platform for COX-2 enzyme detections. Fibers Polym 20:966–974

Assad M, Jackson N (2019) Biocompatibility evaluation of orthopedic biomaterials and medical devices: a review of safety and efficacy models. In: Encyclopedia of biomedical engineering, pp 281–309

Boopathy AV, Mandal A, Kulp DW, Menis S, Bennett NR, Watkins HC, Wang W, Martin JT, Thai NT, He Y (2019) Enhancing humoral immunity via sustained-release implantable microneedle patch vaccination. Proc Natl Acad Sci 116(33):16473–16478

Bussooa A, Neale S, Mercer JR (2018) Future of smart cardiovascular implants. Sensors 18(7):2008

Cao D, Guo W, Cai C, Tang J, Rao W, Wang Y, Wang Y, Yu L, Ding J (2022) Unified therapeutic-prophylactic vaccine demonstrated with a postoperative filler gel to prevent tumor recurrence and metastasis. Adv Funct Mater 32(40):2206084

Chen X, Wang M, Yang X, Wang Y, Yu L, Sun J, Ding J (2019) Injectable hydrogels for the sustained delivery of a HER2-targeted antibody for preventing local relapse of HER2+ breast cancer after breast-conserving surgery. Theranostics 9(21):6080–6098

Chen Y, Zhang J, Liu X, Wang S, Tao J, Huang Y, Wu W, Li Y, Zhou K, Wei X (2020) Noninvasive in vivo 3D bioprinting. Sci Adv 6(23):eaba7406

Chen J, Hu G, Li T, Chen Y, Gao M, Li Q, Hao L, Jia Y, Wang L, Wang Y (2021) Fusion peptide engineered "statically-versatile" titanium implant simultaneously enhancing anti-infection, vascularization and osseointegration. Biomaterials 264:120446

Chen Y, Cai J, Liu D, Liu S, Lei D, Zheng L, Wei Q, Gao M (2022) Zinc-based metal organic framework with antibacterial and anti-inflammatory properties for promoting wound healing. Regener Biomater 9:rbac019

Dhanasekaran P, Uddin M, Wooley P, Asmatulu R (2020) Fabrication and biological characterization of highly porous PEEK bio-nanocomposites incorporated with carbon and hydroxyapatite nanoparticles for scaffold applications. Molecules 25(16):3572

Ding J-D (2022) Biomedical polymer materials. Science Press, Beijing

Ding X, Gao J, Yu X, Shi J, Chen J, Yu L, Chen S, Ding J (2022) 3D-printed porous scaffolds of hydrogels modified with TGF-β1 binding peptides to promote in vivo cartilage regeneration and animal gait restoration. ACS Appl Mater Interfaces 14(14):15982–15995

Ferraris S, Yamaguchi S, Barbani N, Cazzola M, Cristallini C, Miola M, Vernè E, Spriano S (2020) Bioactive materials: in vitro investigation of different mechanisms of hydroxyapatite precipitation. Acta Biomater 102:468–480

Gao J, Ding X, Yu X, Chen X, Zhang X, Cui S, Shi J, Chen J, Yu L, Chen S (2021) Cell-free bilayered porous scaffolds for osteochondral regeneration fabricated by continuous 3D-printing using nascent physical hydrogel as ink. Adv Healthc Mater 10(3):2001404

Gao Y, Kang K, Luo B, Sun X, Lan F, He J, Wu Y (2022) Graphene oxide and mineralized collagen-functionalized dental implant abutment with effective soft tissue seal and romotely repeatable photodisinfection. Regener Biomater 9:rbac024

Gao C-Y, Wang G, Wang L, Wang Q-S, Wang H-C, Yu L, Liu J-X, Ding J-D (2023) A biosurfactant-containing TSD strategy to modify bovine pericardial bioprosthetic valves for anticalcification. Chin J Polym Sci 41(1):51–66

García JA, Rivero PJ, Ortiz R, Quintana I, Rodríguez RJ (2018) Advanced surface treatments for improving the biocompatibility of prosthesis and medical implants. In: Advanced surface engineering research, pp 199–219

Gautam S, Bhatnagar D, Bansal D, Batra H, Goyal N (2022) Recent advancements in nanomaterials for biomedical implants. Biomed Eng Adv 3:100029

Gu Z, Wang J, Fu Y, Pan H, He H, Gan Q, Liu C (2023) Smart biomaterials for articular cartilage repair and regeneration. Adv Funct Mater 33(10):2212561

Han Y-H, Liu C-G, Chen B-Q, Fu C-P, Kankala RK, Wang S-B, Chen A-Z (2022) Orchestrated tumor apoptosis (Cu2+) and bone tissue calcification (Ca2+) by hierarchical Copper/Calcium-ensembled bioactive silica for osteosarcoma therapy. Chem Eng J 435:134820

Hu K, Li Y, Ke Z, Yang H, Lu C, Li Y, Guo Y, Wang W (2022) History, progress and future challenges of artificial blood vessels: a narrative review. Biomater Transl 3(1):81–98

Ijaola AO, Subeshan B, Pham A, Uddin MN, Yang S-Y, Asmatulu E (2023) Fabrication, characterization, and in vitro cytotoxicity assessment of tri-layered multifunctional scaffold for effective chronic wound healing. Bioengineering 10(10):1148

Javaid M, Haleem A (2019) Current status and applications of additive manufacturing in dentistry: a literature-based review. J Oral Biol Craniofac Res 9(3):179–185

Jeong B, Bae YH, Lee DS, Kim SW (1997) Biodegradable block copolymers as injectable drug-delivery systems. Nature 388(6645):860–862

Jin Y, Wang H, Yi K, Lv S, Hu H, Li M, Tao Y (2021) Applications of nanobiomaterials in the therapy and imaging of acute liver failure. Nano-Micro Lett 13:1–36

Joyce K, Fabra GT, Bozkurt Y, Pandit A (2021) Bioactive potential of natural biomaterials: identification, retention and assessment of biological properties. Signal Transduct Target Ther 6(1):122

Kim JY, Lee Y, Won HJ, Jeong H, Park JH, Kim MJ, Tchah H (2020) Diffractive multifocal intraocular lens implantation in patients with monofocal intraocular lens in the contralateral eye. Int J Ophthalmol 13(5):737–743

Kumar K, Gill R, Batra U (2018) Challenges and opportunities for biodegradable magnesium alloy implants. Mater Technol 33(2):153–172

Li F, Yan Y, Wang Y, Fan Y, Zou H, Liu H, Luo R, Li R, Liu H (2021) A bifunctional MXene-modified scaffold for photothermal therapy and maxillofacial tissue regeneration. Regener Biomater 8(6):rbab057

Liang X, Zhang S, Gadd GM, McGrath J, Rooney DW, Zhao Q (2022) Fungal-derived selenium nanoparticles and their potential applications in electroless silver coatings for preventing pin-tract infections. Regener Biomater 9:rbac013

Liu Y, Luo J, Chen X, Liu W, Chen T (2019) Cell membrane coating technology: a promising strategy for biomedical applications. Nano-Micro Lett 11:1–46

Liu Y, Rui Z, Cheng W, Song L, Xu Y, Li R, Zhang X (2021) Characterization and evaluation of a femtosecond laser-induced osseointegration and an anti-inflammatory structure generated on a titanium alloy. Regener Biomater 8(2):rbab006

Luo Q, Li X, Zhong W, Cao W, Zhu M, Wu A, Chen W, Ye Z, Han Q, Natarajan D (2022) Dicalcium silicate-induced mitochondrial dysfunction and autophagy-mediated macrophagic inflammation promotes osteogenic differentiation of BMSCs. Regener Biomater 9:rbab075

Mahmood A, Patel D, Hickson B, DesRochers J, Hu X (2022) Recent progress in biopolymer-based hydrogel materials for biomedical applications. Int J Mol Sci 23(3):1415

Moura BS, Monteiro MV, Ferreira LP, Lavrador P, Gaspar VM, Mano JF (2022) Advancing tissue decellularized hydrogels for engineering human organoids. Adv Funct Mater 32(29):2202825

Najdanović J, Rajković J, Najman S (2018) Bioactive biomaterials: potential for application in bone regenerative medicine. In: Biomaterials in clinical practice: advances in clinical research and medical devices, pp 333–360

Najibi AJ, Shih T-Y, Mooney DJ (2022) Cryogel vaccines effectively induce immune responses independent of proximity to the draining lymph nodes. Biomaterials 281:121329

Poomathi N, Singh S, Prakash C, Patil RV, Perumal P, Barathi VA, Balasubramanian KK, Ramakrishna S, Maheshwari N (2019) Bioprinting in ophthalmology: current advances and future pathways. Rapid Prototyp J 25(3):496–514

Qi Y, Li X, He Y, Zhang D, Ding J (2019) Mechanism of acceleration of iron corrosion by a polylactide coating. ACS Appl Mater Interfaces 11(1):202–218

Qin Y, Yang H, Liu A, Dai J, Wen P, Zheng Y, Tian Y, Li S, Wang X (2022) Processing optimization, mechanical properties, corrosion behavior and cytocompatibility of additively manufactured Zn-0.7 Li biodegradable metals. Acta Biomater 142:388–401

Richbourg NR, Ravikumar A, Peppas NA (2021) Solute transport dependence on 3D geometry of hydrogel networks. Macromol Chem Phys 222(16):2100138

Rodriguez-Gabella T, Voisine P, Puri R, Pibarot P, Rodés-Cabau J (2017) Aortic bioprosthetic valve durability: incidence, mechanisms, predictors, and management of surgical and transcatheter valve degeneration. J Am Coll Cardiol 70(8):1013–1028

Rokaya D, Srimaneepong V, Sapkota J, Qin J, Siraleartmukul K, Siriwongrungson V (2018) Polymeric materials and films in dentistry: an overview. J Adv Res 14:25–34

Sharma S, Sudhakara P, Singh J, Ilyas R, Asyraf M, Razman M (2021) Critical review of biodegradable and bioactive polymer composites for bone tissue engineering and drug delivery applications. Polymers 13(16):2623

Shen Y, Zhang W, Xie Y, Li A, Wang X, Chen X, Liu Q, Wang Q, Zhang G, Liu Q (2021) Surface modification to enhance cell migration on biomaterials and its combination with 3D structural design of occluders to improve interventional treatment of heart diseases. Biomaterials 279:121208

Shi J, Yu L, Ding J (2021) PEG-based thermosensitive and biodegradable hydrogels. Acta Biomater 128:42–59

Todros S, Todesco M, Bagno A (2021) Biomaterials and their biomedical applications: from replacement to regeneration. Processes 9(11):1949

Toong DWY, Toh HW, Ng JCK, Wong PEH, Leo HL, Venkatraman S, Tan LP, Ang HY, Huang Y (2020) Bioresorbable polymeric scaffold in cardiovascular applications. Int J Mol Sci 21(10):3444

Uddin MN, Dhanasekaran PS, Asmatulu R (2019) Mechanical properties of highly porous PEEK bionanocomposites incorporated with carbon and hydroxyapatite nanoparticles for scaffold applications. Prog Biomater 8:211–221

Uddin MN, Desai F, Asmatulu E (2020) Engineered nanomaterials in the environment: bioaccumulation, biomagnification and biotransformation. Environ Chem Lett 18(4):1073–1083

Urciuolo A, Poli I, Brandolino L, Raffa P, Scattolini V, Laterza C, Giobbe GG, Zambaiti E, Selmin G, Magnussen M (2020) Intravital three-dimensional bioprinting. Nat Biomed Eng 4(9):901–915

Wang X, Lei X, Yu Y, Miao S, Tang J, Fu Y, Ye K, Shen Y, Shi J, Wu H (2021) Biological sealing and integration of a fibrinogen-modified titanium alloy with soft and hard tissues in a rat model. Biomater Sci 9(15):5192–5208

Wang J, Tang Y, Cao Q, Wu Y, Wang Y, Yuan B, Li X, Zhou Y, Chen X, Zhu X (2022) Fabrication and biological evaluation of 3D-printed calcium phosphate ceramic scaffolds with distinct macroporous geometries through digital light processing technology. Regener Biomater 9:rbac005

Wu Q, Xu S, Wang F, He B, Wang X, Sun Y, Ning C, Dai K (2021) Double-edged effects caused by magnesium ions and alkaline environment regulate bioactivities of magnesium-incorporated silicocarnotite in vitro. Regener Biomater 8(6):rbab016

Wu Y, Deng Z, Wang X, Chen A, Li Y (2022) Synergistic antibacterial photocatalytic and photothermal properties over bowl-shaped TiO2 nanostructures on Ti-19Zr-10Nb-1Fe alloy. Regener Biomater 9:rbac025

Xie K, Wang L, Guo Y, Zhao S, Yang Y, Dong D, Ding W, Dai K, Gong W, Yuan G (2021) Effectiveness and safety of biodegradable Mg-Nd-Zn-Zr alloy screws for the treatment of medial malleolar fractures. J Orthop Transl 27:96–100

Xu X, Lu Y, Zhou L, He M, Zhuo J, Zhong Q, Luo K, Lin J (2021) Tuning osteoporotic macrophage responses to favour regeneration by Cu-bearing titanium alloy in Porphyromonas gingivalis lipopolysaccharide-induced microenvironments. Regener Biomater 8(1):rbaa045

Yang X, Chen X, Wang Y, Xu G, Yu L, Ding J (2020) Sustained release of lipophilic gemcitabine from an injectable polymeric hydrogel for synergistically enhancing tumor chemoradiotherapy. Chem Eng J 396:125320

Yang X, Wang Y, Mao T, Wang Y, Liu R, Yu L, Ding J (2021) An oxygen-enriched thermosensitive hydrogel for the relief of a hypoxic tumor microenvironment and enhancement of radiotherapy. Biomater Sci 9(22):7471–7482

Yang W, Zhang M, He J, Gong M, Sun J, Yang X (2022) Central nervous system injury meets nanoceria: opportunities and challenges. Regener Biomater 9:rbac037

Yu X, Li G, Zheng Y, Gao J, Fu Y, Wang Q, Huang L, Pan X, Ding J (2022) 'Invisible'orthodontics by polymeric 'clear'aligners molded on 3D-printed personalized dental models. Regener Biomater 9:rbac007

Zhai P, Peng X, Li B, Liu Y, Sun H, Li X (2020) The application of hyaluronic acid in bone regeneration. Int J Biol Macromol 151:1224–1239

Zhang S, Sun X, Kang C, Yang M, Zhao Y, Wang C (2020) Study on repairing canine mandibular defect with porous Mg–Sr alloy combined with Mg–Sr alloy membrane. Regener Biomater 7(3):331–336

Zhang J, Shang Z, Jiang Y, Zhang K, Li X, Ma M, Li Y, Ma B (2021) Biodegradable metals for bone fracture repair in animal models: a systematic review. Regener Biomater 8(1):rbaa047

Zhang X, Liu H, Li L, Huang C, Meng X, Liu J, Bai X, Ren L, Wang X, Yang K, Qin L (2022) Promoting osteointegration effect of Cu-alloyed titanium in ovariectomized rats. Regener Biomater 9. https://doi.org/10.1093/rb/rbac011

Zhao X, Chen X, Yuk H, Lin S, Liu X, Parada G (2021) Soft materials by design: unconventional polymer networks give extreme properties. Chem Rev 121(8):4309–4372

Zheng D-W, Deng W-W, Song W-F, Wu C-C, Liu J, Hong S, Zhuang Z-N, Cheng H, Sun Z-J, Zhang X-Z (2022) Biomaterial-mediated modulation of oral microbiota synergizes with PD-1 blockade in mice with oral squamous cell carcinoma. Nat Biomed Eng 6(1):32–43

Zhi W, Wang X, Sun D, Chen T, Yuan B, Li X, Chen X, Wang J, Xie Z, Zhu X (2022) Optimal regenerative repair of large segmental bone defect in a goat model with osteoinductive calcium phosphate bioceramic implants. Bioact Mater 11:240–253
Zhou L, Li X, Wang K, Shen F, Zhang L, Li P, Shang T, Wang J, Huang N (2020) Cu||-loaded polydopamine coatings with in situ nitric oxide generation function for improved hemocompatibility. Regener Biomater 7(2):153–160
Zhu L, Luo D, Liu Y (2020) Effect of the nano/microscale structure of biomaterial scaffolds on bone regeneration. Int J Oral Sci 12(1):6

Asib Ridwan is a writer and researcher specializing in biomaterials, nanomaterials, nanotechnology, and surface engineering. Asib is pursuing a Master's degree in Mechanical Engineering at Arkansas State University, where he also contributes as a Graduate Research Assistant. His academic endeavors are complemented by his professional experience, providing a unique perspective that blends rigorous research with practical application. Outside of his research, Asib enjoys exploring new developments in science and technology, contributing to academic discourse, and engaging with the engineering community to foster collaboration and knowledge sharing.

Dr. Md. Nizam Uddin is an Assistant Professor of Mechanical Engineering, at Texas A&M University-Texarkana, TX, USA. He holds an MPhil in Mechanical Engineering (Nanotechnology concentration) from Hong Kong University of Science and Technology, Hong Kong, and a Ph.D. from Wichita State University, KS, USA. He also worked as an Assistant Professor at the Department of Mechanical Engineering, Khulna University of Engineering & Technology, Bangladesh. His research interests are design and advanced manufacturing, nanomaterials, polymer-based nanocomposites, and renewable energy and environmental systems. Throughout his studies, he has published 30 journal papers and 22 conference proceedings, authored 1 book & 7 book chapters, and received 20 honors/awards. He has been working as a reviewer of several journals in the field of mechanical engineering and materials science. He has been serving as an editorial board member; and guest editor of several journals. He is a PI, Co-PI, or Senior Personnel for several grants sponsored by the state of Texas and local agencies. To date, his scholarly activities have been cited more than 1618 times, according to the Google Scholar citation.

Asif Karim Neon is a dedicated mechanical engineer with a strong academic background and practical experience in the engineering field. He holds a Bachelor of Science degree in Mechanical Engineering from Khulna University of Engineering and Technology (KUET), Bangladesh.

Md. Aliahsan Bappy received a B.Sc. degree from the faculty of the Mechanical Engineering department at Rajshahi University of Engineering & Technology one of the top engineering universities in Bangladesh in 2012. He is pursuing a Master's in Mechanical Engineering Department at Lamar University Beaumont, Texas, USA. His research interests are machine learning, product design, development and modification, nanoparticles, energy-efficient engines, solar energy, etc.

Index

R. Malviya, S. Sundram (eds.), *Sustainable Nanocomposites with Green Biomaterials*, Biomaterials, Bioengineering and Sustainability 2,
https://doi.org/10.1007/978-3-031-79110-9

N

O

P

Q

R

S

The manufacturer's authorised representative in the EU is Springer Nature Customer Service Centre GmbH, Europaplatz 3, 69115 Heidelberg, Germany. If you have any concerns regarding our products, please contact ProductSafety@springernature.com

Printed and bound by CPI Group (UK) Ltd, Croydon, CR0 4YY
01/10/2026
02225956-0015